International Conference on Materials Engineering, Materials Chemistry and Materials Physics 2023

Journal of Physics: Conference Series Volume 2765

Bhimavaram, India
25-26 September 2023

ISBN: 978-1-7138-9838-2
ISSN: 1742-6588

Printed from e-media with permission by:

Curran Associates, Inc.
57 Morehouse Lane
Red Hook, NY 12571

Some format issues inherent in the e-media version may also appear in this print version.

Institute of Physics
Dirac House, Temple Back
Bristol BS1 6BE UK

Phone: 44 1 17 929 7481
Fax: 44 1 17 920 0979

techtracking@iop.org

Additional copies of this publication are available from:

Curran Associates, Inc.
57 Morehouse Lane
Red Hook, NY 12571 USA
Phone: 845-758-0400
Fax: 845-758-2633
Email: curran@proceedings.com
Web: www.proceedings.com

TABLE OF CONTENTS

Preface

In the creation of this volume, we find the culmination of the International Conference on Materials Engineering, Materials Chemistry, and Materials Physics (MECAP-2023), a meticulously orchestrated event by the Nanotechnology Research Centre at Sagi Rama Krishnam Raju Engineering College, Bhimavaram, India, spanning from September 25 to 26, 2023.

The primary objective of the conference was to forge a vibrant platform for the convergence of experts, scholars, and students, fostering the exchange of research ideas and technological advancements. With a distinct emphasis on the cutting-edge trends and future horizons in Material Engineering and Materials Science, the conference aimed to stimulate intellectual discourse and collaboration.

A pivotal role in this intellectual tapestry was played by the keynote speeches, where luminaries from esteemed institutions such as the Indian Institutes of Technology (IITs), San Sebastian University in Chile, Northwestern Polytechnical University in China, and New York University in Abu Dhabi, showcased their profound contributions. Their presentations not only illuminated significant advancements in Material Engineering and Materials Science but also ignited fervent discussions that reverberated through both conference sessions and breaks, creating an atmosphere of dynamic exchange and exploration of key topics within the field.

The meticulously curated proceedings of MECAP-2023 encompass a collection of accepted papers that domains such as Materials Engineering,Materials Chemistry, and Materials Physics. Subjected to a rigorous peer review process, these papers now grace the pages of this volume, adhering to the professional standards upheld by the Journal of Physics: Conference Series.

Our heartfelt appreciation extends to the committed members of the Conference Advisory Committee, whose unwavering support and insightful suggestions have played an instrumental role in ensuring the resounding success of the conference. Gratitude is also extended to the keynote speakers, whose invaluable contributions and profound ideas have significantly enriched the conference's success. Finally, we extend our acknowledgment to the Journal of Physics: Conference Series team for their meticulous efforts, guiding us through every stage of the publication process.

Editor,
Dr. Venu Reddy

IOP Conf. Series: Journal of Physics: Conf. Series **2765** (2024) 011001 doi:10.1088/1742-6596/2765/1/011001

Advisory Committee

Prof. K. CheolGi, DIGST, Republic of Korea
Dr. S. AnandaKumarSarella, INanoBio, USA
Dr. A. Ali, Verginia Tech, USA
Dr. R. Abedini-Nazzab, TarbiatModares University, Iran
Dr. X. Hu, Max Planck Institute, Germany
Dr. T. Sri Ramulu, Old Dominion University, USA
Dr. T. Quang Hung, eV-Technologies, France
Dr. K. Ramam, Universidad de Concepción, Chile
Dr. K. Durga Prasad, New York University, Abu Dhabi
Dr. K. Varaprasad, San Sebastián University, Chile
Dr. Jay Singh, Banaras Hindu University, India
Dr. K. Prathap, University of Hyderabad, India
Dr. A. Bharath Kumar, Rajiv Gandhi University, India
Dr. M. Kumar Gupta, CSIR-Advanced Material Processes and Research Institute, India
Dr. Ch. Kiran Kumar, Department of Science and Technology, India
Dr. Vijaya Krishna Kari, Indian Institute of Technology Bhubaneswar, India
Dr. P. Sampath Kumar, National Institute of Technology Warangal, India
Dr. V. Srinivasarao, Assistant Director, Lupin, India
Prof. Y L N Murthy, Andhra University, India
Dr. D. Rama Sekhar Reddy, Krishna University, India

Organizing Committee

Prof. PARK Raju, SagiRamakrishnam Engineering College, India
Prof. Venu Reddy, SagiRamakrishnam Engineering College, India
Prof. K. Brahma Raju, SagiRamakrishnam Engineering College, India
Prof. P. Bhavani, SagiRamakrishnam Engineering College, India
Prof. K. V. Ramana Murthy, SagiRamakrishnam Engineering College, India
Dr. B. S. Diwakar, SagiRamakrishnam Engineering College, India
Dr. D. Chandra Sekhar, SagiRamakrishnam Engineering College, India

IOP Conf. Series: Journal of Physics: Conf. Series **2765** (2024) 011002 doi:10.1088/1742-6596/2765/1/011002

Peer Review Statement

All papers published in this volume have been reviewed through processes administered by the Editors. Reviews were conducted by expert referees to the professional and scientific standards expected of a proceedings journal published by IOP Publishing.

- **Type of peer review:** Double Anonymous
- **Conference submission management system:** Morressier
- **Number of submissions received:** 45
- **Number of submissions sent for review:** 41
- **Number of submissions accepted:** 39
- **Acceptance Rate (Submissions Accepted / Submissions Received × 100):** 86.7
- **Average number of reviews per paper:** 2
- **Total number of reviewers involved:** 66
- **Contact person for queries:**
  Name: Dr. Venu Reddy
  Email: rvenu8@srkrec.ac.in
  Affiliation: Sagi Ramakrishnam Raju Engineering College

IOP Conf. Series: Journal of Physics: Conf. Series **2765** (2024) 012001 doi:10.1088/1742-6596/2765/1/012001

Suitability of AA 7075 and Glass Fiber Materials for Aerospace Applications

K.B. Rama Krishna[1], Rajesh Siriyala[1], P. Rama Murty Raju[1], I Rama Pavan Kumar Varma[1], N. Harsha[1]

[1]Department of Mechanical Engineering, S.R.K.R. Engineering College (A), China Amiram, Bhimavaram-534204, Andhra Pradesh, India

Corresponding author: rajeshsiri.mech@gmail.com

Abstract. The primary thing considered while designing an aircraft is to find the right proportion of the vehicle weight and payload. Spars, ribs, and stringers are the main structural components of the aircraft wing. In this paper, the rib is designed using the CATIA solid modelling software with different geometrical optimization and is then analyzed in ANSYS. Aluminium Alloy 7075 and Glass fiber were the materials considered during the analysis. Modal, Structural and Fatigue analysis were performed on the object to determine the natural frequencies, stresses developed, and max. number of cycles a material can withstand respectively.

1. Introduction

Wing system is one of the basic components of an aircraft as it helps to fly, moveable surfaces helps to adjust the altitude, tail surfaces to make wings stable. The thrust needed to move vehicle is provided by engine. The provisions must be made in order to support the plane while landing, takeoff along with the rest condition on ground. Spars, ribs, and stringers are the main structural components of the wing. Trusses, beams, tubes, or other devices, including the skin, reinforce them. The wing's form and thickness are determined by the wing ribs (airfoil). They normally run from the leading edge of the wing to the rear spar. The ribs form the wing's cambered shape and helps to deliver stress from the skin to the spars. Most modern aeroplanes have fuel tanks that are either built into the wing structure or are flexible containers mounted inside the wing.

2. Literature Survey

Dharmendra et al. [1] designed and analyzed wing rib without cutout and with different cut-outs such as circular, elliptical, triangular, and rectangular. Equivalent stresses and deformation for those ribs are estimated by designing in CATIA as designing and analysis carried out in ANSYS software. The materials used are aluminium 7075 and carbon epoxy. By comparing stress and deformation in different configurations, the best suitable configuration with the best material between aluminium 7075 and carbon epoxy is selected.

Rohit Kumar et al. [2] determined stress on wing rib by designing it in SOLIDWORKS software with different designs and selected the appropriate design. ANSYS STATIC STRUCTURAL is used for analysis and material selected in this study is Al alloy cast 713. Concluded by selecting best design by comparing analysis results of 5 designs.

Sudhir reddy & Sujatha [3] designed by using conic lofting-based RDS software to create a 3D layout of the aircraft, it was then simulated in Matlab to demonstrate how closely the design adhered to its objectives. Additionally, static stress analysis is carried out in COMSOL software for wing design purposes, and the computed aerodynamic loads are applied to the wing to assess the reliability of the wing.

Ali et al. [4] evaluated the stresses on the National Advisory Committee for Aeronautics (NACA) 4415 ribs with 4 different types of cut-outs and studied the effect of cut-out shape on the stresses developed in the ribs. Modelling is done in SolidWorks software and analysis is carried out in ANSYS software. Material applied is Aluminium 2024-T3. It was found that elliptical is best shape for cut-out in order to reduce weight.

Nirmal Raj et al. [5] focused on weight reduction of a typical wing for business jet aircraft. The wing model was generated in CATIA V5 by obtaining the dimensions from conventional methods and loads are also obtained by utilizing conventional approaches. MSC Nastran and Patran 2010 is utilised to generate FEM. Nastran and Patran results revealed that composite materials have higher structural efficiency and are lighter than isotropic materials.

Sruthi et al. [6] compared material properties of 4 specimens made by stir casting process, having different SiC percentages in Aluminium LM25 (AL) and Silicon Carbide (SiC) material and acquired the optimum proportion of SiC in aluminium and given its properties as input inorder to carryout structural analysis on NACA 4412 wing profile modelled in CATIA. The analysis results from ANSYS for pure AL25 and metal matrix composite (SiC) were compared and concluded that composite material has better material properties and stresses as compared to pure aluminium.

Sanya Maria Gomez & Alice Mathai [7] used FEM to analyse hypersonic aircraft wing components such as ribs, spars, and panels, taking into account both isotropic and composite materials. By varying the orientation sequence in the composites, the optimum ply orientation was obtained through a parametric study using the ANSYS. From which it is estimated that replacement of Aluminium alloy by CFRP saves 34.82% total structural weight of the aircraft wing.

Sureka & Satya Mehar [8] created a CAD model of an A300 wing with spare parts and ribs in CATIA. Later, ANSYS WORKBENCH was used to perform a modal and structural analysis on the wing's Skeleton structure for 2 different materials (AL alloy and AL alloy 7068). Similarly, Mounika et. Al [9] also designed and performed finite element analysis (static structural and modal analysis) on aircraft wing by considering aluminum and titanium alloys.

Similarly, Mounika et al. [9] also designed and performed finite element analysis (static structural and modal analysis) on aircraft wing by considering aluminum and titanium alloys, and concluded that titanium is best suited material for making of wing.

3. Methodology

In this study the CAD design of wing rib is designed in software Catia V5 using part design module. CATIA - Computer Aided Three-dimensional Interactive Application is one of the significant 3D software utilizing in several industries ranging from aerospace to consumer products.

FEA analysis is done by using software Ansys16.2 on the airbus wing to find out the stress, vibrations and deformation in the static condition of the aircraft. The module tools which are used for this FEA analysis are Static structural analysis to find deformation and stress in the body, Modal analysis to find the frequencies caused by vibrations in the wing and Fatigue to determine maximum number of cycles a material can withstand when boundary conditions and loads are applied on the body. Table 1 shows the geometric parameters used to construct the structure. Figures from 1 to 4 depicts the designs modelled in CATIA V5.

Table 1. Geometrical Parameters of the design

Parameter	Value
Length of the wing	600 mm
First point height of wing (fixed end)	70 mm

Second point height of wing (free end)	30 mm
Curvature radius of the wing	800 mm
Distance between each circle	20 mm
Distance between each rectangle	30 mm
Base of each triangle	60 mm
Distance between each triangle	30 mm

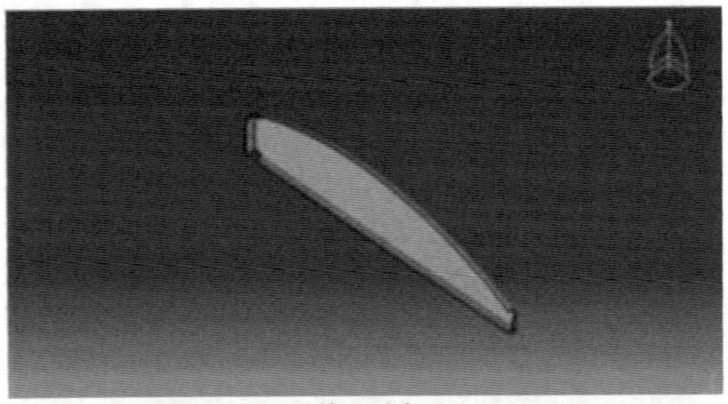

Figure 1. Rib with no cutouts

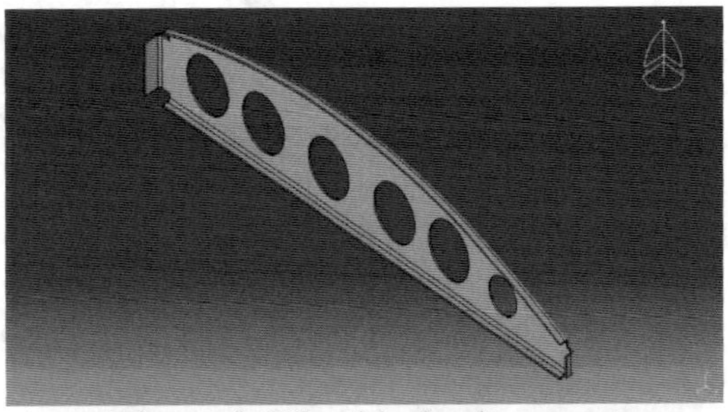

Figure 2. Rib with circular cutout

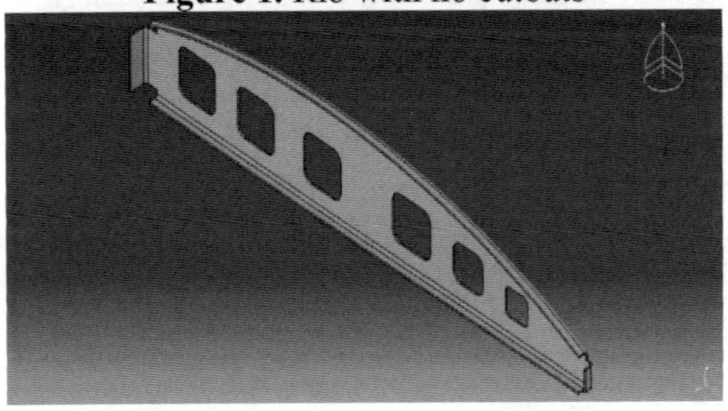

Figure 3. Rib with rectangular cutouts

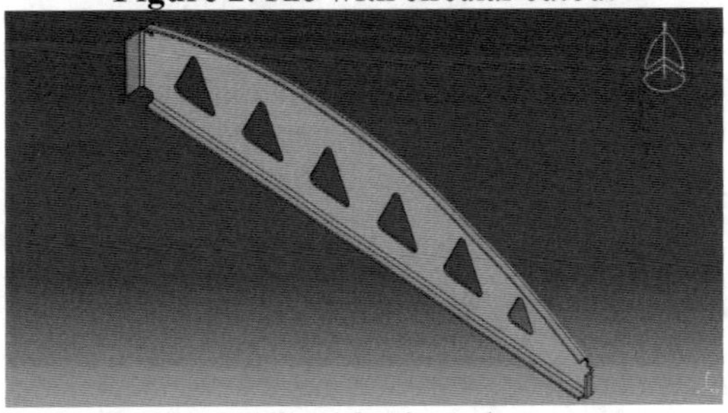

Figure 4. Rib with triangular cutouts

The CATIA file is completed and then saved in the STP format for further analysis. Where STP or STEP indicates standard text for the exchange of product data. The analysis was done using the FE software ANSYS 16.2 workbench. Different analysis such as modal and structural and fatigue analysis were performed by considering the materials Aluminium Alloy 7075 and Glass fibers (S glass fibres). After the application of material to the geometry the process of meshing is started by giving the element size 0.02m (brick element), span angle center as medium and we change the smoothing from course to medium in the details of mesh. The mesh of the wing rib and applied boundary conditions are depicted in figures 5 and 6 respectively.

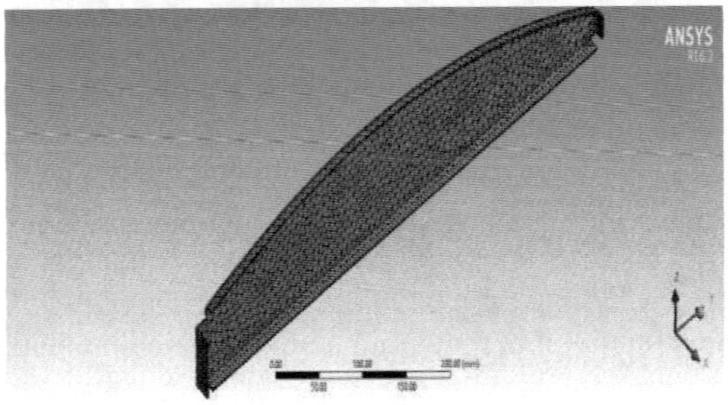

Figure 5. Meshed wing rib

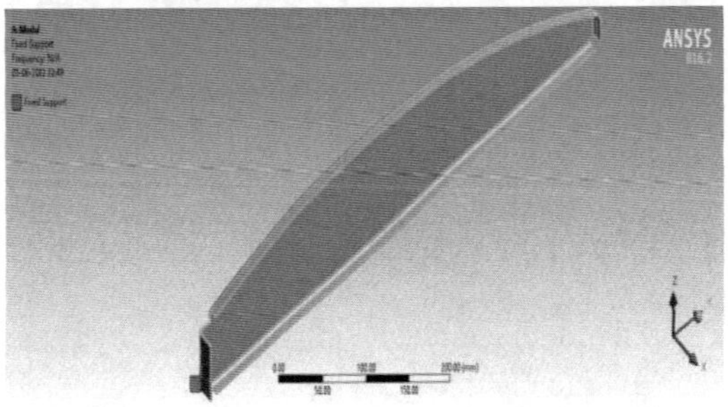

Figure 6. Boundary conditions of the rib

The material properties are obtained from ANSYS library which are given as input in present work are tabulated in below table 2 and table 3.

IOP Conf. Series: Journal of Physics: Conf. Series **2765** (2024) 012001 doi:10.1088/1742-6596/2765/1/012001

Table 2. Properties of the glass fiber

Property	Values
Density	2000 kg/m^3
Young's modulus	75 GPa
Poisson's ratio	0.2
Shear modulus XY	5000 MPa
Shear modulus YZ	3846.2 MPa
Shear modulus XZ	5000 MPa

Table 3. Properties of the aluminum alloy

Property	Values
Density	2770 kg/m^3
Young's modulus	71000 MPa
Poisson's ratio	0.33
Bulk modulus	69608 MPa
Shear modulus	26692 MPa
Tensile yield strength	503 MPa
Compressive yield strength	503 MPa
Tensile ultimate strength	572 MPa

4. Results and Discussion

4.1. Modal Analysis

4.1.1. Modal Analysis of AA 7075 Rib

The modal analysis was done using the material aluminium alloy 7075. The below figures from 7 to 11 shows the 5 mode shapes obtained in modal analysis indicate the mode shapes of the rib without cut outs. Table 4 shows the results of Modal analysis. The highest natural frequency found is 233.7 Hz and low frequency was found at 7.976 Hz. This indicates that the total deformation changes as the natural frequency of the rib changes. The highest natural frequency of circular cut-out rib is found 284.67 Hz and low frequency was found at 25.785 Hz. Similarly, for rectangular cut out it is 156.47 Hz and 20.459 Hz, for triangular cut-out it 174.77 HZ and 12.316 HZ.

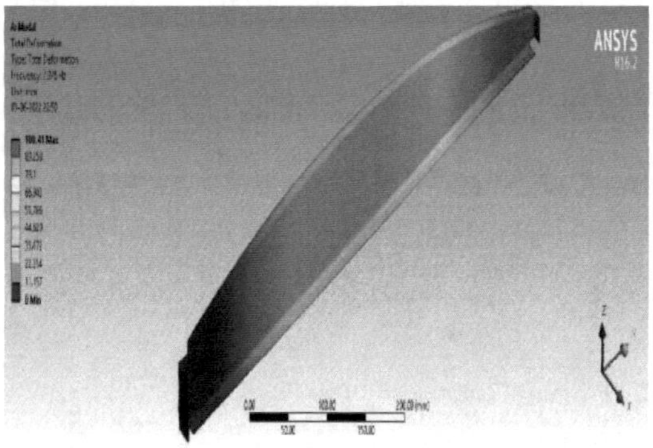

Figure 7. Mode Shape 1

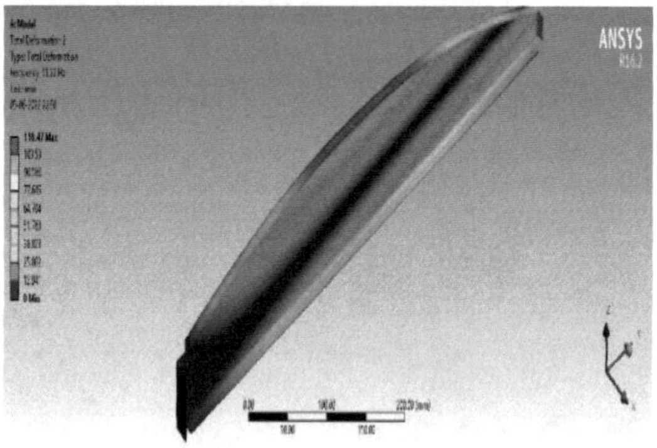

Figure 8. Mode Shape 2

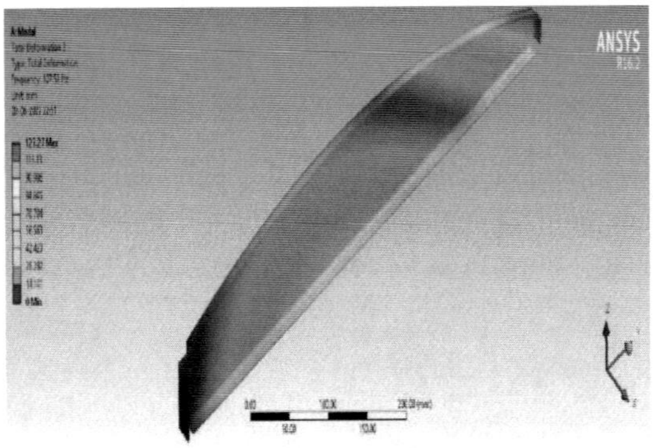

Figure 9. Mode Shape 3

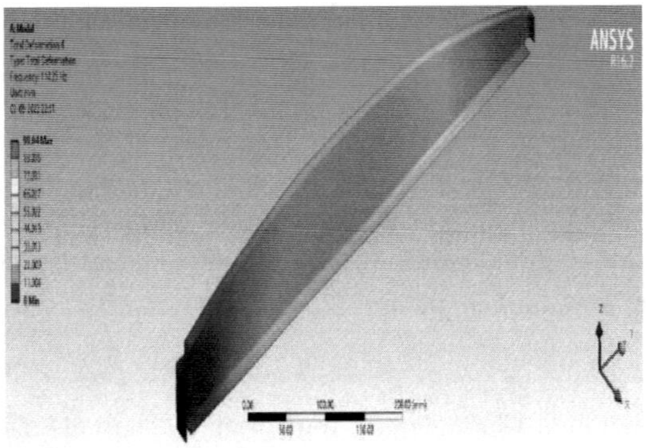

Figure 10. Mode Shape 4

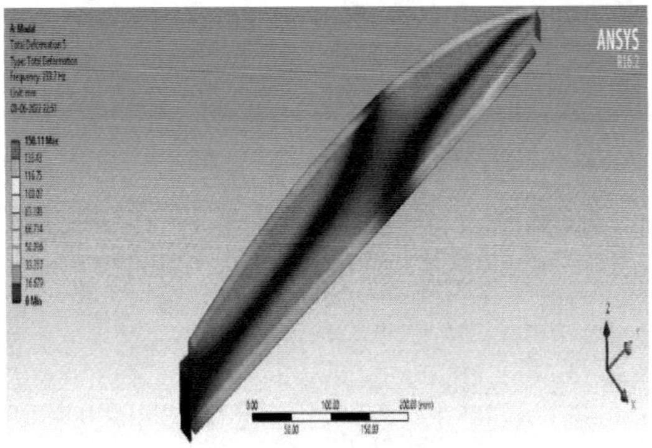

Figure 11. Mode Shape 5

Table 4. Modal Analysis Results of AA 7075 Rib

Mode	Rib with no cut out	Circular Cut	Rectangular cut	Triangular cut
1.	7.976 IIZ	25.785 HZ	20.459 HZ	12.316 HZ
2.	51.22 HZ	69.147 HZ	66.796 HZ	23.285 HZ
3.	107.52 HZ	155.61 HZ	127.74 HZ	43.247 HZ
4.	114.25 HZ	252.49 HZ	136.78 HZ	118.82 HZ
5.	233.7 HZ	284.67 HZ	156.47 HZ	174.77 HZ

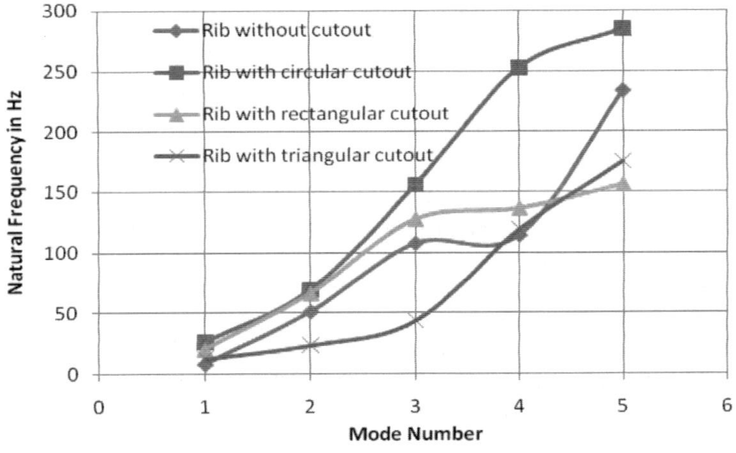

Figure 12. Natural frequencies for various rib configurations for AA 7075 material

4.1.2. Modal Analysis of Glass fibre Rib

The modal analysis was done using the material glass fibres. The below table 5 shows the results obtained from Modal analysis. The highest natural frequency found is 106.67 Hz and low frequency was found at 3.8194 Hz. This indicates that the total deformation changes as the natural frequency of the rib changes. The highest natural frequency of circular cut-out rib is found 127.79 Hz and low frequency was found at 11.621 Hz. Similarly, for rectangular cut out it is 69.729 Hz and 9.3325Hz, for triangular cut-out it 79.752 HZ and 5.6469 HZ. Plots shown in figure 13 reveals that circular cut-out possess highest natural frequency as compared to remaining.

Table 5. Modal Analysis Results of Glass fibre rib

Mode	Rib with no cutout	Circular Cut	Rectangular cut	Triangular cut
1.	3.8194 HZ	11.621 HZ	9.3325 HZ	5.6469 HZ
2.	23.692 HZ	31.674 HZ	30.342 HZ	10.324 HZ
3.	48.283 HZ	70.002 HZ	56.803 HZ	20.284 HZ
4.	57.731 HZ	113.68 HZ	61.903 HZ	54.129 HZ
5.	106.67 HZ	127.79 HZ	69.729 HZ	79.752 HZ

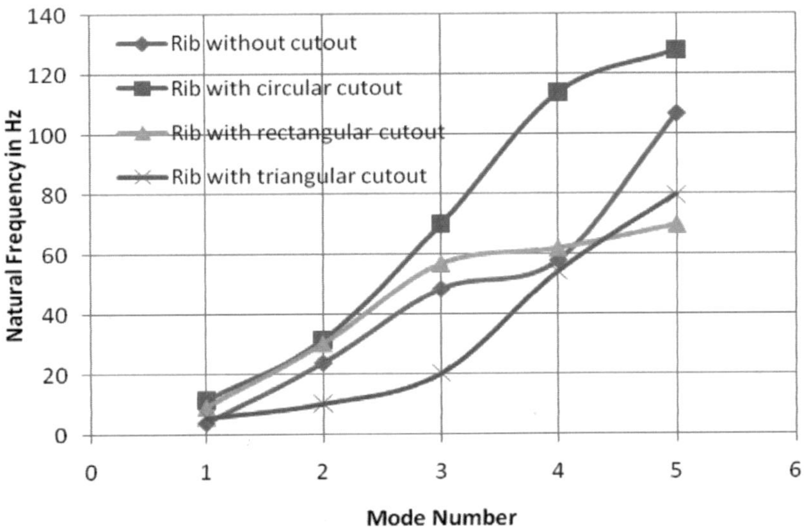

Figure 13. Natural frequencies for various rib configurations for Glass Fibres material

4.2. Structural Analysis

4.2.1. Structural analysis results for AA 7075 rib

The below figure14 and figure 15 indicate the structural analysis of the wing rib without cut-out displaying the total deformation and von misses stress of the rib at a pressure of 0.1 MPa. The change in the wing can be displayed as the deformation produced in the body. This is represented for wing rib without cut outs, rib with circle cut outs, rib with rectangular cut outs and rib with triangular cut outs. The results are tabulated in table 6. The results indicate that rib with aluminium alloy 7075 has 9.531×10^{-2} mm total deformation with von misses stress of 53.558 MPa. And for rib with circular cut outs the total deformation found was 0.1102 mm and von misses stress was found is 53.558 MPa. The rib with rectangular cut outs found total deformation of 0.10443 mm and von misses stress about 52.814 MPa. The rib with triangular cut outs has a total deformation of 9.581×10^{-2} mm and von misses stress of about 53.847 MPa. This indicates that the total deformation is found maximum for the rib with circular cut-out and less for rib without cut outs.

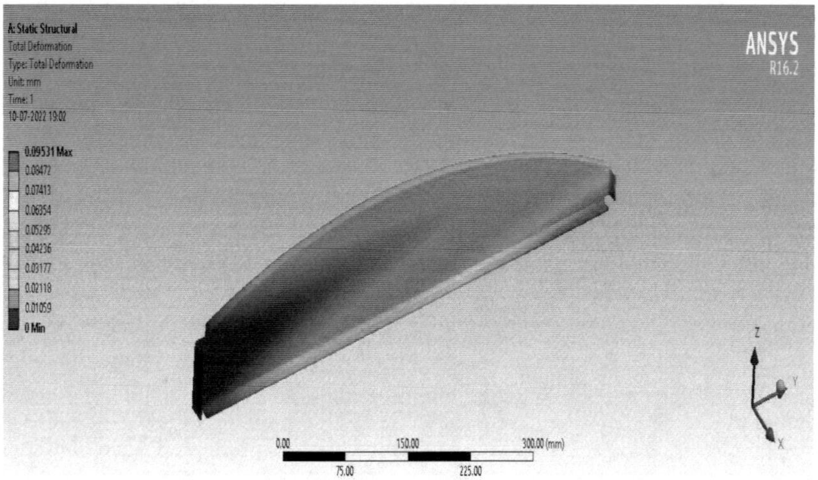

Figure 14. Deformed contour plot of AA 7075 rib without cutouts

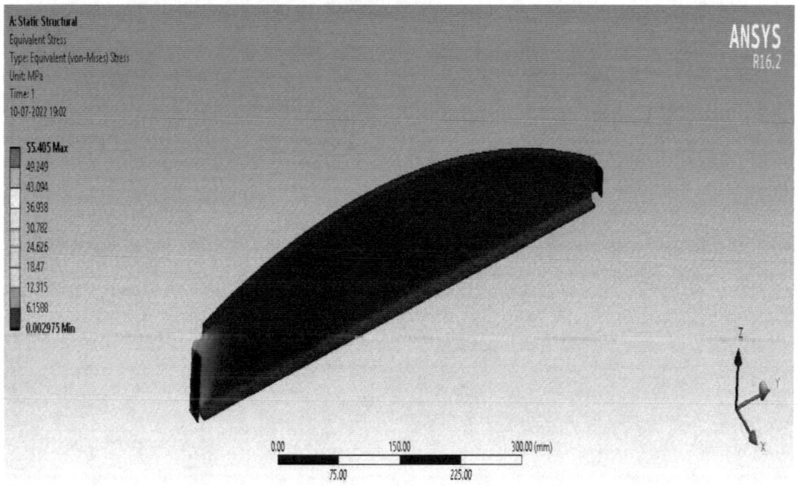

Figure 15. Equivalent stress contour plot of AA 7075 rib without cutouts

Table 6. Structural analysis results for AA 7075 rib

Design	Level	Total Deformation (mm)	Equivalent Stress (MPa)	Maximum Principal Stress (MPa)
Rib with no cut-out	Minimum	0	2.975×10^{-3}	-18.653
	Maximum	9.531×10^{-2}	55.405	54.099
Rib with circular cut-out	Minimum	0	5.2136×10^{-3}	-21.808
	Maximum	0.1102	53.558	57.533
Rib with rectangular cut-out	Minimum	0	3.1456×10^{-3}	-18.821
	Maximum	0.10443	52.814	56.785
Rib with triangular cut-out	Minimum	0	2.955×10^{-3}	-18.501
	Maximum	9.581×10^{-2}	53.847	56.623

4.2.2. Structural analysis results for Glass fibre rib

The below figure 16 and figure 17 indicate the structural analysis of the wing rib without cut-out displaying the total deformation and von misses stress of the rib at a pressure of 0.1 MPa. The change in the wing can be displayed as the deformation produced in the body. This is represented for wing rib without cut outs, rib with circle cut outs, rib with rectangular cut outs and rib with triangular cut outs. The results are tabulated in table 7. The results indicate that rib with glass fibre has 0.85569 mm total deformation with von misses stress of 557.06 MPa. And for rib with circular cut outs the total deformation found was 1.0006 mm and von misses stress was found is 541.99 MPa. The rib with rectangular cut outs found total deformation of 0.94599 mm and von misses stress about 524.11 MPa. The rib with triangular cut outs has a total deformation of 0.85891 mm and von misses stress of about 529.86 MPa. This indicates that the total deformation is found maximum for the rib with circular cut outs and less for rib without cut out.

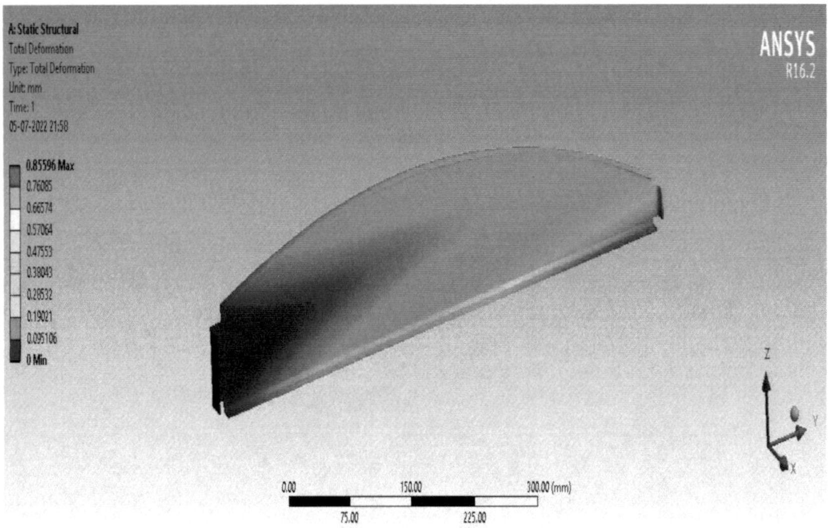

Figure16. Deformed contour plot of s-glass fiber rib without cutouts

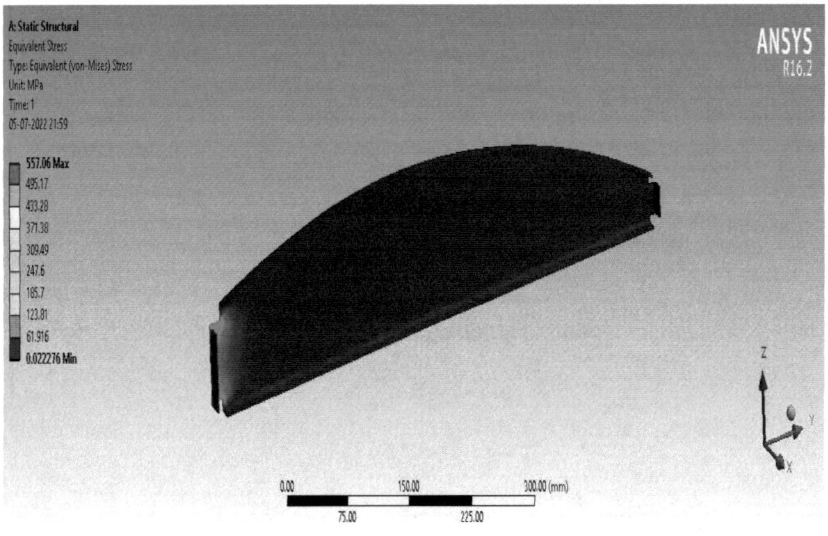

Figure 17. Equivalent stress contour plot of s-glass fiber rib without cutouts

Table 7. Structural analysis results for Glass fibre rib

Design	Level	Total Deformation (mm)	Equivalent Stress (MPa)	Maximum Principal Stress (MPa)
Rib with no cut-out	Minimum	0	2.2276×10^{-2}	-140.05
	Maximum	0.85596	557.06	519.53
Rib with circular cut-out	Minimum	0	2.3296×10^{-2}	-145.33
	Maximum	1.0006	541.99	558.09
Rib with rectangular cut-out	Minimum	0	2.647×10^{-2}	-140.12
	Maximum	0.94599	524.71	526.41
Rib with triangular cut-out	Minimum	0	2.9577×10^{-2}	-134.69
	Maximum	0.85891	529.86	529.27

4.3 Fatigue Analysis

4.3.1. Fatigue Analysis – AA 7075

Fatigue analysis is most commonly used to make a simple and quick estimate of fatigue performance or durability in absence of full time history based loading. The result contour plots shown in below figures 18 to 21 shows the available life for the given fatigue analysis. If loading is of constant amplitude, this represents the number of cycles until the part will fail due to fatigue.

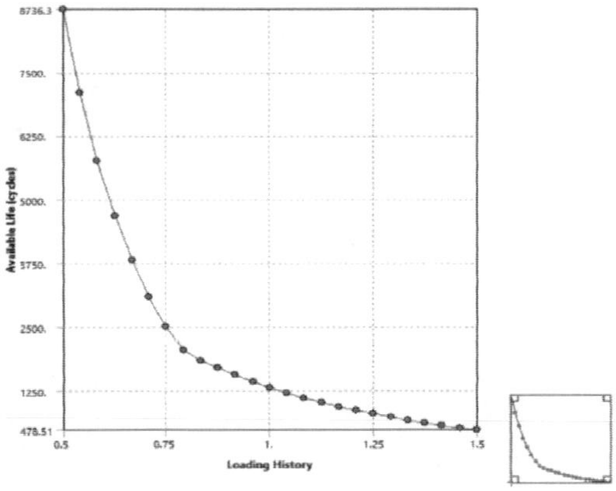

Figure 18. Rib without cutout for AA7075

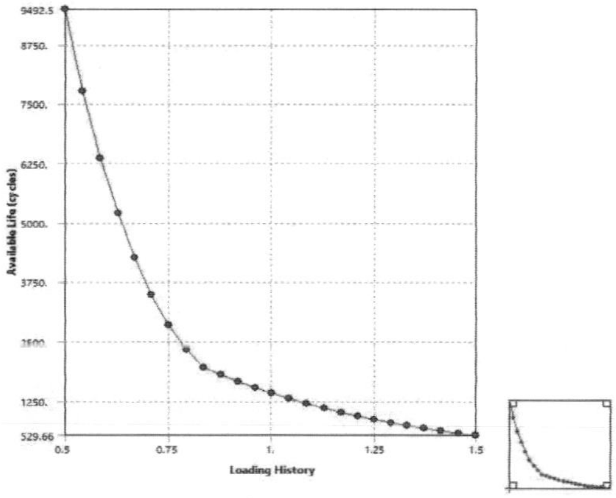

Figure 19. Rib with circular cutout for AA7075

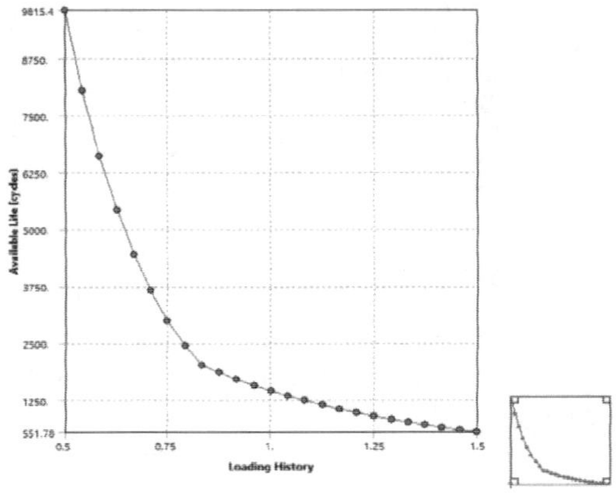

Figure 20. Rib with rectangular cutout for AA7075

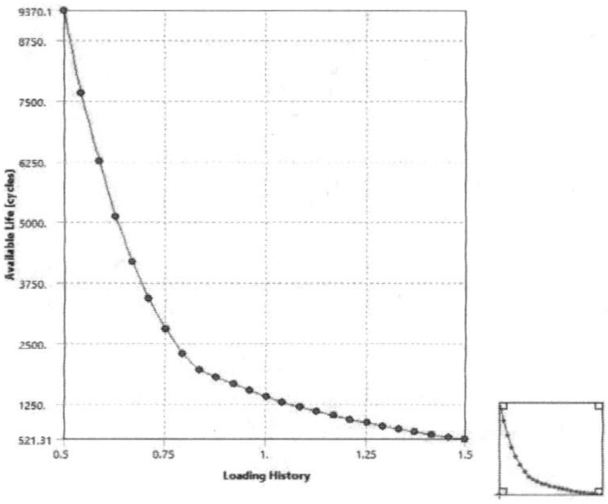

Figure 21. Rib with triangular cutout for AA7075

4.3.2 Fatigue Analysis- Glass fibre Rib

The plots shown in below figures 22 to 25 shows the available life of designed GF rib.

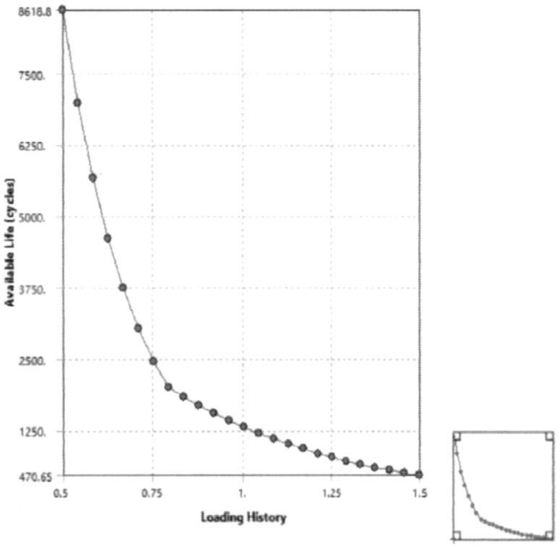

Figure 22. Rib without cutout for s-glass fiber

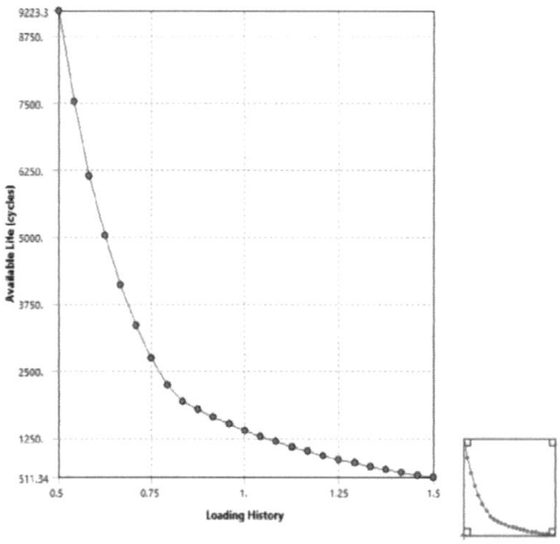

Figure 23. Rib with circular cutout for s-glass fiber

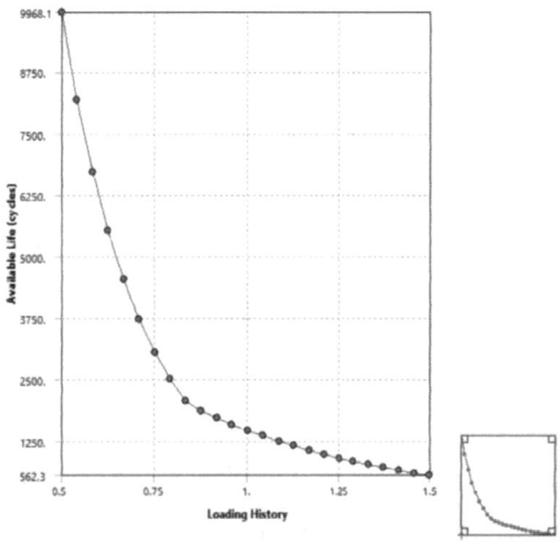

Figure 24. Rib with rectangular cutout for s-glass fiber

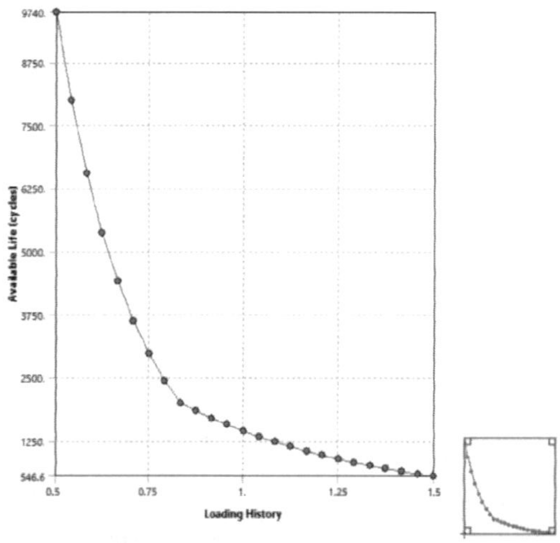

Figure 25. Rib with triangular cutout for s-glass fiber

4.3.3. Comparison of Fatigue Analysis Results among Two Materials

The below figure 26 shows the comparison of fatigue results for both the materials reveals that rectangular cut-out rib withstand for maximum number of cycles as compared to others for both the materials. It also helps to validate the structural analysis in his research as stress developed in rectangular cut-out rib is low as compared to the remaining.

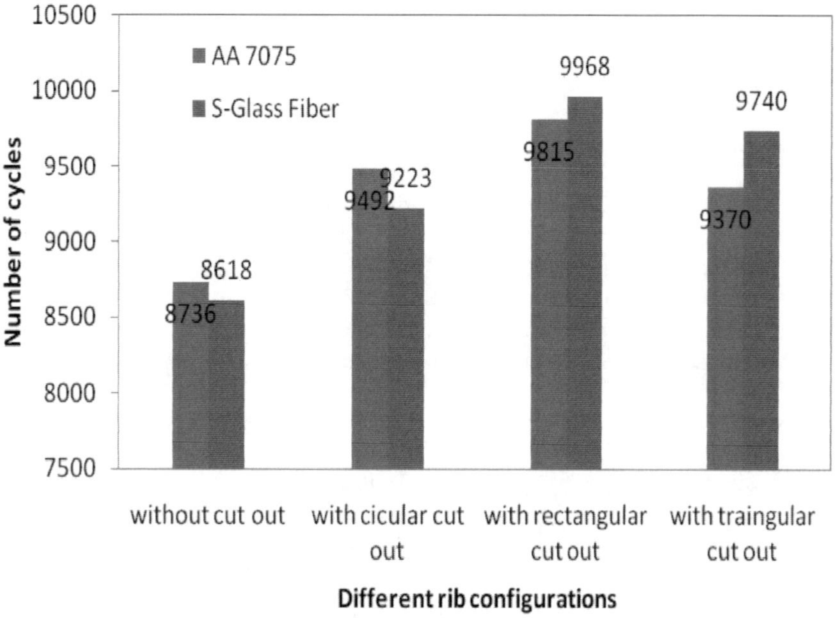

Figure 26. Fatigue analysis results of AA7075 vs Glass fibre

5. Conclusion

In this paper, different types of wing ribs were designed using the CATIA software and were analyzed under boundary conditions using the FE software ANSYS 16.2. The wing ribs considered here are rib with no cut outs, rib with circular cut outs, rib with rectangular cut outs and rib with triangular cut outs. Two different materials such as AA 7075 and Glass fibre (GF) were considered to run the analysis. Analysis such as structural, modal and fatigue were performed on the designed models.

The results indicate that rib with aluminium alloy 7075 has 9.531×10^{-2} mm total deformation with von misses stress of 55.405 MPa. And for rib with circular cut outs the total deformation found was 0.1102 mm and von misses stress was found is 53.558 MPa. The rib with rectangular cut outs found total deformation of 0.10443 mm and von misses stress about 52.814 MPa. The rib with triangular cut outs has a total deformation of 9.581×10^{-2} mm and von misses stress of about 53.847 MPa. This indicates that the total deformation is found maximum for the rib with circular cut outs and less for rib without cut outs.

The results indicate that rib with glass fibre has 0.85569 mm total deformation with von misses stress of 557.06 MPa. And for rib with circular cut outs the total deformation found was 1.0006 mm and von misses stress was found is 541.99 MPa. The rib with rectangular cut outs found total deformation of 0.94599 mm and von misses stress about 524.11 MPa. The rib with triangular cut outs has a total deformation of 0.85891 mm and von misses stress of about 529.86 MPa. This indicates that the total deformation is found maximum for the rib with circular cut outs and less for rib without cut outs.

The results obtained from modal analysis revealed that highest natural frequency is possessed by circular cut-outs in both materials.

From the life cycles obtained in fatigue test it is concluded that the rectangular cut-out rib exhibits better life compared to other designs in both materials. It also helpful to validate the static structural results.

IOP Conf. Series: Journal of Physics: Conf. Series **2765** (2024) 012001 doi:10.1088/1742-6596/2765/1/012001

References

[1] Dharmendra P, Chaithanya K J, Ayesha Sameera, Khyati Kavathiya and Monika K M 2020 *International Research Journal of Engineering and Technology* **7(6)** 180-192.

[2] Rohit Kumar, Syed Mohsin and Ravindra Jaikishan Singh 2020 *Journal of Emerging Technologies and Innovative Research* **7(6)** 1711-1719.

[3] Sudhir Reddy. Konayapalli and Sujatha Y, 2015 *International Journal of Research* **2(10)** 117-123.

[4] Jaffar Syed Mohamed Ali, Wan Muhammad Hafizuddin W. Embong and Abdul Aabid 2021 *CFD Letters*. **13** 87-94.

[5] Nirmal Raj N S, Halesh Koti and Channankaiah, 2014 *International Journal of Engineering Research & Technology* **3(3)** 2300-2303.

[6] K. Sruthi, T. Lakshmana Kishore and M. Komaleswara Rao 2017 *International Journal of Engineering Development and Research* **5(4)** 949-959.

[7] Sanya Maria Gomez and Alice Mathai 2011 *International Journal of Earth Sciences and Engineering* **4(6)** 648-651.

[8] Kakumani Sureka and Satya Mehar R 2015 *International Journal Mechanical Engineering and Robotics Research* **4(2)** 123-130.

[9] Mounika R, Sai Deepthi T, Anil V, Akhil T and Shiva Kumar A 2022 *International Journal of Research Publication and Reviews* **3(7)** 995-1006.

MECHANICAL PROPERTIES OF BASALT / CHOPPED E-GLASS FIBER AND GRAPHITE POWDER REINFORCED HYBRID COMPOSITES

R Bhargavi[1, 3, 5], **V Ajay Kumar**[1, 3], S **Madhusudan**[2, 3] **and S Rajesh**[2, 4]

[1]Associate Professor, Department of Mechanical Engineering.
[2]Professor, Department of Mechanical Engineering.
[3]Usharama College of Engineering and Technology, Telaprolu, Andhra Pradesh, India.
[4]SRKR College of Engineering, Bhimavaram, Andhra Pradesh, India.
Email: rebba.bhar2@gmail.com.

Abstract. In the current investigation an effort has put to study the effect of E-glass chopped mat and graphite powder in variations of 5g, 10g, and 15g on the tensile and flexural behavior of a basalt fiber-reinforced epoxy composite laminate reinforced with E-Glass fiber and Graphite powder. Hand layup technique has been used for the fabrication of the composites. Tensile, flexural, impact, and hardness tests were done on the Basalt/E-glass/Graphite epoxy composite in accordance with ASTM standards. The results indicate significant improvement in tensile strength, impact strength and hardness when E-glass was incorporated in the basalt/epoxy composite. However, basalt/E-glass/10% graphite epoxy composite was shown to have a greater flexural strength. An excellent dispersion of the reinforcements in the polymeric matrix, which elevated surface area for solid interfacial contact and effective load transmission, could possibly be accountable for the improved performance of the laminates that were successfully developed.

1. Introduction

A composite material comprises anything that is produced by combining particles/fibers with a matrix material to form a new material with qualities that differ from the properties of the original material. Composite materials are a promising choice for engineers, material analysts hence; they are being used for great deal of applications. These materials have transformed into the choice of standard fundamental materials. Composites are proved to have phenomenal mechanical properties, low weight, insignificantly expensive and significantly versatile. Basalt fiber is a high-tech fiber that is green, healthful, and ecologically beneficial. Strengthening the investigation and study of the characteristics of basalt fiber and its composite materials is important from both a theoretical and a strategic standpoint. According to published research, basalt fiber outperforms glass, carbon, and aramid fibers in terms of mechanical strength, acid-alkali resistance, electrical performance, high wave permeability, and sound insulation. Seghini et al [1] focused on the mechanical properties of basalt fiber epoxy composites and the association between the impacts of business coupling specialist heated de-estimating treatment and plasma polymerization measure on the fiber Matrix interfacial strength. The surface morphology of the various basalt strands was depicted by FE-SEM perceptions and substance organizations for shaping FT-IR examination. Plasma polymerization results in a homogenous tetra-vinyl-silane (PP-TVS) coating on the surface of basalt filaments, which expands in fiber/Matrix attachment. The AFM has handled the surface discomfort of both treated and untreated basalt.

Maria et al [2] studied the effect of discarded basalt powder on inflexible polyurethane froths. It was revealed that basalt powder causes an increase in the reactivity movement of the polyurethane framework during the frothing cycle, which was also confirmed by the dielectric polarization transforms. Froths being a successful method of minimizing waste administration. The waste filler in basalt powder types catalyzes the shaping system of the polyurethane framework. Danuta et al [3] investigated the mechanical properties of basalt powder epoxy composite. The epoxy matrix slows the rate of degradation of basalt powder, which increases the solidity and hardness of the composite and improves the thermo mechanical properties of the composite with the increase in basalt powder content.

Chen et al [4] conducted testing on basalt fiber supported polymer to know the effect of stacking on mechanical properties. It is discovered that composites become more elastic when strain rate increases, especially when the strain rate is greater than 120 S-1. The quasi-static rigidity of a polymer made of basalt fibers is around 1642.2 MPa, and with a strain rate of 259 S-1, it rises to 3383.3 MPa. The flexible modulus rises from 77.9 GPa to 111.7 GPa with a 43.4 percent increase as the strain rate reaches 259 per second. The mechanical properties of basalt fiber and E-glass fiber built-up plastics made utilizing a vacuum pack process were examined by Lopresto et al [5]. Dorigato et al [6] focused on the impact of Fatigue Obstruction of Basalt Fibers Supported Covers. Hand layup is employed to install the epoxy overlays, which have Carbon, basalt, and E-glass adjustable weaving textures. According to the findings, basalt fiber composites have a greater flexible modulus than matching glass fiber covers, but their rigidity is similar to that of comparing carbon strand covers. Basalt Fiber overlays outperformed glass Fiber covers in terms of firmness persistence at low exhaustion stacks and damping properties, in accordance to the investigation on wear behavior. It was contended that basalt Fiber possessed the potential to replace glass strands in the development of elementary composites that had outstanding mechanical characteristics with interesting energy dissipation capabilities. Characterization of Glass/Vinyl Ester, Basalt/Vinyl Ester, and Basalt/Epoxy Fiber Reinforced Polymer were the main topics of BrahimBenmokrane et al's study [7].

Drilling Analysis on Basalt/Sisal Reinforced Polymer Composites was studied by Sakthivel et al. [8]. Using an HSS drill, ANOVA and regression models were applied to composites made of sisal and basalt. The "Minitab 17" code is then used to prepare the data that was collected by penetration. The effect of heat treatment on the tensile properties of basalt fibers was the primary objective of Sabet et al's study [9]. Authors studied about the rigidity of basalt filaments at room temperature as well as after being exposed to 300°C, 350°C, 400°C, 450°C and 500°C with respect to time. The results demonstrate the optimal conditions for preparing basalt filaments and composites. Alexander et al [10] investigated the combined effects of heat and microwave curing on the mechanical characteristics and surface hardness of basalt/epoxy composites. The results reveal that the surface suits are further developed in tiny wave restored composites as a result of the miniature wave post relieving treatment. In addition to the above, the strength and solidity properties are also improved. The thermo mechanical characteristics of fly ash and glass fiber-reinforced epoxy resin composites have been described by Baheti et al. [11].

The composite was created by hand-laying it up. Fabric and epoxy glue have been incorporated 50:50. The quantities of dry fly ash used were 1, 3, 5, and 10%. When the fly ashes were included in the right weight proportion, the amount of epoxy resin was altered as necessary. The constructed samples were put through tensile testing in accordance with ASTM D638, flexural testing in accordance with ASTM D790, Charpy impact testing in accordance with ASTM D256, a test for resistance to knife penetration, DMA, and fracture surface analysis. In a distinct research, Srivastava and Pawar (2006) developed composites via combining epoxy resin with fly ash particles in a 4:1 ratio with 105 m-sized fly ash. In addition, fatigue strength-weight ratios and moreover tolerance to fatigue damage of many composite laminates are remarkable.

In the current investigation an effort has been made to investigate the effect of E-glass fiber and graphite powder which has been incorporated in variations of 5g, 10g, and 15g on the tensile and flexural characteristics of the basalt fiber-reinforced epoxy resin composite laminas produced through the hand layup technique.

2. Experimentation

2.1 Materials Used

In the current investigation, LY556 type epoxy resin and HY-951 hardener were employed for producing composite elements. Basalt, E-glass, and graphite powders are the reinforcing materials used in the present investigation. These are mixed in various amounts and combinations.

2.1.1 Epoxy Resin

Epoxy is a thermoset polymer principally used in aerospace and aircraft applications. Epoxy LY556 is employed as the lattice material in the current investigation for producing half-and-half fiber epoxy composites. Since epoxy LY556 is a widely utilized material with excellent mechanical properties, low shrinkage, easy creation, amazing substance and dampness obstruction, and high wet capacity, it was chosen. Epoxy pitches are the thermoset plastic most frequently utilized in polymer lattice composites. Epoxy gums are a type of thermoset plastic substance that does not emit reaction products when subjected to curing, resulting in little cure shrinkage. They also have excellent adhesion to a variety of materials, excellent synthetic and environmental obstructive qualities, and excellent protective properties.

2.1.2 Basalt

Basalt fiber, a bio-derived mineral fiber from volcanic rocks, offers numerous of benefits, notably strong mechanical and chemical resistance [12–17]. In accordance with Table 1, basalt fiber demonstrates tensile attributes that are analogous to those of glass fibers and possess a greater maximum service temperature than carbon and glass fibers. This makes basalt fiber composites an efficient and ecologically conscious alternative for glass fiber composites in structural applications. Furthermore, it has a better maximum service temperature as well as superior chemical resistance, making it ideal to use in challenging regions like chemical storage tanks or composite pipes for the oil and gas sectors [13]. Basalt fiber, which originates from volcanic rocks and has mechanical characteristics equivalent to glass fiber, is produced.

2.1.3 E-Glass

Glass fibers are typically employed as reinforcements in polymeric (plastic) matrix composites (PMC). Glass fibers have several advantages, including low cost, great tensile strength, chemical resistance, and good insulating characteristics. E-glass, high-strength (HS)-glass, and corrosion-resistant (CR)-glass are the three most prevalent forms of glass fibers. The first significant synthetic composite reinforcement, E-glass fibers (thus the "E" in its name), was first created for electrical insulation purposes. The sizes of the filaments range from 3 to 20 m. The typical characteristics of E-glass fibers are shown in Table1. When compared to alternative reinforcements, the elastic moduli of e-glass fibers are comparably low. Additionally, creep and creep (stress) rupture are risks for E-glass fibers.

2.1.4 Graphite

Graphite is a carbon allotrope. Graphite is a particularly stable crystalline form of carbon found in nature. Graphite is a black substance. It exhibits the shine of metals and is opaque. The carbon atoms in graphite form a hexagon shape, and the hexagons are stacked in layers. It might be crystalline, amorphous, lumpy, or graphite fiber. The graphite powder has a high electrical conductivity.

Table 1. Properties of Basalt and E-Glass Fiber

Material	Density (g/cm3)	Tensile strength (GPa)	Specific strength	Elastic Modulus (GPa)	Specific Modulus
Epoxy	1.14	0.05-0.13	0.05-0.008	34.7	31
Basalt fiber	2.65	2.9 – 3.1	1.09 – 1.17	85-87	32.1 – 32.8
E- Glass	2.6	2.5	0.962	76	29.2

3. Fabrication of composite specimens (Hand layup)

3.1 Basalt- Epoxy resin composite

The procedures in the processing are really basic. According to the data, the ASTMD790M-86 standard test protocol for mechanical characteristics of fiber-resin composites is used. To prevent polymer from adhering to the surface of the mold, a release gel is first sprayed on it. At the top and bottom of the mold plate, thin plastic sheets are used to provide a high-quality surface finish for the finished product. On top of a thin plastic layer, long fibers formed of the reinforcement (basalt) are placed down in a straight line after being trimmed to meet the mold's specifications. After that, a prescribed hardener (curing agent) is mixed completely with the liquid thermosetting polymer before being placed on top of transparent. A brush is used to distribute the polymer evenly. A second coating of fiber is then added to the surface of the polymer, followed by another layer of polymer, and finally, this is sealed with a further thin plastic sheet after the air is removed by gently moving the squeezer over the thin plastic sheet. The resulting mold is allowed to dry at room temperature for 24 hours.

3.2 E-Glass - Epoxy resin composite

Using the same technique as before, later samples with reinforcements made of basalt and E-Glass fiber are created. Long fiber reinforcement (50 percent basalt and 50 percent e-glass) is cut to the mold size and laid down in a straight line on top of the thin plastic sheet. After that, a prescribed hardener (curing agent) is mixed completely with the liquid thermosetting polymer before being placed on top of transparent. A brush is used to distribute the polymer evenly. A second applying of fiber is then added to the surface of the polymer, followed by another layer of polymer, and finally, this is sealed with a further thin plastic sheet after the air is removed by gently moving the squeezer over the thin plastic sheet. The resulting mold is allowed to dry at room temperature for 24 hours. In this interaction, 100 grams of Epoxy (LY556), which serves as the composite's lattice, is mixed with 8 sheets of 300 GSM E-glass fiber, hacked mate (230/300 mm), and 10 grams of hardener (HY951).

3.3 Basalt/E-glass/Graphite Epoxy resin composite

The graphite powder illustrated in Figure 1 is combined with epoxy resin and binder and thoroughly mixed using a stirrer for at least 10 minutes to create the basalt/ E-Glass/ Graphite epoxy composites. The fly ash and basic plastic need to be combined in the right proportions to achieve the optimum results. Therefore, determining the required amount of graphite is crucial. A variety of samples were created using fly ash in weights ranging from 5 to 15%. By altering the weight percentages of the graphite powders in the basalt/E-glass fiber epoxy resin composite by 5%, 10%, and 15%, samples were created. For tensile testing, the samples were created according to ASTM standards, which call for them to be 165 mm long, 12.5 mm wide, and 4 mm thick. For flexural testing, shapes of 100 mm long, 25 mm wide, and 4 mm thick are constructed. Impact testing uses materials that are 6 mm thick, 12.36 mm wide, and 63.5 mm long. The stepwise manufacturing process used to create the specimens is shown in Figure 2. The steps include; 1. Taking the materials 2 &3. Weighing them to the required quantity 4. Mould preparation 5. Laying a layer of epoxy resin 6. Laying a layer of E-glass and Basalt fiber reinforcement 7. Pouring another layer of resin 8. Drying and demoulding the specimen.

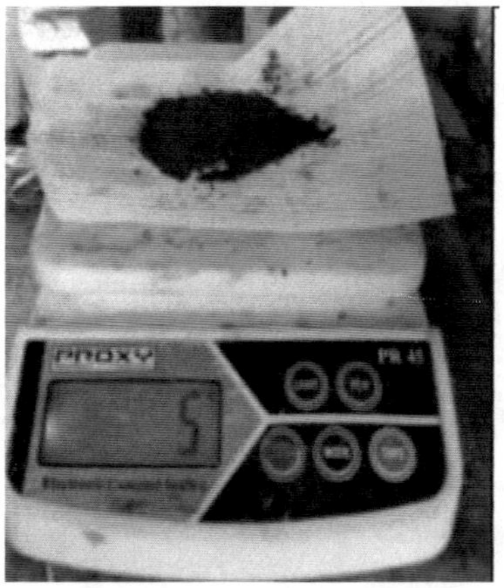

Figure 1. Graphite powder

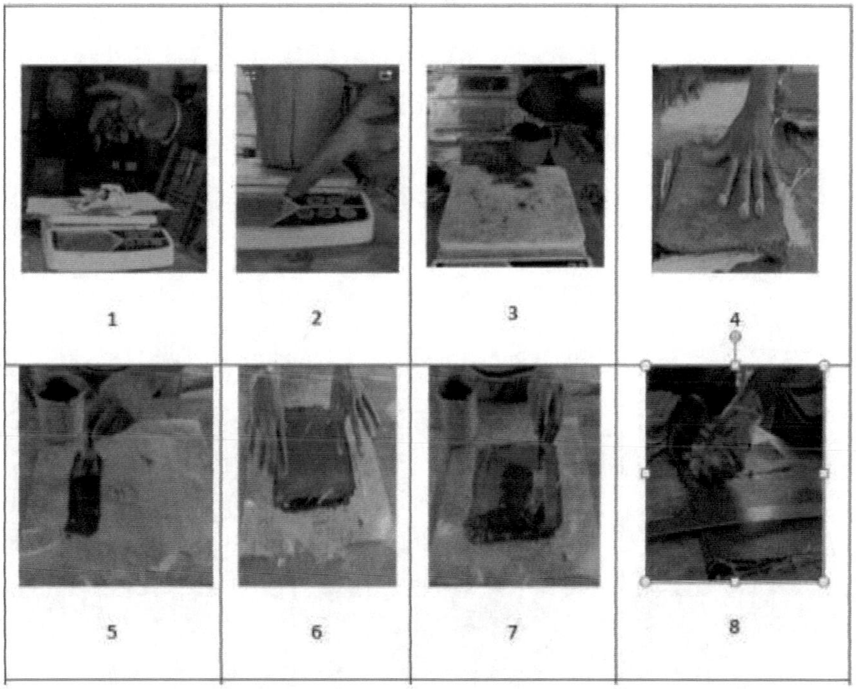

Figure2. Sequential procedure of fabrication of the composites

4. Experimental testing of composites specimens

4.1 Tensile testing of composites

An electronic tensometer of 2 ton capacity, METM 2000 ER-1 model is used to perform the tensile strength test of the composites. For testing the composite, a 2 ton weight cell is used. The electronic tensometer used in the current work is equipped with weight and increase markers, each with an insignificant count of 0.01 kg and 0.01mm. Specimens for the test are 165 mm long, 12.5 mm broad and 4 mm thick.Figure3 shows specimens after tensile test.

IOP Conf. Series: Journal of Physics: Conf. Series **2765**(2024) 012002 doi:10.1088/1742-6596/2765/1/012002

4.2 Flexural testing of composites

According to ASTM-D790M-86 test standard, a three point bending test is carried out. Specimens of 100 mm long, 25 mm wide, and 4 mm thick are employed for this purpose. The oppression of two identical occurrences is used for flexural testing. The test was run at a strain rate of 0.2 mm/min with the outer rollers distanced by 70 mm.Figure4 shows failed specimens after flexural test.

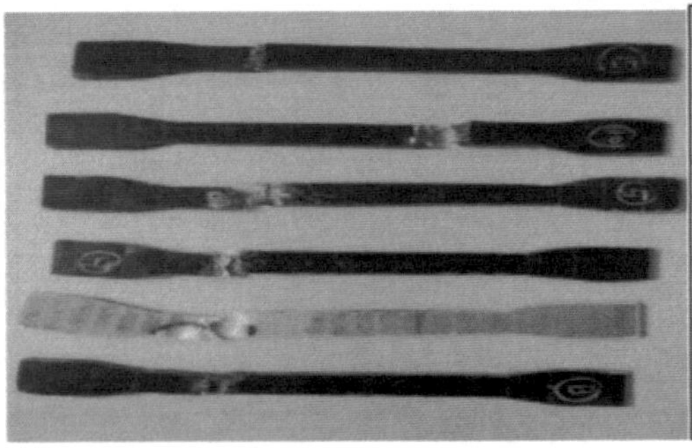
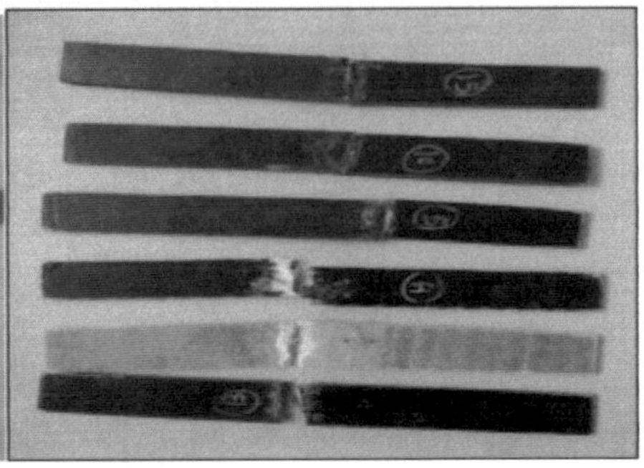

Figure 3. Specimens after tensile test**Figure 4.** Specimens after flexural test

4.3 Impact testing of composites

The charpy v indent test, was performed for testing the impact strength of the fabricated composite. A set of 3 samples were tested in each category and the average results were reported. Figure5 shows the specimens after impact test.

4.4 Hardness

ASTM D 2583 test standard specimens are prepared to determine the hardness of the specimens. Brinnel hardness tester of model GYZJ- 934-1 was used for testing. The Colman BarcolImpressor was used for hardness testing in accordance with determinations. The developed composites were cut into specimens of 12.7 x 12.7 x 3 mm with a length, breadth, and thickness respectively.Figure6 shows the tested specimens.

Figure 5. Specimens after impact test

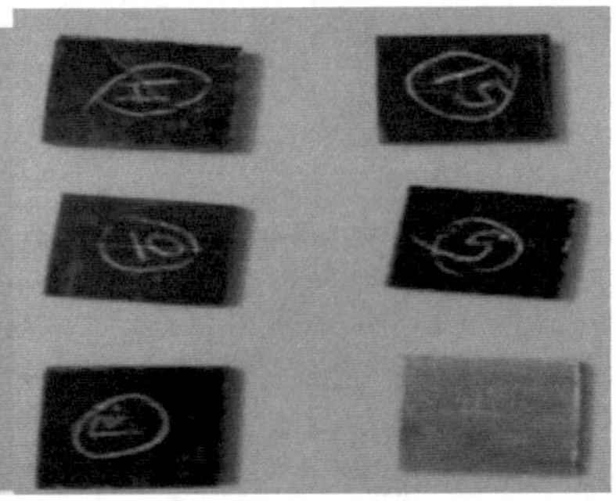

Figure 6. Specimens after hardness testing

5. Results &Discussion

5.1 Tensile strength

Table2.shows the summary of tensile strength for various composites performed as per ASTM D638 standard. Figure 7 shows the tensile load Vs. elongation comparison for Basalt, E glass with chopped mat, Basalt with E glass chopped mat, Basalt and E glass chopped mat 5% graphite powder, Basalt and E glass chopped mat 10% graphite powder, Basalt and E glass chopped mat 15% graphite powder composites. It can be observed that the load Vs elongation for B+E and B+E+10% G has shown higher load withstanding capability as compared to other combinations. Pure B and B+E+15% G composites has shown much inferior tensile strength values. Higher elongation values are also observed for B+E and B+E+10% G composites. Pure B and B+E+15% G composites has shown much inferior elongation to failure values. The variation in the tensile strength exhibited by the samples could be attributed for the factors like uniform distribution of fiber and graphite powder in the resin matrix, bonding between fiber and matrix [3].

Table. 2 Tensile strength for various composites

S.NO	Composite	Load (N)	Elongation (mm)	Tensile strength (N/mm^2)	Elongation (%)
1	B	6890	8.4	11.13	5.51
2	E	8720	16.1	13.90	10.20
3	B+E	9340	22.2	14.84	13.84
4	B+E+G5	4570	8.2	7.62	5.39
5	B+E+G10	9150	20.3	14.56	12.76
6	B+E+G15	8350	9.6	13.34	6.24

5.2 Flexural strength

Table.3 shows the summary of flexural strength for various composites. Figure 8 shows the flexural strength variation for Basalt, E glass with chopped mat, Basalt with E glass chopped mat, Basalt and E glass chopped mat 5% graphite powder, Basalt and E glass chopped mat 10% graphite powder, Basalt and E glass chopped mat 15% graphite powder. It can be observed that E and B+E+10% G composites have shown superior flexural strength values as compared to other composites. Both B (without graphite powder) and B+E (without graphite powder) composite has shown inferior properties, while addition of 10% graphite powder to B+E composite has shown improved flexural strength as compared to E. This was further reduced upon the addition of 15% graphite powder. This can be attributed to brittle nature of composite as result of addition of higher graphite content [3].

Table 3. Flexural strength for various composites

S.NO	Composite	Load (N)	Elongation (mm)	Flexural strength (N/mm^2)
1	B	1170	6.8	307.125
2	E	1450	8.5	475.125
3	B+E	1810	8.0	380.625
4	B+E+G5	1400	7.0	367.5
5	B+E+G10	1960	8.6	514.5
6	B+E+G15	1550	7.3	406.875

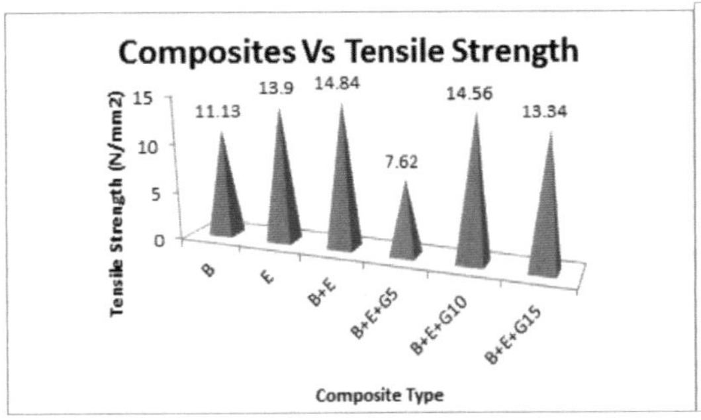

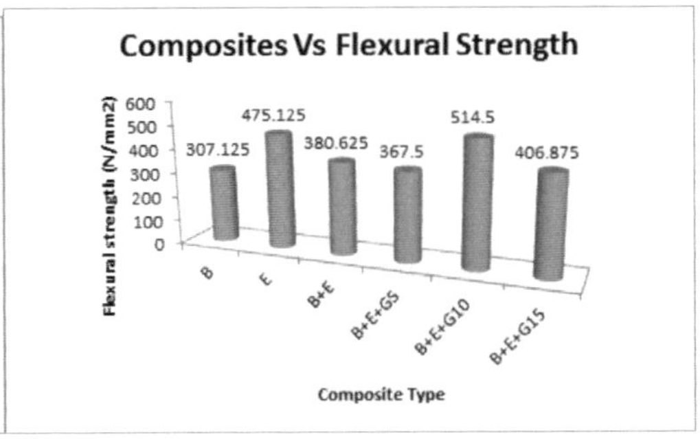

Figure 7. Tensile strength variation for composites**Figure 8.** Flexural strength vs. composites

5.3 Impact strength
The impact strength values are summarized in Table 4. Figure9 shows the variation in impact strength values for plain woven basalt, plain woven E-Glass hacked mate fiber and basalt/E-glass cleaved mate, basalt/E-glass slashed mate with Graphite power variety 5%, 10%, 15% composites. It can be observed that pure E and B + E have shown better impact strength values as compared to other composites. It is true that addition of Graphite powder makes the composite brittle and matrix becomes discontinues there by resulting in reduced impact strength values [4].

Table 4. Impact strength for various composites

S.NO	COMPOSITE	JOULES J/m
1	B	3.9
2	E	5.5
3	B+E	5.7
4	B+E+G5	3.7
5	B+E+G10	5.47
6	B+E+G15	5.4

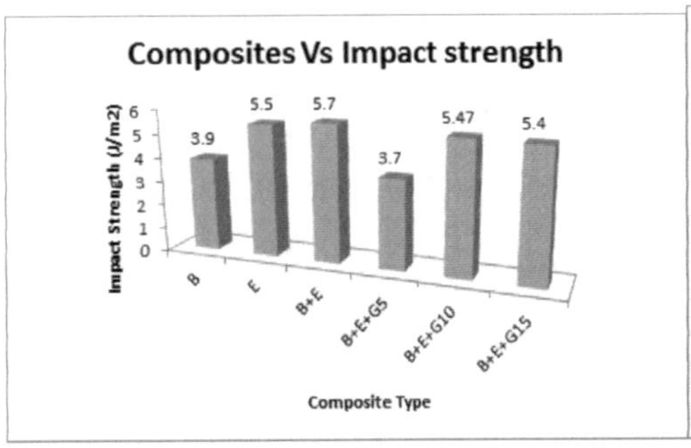

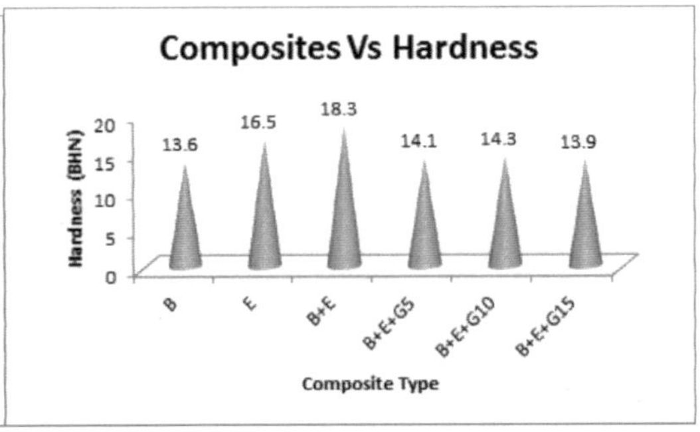

Figure 9. Impact strength vs. composites**Figure 10.** Hardness number vs. composites.

5.4 Hardness number
After testing, the hardness values are summarized in Table 5. Figure10 shows the variation in hardness values for Basalt, E glass with chopped mat, Basalt with E glass chopped mat, Basalt and E glass chopped mat 5% graphite powder, Basalt and E glass chopped mat 10% graphite powder, Basalt and E

glass chopped mat 15% graphite powder. It can be observed that pure E and B+ E have shown higher hardness values as compared to other composites. Since E glass fiber possesses higher hardness as compared to basalt the result for B + E is on higher side. It is interesting and contradicting that addition of graphite powder not yielded in improvement of hardness values. The existence of voids is thought to be the cause of variation in the observed behavior. It is nearly impossible to prevent the production of air bubbles and voids. The voids serve as stress raisers, which start the fractures, in addition to reducing the stress carrying area [3].

Table5. Brinell hardness number

S.NO	COMPOSITE	BRINELL HARDNESS NUMBER
1	B	13.6
2	E	16.5
3	B+E	18.3
4	B+E+G5	14.1
5	B+E+G10	14.3
6	B+E+G15	13.9

6. Conclusion

Basalt/E-glass/Graphite Epoxy resin composites were successfully fabricated by using the basic hand lay-up process. Experiments were carried out to study the effect of E-glass and graphite on the mechanical properties of basalt epoxy resin composites. The following observations have been derived from these studies:

1. Basalt/E-Glass chopped mat composite have shown the best result corresponding to the tensile strength when compared to pure basalt and pure E-glass epoxy resin composite samples.
2. Superior flexural strength has been observed for basalt/E-Glass + 10% graphite composite in comparison with the other composite samples.
3. Though the impact behavior for pure E glass, basalt/E-Glass + 10% graphite, basalt/E-Glass chopped mat + 15% graphite composites were almost closer, yet Basalt/E-Glass chopped mat shown better among them.
4. The Brinell hardness number was found to be higher for Basalt/E-Glass composite as compared to other remaining composites.

References:

[1] M.C.Seghini, F.Touchard, F.Sarasini, V.Cech, L.Chocinski-Arnault, D.Mellier, J.Tirillò, M.P.Bracciale, M.Zvonek, July 2019, "Engineering the interfacial adhesion in basalt/Epoxy composites by plasma polymerization", vol.122, pp. 67-76,

[2] MariaKurańska, July 2019, "Fabrication and Testing of chopped mate e glass and basalt fibers with iron ore powder using hand layup technique"vol.76, pp.90-100.

[3] Danuta, Matykiewicz, MateuszBarczewski, SławomirMichałowski, May 2019, "Basalt powder as an eco-friendly filler for epoxy composites: Thermal and thermo-mechanical properties assessment", vol.164, pp.272-279.

[4] Chen, Wensu, Hong Hao, Michael Jong, Jian Cui, YanchaoShi, Li Chen, and Thong M. Pham, September 2017, " Quasi static and dynamic tensile properties of basalt fiber reinforced polymer" Composites Part B: Engineering, Vol. 125, pp.123-133.

[5] V. Lopresto, C. Leone, and I. De Iorio, June 2011, "Mechanical characterisation of basalt fiber reinforced plastic," Compos. Part B Eng., vol. 42, no. 4, pp. 717–723.

[6] Dorigato and Pegoretti, April 11, 2013, "Flexural and impact behavior of carbon/basalt fibers hybrid laminates", J. Compos. Mater. vol. 48, no. 9, pp. 1121–1130.
[7] Benmokrane B, Elgabbas F, Ahmed E, et al. Dec 2015,"Characterization and comparative durability study of glass/vinylester, basalt/vinylester, and basalt/epoxy FRP bars", *ASCE J Compos Construct*,doi: 10.1061/(ASCE)CC.1943-5614.0000564. pp. 04015008:1-12.
[8] M. Sakthivel, 2015, "Drilling Analysis on Basalt / Sisal Reinforced Polymer Composites Using ANOVA and Regression Model," vol. 9, no. 66, pp. 3285–3290.
[9] S. M. M. Sabet and F. Akhlaghi, February 20, 2015, "The effect of thermal treatment on tensile properties of basalt fibers", J.Ceram. Sci., vol. 248, pp. 245–248.
[10] J. Alexander and B. S. M. Augustine, 2015, "The Effect of Combined Thermal and Microwave Curing on Mechanical Properties and Surface Hardness of Basalt / epoxy Composites," vol. 23, no. 2, pp. 347–349.
[11] Baheti, Vijay &Militky, Jiri & Mishra, Rajesh &Behera, B.K., (2015). "Thermo-mechanical properties of glass fabric/epoxy composites filled with fly ash", Composites Part B: Engineering. 85. 268-276. 10.1016/j.compositesb.2015.09.049.
[12] Wei B., Cao H., Song S., October 2010, "Tensile behavior contrast of basalt and glass fibers after chemical treatment", *Mater. Des, Vol.* 31, Issue 9, pp. 4244–4250.
[13] Fiore V., Scalici T., Di Bella G., Valenza A., June 2015, "A review on basal fiber and its composites", *Compos. Part B Eng.*;74:74–94.
[14] Monaldo E., Nerilli F., Vairo G. 15 April, 2019, " Basalt-based fiber- reinforced materials and structural applications in civil engineering", *Compos. Struct.*; 214:246–263.
[15] Kufel A., Praa S., Kuciel S., 25 Sep 2021,"Basalt/Glass Fiber Polypropylene Hybrid Composites: Mechanical Properties at Different Tem-peratures and under Cyclic Loading and Micromechanical Modelling", *Materials.*14:5574. doi: 10.3390/ma14195574.
[16] Li Z., Ma J., Ma H., Xu X. 2018, "Properties and Applications of Basalt Fiber and Its Composites", *IOP Conf. Ser. Earth Environ. Sci.*; 186:012052.
[17] Technical Datasheet for Basalt Fibers. Lance Brown Import-Export (LBIE) [(accessed on 5 January 2021)].

Impact of Exponential Heat Source and Thermal Radiation on the flow of Hybrid Nanofluid across a Bi-Directional Stretching Surface with Activation Energy

Rajavath Narayana Naik[1], S. Suneetha[2], K.S. Srinivasa Babu[3,*] and M. Jayachandra Babu[4]

[1,2]Department of Applied Mathematics, Yogi Vemana University, Kadapa-516005, India
[3]Department of EM&H, S.R.K.R. Engineering College, Bhimavaram-534204, India
[4]Department of Mathematics, Government Degree College, Rajampeta-516126, India

*Corresponding author email: kssb@srkrec.ac.in

Abstract. The study focuses on how activation energy and exponential heat source affects the radiative motion of a hybrid nanofluid $(EG+ZnO+TiO_2)$ over a bidirectional elongating sheet. Convective boundary condition is assumed. The bvp4c, an inherent function in MATLAB, is used to unravel the altered system deduced from those equations which describe the current work in the two scenarios, i.e., binary ($EG+ZnO+TiO_2$) and mono ($EG+ZnO$) nanofluids. Added a little quantitative discussion regarding engineering parameters including Sherwood number. The most notable findings from this study are mentioned here. Enhanced volume fraction of ZnO lessens the fluid velocity and increasing heat source parameter step-up the fluid temperature. It was noticed that with a raise in magnetic field and the volume fraction of ZnO lessens the friction factor. The rate of mass transfer is lowered with the enhanced activation energy parameter and the same enhances as the reaction rate parameter is amplified. The increment quantities in the Sherwood number against the reaction rate parameter (Γ) are 0.35151 ($EG+ZnO+TiO_2$) and 0.35125 ($EG+ZnO$), when Γ is set to $0 \le \Gamma \le 0.9$. Furthermore, the decrement rates in mass transfer rate against the activation parameter (E_n) are observed as 0.02681 ($EG+ZnO+TiO_2$) and 0.02687 ($EG+ZnO$), when E_n is set to $0 \le E_n \le 2.5$. It is detected that there is an amelioration in the heat transmission rate with the raise in Biot number (Bi). It has been noticed that, when Bi fixed to be $0 \le Bi \le 0.8$, Nusselt number is lowered by 0.736332 ($EG+ZnO+TiO_2$) and 0.621826 ($EG+ZnO$). Our findings are checked against previous findings for validity. A reasonable compromise has been uncovered.

Keywords: Hybrid nanofluid, Bidirectional stretching sheet, Activation energy, Heat source, bvp4c.

1. Introduction

Nanofluids becomes a novel and intriguing class of heat transfer fluids which can be used in place of more conventional options. Heat exchangers, Fuel cells and pharmaceutical procedures are few examples for them. Saba et al. [1] scrutinized the radiative nanofluid flow (Water + SWCNT/MWCNT) by a curled elongating surface together with heat sink/source. Prabavathi et al. [2] considered the same nanofluid and inspected the chemically reactive and radiative Maxwell fluid flow by a vertical cone using FEM (finite element method). Upsurges in the Maxwell parameter is observed to result in an increase in the fluid's temperature. Basha et al. [3] elucidated the radiative nanofluid (water+EG/Diamond and water+EG/SWCNH) flow via a plate/wedge. Saranya and Al-Mdallal [4] carried out a study on various shapes of nanoparticles in the radiative MHD nanofluid (Silicon oil + Al_2O_3) flow by a revolving disk. The heat transmission rate of nanoparticles with a blade form is measured to be significantly higher than that of other shapes. Newly, different researchers [5-7] assumed distinct geometries and analyzed different nanofluid flows including $Ag - Water$ nanofluid. Devi and Devi [8] discussed the features of heat transfer in a HNF ($Cu - Al_2O_3 / Water$) by an elongating sheet. They reported that the enhancement in the heat transmission rate is nearly 17.3% high equated to pure water for nano particle volume fraction. Bioconvective flow of HNF (Water + Copper + Silica) by a revolving disc was modelled mathematically by Jayadevamurthy et al. [9], who utilized a mix of RKF-45 and shooting procedures to unravel the problem. Mabood et al. [10] did an irreversibility analysis on the radiative HNF ($Cu - Al_2O_3 / H_2O$) flow by a melting surface and noticed the increase in heat transmission as the radiation parameter was increased. Khashiie et al. [11] analyzed the same HNF flow by a shrinking/elongating sheet with various conditions including slip with bvp4c function (MATLAB in-built). Acharya and Mabood [12] numerically investigated a hybrid nanofluid (HNF) motion across a slippery bended surface subject to a nonlinear thermal radiation. Freshly, many studies [13- 22] assumed various geometries and examined numerous HNF flows under various conditions, including quadratic velocity. With the increase in nanoparticle volume fraction, it is noticed a 5.5 percent increase in friction factor (2 percent to 4 percent).

Many different industries, including electricity production and solar energy, rely on insights from the heat transmission study of boundary layer flow when radiation is taken into account. Salomatov and Puzyrev [23] provided an analysis for the convective flow of a fluid by a flat surface with thermal radiation in two cases i.e., large and small assesses of radiation. Similar geometry was taken into account by Hossain and Takhar [24], who theoretically explored the radiation-mixed convective flow of a fluid. Researchers have shown that higher levels of radiation lead to a reduction in shear stress. The numerical investigation of the radiative Darcy-Forchheimer flow of an optically dense fluid by a wedge was reported by Al-Odat et al. [25] using the Keller box scheme. It has been noticed that the temperature of the fluid decreases when the heat sink is deployed. Abdel-Wahed [26] modelled the radiative MHD nanofluid flow through a proceeding surface and observed that the magnetic field impact lessens the heat transmission rate. Khan et al. [27] carried out research on the Williamson fluid flow through an elongated surface with the aid of Lie group procedure. Observations show that when the Weissenberg number rises, so does the momentum boundary layer thickness. Sheremet et al. [28] numerically analyzed the radiative convective flow of a nanofluid ($Al_2O_3 + H_2O$) within a tended square cavity. They emphasized that the lean angle of cavity could be a better regulator factor for fluid flow. With the magnetic field effect, on the same fluid, Sheikholeslami et al. [29] studied the radiative convective Darcy flow inside a wavy surface. Later, different researchers [30-40] discussed distinct radiative nanofluid flows through different geometries under various conditions.

On the radiative Hybrid nanofluid flow, ($EG + ZnO + TiO_2$) via a bidirectional elongated sheet, no works were not yet proposed to reveal the impact of activation energy and diverse heat source parameters. SLR (Slope of Linear Regression) and graphs are used to discuss engineering characteristics of relevance, such as skin friction coefficient. The current results are also justified by comparing them with the past results.

2. Formulation

In this study, the radiative hybrid nanofluid motion across a bidirectional elongating surface with activation energy and diverse heat source parameter was investigated. As a base fluid, Ethylene glycol is considered, ZnO and TiO_2 are taken to be nanoparticles for hybrid nanofluid considered. Table-1 displays the numerical values of their thermophysical parameters. Magnetic field with intensity B_0 is enforced in the z - direction (observe Fig. 1). We assumed $U_w(x) = a_1 x$, the elongating velocity of the sheet in x-direction, and it is $V_w(y) = a_2 y$ in y-direction.

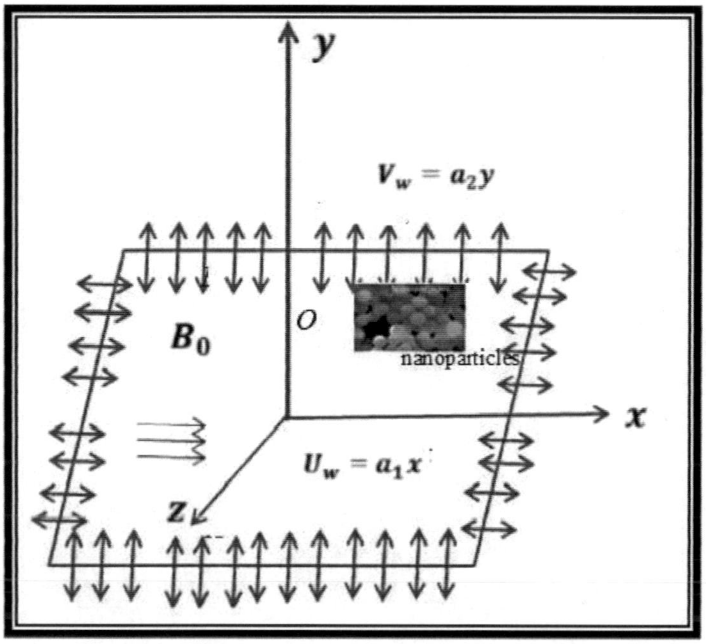

Figure 1. Flow geometry (Gupta and Gupta [41])

Under these premises, governing equations and boundary conditions are presented as (Gupta and Gupta [41], Al-Kouz et al. [42] and Irfan et al. [43]):

$$\frac{\partial}{\partial x}(u)+\frac{\partial}{\partial y}(v)+\frac{\partial}{\partial z}(w)=0 \tag{1}$$

$$\rho_{hnf}\left(u\frac{\partial}{\partial x}(u)+v\frac{\partial}{\partial y}(u)+w\frac{\partial}{\partial z}(u)\right)=\mu_{hnf}\frac{\partial}{\partial z}\left(\frac{\partial}{\partial z}(u)\right)-\sigma_{hnf}B_0^2u-u\frac{\mu_{hnf}}{\kappa} \tag{2}$$

$$\rho_{hnf}\left(u\frac{\partial}{\partial x}(v)+v\frac{\partial}{\partial y}(v)+w\frac{\partial}{\partial z}(v)\right)=\mu_{hnf}\frac{\partial}{\partial z}\left(\frac{\partial}{\partial z}(v)\right)-v\sigma_{hnf}B_0^2-\frac{\mu_{hnf}}{\kappa}v \tag{3}$$

$$\left(\rho C_p\right)_{hnf}\left(u\frac{\partial T}{\partial x}+v\frac{\partial T}{\partial y}+w\frac{\partial T}{\partial z}\right)=\left(k_{hnf}+\frac{16}{3}\frac{\sigma^* T_\infty^3}{k^*}\right)\frac{\partial^2 T}{\partial z^2}$$
$$+q_e\left(T_w-T_\infty\right)\exp\left(-n\sqrt{\frac{a}{\upsilon}}z\right)+q_t\left(T-T_\infty\right) \quad (4)$$

$$u\frac{\partial C}{\partial x}+v\frac{\partial C}{\partial y}+w\frac{\partial C}{\partial z}=D_m\frac{\partial^2 C}{\partial z^2}-k_0^2\left(\frac{T}{T_\infty}\right)^m\exp\left(-\frac{E^*}{k_1 T}\right)\left(C-C_\infty\right) \quad (5)$$

$$\left.\begin{array}{l} at\ z=0:\ u=U_w, v=V_w, w=0, -k_{hnf}\dfrac{\partial T}{\partial z}=h_f\left(T_w-T\right), C=C_w \\ as\ z\to\infty:\ u\to 0, v\to 0, T\to T_\infty, C\to C_\infty \end{array}\right\} \quad (6)$$

Gupta and Gupta [41] proposed the usage of the similarity variables listed below in the transformation of governing equations:

$$\left.\begin{array}{l} \eta=z\sqrt{\dfrac{U_w}{\upsilon x}}, u=a_1 x f', w=-\sqrt{a_1\upsilon}\left(f(\eta)+g(\eta)\right), v=a_1 y g', \\ \theta(\eta)=\dfrac{T-T_\infty}{T_w-T_\infty}, \Phi(\eta)=\dfrac{C-C_\infty}{C_w-C_\infty}. \end{array}\right\} \quad (7)$$

2.1. Hybrid nanofluid (HNF) thermophysical characteristics (Hanif et al. [44])

$$\left.\begin{array}{l} \left(\rho C_p\right)_{hnf}=\left[\left(1-\phi_1\right)\left(\rho C_p\right)_f+\phi_1\left(\rho C_p\right)_{s_1}\right]\left(1-\phi_2\right)+\left(\rho C_p\right)_{s_2}\phi_2, \\ \mu_{hnf}=\dfrac{\mu_f}{\left(1-\phi_1\right)^{2.5}\left(1-\phi_2\right)^{2.5}}, \rho_{hnf}=\left[\left(1-\phi_1\right)\rho_f+\phi_1\rho_{s_1}\right]\left(1-\phi_2\right)+\rho_{s_2}\phi_2, \\ k_{hnf}=k_{nf}\times\dfrac{k_{s_2}+2k_{nf}-2k_{nf}\phi_2+2k_{s_2}\phi_2}{k_{s_2}+2k_{nf}+k_{nf}\phi_2-k_{s_2}\phi_2}, k_{nf}=k_f\times\dfrac{k_{s_1}+2k_f-2k_f\phi_1+2k_{s_1}\phi_1}{k_{s_1}+2k_f+k_f\phi_1-k_{s_1}\phi_1}, \\ \sigma_{hnf}=\sigma_{nf}\times\dfrac{\sigma_{s_2}+2\sigma_{nf}-2\sigma_{nf}\phi_2+2\sigma_{s_2}\phi_2}{\sigma_{s_2}+2\sigma_{nf}+\sigma_{nf}\phi_2-\sigma_{s_2}\phi_2}, \sigma_{nf}=\sigma_f\times\dfrac{\sigma_{s_1}+2\sigma_f-2\sigma_f\phi_1+2\sigma_{s_1}\phi_1}{\sigma_{s_1}+2\sigma_f+\sigma_f\phi_1-\sigma_{s_1}\phi_1}. \end{array}\right\}$$

Next, (7) makes (1) true, rewrites (2-5) and (6) as:

$$\frac{1}{B_1 B_2}f'''+ff''+gf''-f'^2-\left(\frac{B_3}{B_2}M+\frac{\delta}{B_1 B_2}\right)f'=0 \quad (8)$$

$$\frac{1}{B_1 B_2}g'''+fg''+gg''-g'^2-\left(\frac{B_3}{B_2}M+\frac{\delta}{B_1 B_2}\right)g'=0 \quad (9)$$

$$\left(\frac{B_4}{B_5}\frac{1}{\Pr}+\frac{1}{B_5}\frac{Ra}{\Pr}\right)\theta''+\left(f+g\right)\theta'+\frac{Q_T}{B_5}\theta+\frac{1}{B_5}Q_E\exp\left(-n\eta\right)=0 \quad (10)$$

$$\frac{1}{Sc}\Phi''+(f+g)\Phi'-\Gamma(1+\Lambda\theta)^m\exp\left(-\frac{E_n}{1+\Lambda\theta}\right)\Phi=0 \tag{11}$$

$$\left.\begin{array}{l}\text{at } \eta=0: f(\eta)=g(\eta)=0, f'(\eta)=1, g'(\eta)=\lambda, \theta'(\eta)=-Bi(1-\theta(\eta)), \Phi(\eta)=1\\ \text{as } \eta\to\infty: f'(\eta)\to 0, g'(\eta)\to 0, \theta(\eta)\to 0, \Phi(\eta)\to 0\end{array}\right\} \tag{12}$$

Note that

$$\left.\begin{array}{l}B_1=(1-\phi_1)^{2.5}(1-\phi_2)^{2.5}, B_4=\dfrac{k_2+2B_{41}k_f-2\phi_2(B_{41}k_f-k_2)}{k_2+2B_{41}k_f+\phi_2(B_{41}k_f-k_2)},\\ B_3=\dfrac{\sigma_2+2B_{31}\sigma_f-2\phi_2(B_{31}\sigma_f-\sigma_2)}{\sigma_2+2B_{31}\sigma_f+\phi_2(B_{31}\sigma_f-\sigma_2)}, B_{41}=\dfrac{k_1+2k_f-2\phi_1(k_f-k_1)}{k_1+2k_f+\phi_1(k_f-k_1)},\\ B_{31}=\dfrac{\sigma_1+2\sigma_f-2\phi_1(\sigma_f-\sigma_1)}{\sigma_1+2\sigma_f+\phi_1(\sigma_f-\sigma_1)}, B_2=(1-\phi_2)\left[(1-\phi_1)+\phi_1\dfrac{\rho_1}{\rho_f}\right]+\phi_2\dfrac{\rho_2}{\rho_f},\\ B_5=(1-\phi_2)\left[(1-\phi_1)+\phi_1\dfrac{(\rho C_p)_1}{(\rho C_p)_f}\right]+\phi_2\dfrac{(\rho C_p)_2}{(\rho C_p)_f}.\end{array}\right\}$$

and

$$\left.\begin{array}{l}\delta=\dfrac{\upsilon_f}{a_1\kappa}, \Lambda=\dfrac{T_w-T_\infty}{T_\infty}, Ra=\dfrac{16}{3}\dfrac{\sigma^* T_\infty^3}{k_f k^*}, \Gamma=\dfrac{k_0^2}{a_1}, Q_E=\dfrac{q_e}{a_1(\rho C_p)_f}, \lambda=\dfrac{a_2}{a_1},\\ Q_I=\dfrac{q_t}{a_1(\rho C_p)_f}, M=\dfrac{\sigma_f B_0^2}{\rho_f a_1}, E_n=\dfrac{E^*}{k_1 T_\infty}, Sc=\dfrac{\upsilon}{D_m}, Bi=\dfrac{h_f}{k_f\sqrt{\dfrac{a_1}{\upsilon}}}.\end{array}\right\}.$$

In this work, we assume the condition (without losing generality) $(f+g)(0)=0$ as $f(0)=g(0)=0$.

Table 1 Base fluid and nanomaterial thermophysical property values (Jamshed and Aziz [45])

S. No.	Properties	EG (f)	ZnO (s_1)	TiO_2 (s_2)
1	σ	0.0000055	0.01	2380000
2	k	0.252	13	8.953
3	ρ	1114	5660	4250
4	C_p	2415	495.2	686.2

Skin friction coefficient, transfer rates (heat, mass) are defined as:

$$Cf_x = \frac{\tau_{zx}}{\frac{1}{2}\rho_f u_w^2}\Bigg|_{z=0}, Cf_y = \frac{\tau_{zy}}{\frac{1}{2}\rho_f v_w^2}\Bigg|_{z=0}, Nu_x = \frac{xq_w}{k_f\left(T_w - T_\infty\right)}\Bigg|_{z=0}, Sh_x = \frac{xs_w}{D_m\left(C_w - C_\infty\right)}\Bigg|_{z=0} \quad (13)$$

where

$$\tau_{zx} = \mu_{hnf}\frac{\partial u}{\partial z}, \tau_{zy} = \mu_{hnf}\frac{\partial v}{\partial z}, q_w = -\left(k_{hnf} + \frac{16}{3}\frac{\sigma^* T_\infty^3}{k^*}\right)\frac{\partial T}{\partial z}, s_w = -D_m\frac{\partial C}{\partial z}.$$

Terms in (13) can be recast using (7) as:

$$\begin{aligned}&\left(\mathrm{Re}_x\right)^{0.5} Cf_x = \frac{2}{B_1} f''(0), \left(\mathrm{Re}_y\right)^{0.5} Cf_y = \frac{2}{B_1}\frac{1}{\sqrt{\lambda}} g''(0),\\ &\left(\mathrm{Re}_x\right)^{-0.5} Nu_x = -B_4\left(1 + Ra\right)\theta'(0), \left(\mathrm{Re}_x\right)^{-0.5} Sh_x = -\Phi'(0)\end{aligned} \quad (14)$$

3. Numerical Procedure

For obtaining the solution of the system (8 - 11) subject to the conditions (12), we employed bvp4c, an inherent function in MATLAB. The flowchart for the bvp4c numerical technique is presented in Fig.2.

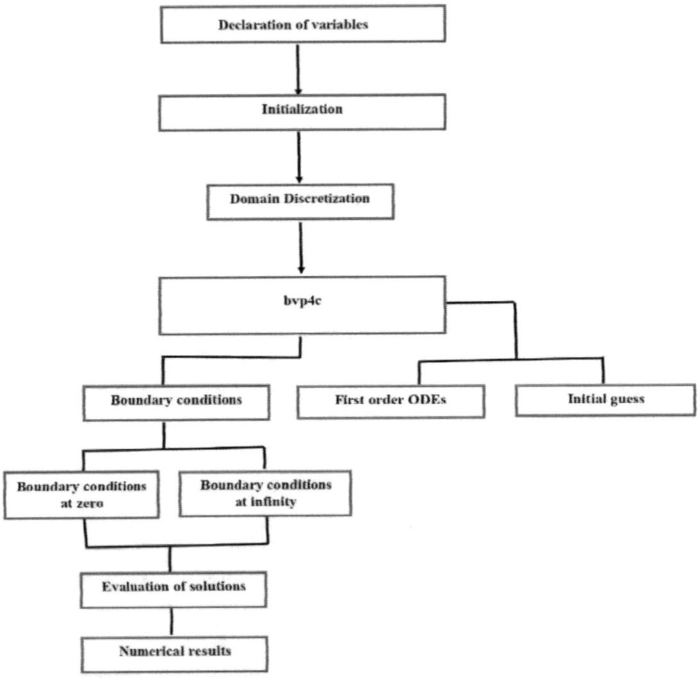

Figure 2. Schema chart for bvp4c (Faiz et al. [46])

4. Validation

Table 2 compares our findings to those of other studies under special situations and reveals a very close agreement.

Table 2 Verification of current findings by referencing previously obtained data for $-\theta'(0)$ when $\phi_1 = \phi_2 = M = \delta = Ra = Q_T = Q_E = \Gamma = 0, \lambda = 1$.

Pr	Khan and Pop [47]	Gupta and Gupta [41]	Current results
0.07	0.066	0.06632019	0.066319
0.20	0.169	0.16831182	0.168314
0.70	0.454	0.45423652	0.454228
2.0	0.911	0.91148210	0.911480

5. Discussion of the outcomes

Results are displayed in the form of graphs for two instances i.e., $EG+ZnO+TiO_2$ and $EG+ZnO$. Figure 3 depicts streamlines representing the current motion of the fluid.

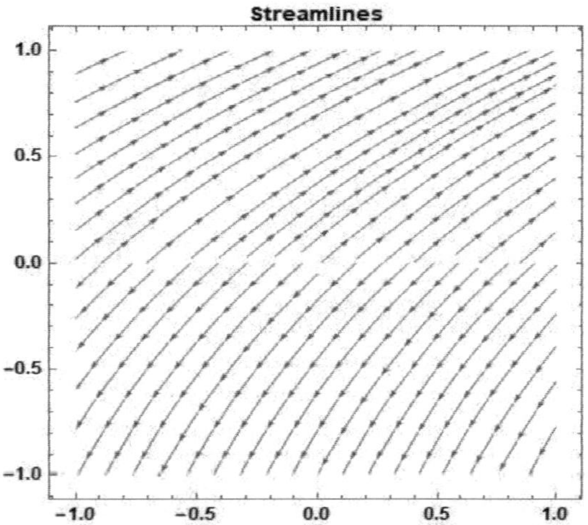

Figure 3. Streamlines for the present flow study

When there is a magnetic field, Lorentz forces are typically generated by the motion of an electrically conducting liquid. Because of the slowing effect of Lorentz forces, fluid velocities drop as magnetic field is increased (Figs. 4 - 5). As the nanoparticle volume percentage rises, the fluid becomes more viscous and becomes more difficult to move through. Therefore, when ϕ_1 grows, velocity fall (Figs. 6 - 7)

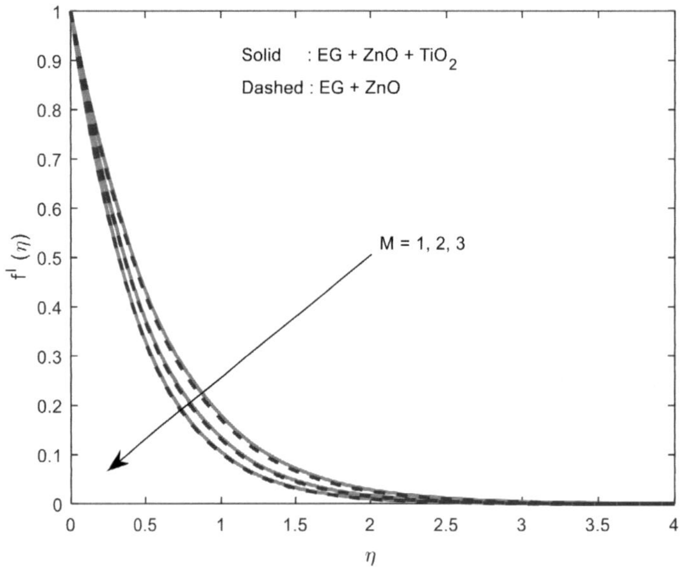

Figure 4. Behavior of $f'(\eta)$ when M is applied

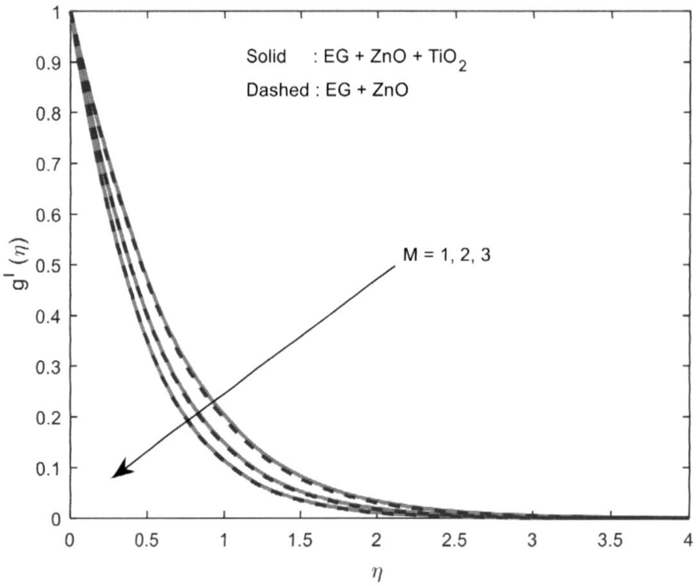

Figure 5. Result of M on $g'(\eta)$

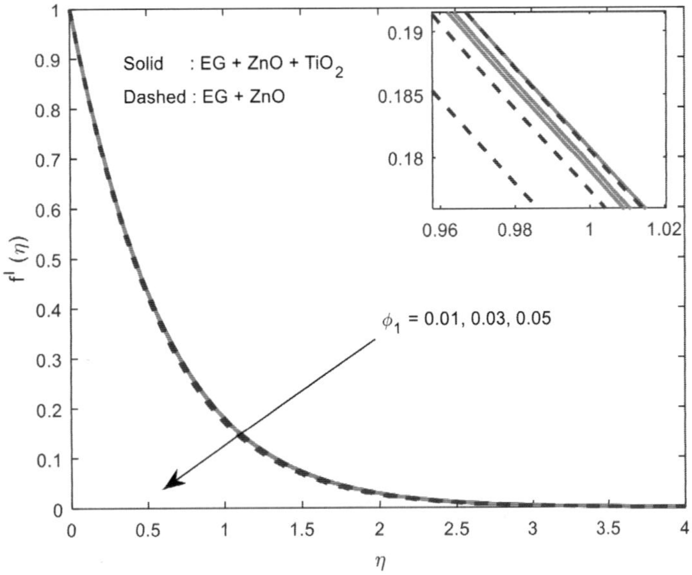

Figure 6. Result of ϕ_1 on $f'(\eta)$

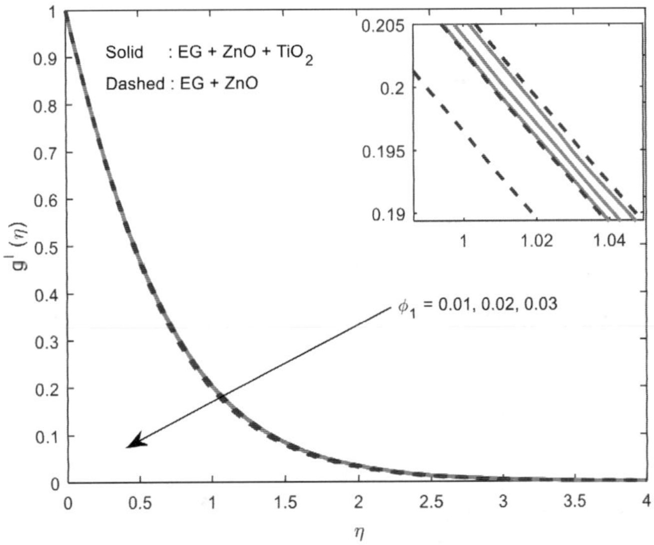

Figure 7. Behavior of $g'(\eta)$ when ϕ_1 is applied

Under the influence of thermal radiation on the flow, amount of heat is transferred into the fluid. If that value is enhanced, there is a huge amount of transfer of heat energy is noticed from source to fluid region. As a result, as illustrated in Fig.8, due to the enhancement in temperature causes to the absorbance of heat energy in fluid particles. More fluid particle collisions take place when the nanoparticles' volume percentage rises. In turn, this causes the fluid's temperature to increase (see Fig. 9). From Fig. 10, it was noticed that, there is an enhancement in fluid temperature due to the increment in Q_T. Fig. 11 shows the acceleration of temperature profile with a rise in Q_E in both scenarios. The ESHS mechanism (Heat Source connects to Exponential space) is the reason for a rise in fluid temperature while enhancing the heat source parameter, which leads to transfer of more energy into the liquid. The rate of heat transmission via convection in a fluid rises in proportion to its Biot number. As a reason, temperature of the fluid is augmented (as depicted in Fig. 12).

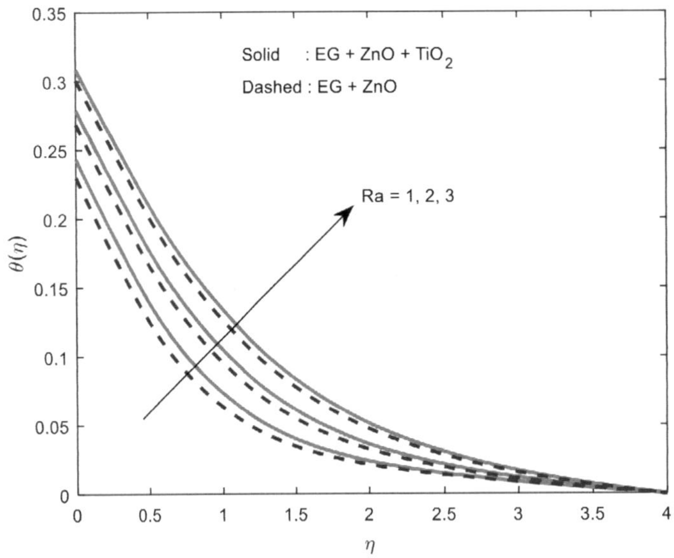

Figure 8. Profile of $\theta(\eta)$ due to impact of Ra

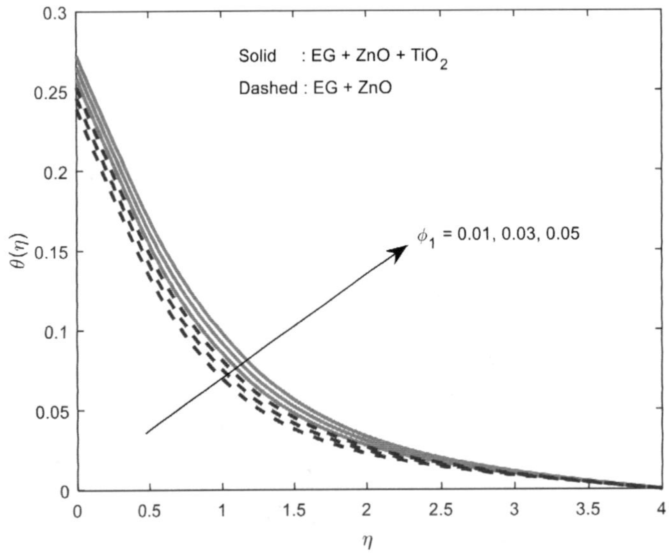

Figure 9. Profile of $\theta(\eta)$ due to impact of ϕ_1

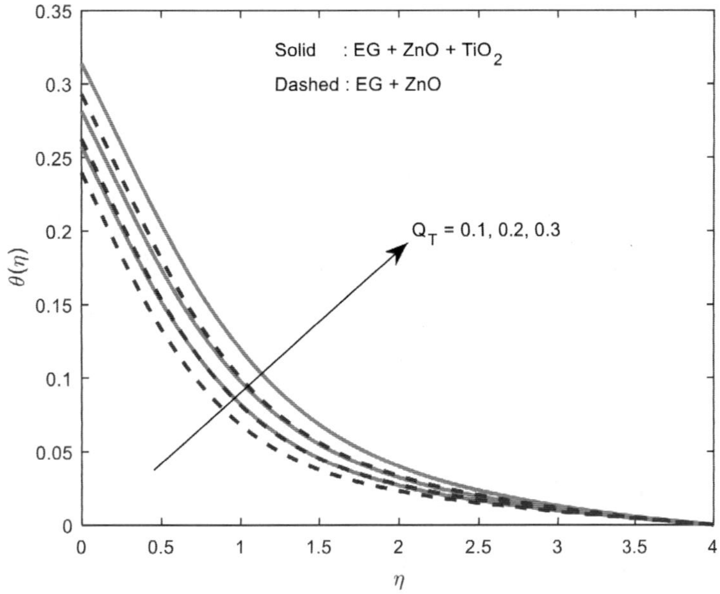

Figure 10. Profile of Q_T due to impact of $\theta(\eta)$

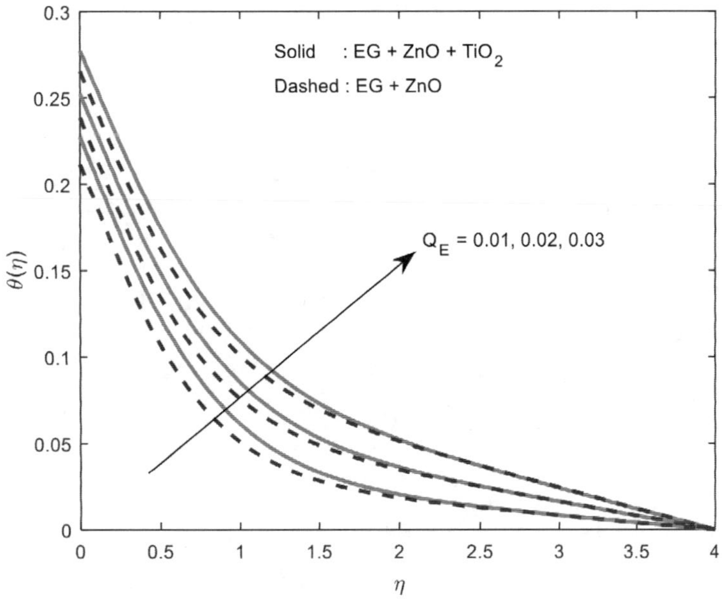

Figure 11. Profile of Q_E due to impact of $\theta(\eta)$

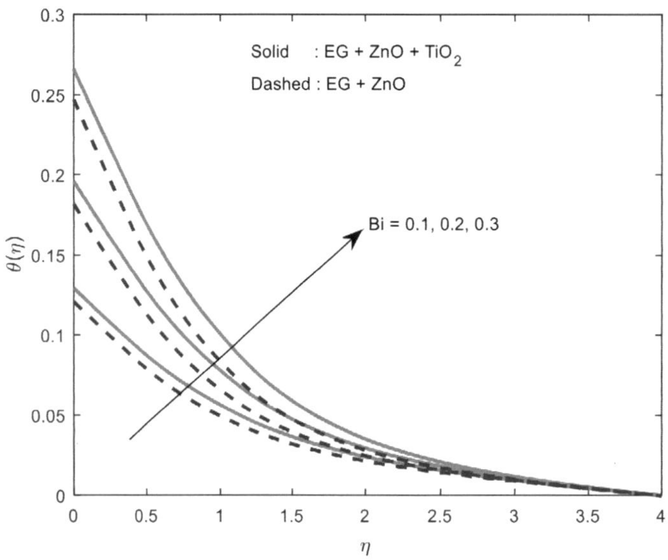

Figure 12. Temperature profile when Bi is applied

By tracking the decline in the threshold energy caused by a rise in activation energy, one can determine the average kinetic energy of a fluid. From this, we can infer a drop in average kinetic energy. If there is no diffusion, there is a high fluid concentration (see Fig.13). As the reaction rate parameter rises, the fluid concentration seems to decrease (see Fig.14). Lower molecular diffusivity is caused by a greater temperature difference. Accordingly, the fluid concentration drops with increasing temperature difference parameter (see Fig. 15).

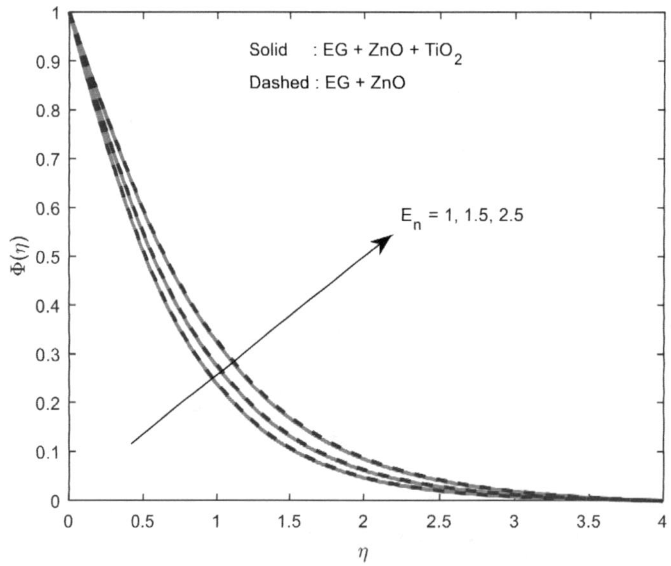

Figure 13. Profile of $\Phi(\eta)$ due to impact of E_n

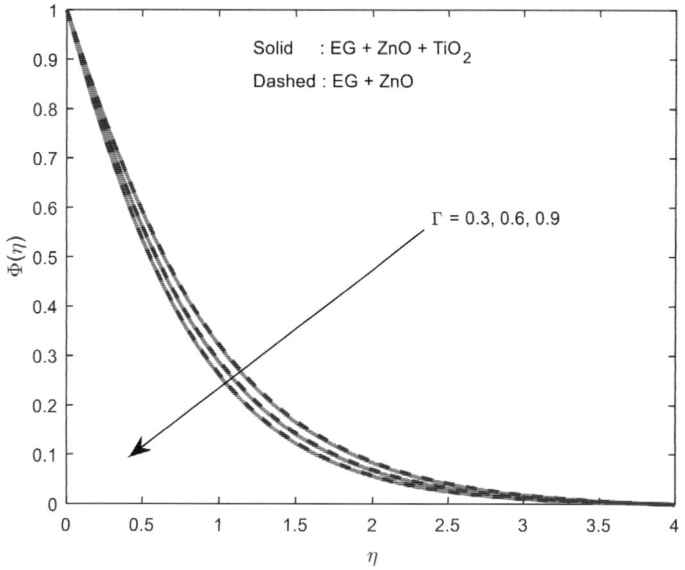

Figure 14. Profile of $\Phi(\eta)$ due to impact of Γ

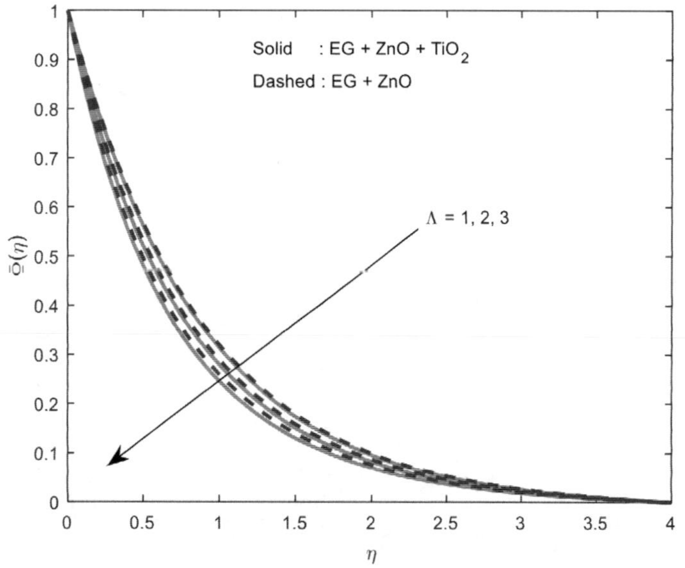

Figure 15. Profile of $\Phi(\eta)$ due to impact of Λ

From Figs. 16-17, it is clear that both M and ϕ_1 are useful to reduce the friction factor in both situations namely, Hybrid Nanofluid ($EG+ZnO+TiO_2$) and Mono Nanofluid ($EG+ZnO$). It can be shown in Fig. 18 that when Q_T upsurges, the heat transmission rate declines. Reduction quantities in the heat transfer rate against Q_T are noticed as 0.15677 ($EG+ZnO+TiO_2$), 0.11564 (MNF). Therefore, it is worthy to mention that reduction rate is higher in case of $EG+ZnO$ compared to $EG+ZnO+TiO_2$. Fig. 19 exhibits the fact that there is an amelioration in the Nusselt number with the raise in Bi. Increment quantity is detected as 0.736332 ($EG+ZnO+TiO_2$) and 0.621826 ($EG+ZnO$

IOP Conf. Series: Journal of Physics: Conf. Series **2765** (2024) 012003 doi:10.1088/1742-6596/2765/1/012003

).From Figs. 20-22, it is discovered that the activation energy parameter diminishes the Sherwood number and Γ, Λ escalate the same. The increment quantities against Γ, Λ are 0.35151 ($EG+ZnO+TiO_2$), 0.35125 ($EG+ZnO$), 0.04911 ($EG+ZnO+TiO_2$) and 0.04300 ($EG+ZnO$) respectively. Furthermore, the decrement in mass transfer rate against the activation parameter is observed as 0.02681 ($EG+ZnO+TiO_2$) and 0.02687 ($EG+ZnO$).

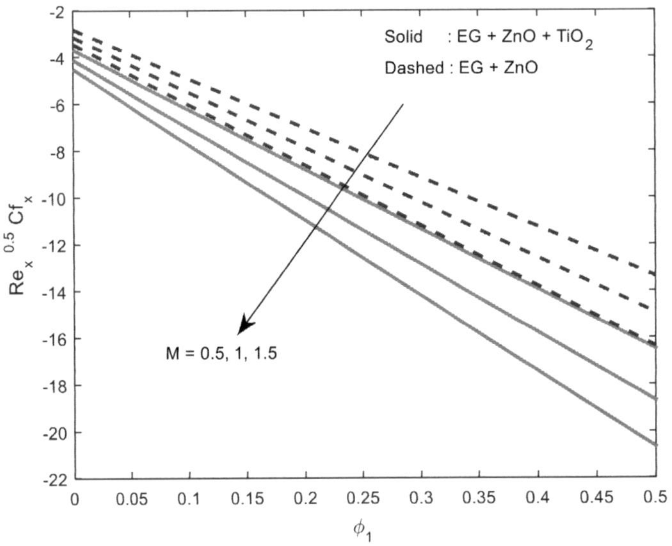

Figure 16. Behavior of skin friction coefficient when M is applied

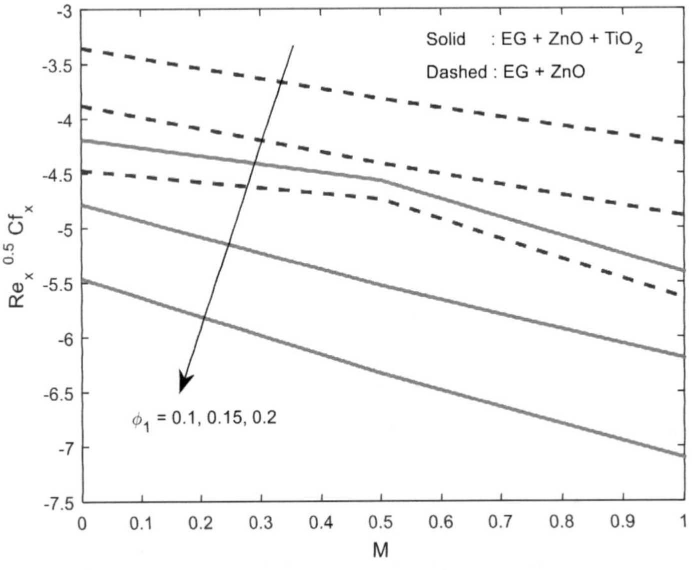

Figure 17. Behavior of skin friction coefficient when ϕ_1 is applied

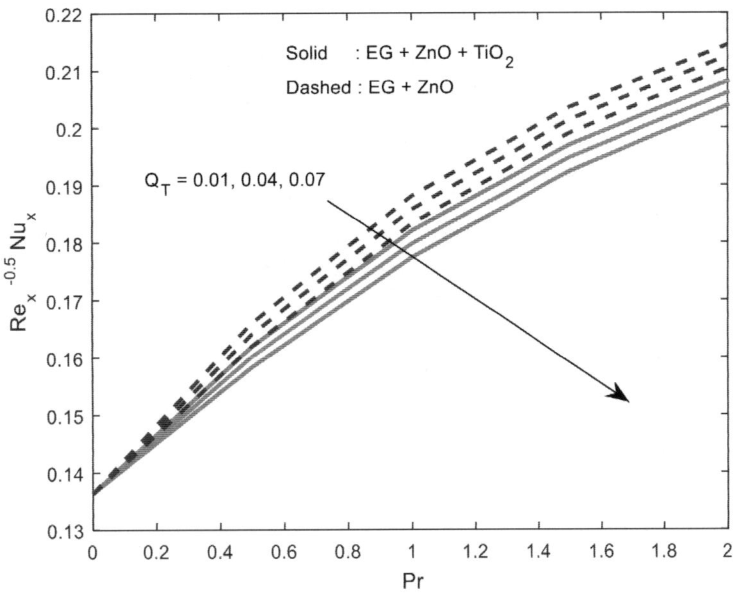

Figure 18. Behavior of heat transmission rate when Q_T is applied

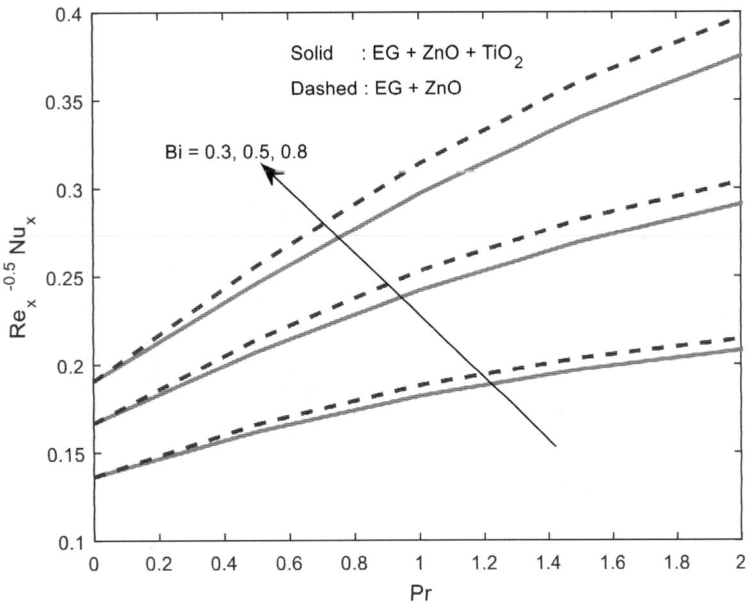

Figure 19. Behavior of heat transmission rate when Bi is applied

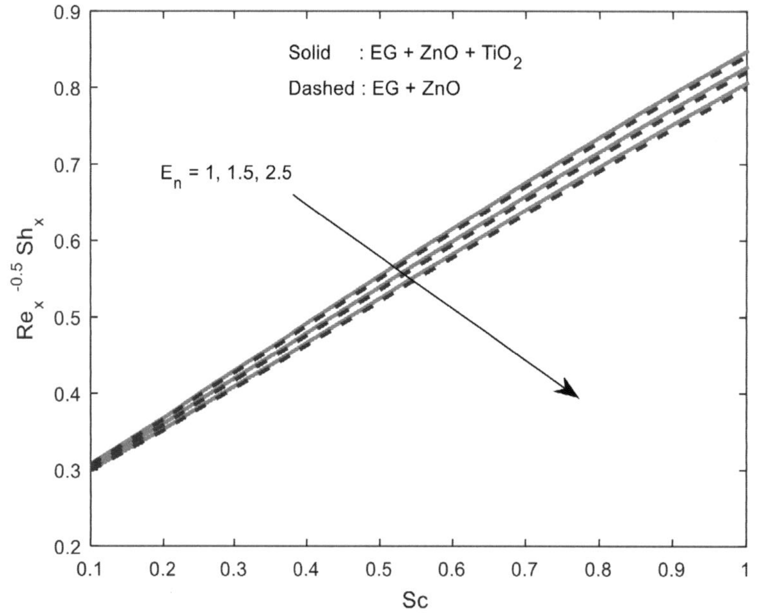

Figure 20. Behavior of mass transmission rate when E_n is applied

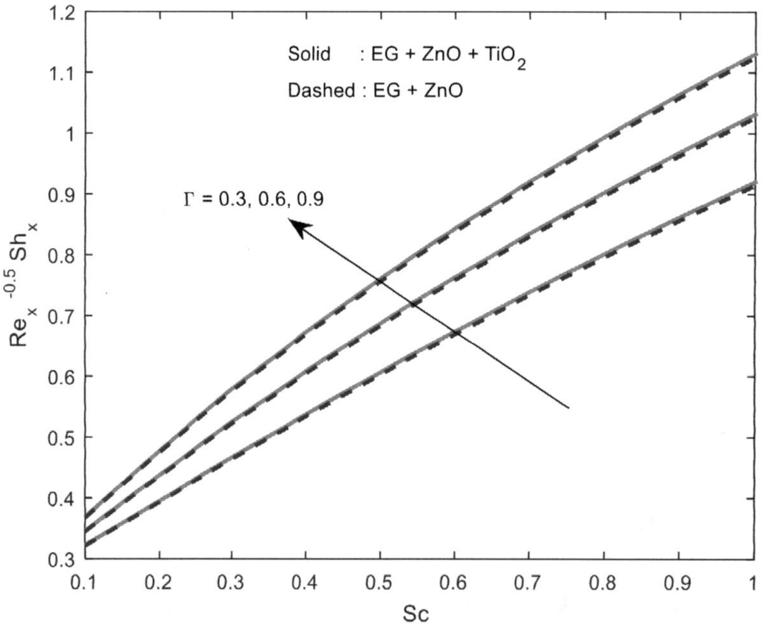

Figure 21. Behavior of mass transmission rate when Γ is applied

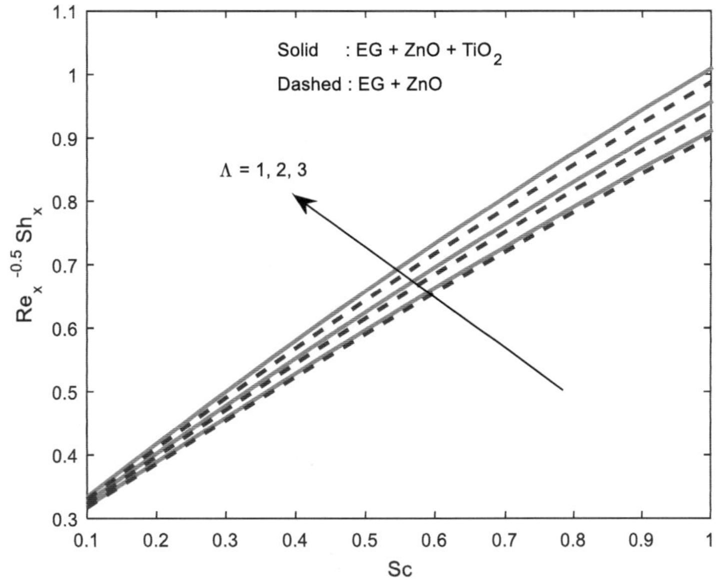

Figure 22. Behavior of mass transmission rate when Λ is applied

6. Conclusion

In this study, radiative hybrid nanofluid ($EG+ZnO+TiO_2$) flow by a bidirectional elongating sheet with different heat source parameters and activation energy is analyzed theoretically. Convective boundary condition is assumed. The results of this investigation were obtained using a pre-existing function in MATLAB called bvp4c. Added a little quantitative discussion regarding engineering parameters inclusive of Sherwood number. The concluding remarks of this study are briefly summarized below.

- Enhanced volume fraction of ZnO lessens the fluid velocity and increasing the heat source parameter step-up the fluid temperature.
- It was observed that, due to the occurrence of magnetic field as well as the volume fraction of ZnO is beneficial in lowering the skin friction coefficient in both scenarios.
- It is discovered that the activation energy parameter lowers the Sherwood number. The decrement rates in the Sherwood number against the activation parameter (E_n) are observed as 0.02681 ($EG+ZnO+TiO_2$) and 0.02687 ($EG+ZnO$), when E_n is set to $0 \le E_n \le 2.5$.
- It is detected that Γ and Λ are helpful to enhances the mass transfer rate. Sherwood number is increased with a rate of 0.35151 ($EG+ZnO+TiO_2$) and 0.35125 ($EG+ZnO$), when Γ is set to $0 \le \Gamma \le 0.9$.
- There is an amelioration in the heat transmission rate with the raise in Bi. It was perceived that, when Bi is set to $0 \le Bi \le 0.8$, Nusselt number is reduced to 0.736332 ($EG+ZnO+TiO_2$) and 0.621826 ($EG+ZnO$).

References

[1] F. Saba, N. Ahmed, S. Hussain, U. Khan, S.T. Mohyud-Din and M. Darus, "Thermal analysis of nanofluid flow over a curved stretching surface suspended by carbon nanotubes with internal heat generation", *Appl. Sciences*, Vol.8, No.3, 395, March 2018.

DOI: **https://doi.org/10.3390/app8030395**.

[2] B. Prabhavathi, P.S. Reddy and R.B. Vijaya, "Heat and mass transfer enhancement of SWCNTs and MWCNTs based Maxwell nanofluid flow over a vertical cone with slip effects", *Powder tech.*, Vol. 340, pp. 253-263, Dec. 2018.
DOI: https://doi.org/10.1016/j.powtec.2018.08.089.

[3] H.T. Basha, R. Sivaraj, A.S. Reddy and A.J. Chamkha, "SWCNH/diamond-ethylene glycol nanofluid flow over a wedge, plate and stagnation point with induced magnetic field and nonlinear radiation–solar energy application", *The European Phy. J. Special Topics*, Vol. 228, No.12, pp. 2531-2551, 2019.
DOI: https://doi.org/10.1140/epjst/e2019-900048-x.

[4] S. Saranya, Q.M. Al-Mdallal, "Computational study on nanoparticle shape effects of Al_2O_3-silicon oil nanofluid flow over a radially stretching rotating disk", *Case Studies in Therm. Engg.*, Vol. 25, 100943, June 2021.
DOI: https://doi.org/10.1016/j.csite.2021.100943.

[5] A. Mishra, A. Kumar Pandey and M. Kumar, "Thermal performance of Ag–water nanofluid flow over a curved surface due to chemical reaction using Buongiorno's model", *Heat Transf.*, Vol. 50, No.1, pp. 257-278, January 2021.
DOI: https://doi.org/10.1002/htj.21875.

[6] A. Shafiq, A.B. Çolak, and T.N. Sindhu, "Designing artificial neural network of nanoparticle diameter and solid fluid interfacial layer on SWCNTs/EG nanofluid flow on thin slendering needles", *Int. J. for Num. Methods in Fluids*, Vol. 93, No. 12, pp. 3384-3404, Dec. 2021. DOI: https://doi.org/10.1002/fld.5038.

[7] M. Bahiraei, N. Mazaheri, M.R. Daneshyar, A. Mwesigye, "Two-phase simulation of irreversibilities for Ag–water nanofluid flow inside an elliptical pin-fin heat sink: Entropy generation and exergy considerations" *Powder Tech.*, *409*, 117723, 2022.
DOI: https://doi.org/10.1016/j.powtec.2022.117723.

[8] S.U. Devi, S.P. Anjali Devi, "Heat transfer enhancement of Cu-Al_2O_3/Water hybrid nanofluid flow over a stretching sheet", *J. of the Nigerian Math. Society*, Vol. 36, No.2, pp. 419-433, Aug. 2017.

[9] P.G.R. Jayadevamurthy, N.K. Rangaswamy, B.C. Prasannakumara, and K.S. Nisar, "Emphasis on unsteady dynamics of bioconvective hybrid nanofluid flow over an upward–downward moving rotating disk". *Num. Methods for Partial Diff. Eqns.,* pp. 1-22, 2020, DOI: https://doi.org/10.1002/num.22680.

[10]F. Mabood, T.A. Yusuf, and W.A. Khan, "Cu-Al_2O_3-H_2O hybrid nanofluid flow with melting heat transfer, irreversibility analysis and nonlinear thermal radiation", *J. of Therm. Analysis and Calorimetry*, Vol. 143, No. 2, pp. 973-984, 2021. DOI: https://doi.org/10.1007/s10973-020-09720-w.

[10] N.S. Khashi'ie, N.M. Arifin, I. Pop, R. Nazar, E.H. Hafidzuddin and N. Wahi, "Three-dimensional hybrid nanofluid flow and heat transfer past a permeable stretching/shrinking sheet with velocity slip and convective condition", *Chinese J. of Phy.*, Vol. 66, pp. 157-171, Aug. 2020.
DOI: https://doi.org/10.1016/j.cjph.2020.03.032.

[11] N. Acharya, F. Mabood, "On the hydrothermal features of radiative Fe3O4–graphene hybrid nanofluid flow over a slippery bended surface with heat source/sink", *J. of Therm. Analysis and Calorimetry*, Vol. 143, No. 2, pp. 1273-1289, 2021.
DOI: https://doi.org/10.1007/s10973-020-09850-1.

[12] A.K. Alzahrani, M.Z. Ullah, A.S. Alshomrani and T. Gul, "Hybrid nanofluid flow in a Darcy-Forchheimer permeable medium over a flat plate due to solar radiation" *Case Studies in Therm. Engg.*, Vol. 26, 100955, Aug. 2021.
DOI: https://doi.org/10.1016/j.csite.2021.100955.

[13] N.A. Zainal, R. Nazar, K. Naganthran and I. Pop, "Stability analysis of MHD hybrid nanofluid flow over a stretching/shrinking sheet with quadratic velocity", *Alexandria Engg. J.*, Vol. 60, No.1, pp. 915-926, 2021.

DOI: https://doi.org/10.1016/j.aej.2020.10.020.
[14] N. Acharya, "On the flow patterns and thermal control of radiative natural convective hybrid nanofluid flow inside a square enclosure having various shaped multiple heated obstacles", *The Eur. Phy. J. Plus*, Vol. 136, No. 8, pp. 1-29, Aug. 2021. DOI: https://doi.org/10.1140/epjp/s13360-021-01892-0.
[15] N. Abbas, S. Nadeem, A. Saleem, M.Y. Malik, A. Issakhov and F.M. Alharbi, "Models base study of inclined MHD of hybrid nanofluid flow over nonlinear stretching cylinder", *Chinese J. of Phy.*, Vol. 69, pp. 109-117, Feb. 2021. DOI: https://doi.org/10.1016/j.cjph.2020.11.019.
[16] J.K. Madhukesh R.N. Kumar, R.P. Gowda, B.C. Prasannakumara, G.K. Ramesh, Khan, M. Ijaz Khan, Sami Ullah Khan and Y.M. Chu, "Numerical simulation of AA7072-AA7075/water-based hybrid nanofluid flow over a curved stretching sheet with Newtonian heating: A non-Fourier heat flux model approach", *J. of Mol. Liquids*, Vol. 335, 116103, Aug. 2021. DOI: https://doi.org/10.1016/j.molliq.2021.116103.
[17] J.K. Madhukesh, M.M. Alam, R.S. Varun Kumar, A. Arasaiah, I. Ahmad, M.R. Gorji and B.C. Prasannakumara, "Exploring magnetic dipole impact on Maxwell hybrid nanofluid flow over a stretching sheet", *Proceedings of the Instt. of Mech. Engineers, Part E: J. of Process Mech. Engg.*, 2022. DOI: https://doi.org/10.1177/09544089211073267.
[18] M. Kumar, P.K. Mondal, "Irreversibility analysis of hybrid nanofluid flow over a rotating disk: Effect of thermal radiation and magnetic field", *Colloids and Surfaces A: Physicochem. and Engg. Aspects*, Vol. 635, 128077, 2022. DOI: https://doi.org/10.1016/j.colsurfa.2021.128077.
[19] H. Gul, M. Ramzan, K.S. Nisar, R.N. Mohamed, H.A.S. Ghazwani, "Performance-based comparison of Yamada–Ota and Hamilton–Crosser hybrid nanofluid flow models with magnetic dipole impact past a stretched surface", *Sci. Reports*, Vol. 12, No.29, pp. 1-11, 2022. DOI: https://doi.org/10.1038/s41598-021-04019-8.
[20] G. Revathi, S. Avadapu, C.S.K. Raju, M.J. Babu, A.M. Zidan, M.K. Alaoui, N.A. Shah, and J.D. Chung, "Dynamics of lorentz force and cross-diffusion effects on ethylene glycol based hybrid nanofluid flow amidst two parallel plates with variable electrical conductivity: A multiple linear regression analysis", *Case Studies in Therm. Engg.*, Vol. 41, 102603, Jan. 2023. DOI: https://doi.org/10.1016/j.csite.2022.102603.
[21] G. Revathi, V.S. Sajja, M.J. Babu, K.S. Srinivasa Babu, A. Suneel Kumar, C.S.K. Raju and S.J. Yook, "Multiple linear regression analysis on the flow of ternary hybrid nanofluid by a quadratically radiated stretching surface with and second order slip". *Waves in Rand. and Com. Media*, pp. 1-18, 2023. DOI: https://doi.org/10.1080/17455030.2023.2181645.
[22] V.V. Salomatov, E.M. Puzyrev, "Influence of thermal radiation on the laminar boundary layer of a non-absorbing fluid", *J. of engg. Phy.*, Vol. 20, No.6, pp. 720-724, 1971.
[23] M.A. Hossain, H.S. Takhar, "Radiation effect on mixed convection along a vertical plate with uniform surface temperature", *Heat and Mass trans.*, Vol.31, No.4, pp. 243-248, 1996.
[24] M.Q. Al-Odat, F.M.S. Al-Hussien, R.A. Damseh, "Influence of radiation on mixed convection over a wedge in non-Darcy porous medium". *Forschung im Ingenieurwesen*, Vol. 69, No.4, pp. 209-215, 2005. DOI: https://doi.org/10.1007/s10010-005-0004-2.
[25] M.S. Abdel-Wahed, "Nonlinear Rosseland thermal radiation and magnetic field effects on flow and heat transfer over a moving surface with variable thickness in a nanofluid", *Canadian J. of Phy.*, Vol. 95, No. 3, pp. 267-273, March 2017. DOI: https://doi.org/10.1139/cjp-2016-0345.
[26] M. Khan, A. Hamid, "Influence of non-linear thermal radiation on 2D unsteady flow of a Williamson fluid with heat source/sink", *Results in phy.*, Vol. 7, pp. 3968-3975, 2017. DOI: https://doi.org/10.1016/j.rinp.2017.10.014.

[27] M.A. Sheremet, I. Pop, A.V. Rosca, "The influence of thermal radiation on unsteady free convection in inclined enclosures filled by a nanofluid with sinusoidal boundary conditions", *Int. J. of Num. Methods for Heat & Fluid Flow*, Vol. 28, No. 8, pp. 1738-1753, Oct. 2018. DOI: https://doi.org/10.1108/HFF-09-2017-0375.

[28] M. Sheikholeslami, M.A. Sheremet, A. Shafee, I. Tlili, "Simulation of nanoliquid thermogravitational convection within a porous chamber imposing magnetic and radiation impacts". *Physica A: Statistical Mechanics and its Appl.*, Vol.550, 124058, July 2020. DOI: https://doi.org/10.1016/j.physa.2019.124058.

[29] S.E. Ghasemi, M. Hatami, "Solar radiation effects on MHD stagnation point flow and heat transfer of a nanofluid over a stretching sheet". *Case Studies in Therm. Engg.*, Vol. 25, 100898, June 2021. DOI: https://doi.org/10.1016/j.csite.2021.100898.

[30] N. Iftikhar, M.R. Bilal, A.A. Zafar, S.M. Husnine, "Nanofluid with nonlinear Rosseland thermal radiation and mixed convection", *Advances in the Theory of Nonlinear Analy. and its Appl.*, Vol. 5, No.3, pp. 412-420, 2021. DOI: https://doi.org/10.31197/atnaa.752667.

[31] Dongmin Yu, and Rijun Wang, "An optimal investigation of convective fluid flow suspended by carbon nanotubes and thermal radiation impact." *Mathematics* Vol.10, No. 9, 1542, 2022. DOI: **https://doi.org/10.3390/math10091542**

[32] U.S. Mahabaleshwar, K. N. Sneha, A. Chan, and Dia Zeidan, "An effect of MHD fluid flow heat transfer using CNTs with thermal radiation and heat source/sink across a stretching/shrinking sheet" *Int. Comm. in Heat and Mass Transf.*, Vol. 135, 106080, 2022. DOI: https://doi.org/10.1016/j.icheatmasstransfer.2022.106080

[33] S.G. Bejawada, M.M. Nandeppanavar, "Effect of thermal radiation on magnetohydrodynamics heat transfer micropolar fluid flow over a vertical moving porous plate", *Exp. and Comp. Multiphase Flow*, Vol. 5, No.2, pp.149-158. 2023. DOI: https://doi.org/10.1007/s42757-021-0131-5

[34] S. Nadeem, B. Ishtiaq, N. Abbas, "Impact of thermal radiation on two-dimensional unsteady third-grade fluid flow over a permeable stretching Riga plate", *Int. J. of Modern Phy. B*, Vol.37, No.1, 2350009, 2023. DOI: https://doi.org/10.1142/S0217979223500091.

[35] Revathi, G., Gopinath, V., Jayachandra Babu, M., Srinivasa Babu, K.S., Suneel Kumar, A., "Darcy–Forchheimer flow of power-law (Ostwald-de Waele type) nanofluid over an inclined plate with thermal radiation and activation energy: an irreversibility analysis", *Int. J. of Ambient Energy*, Vol. 44, No.1, pp. 1980-1989. April 2023. DOI: 10.1080/01430750.2023.2200434.

[36] Rajakumar, K.V.B., Govinda Rao, T., Umasankara Reddy, M. *et al.* Influence of Dufour and thermal radiation on unsteady MHD Walter's liquid model-B flow past an impulsively started infinite vertical plate embedded in a porous medium with chemical reaction, Hall and ion slip current. *SN Appl. Sci.* Vol. 2, 742, March 2020. DOI: https://doi.org/10.1007/s42452-020-2484-y.

[37] Rajkumar, K.V.B., Srinivasa Raju, R., Govinda Rao, T., Balamurugan, K.S., "Influence of viscous dissipation and spanwise cosinusoidally fluctuating temperature on MHD free convective boundary layer flow with radiation absorption and chemical reaction", *Heat Transfer,* Vol. 51, No. 5, pp. 4714-4740. July 2022. DOI: **https://doi.org/10.1002/htj.22520**.

[38] Rajakumar, K.V.B., Balamurugan, K.S., Govinda Rao, T., Satyanarayana Reddy, B. "Influence of Radiation Absorption and Thermal Diffusion on Laminar Flow of a Chemically Reacting Fluid Past a Porous Plate in Conducting Field with Rotation", Recent Advances in Applied Mathematics and Applications to the Dynamics of Fluid Flows. *Lecture Notes in Mechanical Engineering*. Springer, Singapore, pp. 77-87, 2023. DOI: https://doi.org/10.1007/978-981-19-1929-9_7.

[39] Rajakumar, K.V.B., Murthy, P.P., Govinda Rao, T., Ranganath, N., "Significance of Casson dissipative fluid and spanwise cosinusoidally fluctuating temperature on unsteady MHD free

convective flow past a hot vertical permeable plate with radiation absorption and chemical reaction" *Heat transfer*, Vol. 51, No.8, pp. 7725-7752. Dec 2022. DOI: https://doi.org/10.1002/htj.22663.

[40] S. Gupta, S. Gupta, "MHD three dimensional flow of Oldroyd-B nanofluid over a bidirectional stretching sheet: DTM-Padé Solution", *Nonlinear Engg.*, Vol.8, No.1, pp. 744-754, 2019. DOI: https://doi.org/10.1515/nleng-2018-0047.

[41] W. Al-Kouz, B. Mahanthesh, M.S. Alqarni, K. Thriveni, "A study of quadratic thermal radiation and quadratic convection on viscoelastic material flow with two different heat source modulations" *Int. Comm. in Heat and Mass Transf.,* Vol. 126, 105364, July 2021. DOI: https://doi.org/10.1016/j.icheatmasstransfer.2021.105364.

[42] M. Irfan, M. Khan, W.A. Khan, L. Ahmad, "Influence of binary chemical reaction with Arrhenius activation energy in MHD nonlinear radiative flow of unsteady Carreau nanofluid: dual solutions", *Appl. Phy. A*, Vol.125, No.3, 179, 2019. DOI: https://doi.org/10.1007/s00339-019-2457-4.

[43] H. Hanif, I. Khan, S. Shafie, "Heat transfer exaggeration and entropy analysis in magneto-hybrid nanofluid flow over a vertical cone: a numerical study", *J. of Therm. Analysis and Calorimetry*, Vol.141, No.5, pp. 2001-2017, 2020. DOI: https://doi.org/10.1007/s10973-020-09256-z.

[44] W. Jamshed, A. Aziz, "Cattaneo–Christov based study of $TiO_2 - ZnO / EG$ Casson hybrid nanofluid flow over a stretching surface with entropy generation", *Appl. Nanosci.*, Vol.8, No.4, pp. 685-698, April 2018. DOI: 10.1007/s13204-018-0820-y.

[45] M. Faiz, D. Habib, I. Siddique, J. Awrejcewicz, W. Pawłowski, S. Abdal, N. Salamat, "Multiple slip effects on time dependent axisymmetric flow of magnetized Carreau nanofluid and motile microorganisms", *Sci. Reports*, Vol. 12, 14259, 2022. DOI: https://doi.org/10.1038/s41598-022-18344-z.

[46] W.A. Khan, I. Pop, "Boundary-layer flow of a nanofluid past a stretching sheet", *Int. J. of heat and mass transf.* Vol. 53, No.11-12, pp. 2477-2483, 2010. DOI: https://doi.org/10.1016/j.ijheatmasstransfer.2010.01.032.

BIOGENIC SYNTHESIS OF MAGNETITE NANOPARTICLES USING THE ROOTS OF *MIRABILIS JALAPA* FOR EFFICIENT REMOVAL OF ELUENT, ANTI-BACTERIAL, AND ANTIOXIDANT: A SUSTAINABLE APPROACH.

B. Madhavi[†,††], S. Prameeladevi[†††], N. Annapurna[†*], M. Padma[†]

*Department of Engineering Chemistry, Andhra University Visakhapatnam, India- 530017

Correspondence mail id: madhavi.ritukartu@gmail.com, purnaaucevizag@gmail.com

Abstract:

Magnetite nanoparticles hold quite immense applications in the fields of drug delivery, environmental remedies, and magnetic resonance imaging as they are biocompatible and can stabilize and reduce. An eco-friendly approach that is viable and sustainable for synthesizing magnetite nanoparticles with controlled properties has new avenues for their applications in various fields. Magnetite nanoparticles are often used in the removal of organic eluents as they can adsorb heavy metals which binds on the surface and cleans the polluted area. Magnetite nanoparticles synthesized from the roots of *Mirabilis Jalapa* have a variety of uses in the treatment of lesion infections. As antioxidants, they reduce the oxidative stress caused by free radicals. Morphology, size, and form of Magnetite nanoparticles are determined by UV-visible spectroscopy, SEM, EDX, TEM, TGA, XRD, and FTIR, respectively.

Keywords: Biocompatible, eluents, antioxidant, lesions, SEM, TEM, XRD.

[†] Department of Engineering Chemistry, Andhra University Visakhapatnam, India- 530017

[††]Department of Chemistry, St. Joseph's college for women, Visakhapatnam, India- 530004.

[†††]Department of Chemistry, S.R.K.R. Engineering College, Bhimavaram, India-534204

1. Introduction

Nanotechnology has a large assortment of unique applications in the fields of chemical, physical, optical, electrical, mechanical, magnetically, and catalytic properties due to their demand in the present world of science and technology [1-3]. The goal of nanotechnology is to prepare better, simpler, more efficient, eco-friendly nanoparticles[4-6]. The nanoparticles form as a cluster with nano-sized having a range of 100 nm[7]. The bio-genic synthesized nanoparticle has become trending as it possesses many biological applications towards anti-cancer, anti-diabetic, anti-microbial, anti-hyperglycaemic, anti-oxidants, and cardioprotective effect due to its abundant in nature and eco-friendly availability[8-9].

Metallic nanoparticles like magnetite illuminate towards catalytic, biosensors, ferrofluids, separation, and purification of bio-molecules due to their non-toxic and bio-compatibility for researchers as they exhibit various oxidation states, remove inorganic and organic heavy metal pollutant that is present in water[10-12]. Various methods are employed for the synthesis of Fe_3O_2NPs such as physical and chemical approaches, co-precipitation approach, hydrothermal, spray laser pyrolysis, etc[13-14]. To determine the morphology of synthesized nanoparticles that may be in the form of nanorods, nanotubes, nanowires, nanocage nanoflowers, etc. these methods may lead to toxicity, are hazardous to the environment, and may cause adverse effects on mother nature. Hence it has become important for researchers to synthesize nanoparticles with high bio-compatibility and harness them to synthesize NPs that are low-cost, renewable, benign reactions, sustainable, and non-toxic[14-15]. Iron oxide nanoparticles transformed the world as a supplement to combat disease and solve environmental issues. Different parts of plant material like fruit, leaves, flowers, roots, seeds, stem, and bark were used to synthesize Iron oxide NPs. This study is concerned with the potential application of biogenic synthesis of magnetite nanoparticles that are harnessed in the utility of natural resources acting as stabilizing, capping, and reducing agents[16-17]. These harness techniques can be innovatively used in various applications. In this study, Fe_3O_2NPs are used as an antibacterial agent in the treatment of harmful infection caused by lesions, as an antioxidant it reduces oxidative stress that arises due to odd unpaired electrons, and as a catalyst, it removes eluent that damages the environment. Fe_3O_2 NPs were prepared using the roots of *Mirabilis Jalapa*, which belongs to the Nyctaginaceae family the marvel of Peru a perennial herbaceous bushy plant that is also called the 4'oclock plant[18-19]. In some areas where the climatic conditions are warm, the roots can grow to more than 18-20 kg. The *Mirabilis Jalapa* roots are made into a paste and used as a medicine in Ayurveda, Siddha, and in some traditional methods for curing diseases like arthritis, joint pains, piles, and inflammation[21-24].

Herewith, we prepared magnetite nanoparticles from the roots of *Mirabilis Jalapa*. The prepared nanoparticles were characterized UV-visible spectroscopy, Scanning Electron microscopy (SEM), Energy dispersive X-ray spectroscopy (EDX), Transmission Electron Microscopy (TEM), Thermogravimetric analysis (TGA), X-ray diffraction (XRD) and Infrared spectroscopy techniques (FT-IR). Further, the prepared nanoparticles catalytic, antioxidant and antibacterial properties are investigated.

2. Materials and Methods

2.1. Materials

Analytical reagents of Iron II sulphate heptahydrate ($FeSO_4.7H_2O$), Iron III Chloride hexa hydrate ($FeCl_3.6H_2O$), NH_3, and Malachite green dye were purchased from Merch, and Milli-Q-water was used for the analysis.

Fig. (1): a). *Mirabilis Jalapa* plant, b). *Mirabilis Jalapa* Root, c). Dried root d). Dried root powder, e). Root extract, f). Various concentrations of *Mirabilis Jalapa* root extract.

2.2. Methods

2.2.1. Preparation of plant extract

From Visakhapatnam's neighborhood, a fresh *Mirabilis Jalapa* plant was procured. The *Mirabilis Jalapa* root was cut free from the plant and properly cleaned under a running tap then with 2D water and immersed for 30 minutes to eliminate dust. The collected roots were peeled, sliced into fine pieces, and dried for 10 to 15 days in the shade. After drying, the root was ground into a fine powder. The powder was used for additional testing after being kept in an airtight container and used for extract preparation (Figure 1&2).

2.2.2. Synthesis of Iron oxide nanoparticles

The magnetite NPs have been synthesized using the co-precipitation method with roots of MJ extract that act as a capping and reducing agent (Figure 2). In this typical procedure, a 1:1 ratio of $FeSO_4{\cdot}7H_2O$ about 2.78 g, and 0.55 g of $FeCl_3{\cdot}6H_2O$ were prepared and 10 ml of each solution was added to 80 ml of Milli-Q water. This solution was then subjected to heated for 50 min at 80 °C stirring constantly on a hot plate. 5 mL of 5 % aqueous root extract of *Mirabilis jalapa* was added to the reaction mixture. 2 mL of 5 % ammonia was added to the reaction and the pH was maintained. A brownish-black precipitate was observed and the reaction was stopped and filtered[25]. The obtained, Fe_3O_2NPs were allowed to centrifugation and washed thrice with Milli-Q water to free from impurities then dried in an oven at 50^0c, and used for further study. To enhance the yield optimisation conditions were employed.

Fig. (2): Preparation of Magnetite nanoparticles

3. Results and Discussions

3.1. Optimization conditions for the synthesis of magnetite NPs

Optimization studies were done to avoid agglomerisation and to enhance the yield of Fe_3O_2NPs by changing different concentrations of $FeSO_4$, and $FeCl_3$, root extract, pH, temperature, and reaction time. The conditions for the synthesis of Magnetite NPs are identified by the appearance of a characteristic band from 390 nm to 405 nm in the UV-visible spectrophotometer.

3.2. Optimizing with FeSO4 and FeCl3 concentration

Different ratios of $FeSO_4$ and $FeCl_3$ solutions were prepared and used to enhance the production of magnetite NPs. The Concentrations were taken in 1:1 ratio, 2:1 ratio, 2:3 ratio, and 1:2 ratios of $FeSO_4$ and $FeCl_3$. 1:2 ratios of ferrous sulphate and ferric chloride were favorable in enhancing the yield and the UV-Visible spectral analysis as shown in Fig. (3)

3.3. Optimizing with root extract concentration

Various percentages of *MJ* root extract (1%,2%,4%.5%) were prepared to synthesize magnetite nanoparticles. It was observed that as the percentage of the extract increased the color of the reaction also turned darker, the SPR band of maximum absorbance was observed at 404 nm which is shown in Fig. (4) when the reaction was performed using 5% root extract.

3.4. Optimization of pH:

P^H studies were done to synthesize Magnetite NPs to enhance yield. 10 ml of 5%v root extract, a 1:2 ratio of $FeSO_4$ and $FeCl_3$ solutions were added, and different pH conditions were maintained (5,7,8,9) by adding NH_3 and HCl. pH-9 was observed to be quite favorable with maximum absorbance at 406 nm shown in Fig. (5)

3.5. Optimization of reaction times

To obtain stable NPs the duration of the reaction mixture is been increased, and various reaction times were carried out for 80 min, 90 min, and 120 min at pH-9, 5% root extract with 1:2 ratios of $FeSO_4$ and $FeCl_3$ was used to increase the yields of magnetite NPs at 70 ^{0}C. The color of the reaction mixture turned yellowish orange to blackish brown and the absorbance was recorded which is shown in Fig. (6).

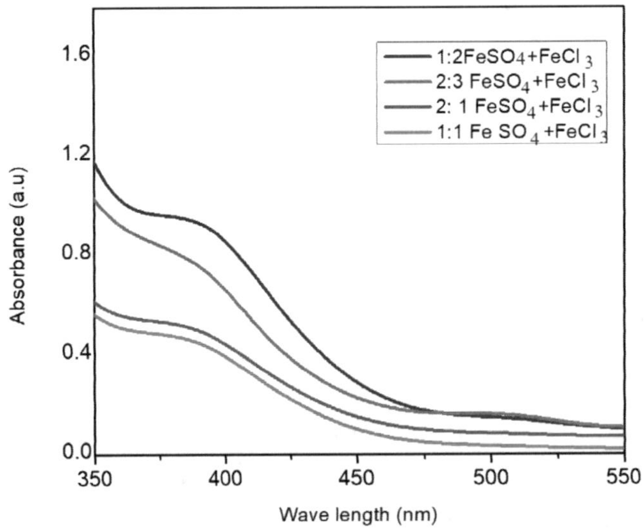

Fig. (3): Different proportions of $FeSO_4$ and $FeCl_3$.

3.6. Optimization of temperature

Temperature plays an important role in the enhancement of Fe_3O_2NPs synthesis to 5 % of 10 ml root extract 1:2 ratios of $FeSO_4$ and $FeCl_3$ solution was added, ammonia solution was added to adjust pH-9 during the course of the reaction, and different temperature conditions were

IOP Conf. Series: Journal of Physics: Conf. Series **2765** (2024) 012004 doi:10.1088/1742-6596/2765/1/012004

maintained in the range of 40^0 -70^0 C to increasing the productivity of magnetite nanoparticles. The SPR band was obtained at 398-409 nm given in Fig. (7) maximum yields were obtained at 70 ^{0}C.

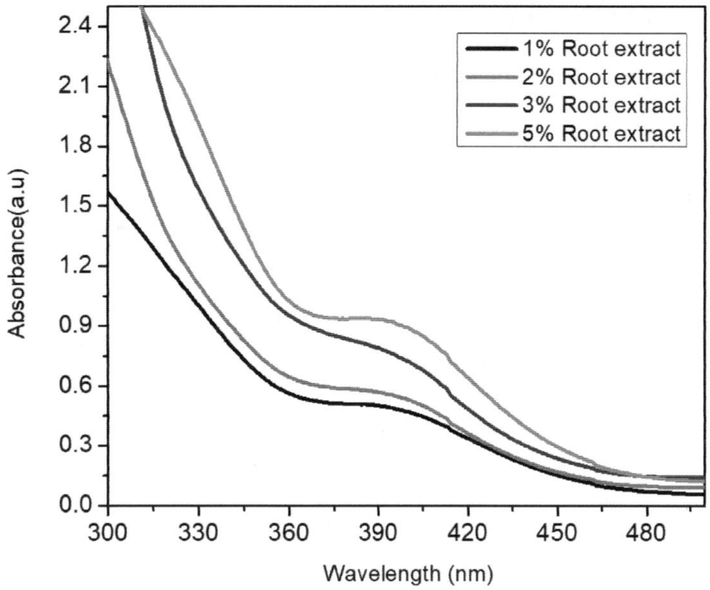

Fig (4): Various percentages of root extract.

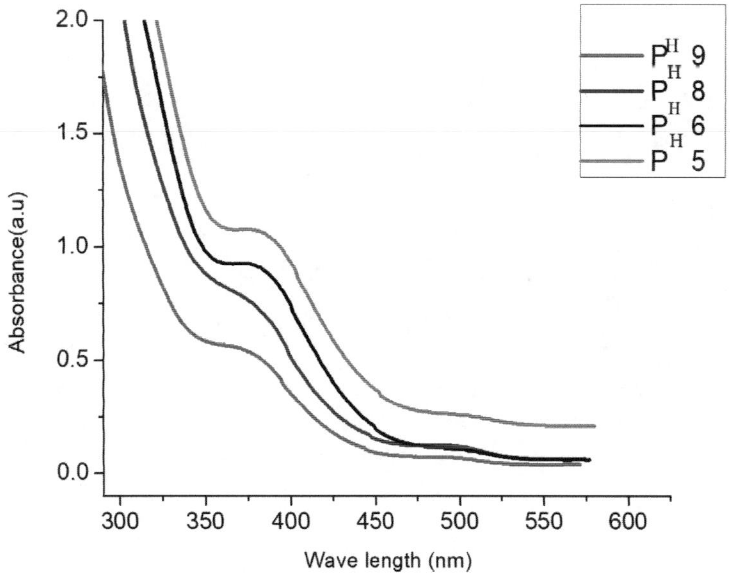

Fig. (5): Various pH.

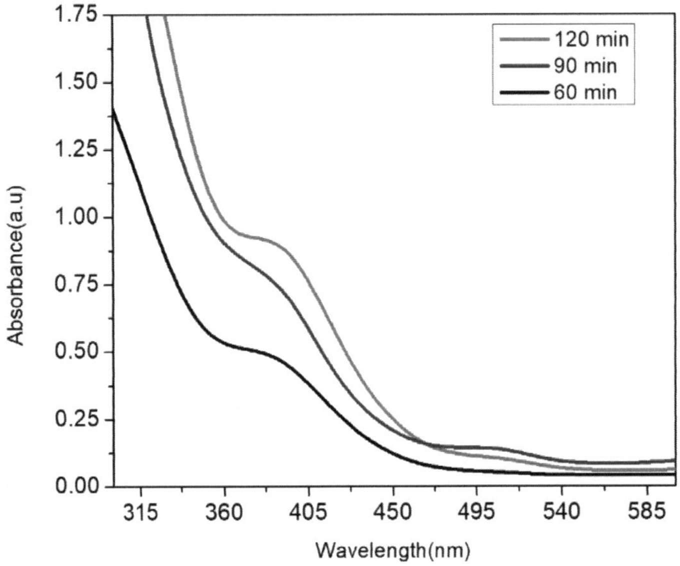

Fig. (6): Various reaction times.

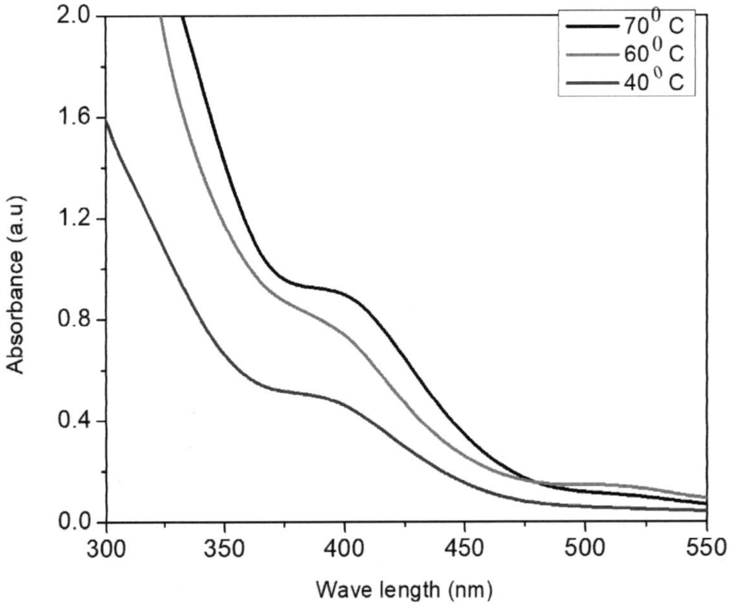

Fig. (7): Various temperature conditions.

3.7. FT-IR studies

The functional groups in synthesized iron oxide nanoparticles were done with FT-IR analysis using the instrument Prestige 21 Shimadzu. These spectroscopy studies enable the identification of various Phyto-chemical constituents / bio-molecules that are capable of reduction and stabilization [26-27] and are shown in Fig (8).

Peaks were identified at 1090 cm^{-1} that corresponds to strong C-O stretching with primary alcohols, 1610 cm^{-1} for strong C=C stretching indicates the presence of α, β -unsaturated ketones, 1710 cm^{-1} for strong C = O stretching with conjugated acids, and 3546 cm^{-1} for strong broad O-H stretching shows the presence of alcohols.

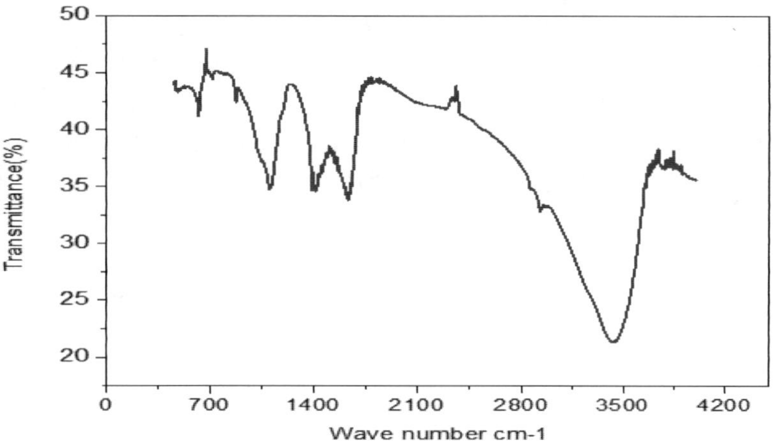

Fig. (8): FTIR spectra of magnetite nanoparticles synthesized using *Mirabilis Jalapa* root extract.

3.8. SEM Studies

The Scanning electron microscope (SEM) enables us to determine the surface morphology, size, and shape of Iron Oxide NPs. The morphological dimensions were assessed by using SEM (ZEISS GEMINI SEM). The SEM images are shown in Fig. (9a-c). The SAED image of prepared nanoparticles is shown in Fig (9d) The average particle size of synthesized iron oxide nanoparticles using aqueous root extract of *M.J* extract was in the range of 59.73nm. The histogram of iron oxide nanoparticles is shown in Fig. (9e) and the particle size of the nanoparticles is less than 100 nm.

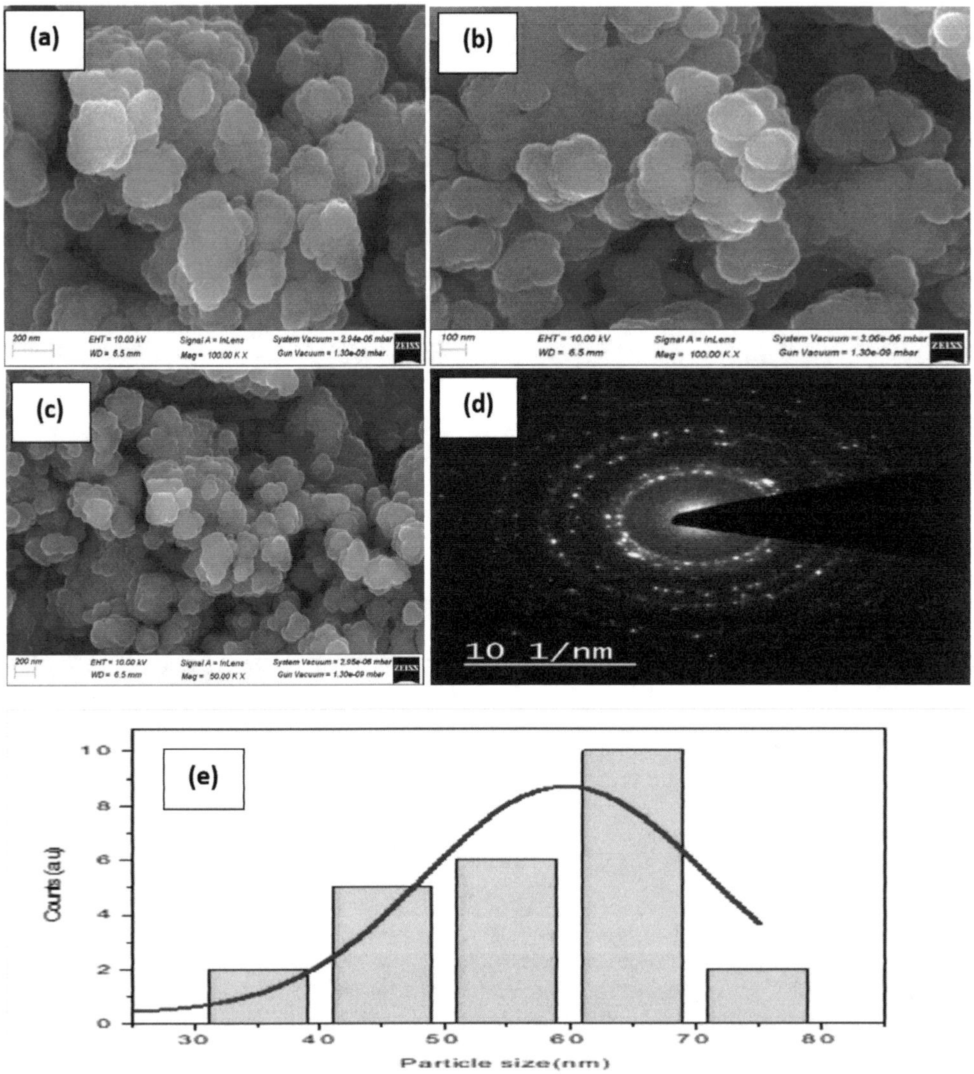

Fig. (9): SEM images of prepared magnetite NPs (a-c) and SAED image of prepared NPs (d) and Histogram of prepared magnetite NPs (e)

3.9. EDX Studies

The chemical composition of the resulting nanoparticles is made possible with the use of EDX analysis. Iron and oxygen show a strong signal on the EDX, and this information makes it possible for the presence of both iron and oxygen in the form of iron oxide nanoparticles with maximum peak intensity at 0.6 e.v for iron with 65.3% and 0.5 ev for oxygen at 32.9 % [28] is displayed in Fig. (10).

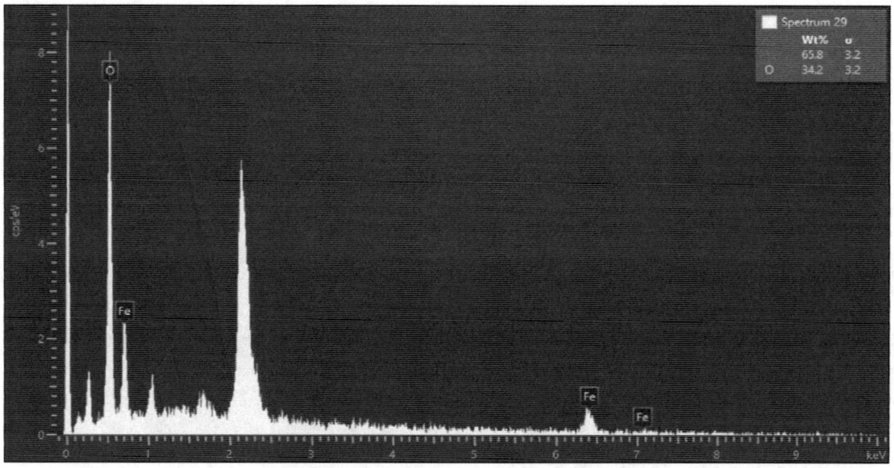

Fig (10): EDX image of Magnetite NPs.

3.10. XRD Studies

X-ray diffraction analysis was performed for Fe $_2O_3$NPs with Bruker AXS D8 equipment at a wavelength of 1.5406, and XRD analysis was carried out at 20^0 - 90^0 at 40 kV with a divergence slit of 10 mm and is shown in Fig. (11). Fe $_2O_3$ NPs peak positions were determined to have 2 θ values at 38.32, 44.49, 64.70, 77.64, and 81.78, which corresponded to (h, k, l) values that matched with JCPDS card numbers 39-1346 and 89-4319. The crystallinity was calculated using the Scherrer equation by the following equation -1 [29].

$$D = (K.\lambda / \beta \cos \theta)\ {}^0A \text{ ----- } (1)$$

Whereas in equation (1) represents K = Scherrer's constant (0.9560), λ = Wavelength of X-ray (1.5406 ^{0}A), β = Full width at half maximum, θ = Bragg's diffraction angle and D = Size of the particle.

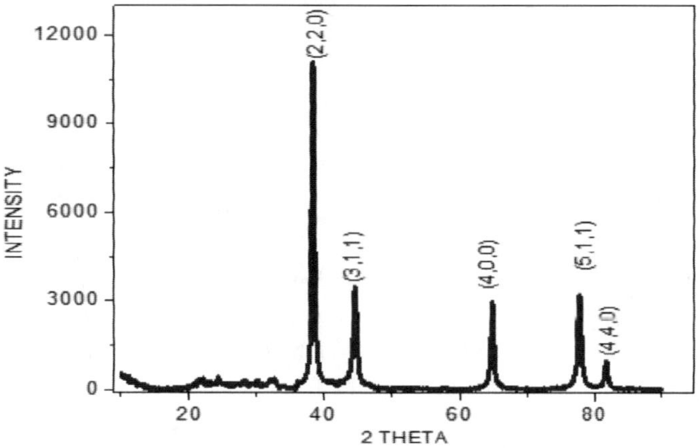

Fig. (11): XRD spectrum of Fe_2O_3 NPs

IOP Conf. Series: Journal of Physics: Conf. Series **2765** (2024) 012004 doi:10.1088/1742-6596/2765/1/012004

3.11. TEM studies

TEM examination of iron oxide nanoparticles was carried out using the FEI Tecnai G2 20 S-Twin. The SAED pattern known as selected area electron diffraction and image were used to study the TEM images of iron oxide nanoparticles that were made utilizing environmentally friendly methods using the roots of *Mirabilis Jalapa*. The produced nanoparticles had been discovered to be crystalline. Iron oxide nanoparticles were found to have a 45 nm particle size. Fig. (12) displays the histogram.

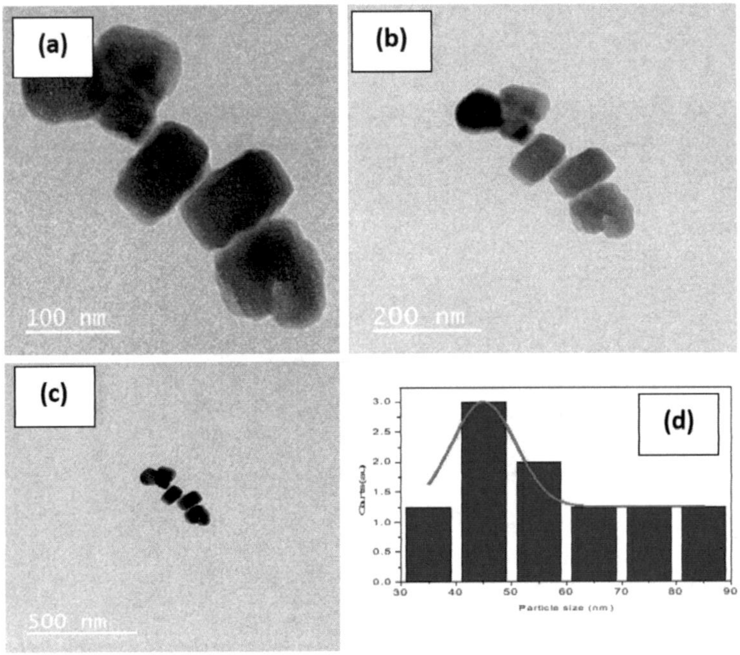

Fig. (12): TEM images (a-c) and average size distribution-Histogram (d)

3.12. TGA studies

TGA study was carried out on synthesized magnetite nanoparticles to incorporate the surface density, identify of nanoparticles composition, determine the effect of the additive present, oxidative assessment and to know the thermal stability as well as purity of the sample. In this process, the temperature of the furnace under which the sample is kept whose weight is measured with the help of an analytical balance provided in it. The loss of weight was observed when the instrument was run from 40-1000^0 c temperature. It was observed that about 18% of total mass was lost by the magnetite NPs and the following graph is shown in Fig 13.

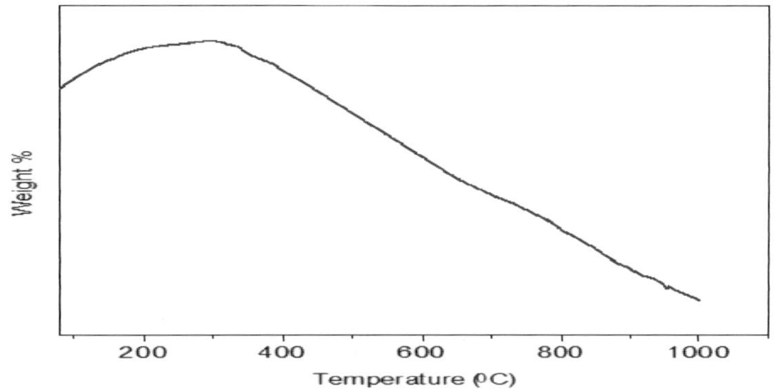

Fig. (13): TGA analysis of magnetite NPs

3.13. Adsorption Studies

Due to the increase in globalization, there is a tremendous spread in trade and industries that enormously increases the pollution of the environment causing serious damage to Mother Nature. The eluents from industries and factories get mixed with water bodies, in the soil and get contaminated. This not only affects the water bodies but also affects the aquatic living system. Dye is the major substance that has to be deprived. Nanoparticles can be used to control these eluents and help the environment. Biogenic synthesized magnetite nanoparticles are proven to be the best pollution deprivers as they are non-toxic and eco-friendly. In this study magnetite nanoparticles that are synthesized using *Mirabilis Jalapa* root extract are used as pollution deprivers to remove Malachite Green dye, a cationic dye.

Adsorption isotherm analyses were done to identify the maximum adsorption capacity of synthesized magnetite NPs. This method enables in evaluation of the performance of adsorbent in the removal of Malachite Green dye that gets contaminated by the water bodies. Various methods of adsorption techniques were performed in the removal of M.G dye such as Langmuir, Freundlich, and Temkin isotherm models [29-31]. Langmuir adsorption assumes monolayer adsorption on homogeneous surfaces without interaction of adsorbate molecules and also provides insight into the nature of adsorption on the surface of magnetite NPs this also helps in depriving MG dye of getting polluted into the environment. A saturated state is formed at which further adsorption does not take place as well and there is no further interaction between the molecules that are adsorbed on neighboring sites. The Freundlich isotherm allows for multi-layer adsorption and is nonuniform energies. This enables us to heterogeneity. The Temkin isotherm involves the interaction between adsorbate over the surface of the adsorptive in a thermodynamical manner. The adsorption isotherms (Fig. 14) parameters R^2 values were calculated from the above techniques are mentioned in **Table** (1). These results in the regression coefficient values of the Langmuir model ($R^2 = 0.99458$), the Freundlich model ($R^2 = 0.98518$), and the Temkin model ($R^2 = 0.94872$) were been evaluated. Adsorption techniques are better fitted to Langmuir isotherm than Freundlich and Temkin isotherms on the surface of Fe_2O_3 is a monolayer. The monolayer adsorption capacity (q_{max}) of Fe_2O_3 NPs was theoretically calculated and found to be (34.0136) mg/g. Fe_2O_3 NPs have comparative adsorption efficiency towards organic eluents. In this study, the adsorption mechanism was done using the kinetics of pseudo-first and second-order reactions (Fig. 15) for obtained experimental data, and the parameters and constants of Fe2O3 NPs are shown in Table 2.

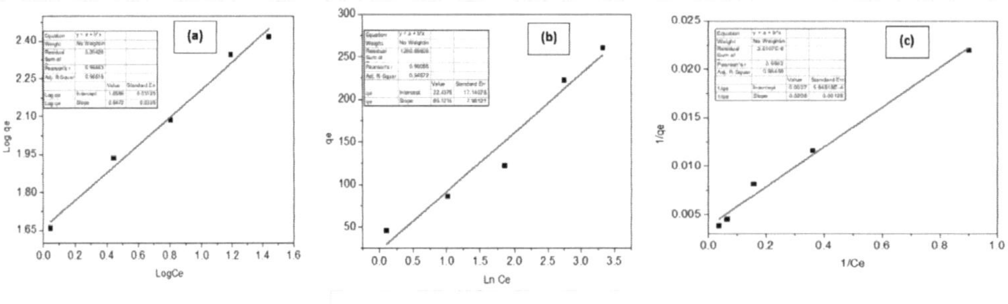

Fig. (14): Langmuir adsorption isotherm (a); Freundlich adsorption isotherm (b) Temkin adsorption isotherm (c)

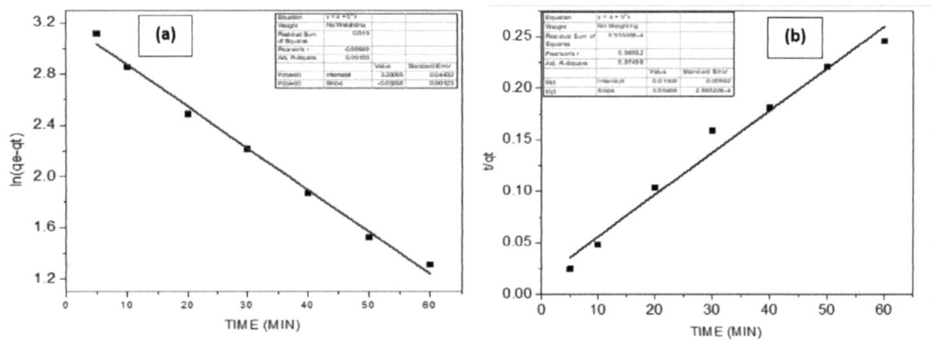

Fig. (15): Pseudo first order reaction (a); Pseudo second order reaction (b).

Table (1): Langmuir, Freundlich, and Temkin isotherms' for adsorption of Fe_2O_3 NPs

Isotherm models	**Parameters**		
Langmuir Isotherm	**qmax (mg·g-¹)**	**KL (L·mg-¹)**	**R^2**
	34.0136	1.14850	0.99458
Freundlich Isotherm	**n**	**K_F**	**R^2**
	1.82505	4.4668	0.98518
Temkin Isotherm	**BT**	**K_T**	**R^2**
	69.1215	2.718282	0.94872

Table (2): Pseudo-first-order and pseudo-second-order kinetics model of Fe3O4 NPs

Kinetic Model	**Parameters**		
Pseudo-first order	**q_e exp (mg/g)**	**K_1 (min-^{1})**	**R^2**
	7.8963	0.000533	0.99159
Pseudo-second order	**q_e exp (mg/g)**	**K_2 (g.mg-1.min-^{1})**	**R^2**
	21.7391	0.0021160	0.97498

3.14. Anti-bacterial studies

Lesions carry bacteria damaging the healthy tissues and increasing the infection as they host the immune system. Lesions can transfer from one organ to many organs as some harmful pathogenic microorganisms present in lesions and delay the healing process causing severe pain. To minimize the infections caused by lesions, antibiotics are prescribed to decrease the infection and proper care must be taken [32-35]. Antibiotics work on infection and kill the growth of bacteria in lesions they do not heal virus or fungus infections. In a few cases, they may need additional interventions such as antibiotics and sometimes surgical debridement. Though antibiotics are necessary for curing diseases improper usage may disrupt the immune system of our body. Biogenic synthesized magnetic nanoparticles target bacteria to reduce contamination and enhance the healing process in the lesion. They can stimulate blood circulation, tissue production, and healing ability. In the present study staphylococcus and E. coli were isolated from the infected lesions and an anti-bacterial assay was performed using disk diffusion / Kerby-Bauer method. The sterile agar growth medium was poured into a petri dish. The infected lesions were collected with a sterile swab and spread on an agar plate. Streaking was done on the entire surface followed by incubation for 48 hours at 37^0 c, the bacteria developed on the agar plate and were identified by the gram staining process. The isolates of staphylococcus aureus and E. coli images are shown in Fig. (16). Agar medium was prepared and known concentrations of Fe_2O_3 NPs were prepared with a suitable solvent and added to the agar medium which was punched with the help of a sterile borer. Ciprofloxacin was used as a reference antibiotic and was allowed for incubation for 24 hours at 37^0 c. A Himedia zone scale was used to measure the inhibition zone. Which were shown in the below Table (3) and the anti-bacterial activity of staphylococcus aureus and E. coli with ciprofloxacin antibiotic on Fe_2O_3 NPs is shown in Fig. (17).

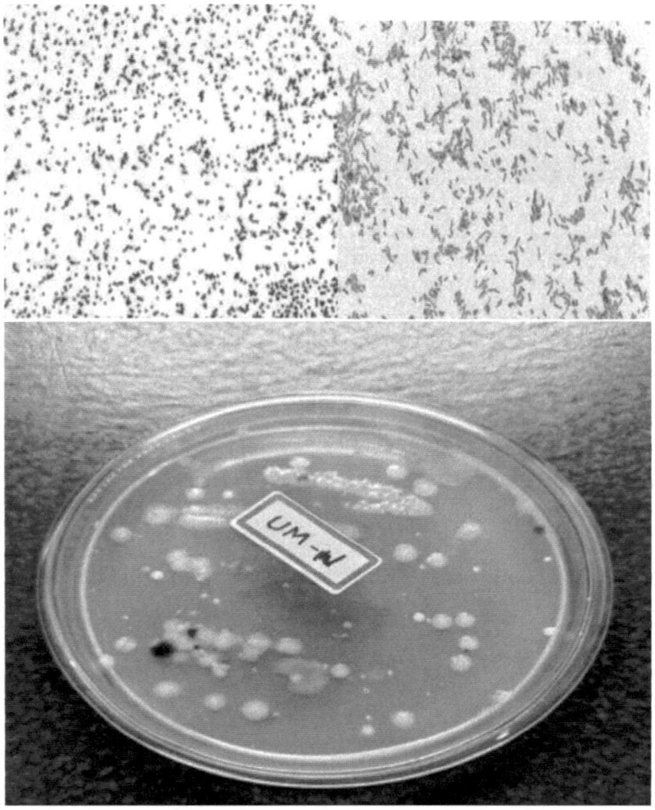

Fig. (16): *Staphylococcus Aureus* and *E.-Coli* bacteria *isolated from the lesion*

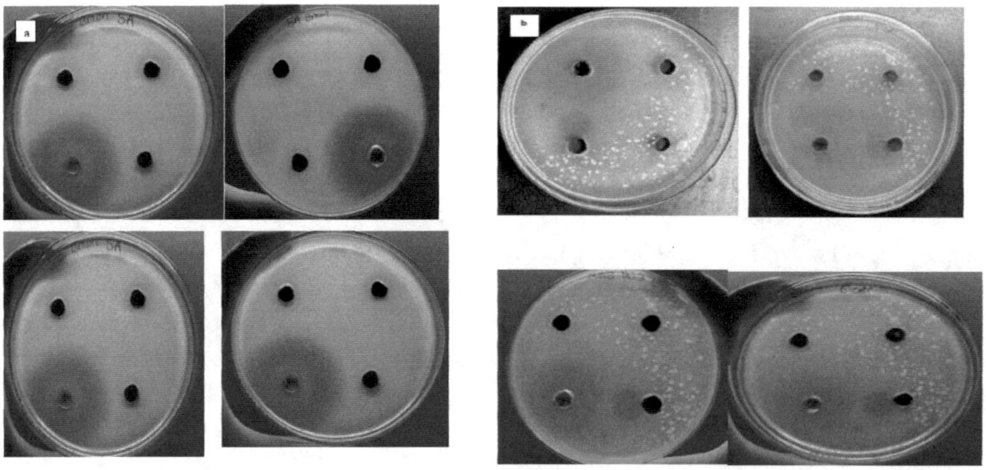

Fig. (17): Anti-bacterial activity of synthesized Fe_2O_3 NPs against *Staphylococcus Aureus* (a) and E. *coli (b)* with different concentrations of Fe_2O_3 NPs, and antibiotic Ciprofloxacin.

Table (3): Zone of inhibition of Fe_2O_3 NPs and zone of inhibition.

Name of the bacteria	**Concentration of Fe_2O_3 NPs**			
Concentration	**15 µg/mL**	**25 µg/mL**	**50 µg/mL**	**Antibiotic (Ciprofloxacin) µg/mL**
ZONE OF INHIBITION				
E-Coli	10	12	14	29
Mean ±S.D	9.66 ± 0.57	11.33 ± 0.57	13.66±0.57	29
Staphylococcus aureus	12	13	14	30
Mean ±S.D	11 ± 0.57	13.33 ± 0.57	14.66±0.57	30

3.15. Anti-oxidant studies

Antioxidant activity can be performed in the laboratory to measure the antioxidant capacity of the desired sample. A substance that stops oxidative damage caused by reactive substances called free radicals and reactive oxygen species. Antioxidants are used in food sciences, medicine, and cosmetics, they can neutralize the free radicals. As food preservers antioxidants can retard the action of oxygen on food, they are more reactive towards oxygen than the food. In this study, DPPH is used in assessing the potential of Fe_2O_3 NPs as it is a rapid simpler, and stable free radical that reacts with antioxidants and turns to violet, and slowly its color becomes pale violet when kept in the dark for half an hour[36]. The degree of color is related to its antioxidant activity. The greater the loss in colour of DPPH greater the antioxidant activity. The antioxidant potential of Fe_2O_3 NPs is measured using a UV-visible spectrophotometer at 517 nm. For the control, Ascorbic acid was used that reduce and neutralize highly reactive oxygen. The percentage inhibition was done using the formula. the antioxidant efficacy is shown in Fig. (18)

$$\% \text{ inhibition} = \{(A_C - A_S)/A_C * 100\}$$

A_C = Absorbance of Ascorbic acid as a control

A_S = Absorbance of the magnetite nanoparticles after performing an antioxidative assay

It was observed that the absorbance of D.P.P.H reduced with a sign of color change from violet to pale yellow color with the increase in the concentration of Fe2O3 NPs.

Mean ± SD was calculated and shown in the table (4)

IOP Conf. Series: Journal of Physics: Conf. Series **2765**(2024) 012004 doi:10.1088/1742-6596/2765/1/012004

Table (4): The antioxidant efficacy using different concentrations of synthesized magnetite NPs against DPPH.

Sample Number	**The concentration of Fe_2O_3 NPs (µg / mL)**	**% of DPPH scavenging activity mean ± SD**
1	50	1.63 ± 0.005
2	100	15.08 ± 0.01
3	150	29.01± 0.001
4	200	40.32 ± 0.015
5	250	49.32 ± 0.015

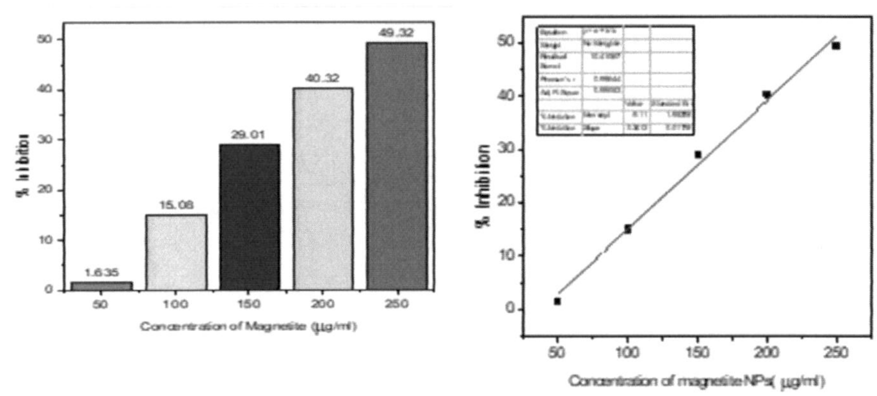

Fig. (18): Antioxidant efficacy of Fe_2O_3 NPs against DPPH.

Conclusions

The Magnetite nanoparticles have been synthesized from the roots of *Mirabilis Jalapa*. The prepared Magnetite nanoparticles characterized using UV-visible spectroscopy, Scanning Electron microscopy, Energy dispersive X-ray spectroscopy, Transmission Electron Microscopy, Thermogravimetric analysis, X-ray diffraction and Infrared spectroscopy techniques were used. The prepared particles are acted as a catalyst in depriving eluent, an anti-oxidant that removes oxidative stress, and as an anti-bacterial agent to cure infective lesions.

References

[1]. M. Ayelén Vélez, M. Cristina Perotti, L. Santiago, A. María Gennaro, E. Hynes, *Nutr. Deliv.*, **6**, 221–250. (2017)

[2]. Bera H. Belhaj, *J. Nat. Gas Sci. Eng.*, **34,** 1284–1309. (2016)

[3]. S Fernandes, A. Benfeito, C. Fonseca, J. Oliveira, E.M. Garrido, et al., *Food Preserv.* **15**, 527–565 (2017).

[4]. L.J. Frewer, N. Gupta, S. George, A.R.H. Fischer, E.L. Giles, D. Coles, *Trends Food Sci. Technol.* **40,** 211–225(2014).

[5]. A.K. Hussein, Sust. *Energ. Rev.* **62,** 767–792 (2016)

[6]. M.K.Sharif, F.H. Shah, M.S. Butt, H.R. Sharif, *Nutr. Deliv.*, **15**, 587–618 (2017)

[7]. McNamara, Karrina, and Syed A. Tofail. *Advances in Physics: X*, 2, 54–88 (2016).

[8]. S. S. Shankar, A. Rai, A. Ahmad, M. Sastry, *J Colloid Interface Sci.*, **275**, 496–502 (2004)

[9]. Beveridge, R. Murray, *J. Bacteriol.*, **141**, 876–887 (1980)

[10] P. Bautista, A. F. Mohedano, J. A. Casas, J. A. Zazo, J.J. Rodriguez, *J. Chem. Technol. Biotechnol.,* **83**, 1323–1338 (2008).

[11] W. Wei, P. Gao, J. Xie, S. Zong, H. Cui, X. Yue, *J. Solid State Chem.,* **204,** 305– 313 (2013).

[12] A. Xu, X. Li, S. Ye, G. Yin, Q. Zeng, *Appl. Catal.* B**, 102,** 37–43. (2011).

[13] P. Kuppusamy, M.M. Yusoff, G.P. Maniam, N. Govindan, *Saudi Pharmaceutical Journa*l, **24**, 473–484, (2016).

[14] P. Raveendran, J. Fu, S.L. Wallen, *J. Am. Chem. Soc.*, **125**, 13940–1394 (2003).

[15] R. Rajan, K. Chandran, S.L. Harper, S.-I. Yun, P. T. Kalaichelvan, *Ind. Crops Prod.*, **70,** 356–373. (2015).

[16] S. Ahmed, M. Ahmad, B. L. Swami and S. Ikram*, Journal of Advanced Research***, 7**, 17–28 (2016).

[17] S. Hao, et al. *Solar energy***, 80**,209-214, (2006).

[18] Kayser O, Quax WJ Eds., Medicinal plant biotechnology: from basic research to industrial applications. Weinheim, Wiley VCH, (2007).

[19]. Daniel M. Medicinal plants chemistry and properties, Taylor and Francis group, (2006)

[20] Kurian J.C, Mutatkar R.K, Samraj E. Plants that heal, Watchman Publishing House, 214-215 (2003).

[21] State Administration of Traditional Chinese Medicine, Chinese materia medica, in Editorial Board, Chinese Materia Medica (II), Shanghai Science and Technology Press, Shanghai, China, 748-749 (1999)

[22] A.L. Bogle, *Journal of the Arnold Arboretum*, **55,** 1-37 (1947)

[23] Watt J.M, Breyer-Brandwijk M.G. The Medicinal and Poisonous Plants of South Africa. The Indian Medial Gazette, 1933.

[24] Bhatia H., Manhas, R.K., Rani M., *J Ethnopharmacol*., **152**, 207-16 (2014).

[25] Lin C, Lee C., Chiu W., *Journal of Colloid and Interface Science***, 291,** 411-20

[26] F. Alam and Q. U. Saqib, *Anci. Sci. Lif.,* **34**, 147 (2015).

[27] R.M. Cornell, U. Schwertmann, Wiley-VCH, Weinheim (2003).

[28] L. Patterson, *Physical Review*, **56**, 978–982 (1939)

[29] Langmuir, *J. Ameri. Chem. Soci.*,**38**, 2221–2295 (1916)

[30] H.M.F. Freundlich, *J. Phys Chem*, **57**, 385–471 (1906)

[31] N. Mohammadi, H. Khani, V.K. Gupta, E. Amereh, S. Agarwal, *J. Colloid Interface Sci.*, **362**, 457–462 (2011).

[32] Anthony, A.A., T.M. Okoh, I. Anthony and J. Steve., *Sci. Res. Essays*, **5,** 3876-3881 (2010).

[33] Bessa, L.J., P. Fazii, M. di Giulio and L. Cellini., *Int. Wound J.*, **12,** 47-52. (2015).

[34] Kenneth, I.E.P. *AASCIT J. Health*, **4,** 63-67 (2017).

[35] Cathy, T.H. *Adv. Skin Wound Care*, **24**, (2011).

[36] A.V. Ramesh, Dharmasoth Rama devi, Ganga Rao, K. Basavaiah, *South African Journal of Chemical Engineering*, **26** 25-34 (2018).

A review on machinability and optimization of machining parameters of metal matrix composites

Tarun Kumar Kotteda[1,3*], Manoj Kumar[1], Pramod Kumar[1], Ajay Gupta[2] and K V Murali Krishnam Raju[3]

[1] Department of Mechanical Engineering, Dr B. R. Ambedkar National Institute of Technology, Jalandhar, Punjab, India
[2] Department of Industrial and Production Engineering, Dr B. R. Ambedkar National Institute of Technology, Jalandhar, Punjab, India
[3] Department of Mechanical Engineering, Sagi Rama Krishnam Raju Engineering College, Bhimavaram, Andhra Pradesh, India

* Corresponding Author: kottedatk.me.21@nitj.ac.in

Abstract. The advancement in the state-of-the-art materials that are harder and difficult to cut till date have led to the upgradation towards cutting-edge technologies for machining them. In recent times, utilization of advanced materials in various domains such as aerospace, automobile industries etc. is significantly increases due to the requirement in the aspect of lighter in weight and hardness. At present study, investigations by many researchers on various machining processes and optimization techniques used so far in several metal matrix composites are mentioned. This includes conventional and unconventional machining methods and the user-friendly optimization techniques such as response surface methodology, artificial neural network, fuzzy logic, and Taguchi approach. In addition to these, this study emphasizes on the modelling of the machining processes in MMCs. Finally, open issues and conclusions drawn on the machinability and optimization of tailor-made composites are illustrated with reference to surface finish.
Keywords: Conventional machining, Machinability, Metal matrix composites, Optimization, Unconventional machining

1. Introduction

Metal matrix composites have induced a discerning attention from the automotive and aeronautic applications undischarged to the possession of excellent mechanical properties and utilization. In the recent years, researchers have disclosed many findings prevailed to these state-of-the-art materials. Over the past two decades, metal matrix composites (MMCs) are extensively used in utmost industries only because of the reason that they are considered as hard materials to machine and having high strength to weight ratio, good surface finish which are required in automotive, marine, aerospace and other similar industries [1-4]. The term metal matrix composites impart most favourable solutions for diverse applications that comprises wide variety of materials ranging from truncated to complex castings. MMCs offering promising solutions in various applications due to its exceptional properties such as high wear resistance, low density material, high thermal stability, elastic modulus, conductivity and magnificent wear resistance [5-7].

Among various composite materials, aluminium matrix with SiC reinforcement is more beneficial and regular. Aluminium, magnesium and titanium alloys were conventionally utilized as matrix form. However, silicon carbide (SiC), boron carbide (B_4C) and alumina (Al_2O_3) are the most favourable group of MMCs reinforcements. Nevertheless, the MMC samples fabricated to net-near structure using various techniques, sub-sequent processing has identified and needed for acquiring the expected size and shape with appropriate surface finish [8-13]. The parent interface of tailor-made composites comprises alumina, silicon carbide particulates, aluminium alloys, magnesium and titanium alloys. Inspite of the fact that, the mechanical behaviour of silicon carbide embedded aluminium composites are of conflictive nature in tandem with non-discrete fibrous composites, they are extremely contending because of effective synthesis tactics, sturdy processing capabilities. In addition to this, the cost-effective factor is of enormous benefit to their widespread use [14,15].

In general, fabrication of composite materials is having a major problem with difficulty in its machinability because of non-uniform shape, abrasive particles, porosity and high hardness of reinforcement form. Factors such as uncontrolled tool wear rate, poor surface finish, lower surface integrity and high cutting temperatures challenged many researchers for machining the composite materials [16]. These materials discover the place in the hard and tough to cut machine group, needs more accuracy and high surface finish prevailing to uneconomical machining cost. The subsistence of ceramic particles categorized them hard to machine that may lead to rapid rise in tool wear rate. Meanwhile, due to these facts both the performance of machining and tool may be affected to get the required shape of the component. Such as during the conventional turning operation of SiC/Al MMCs, very hard, precise and costly cutting enclosures, namely PCD tools are necessary to acquire expected size and shape. Moreover, the mechanical, thermal, electromagnetic and chemical behaviour of these composite materials have become ineffectual in traditional machining processes [17].

Muthu Krishnan et al. [18] noticed the endurance and wear resistance of the PCD cutting tool by performing turning operation on aluminium metal matrix composites (AAMCs) reinforced with SiC particles varying from 10 to 20% with a regular interval of 10%. Investigational output responses showed higher flank wear rate in addition to the various abrasions (includes both two & three-body) whereas, on other hand it was stated that the wt.% of SiC reinforcement is a deciding factor on tool wear rate. It is essential to estimate the cutting tool wear in order to automatize the production and avoiding the break down or any flaws may be occurred due to asymmetry or inadequacy of the machining process. Many researchers have done investigation in several elements such as wear rate, chip formation and deterioration. Transformation of the tailor-made composites is being restricted in various approaches only because of continuous impediments in order to test the machinability, leads to cause extravagant wear of cutting tool on account of existence of coarse-grained fragments [19,20]. On other side, non-traditional machining is accelerated in the contemporary manufacturing era to signify its prominence and acquiring surface finish economically in case of MMCs [21].

In this study, machinability of particle reinforced MMCs using both traditional and non-traditional machining processing has elucidated in detail. More explicit processing operations such as turning, drilling, milling and grinding have been studied to know about the operational behavior of MMCs. With regard to the non-traditional machining, wire-cut electrical discharge machining (WEDM), die sinking and powder mixed electrical discharge machining, electrochemical machining, laser machining, ultrasonic machining (USM) and high-speed machining (HSM) have been studied for recognizing the processing behavior of MMCs. The synopsis of the machining performance is also furnished. As time passes on, most of the researchers are taking interest to employ the modelling and optimization techniques to estimate the machining output responses. Therefore, cutting-edge modelling such as modelling of machinability of MMCs, finite element modelling and simulation whereas artificial intelligence techniques for optimizing the machinability of MMCs that comprises response surface methodology (RSM), adaptive neural networks (ANN), Taguchi technique, fuzzy logic and multi-objective optimization have been studied. Open issues and conclusion for further studies have illustrated in the current survey.

2. Traditional machining processes of MMCs

Turning, drilling, milling and grinding are categorized under traditional machining processes, generally utilized to acquire the required shape and finish to work piece. Because of rapid increase in the usage of MMCs in various engineering applications, lots of work have done so far by many researchers for understanding the machinability of MMCs with traditional approaches. As compared with remaining metals and alloys, MMCs are being selected immensely in automotive and aircraft industrial sectors due to effective cost and feasible metallurgical along with mechanical properties [22]. The properties of MMCs are absolutely preferable; whereas difficulty may arise while machining with conventional technique due to the usage of hard and SiC abrasive reinforcement that may lead to excessive tool wear rate results in tool life shortened. Because of this high tool wear leads to make the machining process of MMCs overpriced [23]. Due to heavy tool wear arising by the inclusion of the hard reinforcement, it is extremely difficult to processing Particle reinforced MMCs using traditional approach. However, the process parameters such as matrix material, reinforcement, tool nomenclature, cutting temperatures, statistical variations etc. may influence the machinability of tailor-made composites. With a view of acknowledging the circumstances, innumerable work has done by many investigators to estimate feasible process variables and cutting implements for machining the composites. Hence, innumerable researchers have committed themselves to elucidate several working conditions of machining the metal matrix composites.

2.1. Turning of MMCs

Among various conventional machining processes, turning is the most widely employed approach to prepare the component in the industry. Especially, for attaining the expected shape and size of the customized and semi-finished rotational component in the market; lots of work have done by many researchers in the area of turning metal matrix composites. As noted from the literature, class of workpiece material and type of cutting tool influences the turning of MMCs in several aspects [24,25]. Literature suggests that cutting forces predominantly identified as one among various notable indicators of tool wear in turning process. It is experimentally proved by various researchers that cutting forces affects the tool wear in turning operation [26,27]. Ravi Sekhar et al. [28] concisely reviewed the research disclosures of various turning mechanisms to make the materials so supercilious. At present, many of the researchers attentive towards MMCs reinforced with nano particles such as single wall & multi wall carbon nanotubes, alumina, graphene etc.

Rashid Ali et al. [29] examined the possessions of substantial recession constraints with respect to inception of machining forces while doing turning operation on Al/SiC_P composite. It is noted from the literature that an explicit rise in feed and depth of cut leads to wear and tear of cutting tool results in poor surface finish. Also, type of cutting tool material influences the surface quality while turning the MMCs; that can be amended by the grain size of tool material [30,31]. Davim et al. [19] explored the relationship in with cutting forces and wear of PCD cutting tool used in turning the MMCs. Tool wear rises with rise in cutting speed whereas two-body and three-body mechanisms identified as governing wear mechanisms. The enhanced rate of surface finish of MMC in hot machining recorded as 60°C [32]. It is a fact that the cutting tools with coatings such as cubic boron nitride, tungsten carbide, polycrystalline diamond, diamond coated carbide etc. helps to improve the surface quality at adequate working conditions [33]. As a result, most of the times, coated tools are employed in turning the composites; whereas abrasive wear is a substantial indemnity with a traditional approach and also a brittle break is examined in this activity [34]. At par from the literature, many researchers preferred PCD cutting tools for processing MMCs. As PCD tools are having good enough physical and chemical properties, they can't possess chemical reaction with MMC material [35]. Moreover, optimal quality of surface finish can be acquired by turning particle reinforced tailor-made composites at elevated turning speeds [36].

2.2. Milling of MMCs

Same as turning, milling is also considered as most satisfactorily familiar and extensive traditional machining technique that is customarily discharged for fabricating the components of tailor-made composites. Milling is the remarkably selected conventional machining operation practiced in the manufacturing world. In general, milling may constitute rotary cutters to withdraw unnecessary material by proceeding a milling cutter towards the workpiece, thereby the desired size and shape of the components are to be processed. Various research articles are referred to understand the operating conditions and outcomes of milling process applied for metal matrix composites. Ozben et al. [37] investigated the out-turn of the milling process variables on the mechanical properties of Al/SiC_p composites in which reinforcement varying from 5 to 15 wt.% with an interval of 5wt.%. Experimental findings revealed that tensile strength, impact strength and hardness were improved with an increase in weight percentage of silicon carbide particles in the composite sample.

Arokiadass et al. [38,39] investigated and analyzed the enhancement in surface quality in end milling process of AMMCs reinforced with Silicon carbide particles. Moreover, rotational speed and cutting rate shows high impact on the surface quality; whereas, depth of cut doesn't exhibit any effect on the surface quality. Authors addressed that higher cutting forces observed at higher depth of cut and cutting rate; Whereas, cutting forces decreases with a gradual increment rotational speed of the milling cutter. However, an excellent surface finish can be attained at higher rotational speeds. On other hand, surface quality may diminish at higher cutting rates and depth of cut values [40]. Many of the researchers noticed that the process parameter named cutting rate desperately influences wear resistance of cutting tool while performing milling operation on metal matrix composites [41,42]. As Polycrystalline diamond (PCD) tools are distinguished for several aspects of cutting process parameters, that may give effective tool life during the milling machining process of MMCs. Also, addressed that chemical vapor deposition carbide and cemented carbide milling cutters doesn't exhibited much attractive results comparatively with PCD tools due to lower tool life in milling operation on composites [43,44]. Reddy et al. [45] examined the comparative analysis on surface integrity along with the degree of subterranean failure of AMMCs reinforced with SiC_p with that of corresponding outcomes from aluminium alloy at various machining constrains on end milling machine.

2.3. Drilling of MMCs

Drilling is one of the conventional machining approaches in which a drill bit can be used to pierce or amplify the holes on the workpiece sample. With reference to the type of component, required dimensions of the hole, processing time etc., miscellaneous cutting tools and strategies are followed for drilling of MMCs. Immense experimental investigations in the area of drilling on tailor-made composites had been carried out from so many years with the aim of enhancing the drill bit life in addition to predict optimum machining parameters. However, drilling is performed in the industries as an exploratory phase for most of the processes such as reaming, chiseling, tapping and boring. Songmene et al. [46] examined various factors such as proportion of matrix and reinforcements, volume fraction, and rigidity modulus of the particle reinforcement influences the life of drill bits operated in the drilling process. Because of exceptional tool wear rate, the above significant factors also influence the formation of chips and its breakage throughout the machining time. Authors addressed that machinability is absolutely influenced by the operating parameter i.e., cutting rate while drilling AMMCs of 56% of SiC particle reinforcement in it [47].

Narutaki et al. [48] reported that majority of research findings covered the impact on tool life with respect to the variation of cutting speeds; but in reality, cutting rate impacts the tool life. Likewise, Rajmohan et al. [49] examined that cutting speed reports insignificant effect on the component material, whereas on other hand, authors revealed that feed rate is the most influencing and deciding factor while drilling a hybrid composite of AA356 matrix and SiC_p reinforcement. It's a fact that the ejection of unwanted portion of a workpiece with the help of a cutting tool subjected to cutting forces exerted may cause wear. In general, three types of mechanisms such as adhesive, abrasive, and fatigue wear are noted to be significant while discussing about the tool life. Paulo Davim et al. [50] addressed that abrasive wear formed on the flank face of the tool is the predominant mechanism and influences the life of drill bit and the surface quality. Also, two types of abrasive wear mechanisms are revealed by the researchers that influences the surface quality of the drill hole such as two-body and three-body abrasions of drilling the tailor-made composites [51,52]. Carbide and cubic boron nitride tool's wear mechanism have been considered for understanding the abrasion and adhesion wear in drilling of titanium metal matrix composites [53].

2.4. *Surface Grinding of MMCs*

Grinding is a traditional processing approach which is usually employed to perform material removal operations on a workpiece with the assistance of a grinding wheel. The rotation of grinding wheels, removes the material from the specimen sample while producing a smooth surface finish in the process. In other words, it can be stated that grinding enhances the feed digestibility (nevertheless the grinding shouldn't be too fine as it deteriorates the digestive characteristics and demands a higher energy expenditure), acceptability, fusibility and pellet ability. Therefore, phenomenon of achieving the damage free and acceptable surface finish is a thrust area for researchers, especially during the usage of grinding the metal matrix composites [54,55]. Factors like cutting speed, speed of the grinding sample, feed rate and depth of cut plays key role that may influence the grinding surface finish of Al/SiC_p tailor-made composites [54]. In traditional processes, elevated volume fraction tailor-made composites are to be machined using grinding process recommended with diamond coated grinding wheels. Generally, the process variables of the any machining decide the accuracy and quality of the finished surface and also reveals whether it meets the industrial demand or not. However, in several cases grinding process is considered as a conventional approach only to remove the unwanted material of the work piece sample of minimum depth varying from 0.25 to 0.5 mm.

Nevertheless, the evaluation of machining characteristics of the MMCs hadn't gained much concentration so far, Thiagarajan et al. [56] examined the impact of grinding parameters on the surface quality of Al/Sic MMCs. As per the authors findings, enhanced surface quality and undamaged with good dimensional accurate surfaces were attained because of less and moderate operating forces with high grinding wheel and object speeds. On other hand, the temperature levels ranging from 740-856 °C irrespective of machining parameter levels. Also, similar proportionality showed by the grinding temperatures with respect to the grinding variables such as cutting speed, work piece speed, depth of cut and rate of feed. Authors have done experimental investigations on the effect of grinding parameters on Al/SiC MMCs and revealed that abrasive particles of grinding wheel and the particle-reinforcement impacts the material removal and surface quality rate in grinding [57]. A detailed comparative analysis had made by the researchers in between the whiskers of polycrystalline diamond compact type and chemical vapor deposition diamond type to evaluate the surface finish of the grinding surface and suggested that grinding wheel of PDC type gives superior edge uniformity [58].

3. Non-traditional machining processes of MMCs

Non-traditional machining processes also termed as an un-conventional machining approach in which the unwanted portion of the work piece material can be removed without make in contact with the cutting tool. Most importantly, unconventional machining processes may come in to existence at a place where traditional process fails. Some of the non-traditional machining processes used for machining the composites are discussed below.

3.1. *Over view of Electrical Discharge Machining*

The working principle of Electric Discharge Machining (EDM) is controlled erosion of the workpiece by means of interrupted electric spark discharge between the tool (cathode) and the workpiece (anode). EDM is an advanced non-traditional machining technique that is commonly employed to machine high-strength-temperature-resistant alloys, MMCs and complex profiles with intricate details. The potential areas of application of EDM include automotive & aerospace industry, medical implants, nuclear research and manufacturing of extrusion dies and blanking punches [59-62]. EDM is the most commonly utilized technique for machining MMCs that are characterized with extreme high hardness which are being extensively selected in several areas such as automotive & aerospace industry, nuclear research [21,63]. EDM is also called spark erosion process whose operation is based on setting up an electric field through a dielectric substance. This electric field imparts high velocity to the free electrons by subjecting them to an electrostatic field. The bombardment of these high velocity electrons raises the temperature enormously at workpiece surface and consequently evaporate the desired portion of the workpiece. A plethora of variants are available in EDM, namely, wire cut EDM (WEDM), die-sinking EDM, rotary disk electrode EDM, dry EDM and Powder mixed EDM.

3.1.1. Die-sinking electrical discharge machining

Die-sinking electrical discharge machining is the most-soughed method for refabricating components such as injection moulding, punching blanks and die-casting [64-66]. The setup of Die-sinking EDM is similar to that of general EDM. But the difference lies in Spark generation which begins with the formation of plasma from the disintegration of the dielectric fluid. With passage of time, due to progressive removal of material, the distance of separation increases prompting the electrode to be lowered without causing any hindrance to the process. It was established through the empirical data pertaining to EDM upon LM24 Al-SiC MMCs obtained by Kathiresan and Sornakumar [67] that there is an increment in the surface roughness and MRR with an increment in the power supply. But, Prabhu and Ramadoss [68] through their work on Al-TiB_2 with a low frequency vibrating tool found an issue with increasing the current i.e., it leads to surface deterioration. Optimization and the analysis of operating behaviour of MMCs along with parallelism between wire-cut EDM and die-sinking EDM is drawn by Garg et al. [69]. Experimental studies [70] arrived at the conclusion that MRR can be increased while reducing tool wear rate by increasing the diameter of the electrode tool. Moreover, it was proven that MRR can also be increased with an increment in the duration of active current supply, but it leads to an accelerated tool wear. The effect of other process parameters such as current supply, peak current along with pulse-on/off time was examined by [71] and the results obtained by them are coherent with previous studies [70]. Experimental studies involving Al/SiC composites reinforced with glass and B_4C by Ahamed et al. [72] revealed that planar imperfections can be minimized by increasing the volume fraction of reinforcements, while having an adverse effect on the MRR. Parametric optimization using TOPSIS in the case of in-situ Al-$CuTiB_2$ composite was done by Prabhu and Vinayagam [73] on the basis of tool wear rate, surface finish and MRR and the findings were in accordance with the runs. Figure 1 depicts the schematic arrangement of Die-sinking EDM test setup.

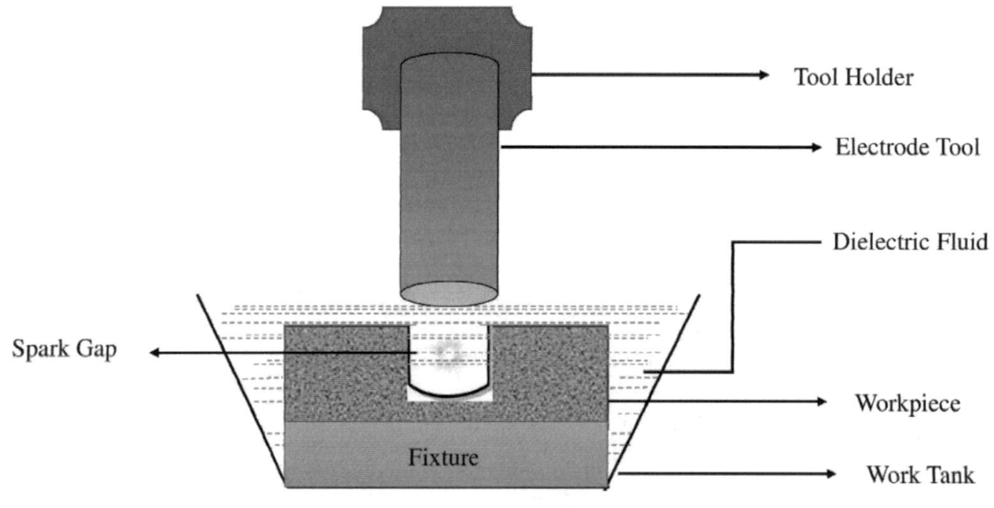

Figure 1: Schematic diagram of Die-sinking EDM

3.1.2. Wire cut electrical discharge machining

Wire cut EDM technique is essentially a thermoelectric process in which material is removed based on the principle of erosion of metals by an interrupted electric spark discharge between the electrode tool and the work-pieced, as the wire is continuously fed. The first EDM machine that used a wire as an electrode became commercially available in 1967. It was developed in the Soviet Union. Wire EDM uses a wire as the tool electrode. A suitable dielectric slurry is continuously forced to maintain a small 'spark gap' between the tool and workpiece, and also to wash away the unwanted material. A numerically controlled electrode with a radius ranged from 0.025 to 0.125 mm is utilized to trace the path fed by the computer to the desired accuracy [74-77]. Figure 2 depicts the schematic arrangement of Wire cut EDM test setup.

During the machining of SiC_p/AA6061 by WEDM, Marigoudar et al. [78] examined the effect of several factors such as wire feed rate, material removal rate (MRR), applied potential difference, interruption interval of current, work piece material, and the width of notch. It is observed that applied potential difference influences MRR and width of notch to a great extent while the influence of interruption interval of current and wire feed rate is trivial. Peak value of MRR and notch width were obtained at lower potential difference and higher pulse interval. The machining of MMC in WEDM with various process parameters such as interruption interval of current, magnitude of current supply while retaining the other parameters constant was investigated by Pramanik [79]. It is evident from the findings that MRR rises with rise in the pulse on time and peak current.

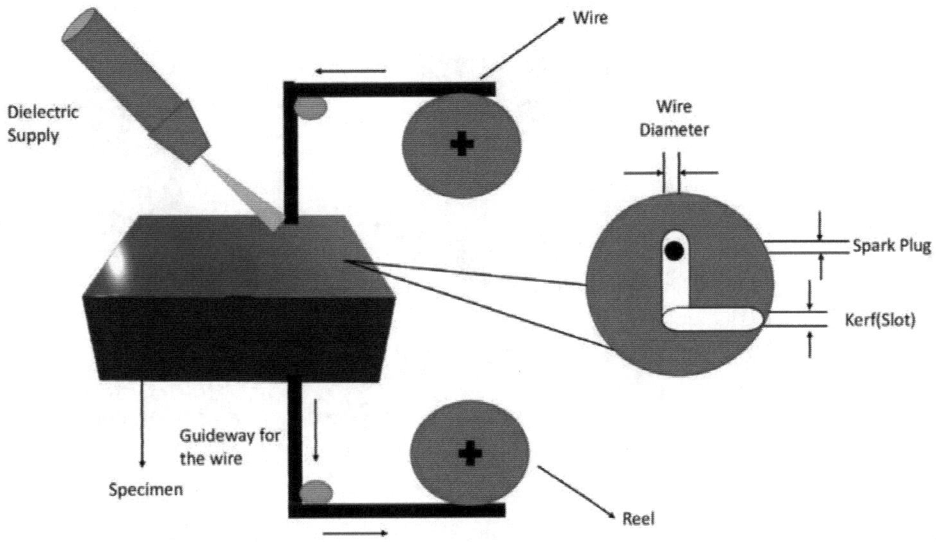

Figure 2: Schematic diagram of Wire cut EDM

3.1.3. Powder mixed EDM

The cutting-edge technological components that are being fabricated now-a-days demand a superior surface finish. This has galvanized researchers globally to devise machining processes to address this challenge. One such process that yields excellent surface finish is powder mixed EDM (PMEDM). This technique employs catalysts to enhance the surface finish. In PMEDM, fine powders are fused with the dielectric medium to not only achieve super-finished surfaces but also to enhance the machining attributes and performance. The introduction of these powders modifies the essential quality of dielectric medium by decreasing the dielectric strength and thereby enlargement of spark gap at specimen and tool interface. This effect has the added benefit of ensuring the effective washing away of the particles removed due to the impinging of high velocity electrons. This in turn ameliorates the surface finish and increases the MRR. Figure 3 depicts the schematic arrangement of Powder mixed EDM test setup.

PMEDM was pioneered by Erden and Bilgin [80] in the late 1980s by introducing powders of Fe, Al and Cu in EDM and have obtained enhanced breakdown attributes. These findings prompted many researchers to investigate the effects of inclusion of various powders that would result in the optimization of EDM. Some of the notable findings are by Narumiya [81] using silicon, graphite and aluminium powders and by establishing a positive relationship between concentration of powders and spark gap. It was shown by Yan and Chen [82] that surface cracks can be minimized through the use of Al_2O_3 powder. Authors addressed through experiments on micro-slit machining of Ti-alloy that the inclusion of aluminium and silicon carbide powders into kerosene dielectric led to an increase in the MRR with a higher depth of cut and spark gap [83]. The influence of the size of powder particles was examined by Tzeng and Lee [84] and arrived at the conclusion that 70-90 nm results in a higher MRR. Authors explored the performance of CH_4N_2O with distilled water in machining of TiN alloy and obtained higher MRR with an accelerated tool wear rate [85]. Table 1 is a compilation of researchers and data pertaining to the powders used.

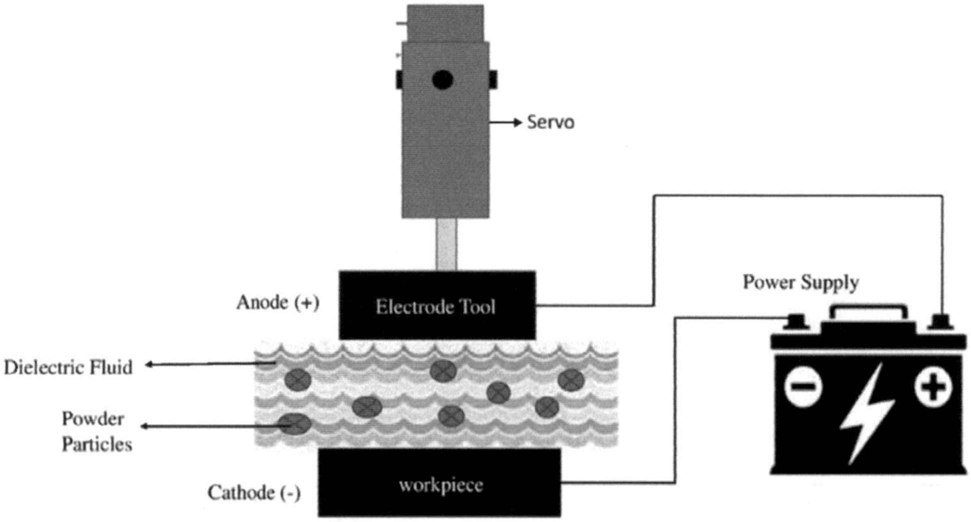

Figure 3: Schematic diagram of Powder mixed EDM

Table 1. Various powders used in PMEDM process by many researchers

Type of Powder used	Dielectric fluid employed	Remarks	Author
Aluminium and Silicon carbide	Kerosene	Both MRR and discharge gap together increases with increase in the size of particle reinforcement. Also, aluminium showed high discharge gap comparatively with SiC	Tzeng et al. [84]
Urea	Distilled water	Showed improvement in friction and wear resistance	Yan et al. [85]
Molybdenum Disulfide	EDM oil	Used ultrasonic vibrations resulted in better surface finish and increased MRR	Prihandana et al. [90]
Chromium	EDM oil	Noticed improvement in machining performance and surface quality	Garg and Ojha [91]

Response surface methodology was utilized by Kansal [86] for optimizing the machining of Al/10 wt.% SiC MMC with aluminium powders. As reported by the researchers, pulse duration has no influence on MRR, but increased with an increment in the peak current. Quantitative analytical technique was utilized to optimize the process variables in PMEDM of $AA6061/Al_2O_3/20_p$ by Singh et al. [87]. It was proved that PMEDM performed better than general EDM through length's technique. Semi-empirical models pertaining to PMEDM of Al/Al_2O_3 MMC using aluminium powder were formulated by Talla et al. [88] and multi-response optimization is performed through grey relational analysis. It was concluded that 4g/L powder concentration with a peak current of 3A on an 85% duty cycle with a pulse-on-time of 150 μs yields best output. A similar approach was followed by Singh and Yeh [89] for PMEDM of $6061Al/Al_2O_3$. The next break-through in PMEDM came with the advent of micro particles. Prihandana et al. [90] exhibited that replacing traditional powders with micro particles and including ultrasonic vibrations resulted in better surface finish and increased MRR through experiments on MoS_2 microparticles. Moreover, usage of graphite powder helps to decrease the machining time by 5 times and increase the vibrational speed of electrode tool by 3 times. Chromium micro powder was used by Ojha et al. [91] in machining EN-8 Steel. In addition to this, researchers also studied the impact of electrode shape on machining.

3.2. Electrochemical Machining

Machining of hard-heat resisting alloys and composites such as MMCs is usually carried out by employing electrochemical machining (ECM) owing to its enhanced durability of tool and high MRR. ECM involves material erosion by deliberated abandonment of the anode of an electrolytic unit. Usually, ECM technique is employed for the metal/materials that are typical to process by mechanical or traditional approaches. Machining responses such as MRR and SR depends only upon factors like electrolyte solution, voltage, cutting speed and inter-electrode gap but not on the intrinsic properties of work material such as toughness and hardness [92]. Senthil kumar et al. [93] investigated the ECM of a specimen having the following composition-15% SiC_p/Al through numerical forecasting of MRR and SR. A judicious compromise between MRR and SR is arrived at by thorough analysis of experimental data. Senthil Kumar et al. [94] have also noticed similar behaviour being exhibited by AMMC sample embedding with 10 wt.% SiC_p reinforcement synthesized via stir casting route. Kumar et al. [95] processed Al/SiC_p composite through ECM based on L27 design of experiments for analyzing the influence of factors like strength of electrolyte, feed and potential difference between the electrodes, with varying percentage of reinforcement in increments of 5% by weight on MRR.

3.3. Laser beam machining

It is an unconventional machining method widely used for machining metal matrix composites. The mechanism as a result of amalgamated melting followed by dissipation, laser beam withdraws the unnecessary material from the workpiece. In which, some gap is to be vindicate in between cutting tool and specimen. Laser beam machining (LBM) is suited only for materials having low heat conductivity or reflectivity. Extremely small holes and complex geometries can be machined with an accuracy of ± 0.1 mm [96]. Manjoth et al. [97] performed experimental investigations on LBM of $Al7075/TiB_2$ composite with emphasis on surface finish, material removal rate and dimensional accuracy while retaining the characteristics like size of the nozzle and power constant while air is utilised as the shielding gas. They established that cutting speed is the decisive factor behind surface finish. On the other hand, the gas pressure and distance of separation has trivial effect on the surface finish. Muller et al. [98] identified the limitations of LBM, the surface finish obtained by LBM is not acceptable because of imperfections like burrs at the tail of the laser cut surface.

3.4. Ultrasonic machining

It is a non-traditional machining process for all sorts of materials regardless of their electrical conductivity and hardness. But it yields low MRR. In ultrasonic machining the tool vibrates in the longitudinal plan with a frequency greater than the audible range with amplitude ranging from 0.01 to 0.06 mm applies load on to the work surface with minimal force so as to cause any distortion. Zhong et al. [99] through empirical data pertaining to MRR in USM and with the assistance of statistical methods have devised a regression model by which they have established that the following parameters in the stated order - grain size of abrasives, feed pressure and feed speed affect the material removal rate. Mohanty et al. [100] studied the MRR in USM while machining MMCs with varying compositions of SiC_p. The MRR depends primarily on the following factors: potential difference applied, concentration of the slurry and hardness of the abrasive particles. Zhong et al. [101] inspected the machinability of SiC_p/Al while performing turning operation with a diamond tool along with the assistance of USM. It is also proved that USM assists in better surface finish.

3.5. High-speed machining

It is also an unconventional process usually preferred when superior surface finish along with high dimensional accuracy is required. The term HSM was coined by Carl Saleman in 1931 [102]. The term

HSM is used only when the operating speed is 5-10 times that of traditional machining methods and this process is usually utilized for materials whose hardness ranges from 45-68 HRC. Great care needs to be exercised while machining materials of high hardness owing to the rapid tool wear and demands the use of tools made of high strength materials [103]. Collins [104] investigated the machining of MMCs at high speed with PCD tools. The empirical data, thus obtained, reveals that with an increase in the grinding wear there is an increase in the requirement of cutting forces. The findings of Jadhav et al. [105] found that improved surface finish can be obtained by preheating the MMCs to a suitable temperature.

Huang et al. [106] performed experimental studies on HSM milling of SiC_p/Al with a high percentage of reinforcements (vol.65%, grain size of 10 μm) with PCD cutting tool to establish a relation between machining parameters and the degree of tool wear. It was evident from the obtained data that with an increase in the cutting speed there is a clear raise in the tool wear with 300 m/min as the limiting value. Sharma et al. [107] observed that cutting speed and cutting forces exhibited an inverse relationship. Brun et al. [108] observed that among all tool materials PCD enhances the tool life while machining 40% (in terms of volume) of SiC_p/Al at low speeds. Wang et al. [109] found that the cutting speeds above 300 m/min are not suitable for practical purposes owing to the abrasive nature of SiC_p and heat generation. A higher feed rate produces better surface finish. One of the crucial factors that affects the tool life is grain size of PCD [110].

3.6. Survey on machinability of tailor-made composites

Among the many aspects that machinability depends upon, chemical composition of the workpiece material and microstructure are the exceptional parameters followed by physical & mechanical properties and machining parameters viz. cutting speed, feed and depth of cut. In addition to these intrinsic factors, imperfections such as surface defects, emerging faults and chip formation mechanism also affect the machinability of a material. In case of metal matrix composites (MMCs) apart from the above-described general aspects, the following factors which are specific to composites also come into play while characterizing the machinability: nature of the reinforcement, spread and type of reinforcement and percentage of reinforcement in the MMC [111]. The inclusion of hard and brittle embedment in matrix alters machinability of the MMC in contrast to the parent metal.

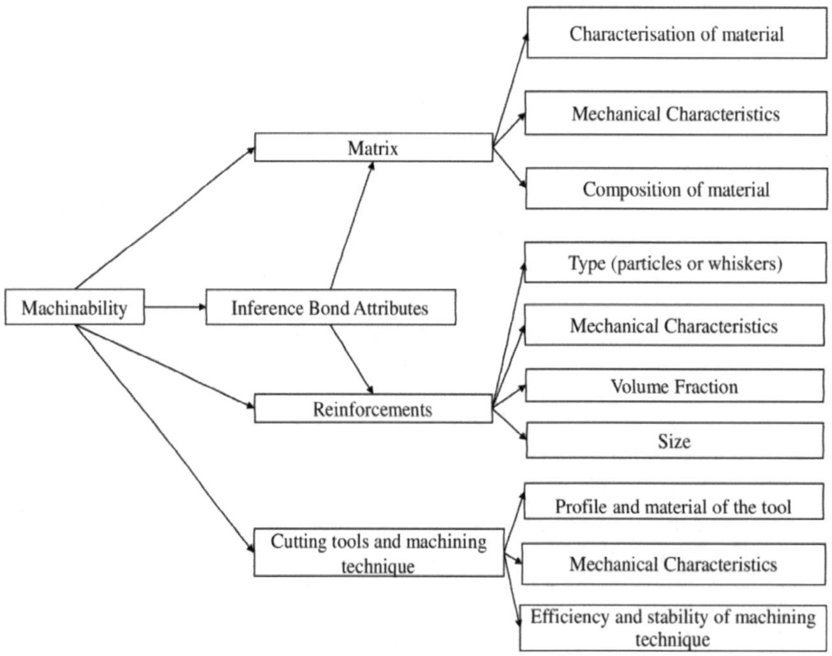

Figure 4: Properties that may influence the machining performance of a material

Machinability can be evaluated by many criteria, with the tool-life between two successive grinds being the major one. This model owes its popularity to the fact that the machinability of two materials with similar mechanical characteristics and chemical compositions can be compared in terms of their tool life when machined in similar conditions. Moreover, the machinability of different materials cannot be compared if the tool material is not invariant. For instance, a comparison cannot be made between HSS tool and carbide tool. The other criteria used to evaluate machinability are surface roughness, material removal rate, requirement of cutting forces, size and form of chips and temperatures induced while machining [44,112]. The above properties are summarized in Fig. 4. In contrast to the traditional machining in non-traditional machining approaches, the removal of material is done through liquefying and devastating the unwanted material of workpiece. In EDM and WEDM, higher current, pulse on time, and increment in the potential difference increases the tool wear, improves surface quality, and MRR [113,114]. If the electrode is made with a material of higher melting point, then the tool wear is gradual. Common type of chips produced are continuous, saw tooth, disintegrated, and other forms. This accounts for the formation of cracks along an extended trail [115].

4. Recent advancements in modelling and optimization techniques in machining of composites

4.1. Modelling of machining process of MMCs

The abundance of experimental studies investigating the machinability of MMCs are not sufficient to completely understand the process of machining by both traditional and non-traditional methods owing to the inherent flaws in the empirical methods. This places emphasis on the need for numerical and analytical method that fit the empirical data. A robust system that predicts the wear mechanism accurately is of immense advantage as it drastically lowers the expenditure and eliminates the need for the inception of a prototype to perform experimental studies upon. Till date, many analytical methods have been formulated to explain various aspects pertaining to machining of MMCs such as the ease with which the material can be machined, its strength, stress analysis, and failure mechanisms among many. The 3 existing models are as listed below:

a) Theory of orthogonal cutting by Ernst-Merchant
b) Tool life equation by F.W. Taylor
c) Theory of elasticity and plasticity

An analytical cutting force model for one dimensional cutting was formulated by Kishway et al. [116] based-on energy considerations and utilized it to forecast the energy expenditure for the deformation. In addition to this, the machining efficiency is modelled as a function of feed and SiC_p/Al composition. By applying the Taylor's equation to Aluminium based MMCs, Hung et al. [117] formulated analytical models to estimate the aggregate wear in tools employed in facing and turning operations. It is evident from findings, thus obtained, that order of various cutting speeds is not affected by tool wear. It was also observed that among all the wear mechanisms grinding wear is prominent. These models were utilized to ameliorate machining operations by enhancing the tool life and thereby reducing the corresponding surface deterioration [118]. Pramanik et al. [119] estimated the cutting force required to machine particle-reinforced MMCs on the basis of orthogonal cutting theory of Merchant and Ernst. The results derived from this model pertaining to ploughing force, particle fracture force and the chip formation force to are in accordance with the empirical findings. However, authors reported that merchant's model failed to estimate the appropriate direction of predominant shear region [120]. While, Majumdar et al. [121] reported that inelastic distortion of the sample with linear orientation depicted as dominated one due to severe resilience of base metal. Whereas the inelastic distortion of the 90° aligned sample depicts failure and plasticity.

An interesting challenge is the geometrical modelling of metal matrix composites with precision of reinforcement geometry specification being the principal hindrance. Li et al. [122] devised a model of tool wear utilizing the correlation based on their geometric features and reinforcement particles displacement. The effects of cutting conditions on the shear angle were investigated by means of statistical methods by Uday et al. [123]. The chip thickness and the corresponding shear angle were taken into consideration in the analytical model in order to perform qualitative analysis. It is evident from it, that the chip formation mechanism is mainly affected by the size of reinforcements and their volume percentage. Moreover, a relationship was established between shear angle and friction angle by Pramanik et al. [124] in order to predict the cutting forces and contact stresses at the tool-chip interface. Tevatia et al. [125] devised the fatigue crack growth live model under the controlled strain conditions for short fibre-reinforced MMC. Modified shear lag theory along with fatigue fracture mechanics theory were utilized to quantify the effect of yield strength. Numerous models were formulated utilizing Johnson cook model, plastic/elastic deformation model, temperature independent hardening and temperature-dependent models [126,127], or Hooke's law to study the machining process of MMCs on material basis. Tsai-Hill model was utilized to model chip formation [128].

4.2. Finite element modelling and simulation

One of the most popular techniques employed in numerical simulation for studying the machining processes is Finite element modelling. The machining behaviour of MMCs can be studied in terms of complex aspects such as metallurgical characteristics, elastic and plastic behaviour of the material, heat transfer, effect of coolants and fracture mechanics with the help of FEM. Moreover, cutting forces estimation, study of deformation process, thermal and stress analysis can be performed by FEM for various machining processes. Mackerle [129,130] compiled the scientific progress in the domain of finite element method applied to machining during the span from 1976 to 2002. A 3-dimensional finite element model for milling with an equivalent homogeneous material (EHM) model was developed by Chen et al. [131] to study the machining process of SiC_p/Al6063/30p based on quasi-static and Split-Hopkinson pressure bar tests with the assistance of ABAQUS software. The predictions of this model were in accordance with the empirical findings with a deviation of 20% in both the components of cutting forces, stating that it is reliable within a selected range of machining parameters.

Machining process of thin wall part SiC_p/Al composite with PCD tools was simulated by Liu et al. [132]. The cutting forces were examined in 3-dimensional simulation of machining process. 2D finite element models of SiC_p/Al were formulated by Fathipour et al. [133] based on ABAQUS/Explicit in order to apply simulation to the process parameters with TiC tools. The hot machining and finite element analysis of SiC_p/Al220 with 10% and 30% of SiC_p respectively, by using tools made of TiAlN coated cemented carbide with due consideration of preheating and machining parameters was presented by Mahesh et al. [134]. The influence of the particles of a tool and machining speed was investigated with the help of a two-dimensional FEM [135]. The high-speed machining of the high-volume fraction SiC_p/Al6063 was examined and simulated by Wang et al. [136] using two-dimensional finite element modeling in ABAQUS software. Multiphase models were employed to estimate Interaction of matrix and it proved instrumental in estimation of stress distribution.

4.3. Optimization techniques

The machining process is optimized by utilizing the following soft computing techniques:

a) Fuzzy logic
b) Taguchi Method
c) Artificial Neural Network (ANN)
d) Response Surface Methodology (RSM)

4.3.1. Fuzzy logic optimization

A wide variety of fuzzy logic theory techniques like Fuzzy expert system which is a combination of fuzzy system and genetic algorithm, and the grey-fuzzy analysis method are indispensable in optimizing the machining of MMCs. Reliable rule-based confident expert systems can be devised with the help of fuzzy logic [137]. Manipulation of discrete data and uncertainty is performed by Grey-fuzzy analysis method. Therefore, the uncertainty in the final system can be minimized by combining both the methods [138]. One of the most commonly used fuzzy logic technique for formulating predictive models is Gray-fuzzy analysis which is discussed at length in [139].

A multi-response predictive model was formulated on basis of Gray-fuzzy logic by Santosh et al [140] while examining the turning process of SiC_p/Al with PCD tools. Thus, fuzzy logic approach is employed to predict the tool wear mechanism. These predictions were in accordance with the experimental findings. The influence of cutting conditions upon the quality of surface finish in milling process of SiC_p/Al356 was examined [141]. Through Statistical analysis, it was found that among all the parameters spindle speed and feed rate have the most influence on surface finish. Low feed rate, high spindle speed, and low depth of cut results in enhancement of surface finish. Based on these findings, an improved fuzzy model (accuracy of 88.44%) was devised to control the process.

4.3.2. Taguchi method

This method is based on statistical analysis and was pioneered by Genichi Taguchi. It is also known as a robust design method. It is advantageous in the sense that it enables us to find the minimum required set of experiments with the accumulated values and variables. Upon culmination of the experimental phase, the results thus obtained are quite often analyzed with ANOVA to quantify the dependence of variables on each other [142]. Taguchi and statistical techniques were used to investigate the turning process of SiC_p/Al with coated carbide tools and it was found that machining is the single most influencing parameter on flank wear and for surface finish it is feed [143]. The Taguchi method was also adopted by Ashok et al. [144] to examine the turning process of SiC_p/Al with carbide tools. The empirical findings and the results obtained from Taguchi method are in good agreement. Sriprateep et al. [145] applied this method to turning A356/20/SiC_p-T6 MMC with PCD insert in order to estimate the surface quality, tool wear and energy expenditure.

4.3.3. Artificial neural network

Artificial Neural Networks can be thought of as a directed weighted graph, with nodes representing the artificial neurons and the relation between output and neuron inputs are exemplified through weighted edges. The features, advantages and functioning of ANN discussed at length in [146-148]. The efficiency of ANN is decided by the number of hidden layers and the neurons comprising each of them along with the choice of algorithm. Santosh et al. [149] examined the turning operation of SiC_p/Al with PCD tool and an ANN based multi-response predictive model is formulated. The performance metrics of the model were good with an error percentage of 7.26% for MRR and 4.46% for surface roughness. Liujie et al. [150] developed the separate ANN machining models for Al-MMC at various feed and machining speeds. Feed forward ANN model was built and trained by using comprehensive data sets. It is established that the cutting speed is the crucial factor that influences the tool wear and energy expenditure.

4.3.4. Response surface methodology

E. P. George and K. B. Wilson developed the response surface methodology in the year 1951 [151]. Response surface methodology (RSM) is based on several mathematical and statistical techniques and is utilized to optimize the response of a process which is affected by several input decision parameters. In RSM, the structure of the designed experiment is utilized to obtain optimality. The first step involved in RSM is to find the optimization model in the design of an experiment.

$Y = F(x_1, x_2, \ldots, x_n)$

Where,

Y is the required function,

F is the response function of variables $(x_1, x_2, \ldots, x_n)$,

$(x_1, x_2, \ldots, x_n)$ represents a combination of machining parameters.

Gaitonde et al. [152] concluded that feed rate is the factor that decides the cutting forces and power expenditure while turning MMCs with PCD tools with the help of RSM technique. A similar approach was adopted by Arokiadass et al. [153] to investigate the effect of SiC composition on flank wear. Al7075/SiCp composites were examined for the influence of machining parameters by RSM mathematical model and revealed that exceptional rate of prediction taken place [154]. The results were correlated through ANOVA. Premnath et al. [155] through his empirical findings pertaining to the milling of Al 6061-aluminum alloy reinforced with varying compositions of Alumina devised RSM model. Tamang et al. [156] examined the mechanism of surface texture with respect to tool life in turning and milling of AMMCs using various cutting tools and experienced feasible predictions by developing forecasting models of ANN and RSM. The process variables of CNC milling of AMMCs were optimized by RSM method with a tolerance of ±5% in accuracy [157]. The RSM approach is employed to evaluate machinability of MMCs for estimating the target responses using the process parameters and experimental outcomes effectively.

5. Observations from literature

Even though, considerable amount of literature available on various machining processes of MMCs, a lot many break-throughs have to be noted for a extensive understanding of machining behaviour of particle reinforced MMCs.

- An interesting challenge is the geometrical modelling of metal matrix composites with precision of reinforcement geometry specification being the principal hindrance.
- Tool wear and failure are the significant complexity confronted in processing the MMCs.
- The phenomena associated with electric discharge machining and electro chemical machining is not yet fully explored in the available literature. Thus, detailed investigation is needed for a comprehensive understanding of the phenomena.
- A judicious compromise needs to be desired between various machining parameters to obtain optimal results.
- The interaction of tool and workpiece with the plasma in non-traditional methods need to investigated to evolve an exceptional perception of the phenomena.
- In laser beam machining, the heat affected zone (HAZ) needs to be examined to arrive at an optimum amount of heat to be supplied in order to prevent the surface deterioration.
- In addition to the machining process, the MMCs should also be modelled.
- An alternative approach needs to be devised for soft-computing to deal with inaccuracies associated with them.

6. Conclusions

The emerging technologies in machining and optimization of the MMCs have been addressed in the current study. In addition, this article covered pertinent research issues and significant challenges involved in various traditional and non-traditional machining techniques applied for processing MMCs. As the tailor-made composites are having exceptional mechanical, metallurgical and thermal properties, they may have huge scope in numerous industrial applications. Considerable investigations are made by several researchers in order to analyze the machining behaviour of various MMCs. Machinability, modelling, and optimization are comprehensively focused in this review.

- As per the literature, most of the works on traditional machining on tailor-made composites such as turning, milling, drilling, and grinding revealed that Polycrystalline diamond tools are better than Polycrystalline cubic boron nitride cutting tools and giving excellent surface finish.
- Metals tend to possess higher hardness due to addition of reinforcement leading to inferior machining performance through traditional ways as noted from literature. Non-traditional machining processes offering promising solutions in processing composites as there is no gap in between the work piece and cutting tool. Electrical discharge machining is one among the advanced, economical and also a conversant approach engaged for making intricate forms.
- Among various unconventional machining operations, Laser machining has been identified as an adverse process for surface quality as it possesses extravagant heat while cutting the material.
- Artificial intelligence tools such as Taguchi for design of experiments, Artificial Neural Networks, Response Surface Methodology, Fuzzy logic and Genetic Algorithm are explicitly explored for predicting output responses based on relevant inputs that will reduce the R&D costs.
- In the aspect of investigating the cutting tool or electrode behaviour on the tailor-made composite material, Finite element modelling and Simulation have been identified as the significant approach.

- As the world is moving towards light weight applications, there is huge demand in the area of studying mechanical as well as thermal properties along with machinability of Nano metal matrix composites.

References

[1] Nicholls C J, Boswell B, Davies I J and Islam M N 2017 Review of machining metal matrix composites *Int. J. Adv. Manuf. Technol.* **90** 2429-41

[2] Maleque M A, Rahman M M and Hossain M S 2011 Conceptual design of aluminium metal matrix composite brake rotor system *Adv. Mat. Res.* **264** 1648-53

[3] Dandekar C R and Shin Y C 2009 Multi-step 3-D finite element modelling of subsurface damage in machining particulate reinforced metal matrix composites *Compos. Part A Appl. Sci. Manuf.* **40** 1231-9

[4] Mukhopadhyay C K, Jayakumar T, Raj B and Venugopal S 2012 Statistical analysis of acoustic emission signals generated during turning of a metal matrix composite *J. Braz. Soc. Mech. Sci. Eng.* **34** 145-54

[5] Wang B B, Xie L J, Wang X B and Chen X L 2014 Simulation studies of the cutting process on SiCp/Al composites with different volume fraction of reinforced SiC particles *Mater. Sci. Forum.* **800** 321-326

[6] Kotteda T K, Kumar M, Kumar P and Chekuri R B 2022 Metal matrix nanocomposites: future scope in the fabrication and machining techniques *Int. J. Adv. Manuf.* (in press)

[7] Kotteda T K, Kumar M and Kumar P 2023 Experimental insights and micrographical investigation on graphene nanoplatelet–reinforced aluminum cast composites *Int. J. Adv. Manuf.* (in press)

[8] Shetty N, Shahabaz S M, Sharma S S and Shetty S D 2017 A review on finite element method for machining of composite materials *Compos. Struct.* **176** 790-802

[9] Bauri R and Yadav D 2017 Metal matrix composites by friction stir processing *Butterworth-Heinemann*

[10] Teng X, Chen W, Huo D, Shyha I and Lin C 2018 Comparison of cutting mechanism when machining micro and nano-particles reinforced SiC/Al metal matrix composites *Compos. Struct.* **203** 636-47

[11] Hung N P and Zhong C H 1996 Cumulative tool wear in machining metal matrix composites Part I: Modelling *J. Mater. Process. Technol.* **58** 109-13

[12] Chandrasekaran M and Devarasiddappa D 2014 Artificial neural network modeling for surface roughness prediction in cylindrical grinding of Al-SiC_p metal matrix composites and ANOVA analysis *Adv. Prod. Eng. Manag.* **9** 59-70

[13] Zhang W 2021 A review of tribological properties for boron carbide ceramics *Prog. Mater. Sci.* **116** 100718

[14] Wang Y, Liao W, Yang K, Chen W and Liu T 2019 Investigation on cutting mechanism of SiC_p/Al composites in precision turning *Int. J. Adv. Manuf. Technol.* **100** 963-72

[15] Li J and Laghari R A 2019 A review on machining and optimization of particle-reinforced metal matrix composites *Int. J. Adv. Manuf. Technol.* **100** 2929-43

[16] Bhattacharyya B and Doloi B 2019 Modern machining technology: Advanced, hybrid, micro machining and super finishing technology *Academic Press*

[17] Ho K H and Newman S T 2003 State of the art electrical discharge machining (EDM) *Int. J. Mach. Tools Manuf.* **43** 1287-300

[18] Muthukrishnan N and Davim J P 2011 An investigation of the effect of work piece reinforcing percentage on the machinability of Al-SiC metal matrix composites *J. Mech. Eng. Res.* **3** 15-24

[19] Davim J P and Baptista A M 2000 Relationship between cutting force and PCD cutting tool wear in machining silicon carbide reinforced aluminium *J. Mater. Process. Technol.* **103** 417-23

[20] Cronjager L and Meister D 1992 Machining of fibre and particle-reinforced *aluminium CIRP Ann.-Manuf. Technol.* **41** 63-6

[21] Popescu I N, Zamfir S, Anghelina V F and Rusanescu C O 2010 Processing by P/M route and characterization of new ecological Aluminum Matrix Composites (AMC) Int. J. Mech. **4** 43-52
[22] Durante S, Rutelli G and Rabezzana F 1997 Aluminum-based MMC machining with diamond-coated cutting tools *Surf. Coat. Technol.* **94** 632-40
[23] Teti R, Jemielniak K, Donnell G and Dornfeld D 2010 Advanced monitoring of machining operations *CIRP annals* **59** 632-40
[24] Das D, Thakur R K, Chaubey A K and Sahoo A K 2018 Optimization of machining parameters and development of surface roughness models during turning Al-based metal matrix composite *Mater. Today: Proc.* **5** 4431-7
[25] Dandekar C R and Shin Y C 2012 Modeling of machining of composite materials: a review *Int. J. Mach. Tools Manuf.* **57** 102-21
[26] Ghani J A, Rizal M, Nuawi M Z, Ghazali M J and Haron C H 2011 Monitoring online cutting tool wear using low-cost technique and user-friendly GUI *Wear* **271** 2619-24
[27] Prasad B S, Prabha K A and Kumar P G 2017 Condition monitoring of turning process using infrared thermography technique-An experimental approach *Infrared Phys. Technol.* **81** 137-47
[28] Sekhar R and Singh T P 2015 Mechanisms in turning of metal matrix composites: a review *J. Mater. Res.* **4** 197-207
[29] Laghari R A, Li J and Mia M 2020 Effects of turning parameters and parametric optimization of the cutting forces in machining SiCp/Al 45 wt% composite *Metals* **10** 840
[30] Tonshoff H K and Karpuschewski B 1999 Manufacturing of magnesium by turning and burnishing operations *Adv. Techology Plast.* **1** 607-12
[31] Tomac N, Tannessen K and Rasch F O 1992 Machinability of particulate aluminium matrix composites *CIRP annals* **41** 55-8
[32] Dabade U A and Jadhav M R 2016 Experimental study of surface integrity of Al/SiC particulate metal-matrix composites in hot machining *Procedia Cirp.* **41** 914-9
[33] Jain V K 2022 Advanced machining science *CRC Press*
[34] Chen T, Zhao B, Lin C, Ding W and Guo G 2023 Machining of titanium metal matrix composites: a short review *Int. J. Adv. Manuf.* (in press)
[35] Jani D 2014 Machining of Sic-metal matrix composite (MMC) by polycrystalline diamond (PCD) tools and effect on quality of surface by changing machining parameters *Int. J. Sci. Res. Dev.* **2** 106-8
[36] Muthukrishnan N, Murugan M and Rao P K 2008 Machinability issues in turning of Al-SiC (10p) metal matrix composites *Int. J. Adv. Manuf.* **39** 211-8
[37] Ozben T, Kilickap E and Cakır O 2008 Investigation of mechanical and machinability properties of SiC particle reinforced Al-MMC *J. Mater. Process. Technol.* **198** 220-5
[38] Arokiadass R, Palaniradja K and Alagumoorthi N 2011 Surface roughness prediction model in end milling of Al/SiCp MMC by carbide tools *Int. J. Eng. Sci. Technol.* **3** 78-87
[39] Arokiadass R, Palaniradja K and Alagumoorthi N 2012 Prediction and optimization of end milling process parameters of cast aluminium based MMC *Trans. Nonferrous Met. Soc. China* **22** 1568-74
[40] Turgut Y, Çinici H, Şahin I and Findik T A 2011 Study of cutting force and surface roughness in milling of Al/Sic metal matrix composites *Sci. Res. Essays* **6** 2056-2062
[41] Songmene V and Balazinski M 2011 Machining of graphitic SiC-reinforced aluminium metal matrix composites with diamond tools *In Proceedings of the CIRP International Seminar on Progress in Innovative Manufacturing Engineering-PRIME, Sestri Levante, Italy* 20-22
[42] Chandrasekaran H and Johansson J O 1997 Influence of processing conditions and reinforcement on the surface quality of finish machined aluminium alloy matrix composites *CIRP Annals* **46** 493-6
[43] Davim J P 2013 Machining composites materials *John Wiley & Sons*
[44] Markopoulos A P, Pressas I S, Papantoniou I G, Karkalos N E and Davim J P 2015 Machining and machining modeling of metal matrix composites-A review *Mod. Manuf. Eng.* 99-141
[45] Reddy N S, Kwang-Sup S and Yang M 2008 Experimental study of surface integrity during end milling of Al/SiC particulate metal–matrix composites *J. Mater. Process. Technol.* **201** 574-9

[46] Songmene V and Balazinski M 1999 Machinability of graphitic metal matrix composites as a function of reinforcing particles *CIRP Annals* **48** 77-80
[47] Huang S T, Zhou L, Chen J and Xu L F 2012 Drilling of SiCp/Al metal matrix composites with polycrystalline diamond (PCD) tools *Mater. Manuf.* **27** 1090-4
[48] Narutaki N, Yamane Y, Hayashi K, Kitagawa T and Uehara K 1993 High-speed machining of Inconel 718 with ceramic tools *CIRP annals* **42** 103-6
[49] Rajmohan T, Palanikumar and Kathirvel M 2012 Optimization of machining parameters in drilling hybrid aluminium metal matrix composites *Trans. Nonferrous Met. Soc. China* **22** 1286-97
[50] Davim JP 2003 Study of drilling metal-matrix composites based on the Taguchi techniques *J. Mater. Process. Technol.* **132** 250-4
[51] Davim J P and Antonio C C 2001 Optimal drilling of particulate metal matrix composites based on experimental and numerical procedures *Int. J. Mach. Tools Manuf.* **41** 21-31
[52] Davim J P and Monteiro Baptista 2001 A Cutting force, tool wear and surface finish in drilling metal matrix composites *Proc. Inst. Mech. Eng. E: J. Process Mech. Eng.* **215** 177-83
[53] Niknam S A, Kamalizadeh S, Asgari A and Balazinski M 2018 Turning titanium metal matrix composites (Ti-MMCs) with carbide and CBN inserts *Int. J. Adv. Manuf. Technol.* **97** 253-65
[54] Patil S S and Bhalerao Y J 2019 Application of NSGA-II for optimisation of cylindrical plunge grinding process parameters *Int. J. Abras. Tech.* **9** 319-29
[55] Monaghan J and Oreilly P 1992 The drilling of an Al/SiC metal-matrix composite *J. Mater. Process. Technol.* **33** 469-80
[56] Thiagarajan C, Sivaramakrishnan R and Somasundaram S 2012 Modeling and optimization of cylindrical grinding of Al/SiC composites using genetic algorithms *J. Braz. Soc. Mech. Sci. Eng.* **34** 32-40
[57] Ilio D A and Paoletti A 2000 A comparison between conventional abrasives and super abrasives in grinding of SiC-aluminium composites *Int. J. Mach. Tools Manuf.* **40** 173-84
[58] Zhang G F, Zhang B and Deng Z H 2009 Mechanisms of Al/SiC composite machining with diamond whiskers *Key Eng. Mater.* **404** 165-75
[59] Khan M A, Rahman M M and Kadirgama K 2014 Neural network modeling and analysis for surface characteristics in electrical discharge machining *Procedia Eng.* **90** 631-6
[60] Khan M A, Rahman M M and Kadirgama K 2014 Electrode wear rate of graphite electrodes during electrical discharge machining processes on titanium alloy Ti-5Al-2.5Sn *Int. J. Automot. Mech. Eng.* **9**
[61] Khan M A and Rahman M M 2014 Surface study in a non-conventional (electrical discharge machining) process for grade 6 titanium material *J. Teknol.* **68** 89-93
[62] Balan A S and Giridharan A 2017 A progress review in wire electrical discharge machining process *Int. J. Automot. Mech. Eng.* **14** 4097-124
[63] Choudhary S K and Jadoun R S 2014 Current advanced research development of electric discharge machining (EDM): a review *Int. J. Res. Advent Technol.* **2** 273-97
[64] Kumar S and Singh R 2010 Investigation of surface properties in manganese powder mixed electrical discharge machining of OHNS and D2 die steels *Int. J. Automot. Mech. Eng.* **2** 181-99
[65] Khan M A, Rahman M M, Noor M M, Kadirgama K and Maleque M A 2011 Current research trends on dry, near-dry and powder mixed electrical discharge machining *Adv. Mat. Res.* **264** 956-61
[66] Khan M A, Rahman M M, Kadirgama K, Maleque M A and Ishak M 2011 Prediction of surface roughness of Ti-6Al-4V in electrical discharge machining: A regression model *J. Mech. Eng. Sci.* **1** 16-24
[67] Kathiresan M and Sornakumar T 2010 EDM studies on aluminum alloy-silicon carbide composites developed by vortex technique and pressure die casting *J. of Minerals and Materials Characterization and Eng.* **9** 79-88
[68] Prabu M, Ramadoss G, Narendersingh P, Christy T V and Eswaran V V 2014 Electrical discharge machining of Al-TiB2 with a low-frequency vibrating tool *Sci. Eng. Compos.* 21 445-52

[69] Garg R K, Singh K K, Sachdeva A, Sharma V S, Ojha K and Singh S 2010 Review of research work in sinking EDM and WEDM on metal matrix composite materials *Int. J. Adv. Manuf.* **50** 611-24

[70] Iosub A, Nagit G and Negoescu F 2009 Tool wear investigation in electric discharge machining of aluminum matrix composite materials *Annals of Dunarea de Jos University of Galati, Fascicle V, Technologies in machine building* **27** 225-8

[71] Nanimina A M, Abdul-Rani A M, Ahmad F, Zainuddin A and Lo S J 2011 Effects of electro-discharge machining on aluminium metal matrix composite *J. Appl. Sci.* 1-5

[72] Ahamed A R, Asokan P and Aravindan S E 2009 EDM of hybrid Al–SiCp–B4Cp and Al–SiCp–glassp MMCs *Int. J. Adv. Manuf.* **44** 520-8

[73] Prabhu S and Vinayagam B K 2016 Multiresponse optimization of EDM process with nanofluids using TOPSIS method and genetic algorithm *Arch. Mech. Eng.* 45-71

[74] Shandilya P, Jain P K and Jain N K 2013 RSM and ANN modeling approaches for predicting average cutting speed during WEDM of SiCp/6061 Al MMC *Procedia Eng.* **64** 767-74

[75] Mouralova K, Benes L, Zahradnicek R, Bednar J, Zadera A, Fries J and Kana V 2020 WEDM Used for machining high entropy alloys *Materials* **13** 4823

[76] Gore A S and Patil N G 2018 Wire electro discharge machining of metal matrix composites: A review *Procedia Manuf.* **20** 41-52

[77] Rao P N 2013 Manufacturing technology: metal cutting and machine tools *Tata McGraw-Hill Education* **2**

[78] Marigoudar R N and Sadashivappa K 2013 Effect of machining parameters on MRR and surface roughness in machining of ZA43/SiCp composite by WEDM *Int. J. Appl. Sci. Eng.* **11** 317-30

[79] Pramanik A 2014 Developments in the non-traditional machining of particle reinforced metal matrix composites *Int. J. Mach. Tools Manuf.* **86** 44-61

[80] Erden A and Bilgin S 1981 Role of impurities in electric discharge machining *Macmillan Education UK* 345-50

[81] Narumiya H 1989 EDM by powder suspended working fluid *Proc. of ISEM-9* **5**

[82] Yan B H and Chen S L 1994 Characteristics of SKD11 by complex process of electrical discharge machining using liquid suspended with alumina powder *J. Jpn. Inst. Met.* **58** 1067-72

[83] Chow H M, Yan B H, Huang F Y and Hung J C 2000 Study of added powder in kerosene for the micro-slit machining of titanium alloy using electro-discharge machining *J. Mater. Process. Technol.* **101** 95-103

[84] Tzeng Y F and Lee C Y 2001 Effects of powder characteristics on electro discharge machining efficiency *Int. J. Adv. Manuf.* **17** 586-92

[85] Yan B H, Tsai H C and Huang F Y 2005 The effect in EDM of a dielectric of a urea solution in water on modifying the surface of titanium *Int. J. Mach. Tools Manuf.* **45** 194-200

[86] Kansal H K, Singh S and Kumar P 2006 An experimental study of the machining parameters in powder mixed electric discharge machining of Al-10% SiCp metal matrix composites *Int. J. Mach. Mach. Mater.* **1** 396-411

[87] Singh S, Maheshwari S, Dey A and Pandey P C 2010 Experimental results and analysis for Electrical Discharge Machining (EDM) of aluminium metal matrix composites with powder-mixed dielectric: Lenth's method *Int. J. Manuf. Technol. Manag.* **21** 67-82

[88] Talla G, Sahoo D K, Gangopadhyay S and Biswas C K 2015 Modeling and multi-objective optimization of powder mixed electric discharge machining process of aluminum/alumina metal matrix composite *Eng. Sci. Technol. an Int.* **18** 369-73

[89] Singh S and Yeh M F 2012 Optimization of abrasive powder mixed EDM of aluminum matrix composites with multiple responses using gray relational analysis *J. Mater. Eng. Perform.* **21** 481-91

[90] Prihandana G S, Mahardika M, Hamdi M, Wong Y S and Mitsui K 2009 Effect of micro-powder suspension and ultrasonic vibration of dielectric fluid in micro-EDM processes-Taguchi approach *Int. J. Mach. Tools Manuf.* **49** 1035-41

[91] Ojha K, Garg R K and Singh K K 2011 Parametric optimization of PMEDM process using chromium powder mixed dielectric and triangular shape electrodes *J. of Minerals and Materials Characterization and Eng.* **10** 1087
[92] Rajurkar K P and Zhu D 1999 Improvement of electrochemical machining accuracy by using orbital electrode movement *CIRP Annals* **48** 139-42
[93] Senthilkumar C, Ganesan G and Karthikeyan R 2012 Electrochemical machining of Al/15% SiCp composites through a response surface methodology-based approach *Int. J. Mater. Res.* **103** 378-82
[94] Senthilkumar C, Ganesan G and Karthikeyan R 2009 Study of electrochemical machining characteristics of Al/SiCp composites *Int. J. Adv. Manuf.* **43** 256-63
[95] Kumar K S, Sivasubramanian R and Kalaiselvan K 2009 Selection of optimum parameters in non-conventional machining of metal matrix composite *Port. Electrochim.* **27** 477-86
[96] Wei C, Guo W, Pratomo E S, Li Q, Wang D, Whitehead D and Li L 2020 High-speed, high-power density laser-assisted machining of Al-SiC metal matrix composite with significant increase in productivity and surface quality *J. Mater. Process. Technol.* **285** 116784
[97] Manjoth S, Keshavamurthy R and Kumar G P 2016 Optimization and analysis of laser beam machining parameters for Al7075-TiB2 in-situ composite *IOP Conf. Ser.: Mater. Sci. Eng.* **149** 012013
[98] Muller F and Monaghan J 2000 Non-conventional machining of particle reinforced metal matrix composite *Int. J. Mach. Tools Manuf.* **40** 1351-66
[99] Zhong G, Xu J, Wu Y and Yang S 2015 Statistical Analyses and Regression Modeling for Influence of Process Parameters on Material Removal Rate in Ultrasonic Machining *Glob. J. Technol. Optim.* **6** 1-6
[100] Mohanty S and Routara B C 2016 A review on machining of metal matrix composites using nanoparticle mixed dielectric in electro-discharge machining *Int. J. Automot. Mech.* **13**
[101] Zhong Z W and Lin G 2006 Ultrasonic assisted turning of an aluminium-based metal matrix composite reinforced with SiC particles *Int. J. Adv. Manuf.* **27** 10771-81
[102] Pasko R, Przybylski L and Slodki B 2002 High speed machining (HSM)–the effective way of modern cutting *Int. workshop CA systems and Technologies* **18** 72-79
[103] Bartarya G and Choudhury S K 2012 State of the art in hard turning *Int. J. Mach. Tools Manuf.* **53** 1 4
[104] Collins J L 2001 High speed dry machining of MMCs with diamond tools *15lh International Plansee Seminar* **2** 711-25
[105] Jadhav M R and Dabade U A 2016 Multi-objective optimization in hot machining of Al/SiCp metal matrix composites *IOP Conf. Ser.: Mater. Sci. Eng.* **114** 012122
[106] Huang S, Guo L, He H and Xu L 2018 Study on characteristics of SiCp/Al composites during high-speed milling with different particle size of PCD tools *Int. J. Adv. Manuf.* **95** 2269-79
[107] Sharma V S, Dhiman S, Sehgal R and Sharma S K 2008 Estimation of cutting forces and surface roughness for hard turning using neural networks *J. Intell. Manuf.* **19** 473-83
[108] Brun M K, Lee M and Gorsler F 1985 Wear characteristics of various hard materials for machining SiC-reinforced aluminum alloy *Wear* **104** 21-9
[109] Wang T, Xie L, Wang X and Ding Z PCD tool performance in high-speed milling of high-volume fraction SiCp/Al composites *Int. J. Adv. Manuf.* **78** 1445-53
[110] Zhou L, Cui C, Zhang P F and Ma Z Y 2017 Finite element and experimental analysis of machinability during machining of high-volume fraction SiCp/Al composites *Int. J. Adv. Manuf.* **91** 1935-44
[111] Jain P K, Soni S C and Baredar P V 2014 Review on machining of aluminium metal matrix composites Mater. Sci. Res. **11** 114-20
[112] Channi A S, Bains H S, Grewal J S, Chidambranathan V S and Kumar R 2023 Tool wear rate during electrical discharge machining for aluminium metal matrix composite prepared by squeeze casting: A prospect as a biomaterial *J. Electrochem. Sci. Eng.* **13** 149-62

[113] Kar C, Surekha B, Jena H and Choudhury S D 2018 Study of influence of process parameters in electric discharge machining of aluminum–red mud metal matrix composite *Procedia Manuf.* **20** 392-9
[114] Kandpal B C, Kumar J and Singh H 2018 Optimisation of Process Parameters of Electrical Discharge Machining of Fabricated AA6061/10% Al2O3 Aluminium Based Metal Matrix Composite *Mater. Today: Proc.* **5** 4413-20
[115] Huang S, Yu X, Wang F and Xu L 2015 A study on chip shape and chip-forming mechanism in grinding of high-volume fraction SiC particle reinforced Al-matrix composites *Int. J. Adv. Manuf.* **80** 1927-32
[116] Kishawy H A, Kannan S and Balazinski M 2004 An energy based analytical force model for orthogonal cutting of metal matrix composites *CIRP Annals* **53** 91-4
[117] Hung N P, Loh N L and Xu Z M 1996 Cumulative tool wear in machining metal matrix composites part II: machinability *J. Mater. Process.* **58** 114-20
[118] Hung N P, Boey F Y, Khor K A, Phua Y S and Lee H F 1996 Machinability of aluminum alloys reinforced with silicon carbide particulates *J. Mater. Process.* **56** 966-77
[119] Pramanik A, Zhang L C and Arsecularatne J A 2006 Prediction of cutting forces in machining of metal matrix composites *Int. J. Mach. Tools Manuf.* **46** 1795-803
[120] Molinari A and Moufki A 2008 The Merchant's model of orthogonal cutting revisited: A new insight into the modeling of chip formation. *Int. J. Mech. Sci.* **50** 124-31
[121] Majumdar B S and Newaz G M 1992 Inelastic deformation of metal matrix composites: plasticity and damage mechanisms. *Phil. Mag.* **66** 187-212
[122] Li X and Seah W K 2001 Tool wear acceleration in relation to workpiece reinforcement percentage in cutting of metal matrix composites *Wear* **247** 161-71
[123] Dabade U A and Joshi S S 2009 Analysis of chip formation mechanism in machining of Al/SiCp metal matrix composites *J. Mater. Process.* **209** 4704-10
[124] Pramanik A, Zhang L C and Arsecularatne J A 2008 Machining of metal matrix composites: effect of ceramic particles on residual stress, surface roughness and chip formation *Int. J. Mach. Tools Manuf.* **48** 1613-25
[125] Tevatia A and Srivastava S K 2015 Modified shear lag theory-based fatigue crack growth life prediction model for short-fiber reinforced metal matrix composites *Int. J. Fatigue* **70** 123-9
[126] Zhou Y, Gu Y, Lin J, Zhao H, Liu S, Xu Z, Yu H and Fu X 2022 Finite element analysis and experimental study on the cutting mechanism of SiCp/Al composites by ultrasonic vibration-assisted cutting *Ceram. Int.* **48** 35406-21
[127] Daphale A M, Patil C S, Ansari M I and Thakur D G 2021 Numerical investigation of the effect of fibre volume fraction on the stress distribution of aerospace grade Al-Li 8090 metal matrix composite *Int. J. Mater. Struct. Integr.* **14** 102-19
[128] Pramanik A, Zhang L C and Arsecularatne J A 2008 Deformation mechanisms of MMCs under indentation *Compos. Sci. Technol.* **68** 1304-12
[129] Mackerle J 1999 Finite-element analysis and simulation of machining: a bibliography (1976-1996) *J. Mater. Process. Technol.* **86** 17-44
[130] Mackerle J 2003 Finite element analysis and simulation of machining: an addendum: A bibliography (1996–2002) *Int. J. Mach. Tools Manuf.* **43** 103-114
[131] Chen X, Xie L, Xue X and Wang X 2017 Research on 3D milling simulation of SiC p/Al composite based on a phenomenological model *Int. J. Adv. Manuf.* **92** 2715-23
[132] Liu Y N, Huang S T, Zhou L and Xu L F 2014 Three-dimensional finite element simulation analysis of cutting force of SiCp/Al composite thin-walled parts *Key Eng. Mater.* **589** 106-10
[133] Fathipour M, Zoghipour P, Tarighi J and Yousefi R 2012 Investigation of reinforced sic particles percentage on machining force of metal matrix composite *Mod. Appl. Sci.* **6** 9-20
[134] Jadhav M R and Dabade U A 2016 Modelling and simulation of Al/SiCp MMCs during hot machining *ASME Int. Mech. Eng. Congress Expos.* **50527** V002T02A023
[135] Liu J, Cheng K, Ding H, Chen S and Zhao L 2018 Simulation study of the influence of cutting speed and tool-particle interaction location on surface formation mechanism in micromachining SiCp/Al composites *Proc. Inst. Mech. Eng. C J. Mech. Eng. Sci.* **232** 2044-56

[136] Wang B, Xie L, Chen X and Wang X 2016 The milling simulation and experimental research on high volume fraction of SiCp/Al *Int. J. Adv. Manuf.* **82** 809-16
[137] Srinivasan A, Arunachalam R M, Ramesh S and Senthilkumaar J S 2012 Machining performance study on metal matrix composites-a response surface methodology approach *Am. J. Appl. Sci.* **9** 478-83
[138] Rajmohan T, Palanikumar K and Prakash S 2013 Grey-fuzzy algorithm to optimise machining parameters in drilling of hybrid metal matrix composites *Compos. B. Eng.* **50** 297-308
[139] Suresh P, Marimuthu K, Ranganathan S and Rajmohan T 2014 Optimization of machining parameters in turning of Al-SiC-Gr hybrid metal matrix composites using grey-fuzzy algorithm *Trans. Nonferrous Met. Soc. China.* **24** 2805-14
[140] Tamang S and Chandrasekaran M 2014 Application of grey fuzzy logic for simultaneous optimization of surface roughness and metal removal rate in turning Al-SiCp metal matrix composites *AIMTDR 2014* 832-1
[141] Kalaichelvi V, Karthikeyan R, Sivakumar D and Srinivasan V 2012 Tool wear classification using fuzzy logic for machining of Al/SiC composite material Model. Num. Simul. Mat. Sci. **2** 1-9
[142] Chandrasekaran M, Das S and Devarasiddappa D 2015 Determining the effect of cutting parameters on surface roughness in end milling of Al-356/SiCp MMC using fuzzy logic International Conference on Precision, Meso, Micro and Nano Engineering
[143] Dhavamani C and Alwarsamy T 2012 Optimization of machining parameters for aluminum and silicon carbide composite using genetic algorithm *Procedia Eng.* **38** 1994-2004
[144] Sahoo A K and Pradhan S 2013 Modeling and optimization of Al/SiCp MMC machining using Taguchi approach *Measurement* **46** 3064-72
[145] Sriprateep K, Patumchat P and Theansuwan W 2011 Application of Taguchi method in the optimization of cutting parameters for turning metal matrix composite *Adv. Mat. Res.* **189** 3056-60
[146] Shetty R, Pai R B, Rao S S and Nayak R 2009 Taguchi's technique in machining of metal matrix composites *J. Braz. Soc. Mech. Sci. Eng.* **31** 12-20
[147] Shabani M O and Mazahery A 2012 Artificial intelligence in numerical modeling of nano sized ceramic particulates reinforced metal matrix composites *Appl. Math. Model.* **36** 5455-65
[148] Tamang S and Chandrasekaran M 2014 Experimental investigation and development of multi response ANN modeling in turning Al-SiCp MMC using polycrystalline diamond tool *Int. J. Curr. Eng. Technol.* **2** 1-8
[149] Muthukrishnan N and Davim J P 2009 Optimization of machining parameters of Al/SiC-MMC with ANOVA and ANN analysis *J. Mater. Process. Technol.* **209** 225-32
[150] Xu L and Davim J P 2008 Modelling cutting power and tool wear in turning of aluminium matrix composites using artificial neural networks *Int. J. Mater. Prod. Technol.* **32** 333-42
[151] Jeyapaul R and Sivasankar S 2011 Optimization and modeling of turning process for aluminium-silicon carbide composite using artificial neural network models *IEEE International Conference on Industrial Engineering and Engineering Management 2011* 773-778
[152] Gaitonde V N, Karnik S R and Davim JP 2009 Some studies in metal matrix composites machining using response surface methodology *J. Reinf. Plast. Compos.* **28** 2445-57
[153] Arokiadass R, Palaniradja K and Alagumoorthi N 2012 A study on tool wear and surface roughness in end milling of particulate aluminum metal matrix composites: Application of response surface methodology *J. Comput. Appl. Res. Mech. Eng.* **2** 1-2
[154] Setia S and Chauhan S R 2021 Comparative analysis and optimization of FEM and RSM based regression model with experimental results for the dry turning of SiCp-Al7075 composite *Silicon* **13** 4681-701
[155] Premnath A A, Alwarsamy T and Rajmohan T 2012 Experimental investigation and optimization of process parameters in milling of hybrid metal matrix composites *Mater. Manuf.* **27** 1035-44

[156] Tamang S K and Chandrasekaran M 2015 Modeling and optimization of parameters for minimizing surface roughness and tool wear in turning Al/SiCp MMC, using conventional and soft computing techniques *Adv. Prod. Eng.* **10** 59
[157] Seeman M, Ganesan G, Karthikeyan R and Velayudham A 2010 Study on tool wear and surface roughness in machining of particulate aluminum metal matrix composite-response surface methodology approach *Int. J. Adv. Manuf.* **48** 613-24

Experimental Investigation on Mechanical and Viscoelastic properties of Kenaf, Basalt and Carbon Fiber reinforced Hybrid Epoxy Polymer Composites

M. INDRA REDDY[a,*], V. MANIKANTH[a], S. MADHAVA RAO[a], P.V.CH.R.K. SANTOSH[a]

[a]Dept. of Mechanical Engineering, SRKR Engineering College, Bhimavaram-534204, India.

*Corresponding Author email: reddy.indra.m@gmail.com

Abstract. In this study, a hybrid polymer composite reinforced with natural fibers such as Kenaf and Basalt, as well as carbon fiber was developed and analyzed for mechanical properties of the composite with varying fiber content and orientation. The composites were made using a hand layup process in which the plies were alternately stacked. Further, specimens were cut from the fabricated laminate according to ASTM standards. For the tensile and flexural test was conducted using a Universal Testing Machine (UTM). It is noticed that a hybrid composite made up of both natural fibers and carbon fiber has shown greater strength than a composite made up of only natural fibers with the same volume percentage of fiber. Furthermore, it is observed that the hybrid fiber composite in equal proportion is shown maximized strength.

1. Introduction

A number of researchers are working on natural fiber composites due to their numerous advantages like biodegradability, less weight, ease of availability, and superior strength. Further, these composites have benefits in terms of technology, society, and the environment but the mechanical properties are poor as comparison to synthetic fiber composites [1]. The vacuum bagging method was primarily used in the production of hybrid composite materials. The attributes of hybrid composites, including as tensile strength, hardness, tensile modulus, and peak load, have been enhanced by increasing the fiber reinforcement content in the matrix material. The peak load, ultimate tensile strength, and yield strength of the composite were significantly enhanced when reinforced with carbon fiber mats. When compared to other types of composites, carbon fiber reinforced ones are more ductile [2].The reason for this is because, compared to most conventional composites and metals utilized in structural applications today, polymer composites provide a greater strength to weight ratio [3].

As compared to conventional composites, hybrid composites have stronger mechanical properties. Jute fibers resist bending stresses, so when fibers are hybridized, better properties can be obtained due to the fibre's supporting existence [4].Carding and needle punching were used to transform the applied fibers into semi-finished materials, which were then further processed by press-moulding. The carding method was meticulously optimized, resulting in homogeneous fabrics made from a variety of fibers (bast, basalt and polypropelyne). Due to the basalt fiber addition, not only it enhanced the composites stiffness and strength, but also improved their energy absorption properties, which is a very crucial in the automobile industry. The fracture surface study revealed that the thermoset composites have greater fiber matrix adhesion than thermoplastic composites. Renewable hybrid polymer composites that have been studied have a lot of potential to replace non renewable fiber-reinforced polymer composites [5]. The best composition of fabricated samples can be achieved by addition of glass fiber (fixed) with 5wt%, natural fiber of 10 wt% with epoxy and vinyl ester composite of the equivalent composition [6].Channabasavanna et al. developed and tested kenaf/glass polymer hybrid composites by using egg shell powder as a filler and found that the mechanical properties were significantly improved [7]. These composites were prepared with a fiber-to-resin weight proportion of 18:82. However, mechanical properties of the jute-ramie fiber reinforced epoxy composite shows 30 percent enhanced properties than the Jute-polyester composite [8].

The composites were made using six distinct formulae of unsaturated polyester resin, basalt, and glass fibers in a manual lay-up process. The tensile and flexural strengths of glass-fiber reinforced unsaturated polyester resin were improved by adding basalt [9]. In addition, it has been discovered that using various types of fibers greatly improves the effectiveness of jute/sisal/glass fiber reinforced polypropylene hybrid composites. With a maximum impact intensity of 44.15 J/m, the identical composition outperforms the virgin sample by over 40%. Furthermore, it was discovered that hybrid and polymer composites with water concentrations below 0.3% (by weight) absorbed less water[10].

Moreover, the results show that a 9-hour treatment with a 10% Sodium Hydroxide concentration results in enhancement of composites properties, which implies by using this form of hybridization for any realistic composites application, is a superior choice for high efficiency [11]. The composite made of 85% jute and 15% coconut has shown enhanced tensile strength because of its higher weight percentage of jute fibre. The composite functions as a semi-brittle and ductile material because it is made up of 50/50 jute and coconut coir fibre [12]. The epoxy matrix was used to fabricate luffa and carbon fiber reinforced composites with weight percentages of 40/0, 20/20, 25/15, and 15/25. The mechanical properties of the tested samples were improved with the increased percentage of carbon fiber and luffa fiber [13]. The type of fibres is the factor that has the second-largest impact on mechanical properties after the volume fraction of the fibre. The kind of chemical treatment, on the other hand, barely makes a difference. The maximum volume percentage of fibre was then employed, and the fibres were treated with sodium bicarbonate NaHCO3, to produce composites with the best mechanical propertiesLastly, when the fiber mix was 80 weight percent jute and 20 weight percent flax, the created hybrid composite performed better than preceding composites when looking at individual fiber composites [14]. Composites bonded with coir fiber and strengthened with palmyra and high-strength sisal fibers improve mechanical characteristics. Using a fiber-to-matrix volume proportion of 40:60, the hybrid composites were produced by a compression molding process. [15]. With regard to flexural strength, tensile modulus, and tensile strength, pre-treated composites outperformed untreated composites by 128 percent, 144 percent, and 132 percent, respectively [16]. Hybrid composites had greater modulus and flexural strength, but flax/PLA had greater tensile strength. When compared to its rivals, the hybrid composite offers superior impact force. Furthermore, as compared to preexisting composites, flax/PLA composites had superior dynamic mechanical characteristics [17].

A hybrid composite was made by impregnating 30 wt% areca leaf stalk with 70 wt% isophthalic resin and then combining it with sisal fibers in a 1:1 ratio. Using a compression mould procedure, continuous sisal/areca leaf stalk fiber reinforced polymer composites with varying percentages of alkali treatment were prepared and studied. The chemical treatment with the alkali greatly enhanced the fibers' mechanical characteristics, according to the findings. Fibers that were treated with 15% NaOH had much higher interracial adherence to the matrix. The study's characteristics consistently showed that fibers treated with 15% alkali outperformed raw fibers and fibers treated with 5% or 10% NaOH [18]. The optimal mechanical properties were observed in hybrid composites reinforced with 20% jute fiber, 10% glass fiber, and 10% tea leaf fiber, according to research by L Prabhu et al. [19]. This blend has the potential to replace pure glass fiber reinforced polymer composites. Mechanical, wear, and water absorption characteristics were studied in relation to the weight percentage of reinforcing particles in composite materials (ranging from 5 to 25 wt percent). Improving the material's tribological and mechanical qualities was achieved by including particles of coconut shell, rice husk, and bagasse [20]. Because the filler interacts with the structure, enhancing the structure's macroscopic mechanical characteristics [21]. Experimental results show that the stacking sequence and fiber orientation of the fiber layers affect the mechanical characteristics of composite materials [22].

The 2g of CaCO3-coir/luffa cylindrical (C/L/C) hybrid composites outperformed the other composite samples in terms of thermal and dynamic mechanical qualities. The Cole-Cole plot also demonstrated strong bonding among the matrix, filler, and fiber. Composites containing 2 g filler material are expected to have better dynamic mechanical characteristics, making them suitable for use in leading engineering applications [23]. The composite with the greatest flexural strength is the one that is made with 15% CSP and 15% RHP filler [24].

The original aim of the hybrid composite was to provide more strength than that of single fiber composites [25]. Fiber reinforced epoxy nano-composites have completely replaced conventional polymer composites due to their many inherent benefits. A combination of fillers with nano and micron sized particles enhanced the composites' characteristics. Incorporating nano and micro fillers into a material may raise its tensile and compressive strengths by 9–23%, improve its storage modulus by 56%, and reduce its loss modulus by 20–40% [26]. Compression molding is used to prepare composite samples from fibers of different lengths (5 mm, 10 mm, 15 mm, 20 mm, 25 mm). The best properties were found in a reinforced composite with a fiber length of 5 mm [27]. The mechanical characteristics, including as tensile, flexural, effects, compression, and inter laminar shear stress (ILSS), were strongest in a composite material that included 45 weight percent bamboo and cotton, thanks to the orientation of the bamboo yarn's weft [28-33].

Composites have clearly been the subject of much research, as is seen from the literature study. Composites have become more popular because to developments in analytical and testing procedures, enhancements to production processes, and novel materials. When compared to other alloys, fiber-reinforced composites provide superior mechanical characteristics and strength-to-weight ratios. Their transparency, strong bonding powers, ease of repairability, and lack of corrosiveness are only a few of the advantages they possess over metals. Based on these benefits, we set out to create and evaluate the mechanical characteristics of an unprecedented fiber reinforced hybrid composite using kenaf, Basalt, and carbon fiber. The greatest qualities of both synthetic and natural fibers may be harnessed by combining them in a single matrix to create hybrid composites. Producing goods with high specific modulus, strength, corrosion resistance, and, often, outstanding thermal stability may be accomplished at a lower cost using hybrid composites.

2. Materials and Methods

2.1. Materials

Because of their superior mechanical and adhesive properties, epoxy resins have played a significant role in polymer matrix composite materials (PMC). They've been used as a structural adhesive as well as a matrix to keep high-performance fibre reinforcement together in composite materials. As a matrix material, we used Epoxy resin (LY-556) in our research. GV Traders in Madurai provides the resin. The resin has a density of 1258 kg/m3, a viscosity of 500 cps at 25°C, and a monomer content of 35%. Hardener is an epoxy or fiber glass curing agent. To begin curing, epoxy resin needs a hardener, also identified as a catalyst, which is the material that hardens the adhesive when combined with resin. The final characteristics and suitability of the epoxy coating for a given environment are determined by the precise selection and combination of epoxy and hardener components.

Table 2.1 Properties of fibers

Fibers	Density (gm/cm3)	Tensile Strength (MPa)	Young's Modulus (GPa)
Kenaf fiber	1.45	930	53
Basalt fiber	2.67	2900-3100	85-87
Carbon fiber	1.57	3500	120

2.2 Fabrication of Composites

The simplest composite processing technique is hand lay-up. The infrastructure needed for this method is also minimal. The steps in the processing are simple. To begin, a release gel is spritzed over the mold's surface to prevent the polymer from sticking. A excellent surface finish on the component is achieved by using thin plastic sheets at the top and bottom of the mould plate. Reinforcement, such as chopped strand mats or woven mats, is trimmed to size and laid on top of the perspex film thereafter. Following the mat's placement in the mold, the thermosetting polymer is mixed with a specified hardener (curing agent) and then poured over the mat's surface.

Using a brush, the polymer is sprayed uniformly. Rolling a roller gently over the mat-polymer layer after applying a second layer of mat to the polymer surface removes the excess polymer and trapped air. Once all of the required layers of mat and polymer have been layered, the procedure is repeated for each subsequent layer. For example, at room temperature, systems based on epoxy usually require about 24 to 48 hours to cure. Composites based on thermosetting polymers are ideal for this technique. Reduced capital and infrastructure needs set it apart from competing methods. The manufacturing rate is reduced and it is challenging to get a high-volume component of reinforcement in processed composites. Aircraft components, automobile parts, boat hulls, diase floors, and decks are just some of the many places you could find the hand lay-up process in action.

2.3 Fiber content in samples

In this work, we are intended to make 9 different samples of composites, in order to find which composition gives out the best results. The fiber content in different samples is tabulated as follows:

Table 2.2 Fiber content in samples

Sample	Fiber content (Volume %)			Total Fiber Content (Volume %)
	Kenaf	Basalt	Carbon	
S1 (K7.5B7.5)	7.5	7.5	-	15
S2 (K15B15)	15	15	-	30
S3 (K20B20)	20	20	-	40

S4 (K5B5C5)	5	5	5	15
S5 (K10B10C10)	10	10	10	30
S6 (K13.3B13.3C13.3)	13.3	13.3	13.3	40
S7 (K15B15C10)	15	15	10	40
S8 (K10B10C20)	10	10	20	40
S9 (K5B5C30)	5	5	30	40

2.4 Mechanical Properties

Tensile test: Tensile test samples were cut into dog bone shapes. The specimen's tensile strength is measured using a UTM (FIE Make, UTES-40) in accordance with ASTM standards. The dimensions and normal tensile strength test method are chosen in accordance with ASTM D3039. The prepared specimens (250253mm) for tensile testing are shown in the diagram below. Tension testing is another name for tensile testing. It is commonly used to determine how a material will behave when exposed to a slowly applied tensile load. Uni-axial tensile testing is the most popular method for determining mechanical properties of materials. Placing the test specimen in the testing machine and adding stress to it until it breaks is the first step in the testing process. As gauge length increases, the tensile force is calculated. When stress is applied, the gauge section's elongation is calculated in relation to the applied force.
Flexural test: The prepared specimens (125-203mm) for flexural test specimens according to ASTM D709 are shown in the figure below. Flexural tests are frequently used to evaluate the flexural modulus or flexibility of a material. Flexural strength is defined as the greatest stress at the outermost fibre on either the compression or tension side of the specimen. The most popular flexural test for polymer composite blend materials is the 3-point flexural bending test. The crosshead location is used to determine the deflection of the specimen. Flexural power and displacement are among the test findings. The testing technique involves inserting the test specimen into the universal testing machine and exerting pressure on it until it splits and fractures.

2.5 Dynamic mechanical analysis

Studying polymer viscosity can benefit greatly from the approach of dynamic mechanical analysis, which is used to analyse and characterise materials. A temperature material sample is subjected to sinusoidal stress, which is then measured. The stress level is frequently changed, and this leads to fluctuations in damping values. To test the viscoelasticity (strain) of a polymer, dynamic mechanical analysis applies a sinusoidal force and measures the ensuing displacement. The viscous fluid's displacement pressure will be entirely postponed in the orthogonal stress-relationship process. During DMA studies, phase delay might occur due to the characteristics of flexible viscous polymers. Loss modulus of heat-radiating energy represents the viscous part of the storage modulus, while storage modulus represents the elastic part. An LVDT displacement sensor measures the voltage change caused by the probe moving a magnetic core; it then monitors the temperature, the temperature control device, and the engine drive furnace (a linear load motor that provides load in response to applied force). The power of the sample engine and clamps will be determined by the support and steering shaft drive. The samples will undergo separate processing and handling for the sample under review, based on their weighting rationale.

3. Results and Discussion

The dispersion and direction of the fibre or fillers in relation to one another, the fibre concentration, the stress-strain conduct of the matrix and fiber phases, and phase volume fractions are all variables that affect the mechanical properties of natural fibre composites. By varying the fiber composition and material, we were able to determine the mechanical, thermal, and viscoelastic properties of various samples.

3.1 Mechanical Properties

3.1.1 Tensile Strength of Composites

Figures 3.1 to 3.3 show the difference in composite mean tensile strength as a function of fiber material. It was apparent that as the fiber content in the epoxy matrix was increased, the tensile strength increased as well. This is owing to the epoxy resin's ability to transmit and distribute applied stress to the sprout fibers, resulting in increased intensity. The composite (K20B20) has a tensile strength of 124.67 MPa at full fiber content (40 percent volume of fiber). At 15 and 30 volume percentage of fiber, respectively, the tensile strength of the Kenaf and Basalt fiber reinforced composites increased by 35.29% and 1.45% over the matrix. In addition, the tensile strength of Kenaf, Basalt, and Carbon fiber reinforced composites is increased by 8.5% and 24.49%, respectively, over 15% and 30% of the fiber content in composite. The composite's tensile strength is found to be 291.3 MPa at full fiber content (K5B5C30). Figure 3.4 depicts the specimens that were tensile strength measured.

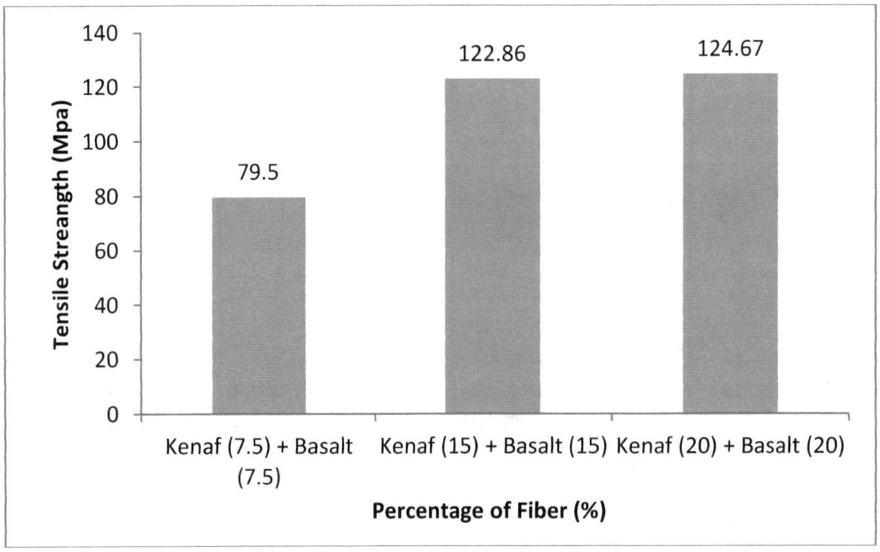

Fig. 3.1 Effect of fiber loading on Tensile Strength of Composite

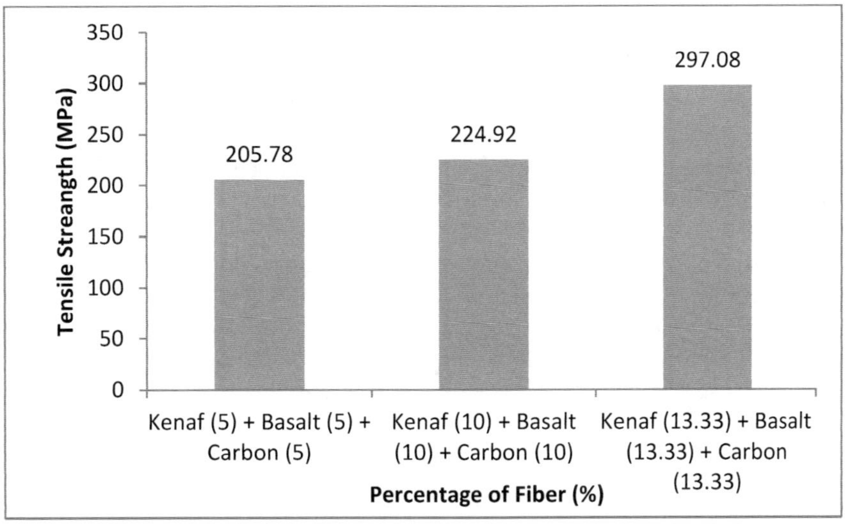

Fig. 3.2 Effect of fiber loading on Tensile Strength of Composite

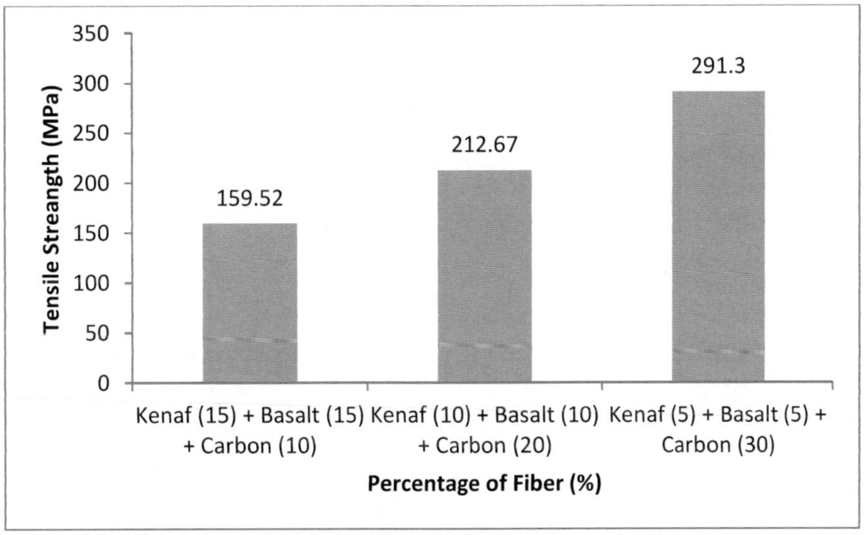

Fig. 3.3 Effect of fiber loading on Tensile Strength of Composite

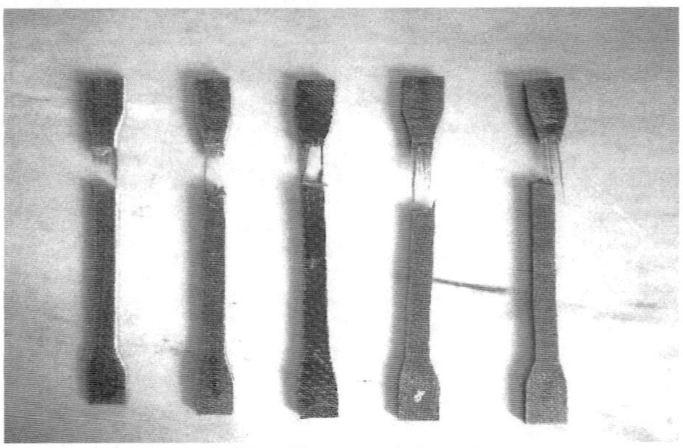

Fig.3.4. Tensile tested Specimens.

3.1.2 Flexural Strength of Composites

Figures 3.5 to 3.7 show the flexural behavior of kenaf, basalt, and carbon fiber reinforced composites. The same cause, as previously noted, is attributed to these plots' comparable pattern of tensile characteristics. The composite's flexural strength at full fiber (K5B5C30 volume of fiber) content is about 331.54 MPa. Over a 15% and 30% volume of fiber, the flexural strength of the kenaf, basalt, and carbon fiber reinforced composite improved by 36.83 and 14.82, respectively. This is due to the fact that the epoxy resin transfers and distributes the applied stress to the fibres, increasing strength. In comparison to virgin epoxy, the composite can withstand a heavier weight before failing. Figure 3.8 depicts the specimens that were subjected to flexural testing.

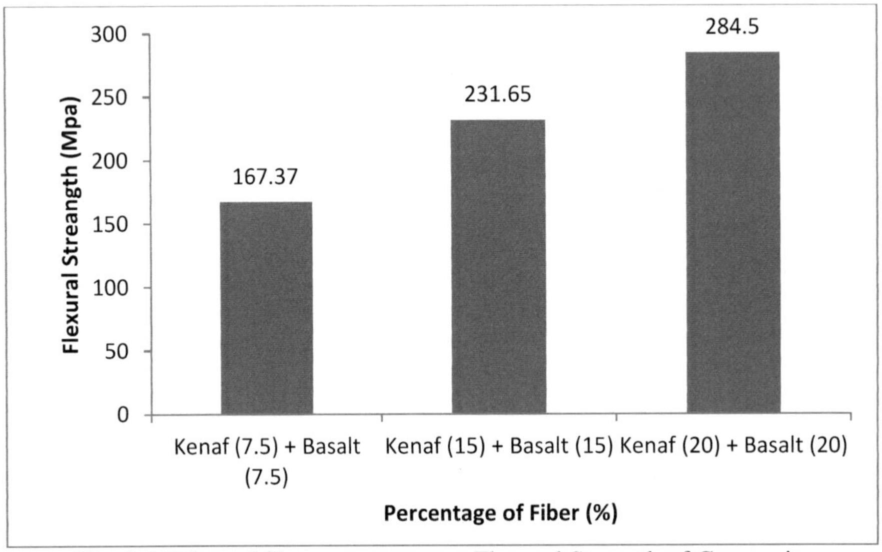

Fig. 3.5 Effect of fiber percentage on Flexural Strength of Composite

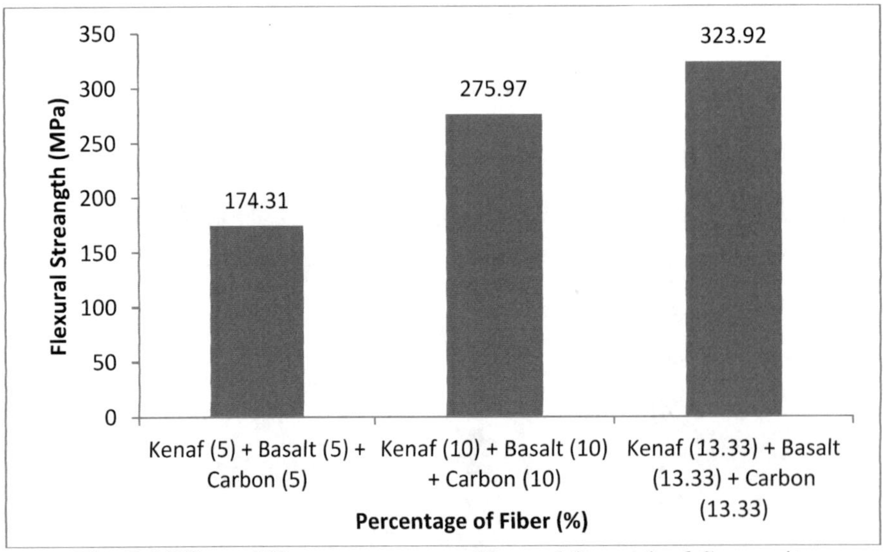

Fig. 3.6 Effect of fiber percentage on Flexural Strength of Composite

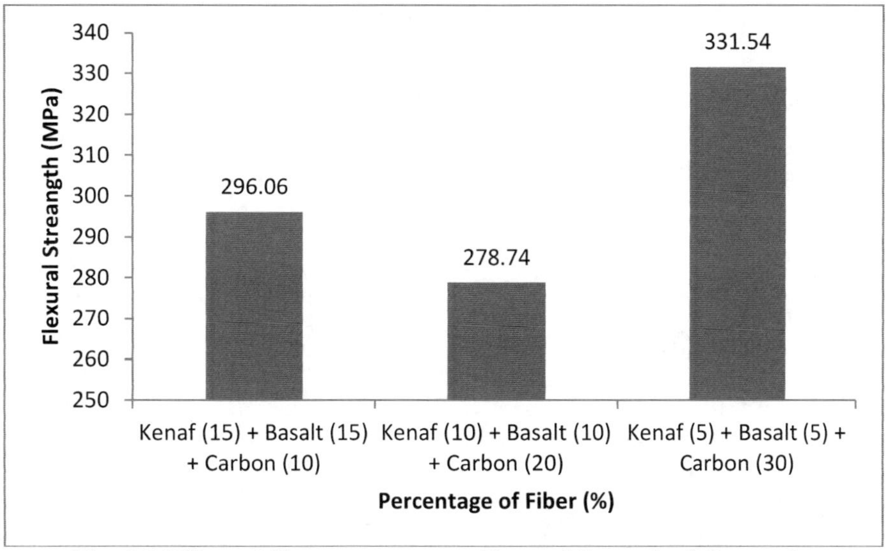

Fig. 3.7 Effect of fiber percentage on Flexural Strength of Composite

Fig. 3.8 Flexural tested Specimens

3.2 Viscoelastic Properties

3.2.1 Storage Modulus

Figure 3.13 shows the storage modulus of kenaf, basalt, and carbon fiber reinforced hybrid composites as a function of temperature at 1 Hz frequency. Storage moduli of hybrid composites are directly proportional to fiber weight percentages up to 40%, then decrease due to poor fiber-to-matrix adhesion. The value of E' was found to be higher for composite K5B5C30, while composite K20B20 had the lowest value. It is well known that as the temperature rises, the value of E' in all composites decreases gradually. The E' of epoxy composites is observed to decrease as the temperature rises due to fiber stiffness loss at high temperatures in these situations. The value of E' rises as the adhesion between fibers and matrix increases. Composite K5B5C30, on the other hand, is inversely proportional to molecular mobility at peak temperature. Furthermore, all of the composites' storage moduli were discovered to be very similar to each other. This is due to the fact that at peak temperature, the fibres have no effect on the materials. In different frequencies, there is no discernible effect on the storage moduli of composites. Composites showed the highest value of storage modulus at 10 Hz frequency and the lowest value at 1 Hz frequency at lower temperature scales. It was also fascinating to note that the influence of frequencies was stronger at lower temperatures than at higher temperatures.

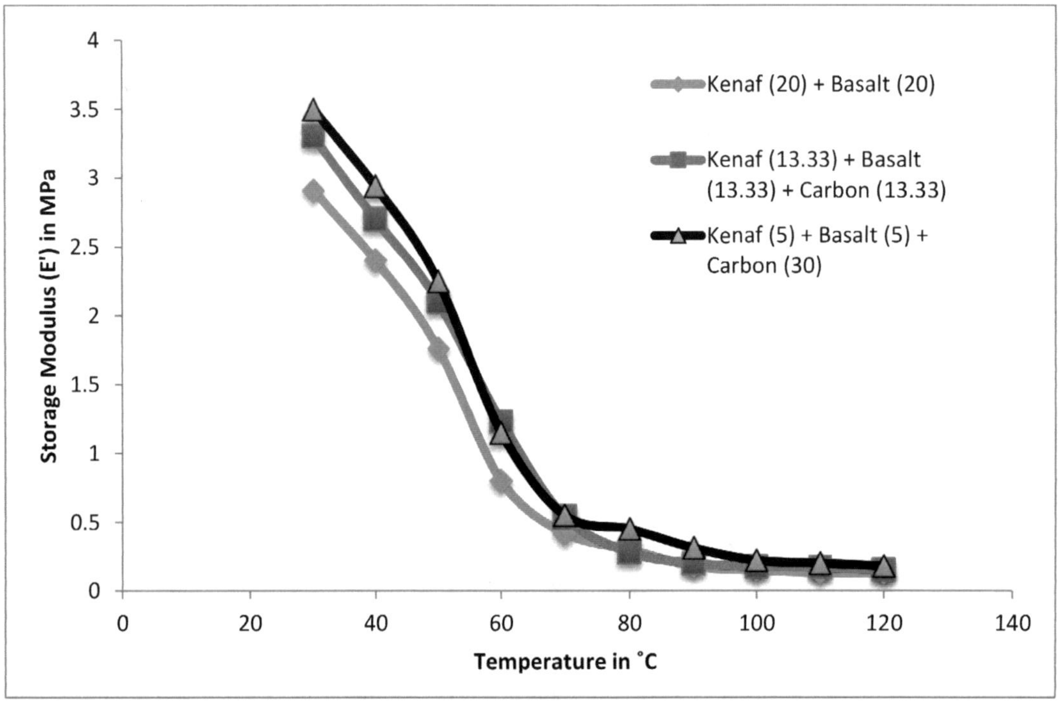

Figure 3.9 Storage modulus of kenaf, basalt, and carbon fiber reinforced hybrid composites at 1 Hz frequency

3.2.2 Loss Modulus

Loss modulus is the highest amount of energy that composite materials emit during deformation. It describes how the mobility of molecules in composites affects viscous materials. Figure 3.14 depicts the relationship between composite loss modulus and temperature at a frequency of 1 Hz. The peak value of a loss modulus curve reflects the Tg of corresponding composites. The values of E" increased until they reached Tg, then decreased as the temperature rose. K5B5C30 had the highest peak on the E" curve. Tg is the value obtained from the loss modulus curve for hybrid composites. The variation in hybrid composite loss modulus with temperature at various frequencies is comparable to storage moduli. The curve's loss modulus values might be increased as molecular mobility improves with increased frequencies.

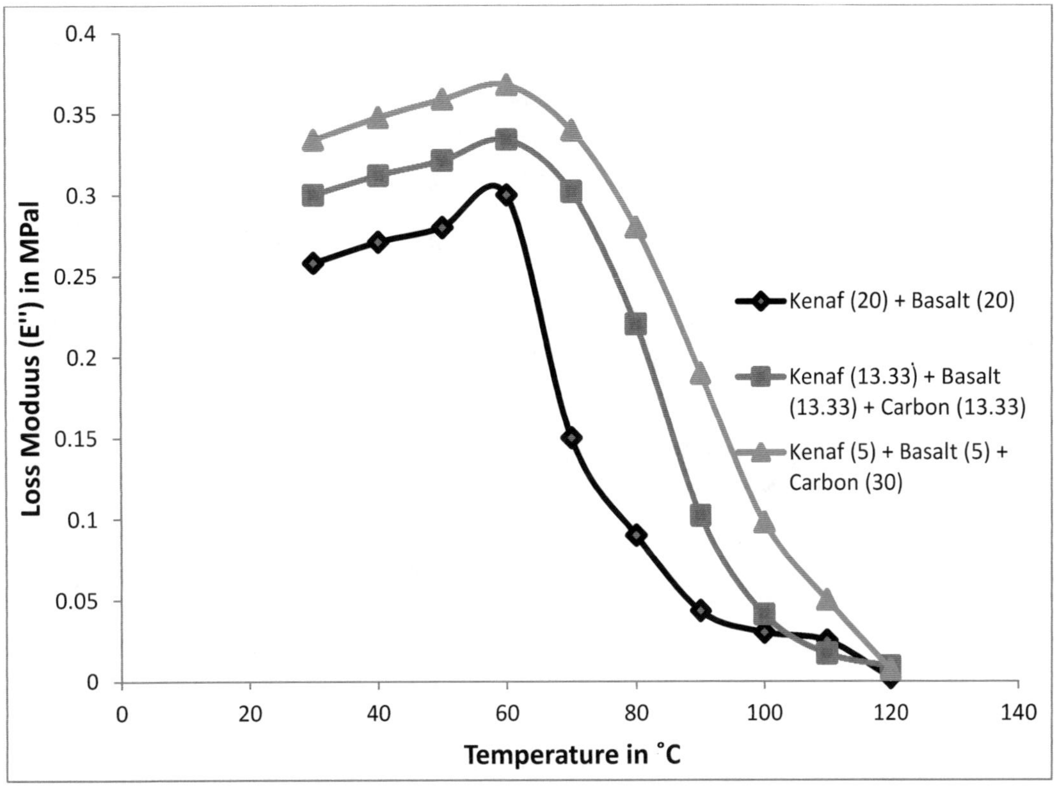

Figure 3.10 Loss modulus of kenaf, basalt, and carbon fiber reinforced hybrid composites at 1 Hz frequency

3.2.3 Damping (Tan δ)

Damping is known as the ratio of loss modulus to storage modulus. Damping is determined by the fibres' and matrix's ability to hold together. Higher damping is linked to poor fibre-matrix sticking properties. This can be shown by the fact that heavy fibre-matrix adhesion can reduce polymer chain mobility, resulting in reduced damping. The lower the damping value, the better the composite's load bearing power. The graph between the damping of composites and temperature is shown in Fig. 3.15 at a frequency of 1 Hz. K13.33B13.33C13.33 has higher values and tends to have greater damping than the other composites in the Tan δ curve. Composite K20B20, on the other hand, had the lowest peak of the Tan curve, indicating better load bearing power. This may be due to the fact that heavy fiber-matrix adhesion results in efficient stress transfer. Tg was observed to migrate towards higher temperatures in composite K13.33B13.33C13.33. This is due to the addition of high-strength natural fibers, which reduces the mobility of the polymer chain. Tg (70°C) was higher in the hybrid composite K13.33B13.33C13.33, indicating greater thermal stability. The maximum thermal stability was found at a lower frequency of 1 Hz, while the lowest thermal stability was found at a higher frequency of 10 Hz. These findings revealed that frequency variation has a significant impact on composite damping.

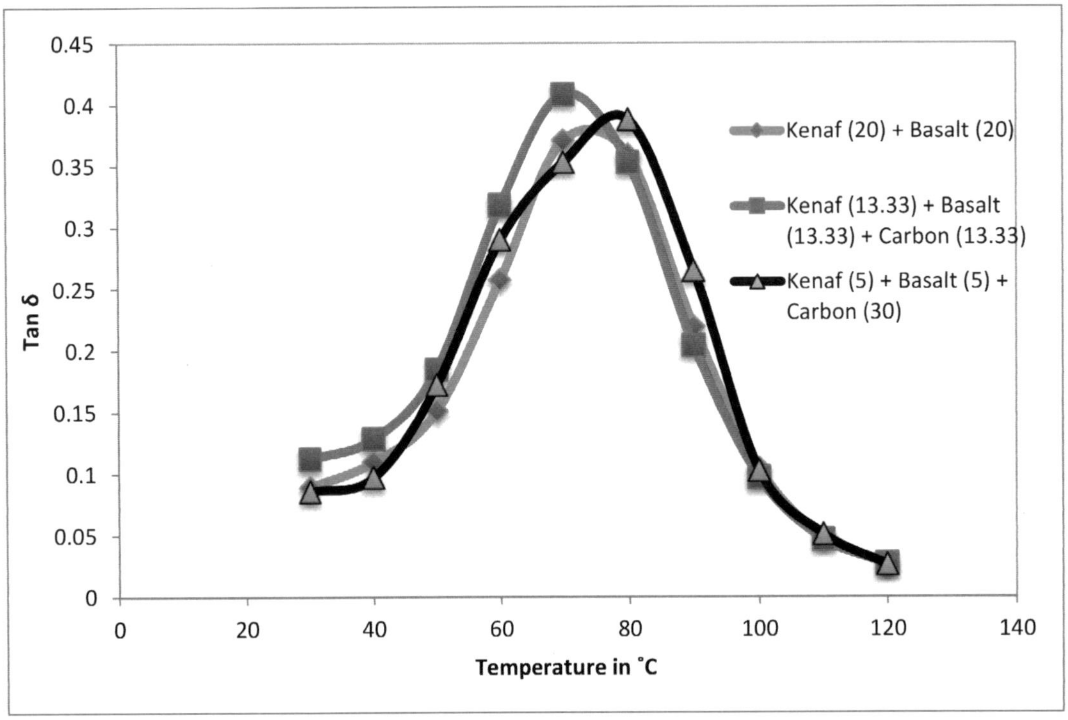

Figure. 3.11. Damping behaviour of hybrid composites at 1 Hz frequency

4. Conclusion

Experiments on kenaf, basalt, and carbon fiber reinforced composites resulted in the following conclusions:

- The tensile strength and flexural strength of the hybrid fiber composite (Kenaf+Basalt+Carbon) are highest when all three fibers are used in the same proportion at 40% fiber volume.
- It has been discovered that the strength of a hybrid composite fabricated with both natural fibers and carbon fiber is greater than a composite fabricated solely with natural fibers at the same volume percentage of fiber.
- These composites have tensile and flexural qualities that are very similar to aluminum alloys. As a result, it implies that it will take the place of several alloy materials currently in use. These composite specimens are also recommended for use in automobile body construction and structural materials.
- Finally, it is concluded that when more strength is needed, synthetic fiber composites are typically used; however, replacing them with hybrid composites, which include both natural and synthetic fibers, provides a number of benefits, including lower material costs, reduced harmful gas emissions, partial eco-friendliness, and good mechanical properties.
- This research also looked at the viscoelastic behavior of hybrid composites.

References

1. Czigány, Tibor. "Basalt fiber reinforced hybrid polymer composites." In Materials Science Forum, vol. 473, pp. 59-66. Trans Tech Publications Ltd, 2005.
2. Jagannatha, T. D., and G. Harish. "Mechanical properties of carbon/glass fiber reinforced epoxy hybrid polymer composites." International Journal of Mechanical Engineering and Robotics Research 4, no. 2 (2015): 131-137.
3. Sinha, Agnivesh Kumar, Harendra Kumar Narang, and Somnath Bhattacharya. "Mechanical properties of hybrid polymer composites: a review." Journal of the Brazilian Society of Mechanical Sciences and Engineering 42, no. 8 (2020): 1-13.
4. Raju, B. S., L. H. Manjunatha, and N. Jagadeeswaran. "Fabrication & characterization of ZnS micro particulate filled glass and jute fibre reinforced hybrid polymer composites." Materials Today: Proceedings 20 (2020): 125-133.
5. Saleem, Anjum, Luisa Medina, and Mikael Skrifvars. "Mechanical performance of hybrid bast and basalt fibers reinforced polymer composites." Journal of Polymer Research 27, no. 3 (2020): 1-13.
6. Khare, Jeetendra Mohan, Sanjeev Dahiya, Brijesh Gangil, and Lalit Ranakoti. "Influence of different resins on Physico-Mechanical properties of hybrid fiber reinforced polymer composites used in human prosthetics." Materials Today: Proceedings 38 (2021): 345-349.
7. Channabasavanna, S. G., T. Venkate Gowda, L. H. Manjunatha, Pavankumar R. Sharvani Shekar, and Shashank M. Chandan Kumar. "Development and testing of glass/kenaf inter-ply hybrid polymer composites with egg shell powder as filler." Materials Today: Proceedings (2020).
8. Santhi, K. Aruna, C. Srinivas, and R. Ajay Kumar. "Experimental investigation of mechanical properties of Jute-Ramie fibres reinforced with epoxy hybrid composites." Materials Today: Proceedings 39 (2021): 1309-1315.
9. Sapuan, S. M., H. S. Aulia, R. A. Ilyas, A. Atiqah, T. T. Dele-Afolabi, M. N. Nurazzi, A. B. M. Supian, and M. S. N. Atikah. "Mechanical properties of longitudinal basalt/woven-glass-fiber-reinforced unsaturated polyester-resin hybrid composites." Polymers 12, no. 10 (2020): 2211.
10. Ray, Krutibash, Hemalata Patra, Anup Kumar Swain, Bibhudatta Parida, Sourabh Mahapatra, Asit Sahu, and Suryakanta Rana. "Glass/jute/sisal fiber reinforced hybrid polypropylene polymer composites: Fabrication and analysis of mechanical and water absorption properties." Materials Today: Proceedings 33 (2020): 5273-5278.
11. Yoganandam, K., P. Ramshankar, P. Ganeshan, and K. Raja. "Mechanical properties of alkali-treated Madar and Gongura fibre-reinforced polymer composites." International Journal of Ambient Energy 41, no. 8 (2020): 849-850.
12. Singh, Chandra Prakash, Raj Vardhan Patel, Mohd Faizul Hasan, Anshul Yadav, Virendra Kumar, and Anil Kumar. "Fabrication and evaluation of physical and mechanical properties of jute and coconut coir reinforced polymer matrix composite." Materials Today: Proceedings 38 (2021): 2572-2577.
13. Ashok, K. G., K. Kalaichelvan, V. Elango, Ajith Damodaran, B. Gopinath, and M. Raju. "Mechanical and morphological properties of luffa/carbon fiber reinforced hybrid composites." Materials Today: Proceedings 33 (2020): 637-641.
14. Benkhelladi, Asma, Hamdi Laouici, and Ali Bouchoucha. "Tensile and flexural properties of polymer composites reinforced by flax, jute and sisal fibres." The International Journal of Advanced Manufacturing Technology 108 (2020): 895-916.

15. Nagarjun, J., J. Kanchana, and G. Rajesh Kumar. "Improvement of mechanical properties of coir/epoxy composites through hybridization with sisal and palmyra palm fibers." Journal of Natural Fibers (2020): 1-10.
16. Yogeshwaran, S., L. Natrayan, Gobikrishnan Udhayakumar, G. Godwin, and L. Yuvaraj. "Effect of waste tyre particles reinforcement on mechanical properties of jute and abaca fiber-epoxy hybrid composites with pre-treatment." Materials Today: Proceedings 37 (2021): 1377-1380.
17. Manral, Ankit, Furkan Ahmad, and Vijay Chaudhary. "Static and dynamic mechanical properties of PLA bio-composite with hybrid reinforcement of flax and jute." Materials Today: Proceedings 25 (2020): 577-580.
18. Hariharashayee, D., M. P. Prabakaran, C. Chelladurai, S. Manikandan, N. S. Mohan, and M. Pradeepa. "Experimental Investigation of Raw and Alkali Treated Areca Leaf Stalk and Sisal Fibres Reinforced Hybrid Polymer Composites."
19. Prabhu, L., V. Krishnaraj, S. Sathish, S. GokulKumar, and N. Karthi. "Study of mechanical and morphological properties of jute-tea leaf fiber reinforced hybrid composites: Effect of glass fiber hybridization." Materials Today: Proceedings 27 (2020): 2372-2375.
20. Suresh, S., D. Sudhakara, and B. Vinod. "Investigation on industrial waste eco-friendly natural fiber-reinforced polymer composites." *Journal of Bio-and Tribo-Corrosion* 6, no. 2 (2020): 1-14.
21. Novak, Nejc, Lovre Krstulović-Opara, Zoran Ren, and Matej Vesenjak. "Mechanical properties of hybrid metamaterial with auxetic chiral cellular structure and silicon filler." *Composite Structures* 234 (2020): 111718.
22. Sujon, Md Abu Shaid, Mohammad Ahsan Habib, and Mohammad Zoynal Abedin. "Experimental investigation of the mechanical and water absorption properties on fiber stacking sequence and orientation of jute/carbon epoxy hybrid composites." Journal of Materials Research and Technology 9, no. 5 (2020): 10970-10981.
23. Krishnudu, D. Mohana, D. Sreeramulu, and P. Venkateshwar Reddy. "A study of filler content influence on dynamic mechanical and thermal characteristics of coir and luffa cylindrica reinforced hybrid composites." Construction and Building Materials 251 (2020): 119040.
24. Sundarababu, Jagannathan, Shanmuga Sundaram Anandan, and Paulius Griskevicius. "Evaluation of mechanical properties of biodegradable coconut shell/rice husk powder polymer composites for light weight applications." Materials Today: Proceedings 39 (2021): 1241-1247.
25. Babu, N., C. S. Rathnasabapathy, and N. Karunakaran. "Evaluation of Mechanical and Morphological Properties of Hybrid Reinforced Polymer Composites." In Advances in Industrial Automation and Smart Manufacturing, pp. 771-783. Springer, Singapore, 2021.
26. Kuruvilla, Santhy P., N. M. Renukappa, and J. Sundara Rajan. "Influence Hybrid Fillers on Electrical and Mechanical Properties of Fiber Reinforced Epoxy Composites."
27. Sajin, J. B., P. Babu Aurtherson, J. S. Binoj, N. Manikandan, MS Senthil Saravanan, and T. M. Haarison. "Influence of fiber length on mechanical properties and microstructural analysis of jute fiber reinforced polymer composites." Materials Today: Proceedings 39 (2021): 398-402.
28. Aruchamy, Karthik, Sampath Pavayee Subramani, Sathish Kumar Palaniappan, Balu Sethuraman, and Gobinath Velu Kaliyannan. "Study on mechanical characteristics of woven cotton/bamboo hybrid reinforced composite laminates." Journal of Materials Research and Technology 9, no. 1 (2020): 718-726.

29. Battina, N. Malleswararao, Varaha Siva Prasad Vanthala, and Hari Krishna Chirala. "Influence of tool pin profile on mechanical and metallurgical behavior of friction stir welded AA6061-T6 and AA2017-T6 tailored blanks." Engineering Research Express 3, no. 3 (2021): 035026.
30. Battina, N. Malleswararao, Varaha Siva Prasad Vanthala, and Hari Krishna Chirala. "Influence of tool pin profile on mechanical and metallurgical behavior of friction stir welded AA6061-T6 and AA2017-T6 tailored blanks." Engineering Research Express 3, no. 3 (2021): 035026.
31. Battina, N. Malleswararao, Chirala Hari Krishna, and Varaha Siva Prasad Vanthala. "Influence of pin profile on formability of friction stir-welded aluminum tailor-welded blanks: an experimental and finite element simulation analysis." Transactions of the Canadian Society for Mechanical Engineering 46, no. 3 (2022): 602-613.
32. Battina, N. Malleswararao, Hari Krishna Chirala, Varaha Siva Prasad Vanthala, Eswara Manikanta Javvadi, Raja Sekhar Nukathoti, and Azaharahmed Mohammed. "Experimental and finite element investigations on formability of friction stir welded tailor welded blanks of AA6061 and AA2017." Materials Today: Proceedings (2023).
33. Battina, N. Malleswararao, Hari Krishna Chirala, Ramu Inala, Obula Reddy Kummitha, Lakshmi Narayana Veeravalli, and Srinivasa Raju Pericherla. "Influence of coefficient of friction between punch-blank interface on formability of friction stir welded aluminum tailor welded blanks–An experimental and finite element simulation investigations." Proceedings of the Institution of Mechanical Engineers, Part C: Journal of Mechanical Engineering Science 236, no. 17 (2022): 9677-9686.

A Review on Energy Storage Technologies: Current Trends and Future Opportunities

Ankammarao Padamurthy[1,*], Manikanth Vatala[2], Praveen Kumar Gandla[3], V Ch R K Santosh P[2], Amar Sheelwant[4], Harinadh Vemanaboina[5]

1 Department of Mechanical Engineering, Sri Venkateswara College of Engineering and Technology, Chittoor, Andhra Pradesh, INDIA
2 Department of Mechanical Engineering, Sagi Rama Krishnam Raju Engineering College, Bhimavaram, Andhra Pradesh, INDIA
3 Department of Mechanical Engineering, G. Pullaiah College of Engineering and Technology, Kurnool, Andhra Pradesh, INDIA
4 Application Scientist, Ducom Instruments, Bengaluru, Karnataka, INDIA
5 VEMU Institute of Technology, Chittoor, Andhra Pradesh, INDIA

ankamarao301@gmail.com

Abstract. Despite the fact that numerous energy storage technologies have already been reviewed in the scientific literature, these innovations are right now at various stages of technological development, with only a few having been demonstrated for use on a large scale. The majority of research articles on energy storage give detailed descriptions of these technologies, but there is little data on how these technologies compared when used for energy storage. This paper provides a comprehensive review on energy storage concepts and also compares the different energy storage technologies in terms of research trends and future opportunities. Among the reviewed energy storage technologies, the thermochemical energy storage (TCES) method stands as an attractive and potential alternative to the conventional heat storage methods like sensible and latent thermal energy storage (STES and LTES) for several reasons that include: high energy storage density, minimal heat losses between the system and surroundings, and long term heat storage.

Keywords. Energy Storage Technologies, Non-Thermal Energy Storage, Thermal Energy Storage, Thermochemical Energy Storage, Energy Storage Density

1. Introduction

The exploration and proper utilization of latest technologies help place the individuals and organizations in leading position among their peers, in this socio-economic competitive world. Using the available technologies, the researchers are continually working hard to provide better infrastructure and amenities like energy, medical care, roads, etc. to their citizens for better living standards. Despite of all efforts, the demand for energy appears to be growing non-stop. The world's total energy consumption is mostly due to the sectors like power, industry, transport, residential and commercial buildings, etc. [1]. The rapid growth in the consumption of energy can be attributed to two reasons, viz. increase in world's population and accelerated habitual practices of mankind. The world's total energy consumption was estimated to be at 13,147 million tonnes of oil equivalent (MTOE) in the year 2015 [2], of which, fossil fuels alone took a major share and account for approximately 85%.

Going by the world energy scenarios, the demand for primary energy sources is likely to increase at least by 50% by the year 2050 [3]. Owing to their limited availability, the fossil fuels are very unlikely to support this growing demand of energy in a sustainable manner in future. Further, the usage of fossil fuels at the current rate shall result in serious environmental issues like greenhouse gas emissions, global warming, acid-rain, depletion of ozone layer, etc. The world's carbon dioxide emissions due to fossil fuels was estimated to be 32.3 billion metric tons for the year 2012 and projected at 43.2 billion metric tons for the year 2040 [1].

It is evident from the above discussion that the gap between energy supply and demand is growing continually, while the fossil fuels are diminishing very rapidly. In the near future, this condition is likely to pose a serious challenge to the mankind in meeting their energy needs. This problem can possibly be addressed by (i) using the available energies effectively, which means that enhancing the systems' performance further to their possible maximum levels, (ii) exploring and utilization of alternate sources of energies including renewable energies, and (iii) storing the energy when it is available in abundance and reusing it during the time of need. Of these three options at hand, the first one i.e. enhancing the systems' performance is always limited by a value, determined by the second law of thermodynamics. It is thus highly difficult to bring out further improvements in the systems that are already delivering their best performance under the given conditions. This option is only relevant for the systems that are yet to mature or establish.

The non-hydro renewable energy share of contribution towards the world's total energy need was very little and accounted for 3.3% in the year 2015, and is projected to reach only 10% by the year 2035 [2]. While the institutions and governments are working aggressively to enhance the renewable energy usage, the gap between total energy supply and demand seems to be not declining, and thus force to look forward to other alternate avenues.

2. Need for Energy Storage

Heat is a form of energy which has been in use over the years, in applications like power plants, process industries, etc. through burning of fuels. Before it reaches the beneficiary in final form, energy heat undergoes several conversions with huge losses at many places. For instance, in the case of power generation, as illustrated in Figure 1, it is understood that the energy undergoes multiple conversions before it being used, resulting in huge losses. The present illustration is a classic example to emphasize the need for minimizing such energy losses during the conversion and the associated opportunities for energy storage.

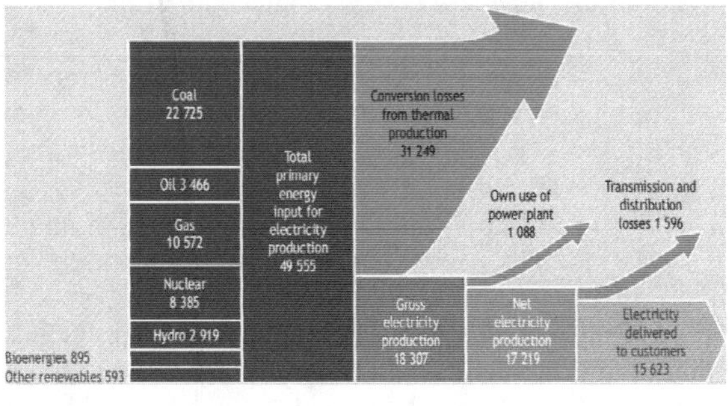

Figure 1. Energy flows (TWh) of the global electricity system [6].

2.1. Energy Storage Methods

Energy storage is a mechanism of storing energy in a medium for future use, either in the same or different form of the energy supplied. Normally, the input energy for energy storage is either waste heat or renewable energy. In some instances, viz. chemical battery, selected chemicals in combination with electrodes are used to derive electrical energy. This sort of energy storage offers many advantages viz. (i) brings down the demand on fossil fuels, (ii) reduces the gaseous emissions due to the fossil fuels, (iii) acts as a stand-alone source of energy, (iv) supplements the peak power demand, (v) helps controlling the overall energy cost. From the literature, it is understood that various types of energy storage methods have been conceived and developed over the decades and they are classified as shown in Figure 2. Subject to the application, a suitable energy storage method is usually opted. Such choice depends on several parameters that includes (i) storage medium/material, (ii) storage technique, (iii) storage capacity, (iv) storage density, (v) storage efficiency, (vi) storage duration, (vii) energy charge and discharge rates, (viii) storage device durability, (ix) initial and running costs, (x) effect on environment.

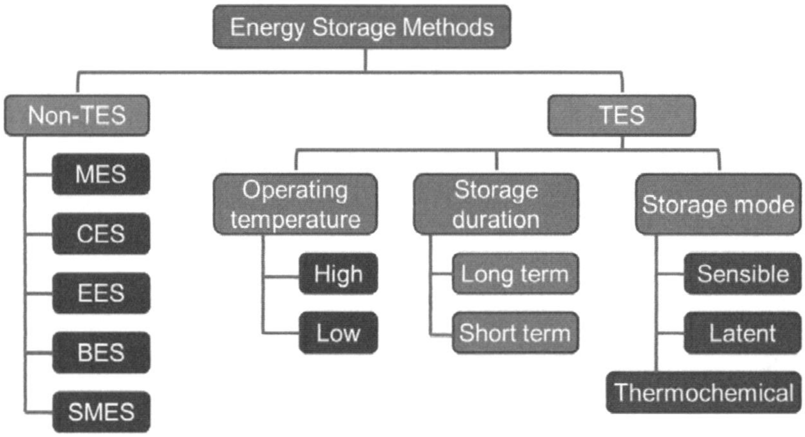

Figure 2. Classification of energy storage methods.

Acknowledging the role of these influencing parameters, a brief outline is presented here for a few prominent energy storage methods which are divided into two groups viz. thermal and non-thermal

3. Non-thermal Energy Storage

Mechanical, chemical, electrostatic, magnetic, and biological are a few conventional methods in use today to store the energy. Of these, mechanical energy storage (MES) and chemical energy storage (CES) are the attractive methods that help storing larger quantities of energy for future use. The MES is further classified into three types viz. pumped hydro storage (PHS), compressed air energy storage (CAES) and flywheel. The PHS runs on the principle of maintaining the desired pressure head through a water reservoir. Such pressure head can be obtained by raising water levels across a dam by either stopping the flowing river or pumping the water from a lower level, when surplus electric power is available. The first PHS was realized in the year 1909 at Schaffhausen in Switzerland [4]. This technology is now fully matured and being used across the globe through 352 installations to a rated power of 183.8 GW [5]. Another form of MES technology is compressed air storage. In this method, the air is compressed and stored in a container when surplus energy is available. This compressed air is used in a gas power plant to generate electricity. It has high power capacity range from 100-300 MW, and low energy storage density 3-12 kWh/m^3. Flywheel is a device that stores the rotational energy of a shaft and returns it when the shaft's momentum is low. A flywheel-powered bus was developed in 1940's by Oerlikon, a Swiss company [7].

In chemical energy storage method, the energy is stored in chemical compounds and retrieved in electrical energy form through chemical reactions [8]. Chemical batteries and fuel cells are a few examples of this category [9]. Chemical battery technology is well matured and hence finds suitability in all scales of applications range from hybrid vehicles to wrist watches. In electrostatic energy storage (EES) method, energy is stored in the form of electric charge between two metallic (conductive) plates of nearly equal size and separated by a dielectric medium (electric insulator) when voltage is applied across them [10]. The direct current supplied to the initial plate induces secondary current in the adjacent plate. Conventional capacitor and super capacitor are the examples for EES method. Compared to the conventional capacitors, super capacitors offer high storage capacity and power output [11].

Superconducting magnetic energy storage (SMES) method is an emerging technology and operates on the principle of electrodynamics. Energy is stored in the form of magnetic field in a superconducting coil when direct current is applied. This method was first introduced in the year 1969 for an electric power application [12]. SMES systems offer significantly high energy storage densities when compared to capacitors. These systems are commercially available in the power range of 0.1 – 10 MW [13].

Bio-batteries are best example of biological energy storage (BES) method where energy is stored by breaking down glucose molecules by enzymatic action. Glucose acts as a medium for energy storage and examples include natural sugar, soda and old papers. The design of bio-battery consists of an electrolyte, separator, anode and cathode, same as that of a conventional battery [14].

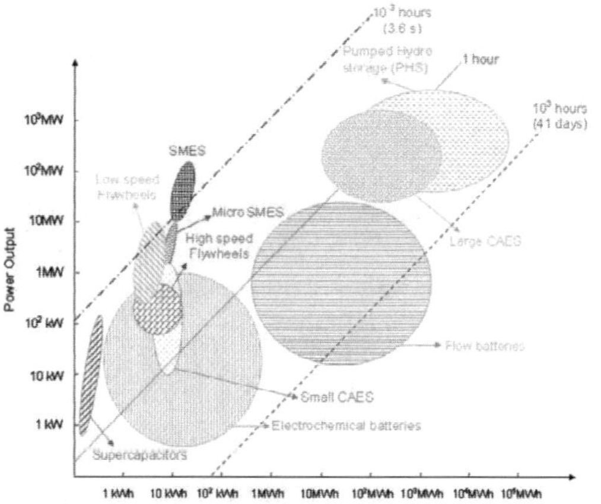

Figure 3. Energy storage capacity and power output of various energy storage technologies [15].

Figure 3 shows a schematic presentation of power output and energy storage capacity of different types of energy storage technologies. It is understood that the PHS is the most promising and attractive energy storage method as it offers significant energy storage capacity and power output. Table 1-4 enumerate key features and issues of various non-thermal energy storage methods. According to the global energy database [5], the total power generation by all these energy storage technologies, in the year 2017 was 193.34 GW from the 352 installations, where the PHS alone accounts for 183.8 GW and the rest is covered by electro-chemical, electro-mechanical, thermal storage, compressed air energy storage, etc. Since the hydel power generation largely depends on the available static head across the dams, the scope for building new dams is curbed by the factors viz. the limited resources like flowing rivers, geographical constraints and the associated environmental concerns. On the other hand, the current installed non-hydro energy storage share is meager, and technologies are yet to establish for building the large scale storage capacities. At this backdrop, thermal energy storage systems appear to be a potential, promising and sustainable candidate for future energy demands.

Table 1. Mechanical Energy Storage (MES) – Key features and Issues [7,16–18].

MES Type	Applications	Influencing parameters	Advantages/ Features	Limitations
PHS	Hydroelectric power plant	Reservoir head and its storage capacity	High reliability, Low maintenance cost, High storage capacity, High durability (40-100 years), High power rating (100-5000 MW), Efficiency (70-87 %)	High capital cost, Environmental issues, Limited locations
CAES			Cycle efficiency (40-80 %), High durability (20-100 years), Flexible in size.	High initial cost, Slow response, Limited location, Hazardous for high-pressure air reservoirs
Flywheel	Automobiles, aerospace, space satellite and hybrid electric vehicles	Flywheel material's mechanical properties	Cycle efficiency (80-99 %), Power density (400-1600 W/kg), Low speed flywheels are cheaper, High discharge time (high speed flywheels), High power rating (0.002-20 MW), Durability (15-20 years)	High material cost , Expensive (high speed flywheels), Low energy storage density (low speed flywheel), Low discharge time (low speed flywheel)

Table 2. Chemical Energy Storage (CES) – Key features and Issues [16,19,20].

CES Type	Applications	Influencing parameters	Advantages/ Features	Limitations
Lead Acid Battery	Vehicle industry		Energy storage capacity (80 MWh)	Poor performance at low temperatures, Low durability, Ecological issues due to lead usage.
Nickel Cadmium Battery	Spinning reserve and black start		High discharging capacity, Long life, Temperature tolerance, High energy storage density	Highly toxic, High cost
Nickel Metal Hydride Battery	Electric vehicles	Charge and discharge rates, operating temperature (20-25 °C), and discharging capacity	High energy storage density, Eco friendly system (use of nontoxic materials)	High self-discharge
Lithium-Ion Battery	Small appliances		High efficiency & reliability, High energy storage density, Low self-discharge rate, Low cost, High energy capacity (30 MWh)	Expensive for medium and largescale power generation
Sodium Sulphur Battery	Electric cars and hybrid vehicles		Rapid reversibility, Efficient operation, Provides high pulse power, Long life, Low maintenance	Corrosion problem
Flow Battery			High efficiency (75-85 %), Allows high voltage and current	High capital and running costs

Table 3. Electrostatic Energy Storage (EES) – Key features and Issues [10,11,21–23].

EES Type	Applications	Influencing parameters	Advantages/ Features	Limitations
Capacitor	Vehicle industry	Plate conductivity, plates distance, plate size and Material selection	High energy release at low discharge rate, Deliver and accept high current in a short period	Very low energy density, Energy delivery reduces due to heat losses
Super Capacitor	Electric hybrid vehicles, Synchronous generator, Wind turbines, and Voltages sag regulators		Rapid response, High durability (100,000 cycles), No maintenance, High efficiency (97 %)	Low energy storage density, High self-discharge, High cost

Table 4. Superconducting Magnetic Energy Storage (SMES), Biological Energy Storage (BES).

Energy Storage Type	Applications	Influencing parameters	Advantages/ Features	Limitations
SMES [12,13 24]	Central power stations	Magnetic field intensity and materials permeability	High durability, High reliability, Low maintenance, No self-discharge, High efficiency (85-99 %), Generate high power pulse, Negligible heat losses, Storage of energy in a short duration	High cost, Unsuitable for large scale applications
BES [14]		Quantity or mass of sample (glucose)	Recharges instantly, No environmental issues, Feasible, Nontoxic, Nonflammable	Low rate of retention energy

4. Thermal Energy Storage (TES)

Beginning with the preservation of food, the mankind had started realizing the significance of either storing or preserving several other commodities like water, goods, energy, environment, etc. for future needs. The energy, stored in thermal form, was realized in earlier times in several ways like meat/fish/food preservation, space cooling/heating for comfort living, etc. The storage of thermal energy in sensible form was understood to have begun as many as 350 years ago [25], while the research on conventional thermal energy storage technology has started from the year 1970 [26]. Total power generation capacity on the basis of thermal energy storage stands for 3622 MW from the 206 installations across the globe [3].

Using either solar radiation or waste heat due to the industrial processes, energy is stored in a chosen medium for future use. Such stored energy, mostly in the form of heat, is useful for either seasonal or long-term needs/applications, viz. space heating, water/air/gas heating, and power generation. A TES method constitutes primarily three stages, namely charging, storing and discharging, with heat losses being associated with each of them. The TES methods are broadly classified on the basis of three parameters, viz. (i) mode of energy storage (sensible, latent and thermochemical), (ii) material's operating temperature (low (below 200 °C), and high (above 200 °C)), and (iii) energy storage duration (long term, short term).

Following are the parameters that help selecting a suitable thermal energy storage method and/or system: storage capacity, storage period, storage efficiency, storage system's size, storage material, material cost, material availability, material cyclability, material's charging and discharging rates, charging and discharging temperatures, material's self-discharging rate, heat losses between the system and surroundings, storage system's durability, etc.

4.1. Sensible thermal energy storage (STES)

In STES (Sensible Thermal Energy Storage) involves storing internal energy in a material without changing its phase [27]. The material's temperature adjusts during charging and discharging, following the energy flow direction. Energy storage capacity relies on the material's heat capacity (mcp) and temperature gain during charging. Applications include solar water heating. Common materials are water (85 °C), oil, and molten salts for liquids, and rock (50 °C), metals, and concrete (400 °C) for solids [28]. Figure 4 illustrates the storage capacity of a widely used liquid medium (water) [29].

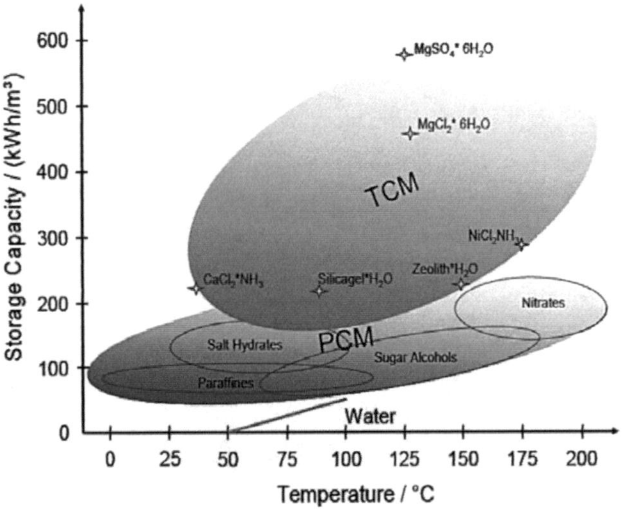

Figure 4. Energy storage capacity and power output of various energy storage technologies [29].

4.2. Latent thermal energy storage (LTES)

In LTES (Latent Thermal Energy Storage), phase change materials (PCMs) undergo a phase change during both charging and discharging, maintaining constant temperature. PCMs like inorganic salt hydrates, fatty acids, eutectics, and paraffin waxes are commonly used [28,30]. Energy storage in LTES depends on PCM mass and latent heat of fusion. Applications include space heating/cooling, water heating/cooling, and cooling systems for various purposes [31]. LTES demonstrates superior storage capacity compared to STES, expanding its versatile applicability [Figure 4].

4.3. Thermochemical energy storage (TCES)

Thermochemical energy storage (TCES) stands out as an innovative and promising thermal energy storage method, utilizing reversible chemical reactions for energy storage and release. During charging, hot air induces reversible reactions in reactive components, breaking intermolecular bonds and producing energy-storing materials [32]. This charged material releases heat during the discharging process, enabling applications such as water heating, space heating, and power generation. Thermochemical materials (TCMs) like salt hydrates [33,34], metal hydroxides [35,36], carbonates, zeolites [37,38], metal organic frameworks [39], metal oxides [40], metal hydrides [41-44] and composites [45,46] exhibit diverse reactions. TCMs offer a significantly higher energy storage density, approximately 8 to 10 times greater than sensible energy storage and twice that of phase change materials (PCMs) [47]. Key features and challenges of various thermal energy storage methods are summarized in Table 5 [29].

Table 5. Comparison of thermal energy storage methods

Energy storage method	Applications with their temperature range	Material/medium with their temperature range	Advantages	Disadvantages	Influencing parameters
STES [26,28,47-49]	Solar water heating (< 100 °C) Space heating for buildings (< 100 °C)	Water (85 °C) Rock (Ground storage) and aquifers (50 °C) Concrete (400 °C)	Reliable, Simple design and construction, Inexpensive storage material, Material's thermal conductivity, Long life, Matured technology, Ecofriendly.	High heat losses, Bigger in size, Low energy storage density (0.2 GJ/m^3), Inability to deliver the energy at constant temperature	System/material size/volume, Specific heat, Temperature difference of the working medium
LTES [47,50,41,52-56]	Space heating/cooling of buildings (5 °C & 30 °C) Domestic hot water production, space heating in buildings and solar heat storage (40-120 °C)	Paraffins (20-70 °C) Salt hydrates (30-80 °C)	Medium energy storage density (0.3-0.5 GJ/m^3), High heat storage capacity, Small temperature gap during the charging and discharging processes, Mixed type PCMs flexible and favorable, Low energy storage volume, Proper utilization of solar energy, High heat of reaction	Not suitable for long distance transport possibility, Low thermal conductivity of PCMs leads to prolong the charging and discharging duration (Paraffin K = 0.2 W/m-k), Material corrosion, High heat losses, Limited life	System/material size/volume, Material's latent heat
TCES [47,57-61]	Low temp. for household applications i.e. < 100 °C Medium temp. for space heating applications i.e. 100-300 °C High temperature for concentric solar power plant applications i.e. > 300 °C	Salts hydrates (<105 °C) Hydroxides (150-300 °C) Metal oxides (> 300 °C)	High energy storage density (0.5-3 GJ/m^3), No heat losses, Energy stored at atmospheric conditions, Long storage period, Compact Transport possibility (long distance), High energy storage capacity (200-600 kWh/m^3)	High initial cost Technically complex Life depends on chemical reactants and side reactions Technically not available (research is ongoing) Material's corrosion	System/material size/volume Charging and discharging temperatures Partial pressure, temperature and mass flow rates of the working fluid

5. Scope for Thermochemical Energy Storage

TCES emerges as the most promising energy storage method due to its high energy storage capacity (200-600 kWh/m^3), extended storage period (3-6 months), capability to attain high temperatures (400-1200 °C), and minimal heat losses to the surroundings [47,60].

Aydin et al. [62] conducted a six-month review on thermal energy storage materials for a 100 m^2 UK domestic building. Thermochemical materials outperformed sensible and latent heat materials consistently in supporting the building's heat load. This finding aligns with Yu et al.'s [63] (Figure 5) conclusion, drawn from experimental data, emphasizing thermochemical materials for their high energy storage densities, ranging from 408 to 1067 Wh/kg across various temperatures.

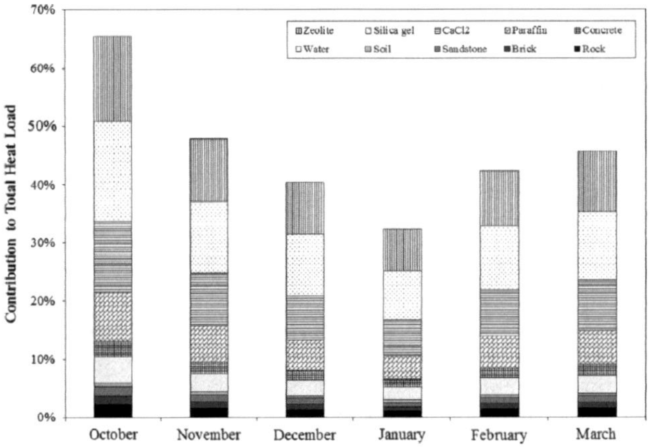

Figure 5. Heat load support from various thermal energy storage materials for a building in UK [63].

Figure 6 shows the publication trend on the subject of thermal energy storage over the years. The data is drawn from the Scopus search tool using 'thermal energy storage' and 'thermochemical energy storage' as the keywords. This data clearly indicates that the subject of thermochemical energy storage has gradually emerged in the last few years and has a good scope to explore and investigate further.

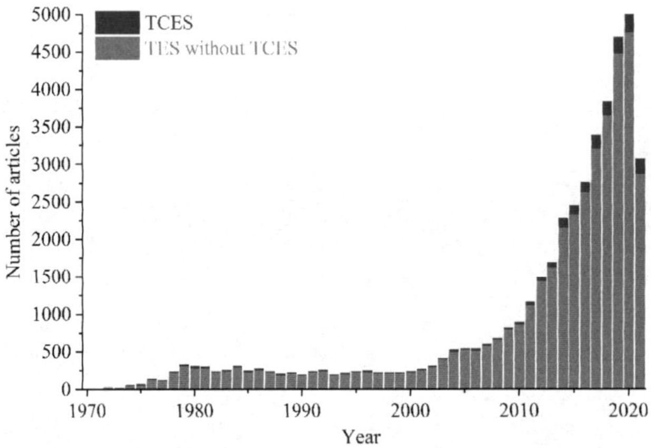

Figure 6. Number of articles published over the years on thermal energy storage (Source: Scopus; Keywords: "thermal energy storage" AND "thermochemical").

6. Conclusions and Future Scope

This work offers a thorough analysis of energy storage technologies and compares various energy storage technologies in terms of current and upcoming research directions. Subject to the application, a suitable energy storage method is usually opted. Such choice depends on several parameters that includes storage medium/material, storage technique, storage capacity, storage density, storage efficiency, storage duration, energy charge and discharge rates, storage device durability, initial and running costs, effect on environment. Among the available non thermal energy storage methods, the PHS is the most promising and attractive energy storage method as it offers significant energy storage capacity and power output. However, the PHS exhibits huge drawback since the hydel power generation largely depends on the available static head across the dams, the scope for building new dams is curbed by the factors that include the limited resources like flowing rivers, geographical constraints and the associated environmental concerns. As per the review reports on TES methods, the TCES method stands as the most promising and attractive energy storage method for several reasons including the high energy storage capacity (200-600 kWh/m^3), long term storage period (3-6 months), possibility of achieving high temperatures (400-1200 °C), minimal heat losses between the system and surroundings. The Scopus data clearly indicates that the subject of thermochemical energy storage is gradually emerging in the last few years and has a good scope to explore and investigate further.

References

[1] International Energy Outlook 2016 with projections to 2040, 2016. https://doi.org/www.eia.gov/forecasts/ieo/pdf/0484(2016).pdf.
[2] BP Energy Outlook 2017 edition, 2017. https://www.bp.com/content/dam/bp/pdf/energy-economics/energy-outlook-2017/bp-energy-outlook-2017.pdf.
[3] World Energy Scenarios: Composing energy futures to 2050, 2013.
[4] M.S. Whittingham, History, Evolution, and Future Status of Energy Storage, Proc. IEEE. 100 (2012) 1518–1534. https://doi.org/10.1109/JPROC.2012.2190170.
[5] DOE Global Energy Storage Database, Sandia Natl. Lab. (2016). http://www.energystorageexchange.org/projects/data_visualization (accessed December 3, 2017).
[6] Co-generation and Renewables: Solutions for a low-carbon energy future, 2011.
[7] A.B. Gallo, J.R. Simoes-Moreira, H.K.M. Costa, M.M. Santos, E. Moutinho dos Santos, Energy storage in the energy transition context: A technology review, Renew. Sustain. Energy Rev. 65 (2016) 800–822. https://doi.org/10.1016/j.rser.2016.07.028.
[8] M.A. Hannan, M.M. Hoque, A. Mohamed, A. Ayob, Review of energy storage systems for electric vehicle applications: Issues and challenges, Renew. Sustain. Energy Rev. 69 (2017) 771–789. https://doi.org/10.1016/j.rser.2016.11.171.
[9] D. Lefebvre, F.H. Tezel, A review of energy storage technologies with a focus on adsorption thermal energy storage processes for heating applications, Renew. Sustain. Energy Rev. 67 (2017) 116–125. https://doi.org/10.1016/j.rser.2016.08.019.
[10] Everything about solar energy, (2017). http://energyprofessionalsymposium.com/?p=32410 (accessed December 3, 2017).
[11] T. Kousksou, P. Bruel, A. Jamil, T. El Rhafiki, Y. Zeraouli, Energy storage: Applications and challenges, Sol. Energy Mater. Sol. Cells. 120 (2014) 59–80. https://doi.org/10.1016/j.solmat.2013.08.015.
[12] C.-S. Hsu, W.-J. Lee, Superconducting magnetic energy storage for power system applications, IEEE Trans. Ind. Appl. 29 (1992) 990–996. https://doi.org/10.1109/28.245724.
[13] X. Luo, J. Wang, M. Dooner, J. Clarke, Overview of current development in electrical energy storage technologies and the application potential in power system operation, Appl. Energy. 137 (2015) 511–536. https://doi.org/10.1016/j.apenergy.2014.09.081.
[14] S. Kalaiselvam and R. Parameshwaran, Thermal Energy Storage Technologies for Sustainability: Systems Design, Assessment and Applications, 2014.

[15] H. Ibrahim, A. Ilinca, J. Perron, Energy storage systems-Characteristics and comparisons, Renew. Sustain. Energy Rev. 12 (2008) 1221–1250. https://doi.org/10.1016/j.rser.2007.01.023.
[16] H.L. Ferreira, R. Garde, G. Fulli, W. Kling, J.P. Lopes, Characterisation of electrical energy storage technologies, Energy. 53 (2013) 288–298. https://doi.org/10.1016/j.energy.2013.02.037.
[17] A.J. Pimm, S.D. Garvey, M. de Jong, Design and testing of Energy Bags for underwater compressed air energy storage, Energy. 66 (2014) 496–508. https://doi.org/10.1016/j.energy.2013.12.010.
[18] Bitterly Jack G., Flywheel technology past, present, and 21st century projections, (n.d.) 2312–2315. https://doi.org/10.1109/62.707557.
[19] Z. Yang, J. Zhang, M.C.W. Kintner-meyer, X. Lu, D. Choi, J.P. Lemmon, Liu Jun, Electrochemical energy storage for green grid, Chem. Rev. 111 (2011) 3577–3613. https://doi.org/10.1021/cr100290v.
[20] N. Chouhan, R.-S. Liu, Electrochemical Technologies for Energy Storage and Conversion, (2012) 1–43. https://doi.org/10.1002/9783527639496.
[21] L.J.O. and T.T. Chen Wenjie, Adnanses Alf Kare, Hansen Jan Fredrik, Super-capacitors based hybrid converter in marine electric propulsion system, in: 19th Int. Conf. Electr. Mach. ICEM, 2010. https://doi.org/10.1109/ICELMACH.2010.5607967.
[22] I.P. Gyuk, EPRI-DOE Handbook of Energy Storage for Transmission & Distribution Applications, 2003. https://doi.org/1001834.
[23] S. Lemofouet, a. Rufer, A Hybrid Energy Storage System Based on Compressed Air and Supercapacitors With Maximum Efficiency Point Tracking (MEPT), IEEE Trans. Ind. Electron. 53 (2006) 1105–1115. https://doi.org/10.1109/TIE.2006.878323.
[24] A. V. Pan, L. MacDonald, H. Baiej, P. Cooper, Theoretical Consideration of Superconducting Coils for Compact Superconducting Magnetic Energy Storage Systems, IEEE Trans. Appl. Supercond. (2016) 1–5. https://doi.org/10.1109/TASC.2016.2533564.
[25] E. Morofsky, Thermal Energy Storage for Sustainable Energy Consumption Fundamentals, Case Studies and Design, in: 2007: pp. 3–22. https://doi.org/10.1007/978-1-4020-5290-3_1.
[26] A.I. Fernandez, M. Martinez, M. Segarra, I. Martorell, L.F. Cabeza, Selection of materials with potential in sensible thermal energy storage, Sol. Energy Mater. Sol. Cells. 94 (2010) 1723–1729. https://doi.org/10.1016/j.solmat.2010.05.035.
[27] S. Khare, M. Dell'Amico, C. Knight, S. McGarry, Selection of materials for high temperature sensible energy storage, Sol. Energy Mater. Sol. Cells. 115 (2013) 114–122. https://doi.org/10.1016/j.solmat.2013.03.009.
[28] A. Abhat, Short term thermal energy storage, Energy Build. 3 (1981) 49–76. https://doi.org/10.1016/0378-7788(81)90005-0.
[29] V.G. Gude, Energy storage for desalination processes powered by renewable energy and waste heat sources, Appl. Energy. 137 (2015) 877–898. https://doi.org/10.1016/j.apenergy.2014.06.061.
[30] P.B. Salunkhe, D. Jaya Krishna, Investigations on latent heat storage materials for solar water and space heating applications, J. Energy Storage. 12 (2017) 243–260. https://doi.org/10.1016/j.est.2017.05.008.
[31] G. Kandalkar, S.J. Deshmukh, R.R. Kolhekar, Latent Heat Storage for Cooling Application : A Review, in: Int. Conf. Electr. Electron. Optim. Tech. - 2016, 2016: pp. 81–84.
[32] J. Yan, C.Y. Zhao, Experimental study of CaO/Ca(OH)2 in a fixed-bed reactor for thermochemical heat storage, Appl. Energy. 175 (2016) 277–284. https://doi.org/10.1016/j.apenergy.2016.05.038.
[33] A. Padamurthy, J. Nandanavanam, P. Rajagopalan, Assessment of selected salt hydrates for thermochemical energy storage applications, Mater. Today Proc. (2022) 1–7. https://doi.org/10.1016/j.matpr.2022.04.048.

[34] A. Padamurthy, J. Nandanavanam, P. Rajagopalan, Evaluation of reaction characteristics of Na2S2O3.5H2O for thermochemical energy storage, Mater. Today Proc. 17 (2019) 239–245. https://doi.org/10.1016/j.matpr.2019.06.425.
[35] M. Kubota, S. Matsumoto, H. Matsuda, Enhancement of hydration rate of LiOH by combining with mesoporous carbon for Low-temperature chemical heat storage, Appl. Therm. Eng. 150 (2019) 858–863. https://doi.org/10.1016/j.applthermaleng.2019.01.049.
[36] K. Wang, T. Yan, R.K. Li, W.G. Pan, A review for Ca(OH)2/CaO thermochemical energy storage systems, J. Energy Storage. 50 (2022). https://doi.org/10.1016/j.est.2022.104612.
[37] A. Padamurthy, J. Nandanavanam, P. Rajagopalan, Sustainable and open sorption system for low-temperature heat storage applications, Int. J. Energy Res. 46 (2022) 20004–20020. https://doi.org/10.1002/er.7867.
[38] A. Padamurthy, J. Nandanavanam, P. Rajagopalan, Thermal stability evaluation of selected zeolites for sustainable thermochemical energy storage, Energy Sources, Part A Recover. Util. Environ. Eff. (2021). https://doi.org/10.1080/15567036.2021.1880502.
[39] A. Padamurthy, J. Nandanavanam, P. Rajagopalan, Preparation and characterization of metal organic framework based composite materials for thermochemical energy storage applications, Appl. Surf. Sci. Adv. 11 (2022) 100309. https://doi.org/10.1016/j.apsadv.2022.100309.
[40] W. Shi, Y. Zhu, C. Shen, J. Shi, G. Xu, X. Xiao, R. Cao, Water sorption properties of functionalized MIL-101(Cr)-X (X=-NH2, –SO3H, H, –CH3, –F) based composites as thermochemical heat storage materials, Microporous Mesoporous Mater. 285 (2019) 129–136. https://doi.org/10.1016/j.micromeso.2019.05.003.
[41] Choudhari M. S. and Sharma V. K., 'Thermodynamic simulation of hydrogen based thermochemical energy storage system'. International Journal of Hydrogen Energy. 46 (2021), 16440-16452. https://doi.org/10.1016/j.ijhydene.2020.09.074.
[42] Choudhari M. S., Sharma V. K. and Paswan M. K., 'Metal hydrides for thermochemical energy storage applications'. International Journal of Energy Research. 45 (2021), 14465-14492. https://doi.org/10.1002/er.6818.
[43] Choudhari M. S. and Sharma V. K., 'Performance investigations on hydrogen-based thermochemical energy storage system through finite volume method and thermodynamic simulation'. International Journal of Energy Research. 45 (2021), 20156-20175. https://doi.org/10.1002/er.7093
[44] Mouli B. C., Tension T. J., Choudhari M. S., Paswan, M. K. and Sharma V. K. (2022), 'Recent advancements in metal hydride based thermochemical energy storage system – A short review'. Materials Today: Proceedings. 51(2022), 1521-1526. https://doi.org/10.1016/j.matpr.2021.10.313.
[45] R. Sutton, E. Jewell, J. Searle, J. Elvins, Discharge performance of blended salt in matrix materials for low enthalpy thermochemical storage, Appl. Therm. Eng. 145 (2018) 483–493. https://doi.org/10.1016/j.applthermaleng.2018.09.052.
[46] M. Gaeini, A.L. Rouws, J.W.O. Salari, H.A. Zondag, C.C.M. Rindt, Characterization of microencapsulated and impregnated porous host materials based on calcium chloride for thermochemical energy storage, Appl. Energy. 212 (2018) 1165–1177. https://doi.org/10.1016/j.apenergy.2017.12.131.
[47] D. Aydin, S.P. Casey, X. Chen, S. Riffat, Novel "open-sorption pipe" reactor for solar thermal energy storage, Energy Convers. Manag. 121 (2016) 321–334. https://doi.org/10.1016/j.enconman.2016.05.045.
[48] O. Ercan Ataer, Storage of Thermal Energy, in: Encycl. Life Support Syst., 2006. https://doi.org/10115582565.
[49] I. Dincer, On thermal energy storage systems and applications in buildings, Energy Build. 34 (2002) 377–388. https://doi.org/10.1016/S0378-7788(01)00126-8.
[50] Y. Zhang, G. Zhou, K. Lin, Q. Zhang, H. Di, Application of latent heat thermal energy storage

in buildings: State-of-the-art and outlook, Build. Environ. 42 (2007) 2197–2209. https://doi.org/10.1016/j.buildenv.2006.07.023.

[51] G. Zhou, Y. Zhang, X. Wang, K. Lin, W. Xiao, An assessment of mixed type PCM-gypsum and shape-stabilized PCM plates in a building for passive solar heating, Sol. Energy. 81 (2007) 1351–1360. https://doi.org/10.1016/j.solener.2007.01.014.

[52] A. Sari, Form-stable paraffin/high density polyethylene composites as solid-liquid phase change material for thermal energy storage: Preparation and thermal properties, Energy Convers. Manag. 45 (2004) 2033–2042. https://doi.org/10.1016/j.enconman.2003.10.022.

[53] D. Zhou, C.Y. Zhao, Y. Tian, Review on thermal energy storage with phase change materials (PCMs) in building applications, Appl. Energy. 92 (2012) 593–605. https://doi.org/10.1016/j.apenergy.2011.08.025.

[54] A. Abhat, Low temperature latent heat thermal energy storage : heat storage materials, Sol. Energy. 30 (1983) 313–332.

[55] J. Jeon, J.H. Lee, J. Seo, S.G. Jeong, S. Kim, Application of PCM thermal energy storage system to reduce building energy consumption, J. Therm. Anal. Calorim. 111 (2013) 279–288. https://doi.org/10.1007/s10973-012-2291-9.

[56] S.-G. Jeong, S. Jin Chang, S. We, S. Kim, Energy efficient thermal storage montmorillonite with phase change material containing exfoliated graphite nanoplatelets, Sol. Energy Mater. Sol. Cells. 139 (2015) 65–70. https://doi.org/10.1016/j.solmat.2015.03.010.

[57] K.E. N'Tsoukpoe, T. Schmidt, H.U. Rammelberg, B.A. Watts, W.K.L. Ruck, A systematic multi-step screening of numerous salt hydrates for low temperature thermochemical energy storage, Appl. Energy. 124 (2014) 1–16. https://doi.org/10.1016/j.apenergy.2014.02.053.

[58] Y. Kato, R. Takahashi, T. Sekiguchi, J. Ryu, Study on medium-temperature chemical heat storage using mixed hydroxides, Int. J. Refrig. 32 (2009) 661–666. https://doi.org/10.1016/j.ijrefrig.2009.01.032.

[59] T. Block, N. Knoblauch, M. Schmücker, The cobalt-oxide/iron-oxide binary system for use as high temperature thermochemical energy storage material, Thermochim. Acta. 577 (2014) 25–32. https://doi.org/10.1016/j.tca.2013.11.025.

[60] L. André, S. Abanades, G. Flamant, Screening of thermochemical systems based on solid-gas reversible reactions for high temperature solar thermal energy storage, Renew. Sustain. Energy Rev. 64 (2016) 703–715. https://doi.org/10.1016/j.rser.2016.06.043.

[61] A.H. Abedin, M.A. Rosen, A Critical Review of Thermochemical Energy Storage Systems, Open Renew. Energy J. 4 (2011) 42–46.

[62] D. Aydin, S.P. Casey, S. Riffat, The latest advancements on thermochemical heat storage systems, Renew. Sustain. Energy Rev. 41 (2015) 356–367. https://doi.org/10.1016/j.rser.2014.08.054.

[63] N. Yu, R.Z. Wang, L.W. Wang, Sorption thermal storage for solar energy, Prog. Energy Combust. Sci. 39 (2013) 489–514. https://doi.org/10.1016/j.pecs.2013.05.004.

The study of rate constants of redox reactions on cobalt (III) complexes of α-hydroxy acids with quinolinium chlorochromate (QCC) in micellar medium

M. V. Raghupathi Raju[1], N. Vijaya Kumar[1], K. Tulasi Bhavani[2], K. Subramani[3], G. Rama Swamy[4], S. Prameela Devi[2], B. Tirumala Rao[4], G. V. Lakshmi Kanth[4]
[1]Department of Chemistry, D.N.R. College (A), Bhimavaram, AP, India-534202.
[2]Department of Chemistry, SRKREngineering College (A), Bhimavaram, AP, India-534204.
[3]Department of Chemistry, Islamiah College, Vaniyambadi, TN, India-655752.
[4]Department of Basic Sciences, VIT (A), Bhimavaram, India -534202.
[1]Corresponding author: nethalavk@gmail.com

Abstract. Complexes of QCC with bound and unbound alpha-hydroxy acids of Co (III) in the micellar medium is unable to obtain some sort of electron transport mode kinetics in the medium. Unbound and bound hydroxy acids are oxidised by QCC, which also oxidizes Cobalt (III). The Cobalt (III) centre cannot receive an electron while the C-C bond is being split. These compounds' oxidation rates rise in direct proportion to their temperature. When the concentration of surfactant is increased, the rate increases. More than ammonium laurylsuphate, DDAC can speed up the oxidation of the process (ALS). A similar pattern was found in the Co (III) complexes containing lactato, glycolato, and mandelato.
Keywords: Quinolinium chlorochromate (QCC), Ammonium Laurylsuphate (ALS), Dimethyl dioctadecyl ammonium chloride (DDAC), Mandelato

1. Introduction

Understanding inorganic reaction mechanisms is crucial for gaining insights in to coordination chemistry, and the significance of metals in biochemical, organometallic chemistry. In this study, that is an important to note the interaction of oxygen with others in the reaction for making bonds is a great concern to get a confirmed structure is a challenging task for the researchers in the field. In 1852, Fremy reported for the first time about the structures of di-cobalt (III) complexes and later in 1910 Werner established the preparation of a wide range of ammine and ethylenediamine complexes, with many bridging groups. It is widely acknowleged in the literature that the addition of an oxygen atom to the respective Co(II) precursors results in the formation of diverse Co(III)-dioxygen complexes. There are two types of oxidation states of di-cobalt complexes with O_2 bridging ligands are observed in general and they are classified as (i) $O_2^{\cdot-}$ 'superoxo' and (ii) O_2^{2-}, peroxo-bridged complexes, respectively.

The investigation focused on examining the oxidation process of α-hydroxy acids, namely Lactic, Mandelic and Glycolic acids, utilizing Quinolinium chlorocromate as both an oxidant and a surfactant[1-3]. With synchronous carbon–carbon bond fission and decarboxilation, electron transport induces approximately 100% reduction at the Co (III) core[4-9].Co (III) Schiff base complexes energies for the oxidation of organic molecules, in the oxidation of phenols, indols and amines a reversible process like Co(III)/ Co(II) redox potential.

The catalytic activity of Co (III) salen (bis(salicylidene)ethylenediamine) complexes describes the þ3 oxidation state of cobalt is catalytically dynamic state. These complexes have been widely utilized to mimic cobalamin (B12) coenzymes instead. Metal complexes of salophen (bis(salicylidene) phenylenediamine) and salen-type ligands also found applications in the embolism of DNA base pairs and basic anions existing in biological and environmental systems. In addition, these ligands have also been utilized as nitrate dectectors. Moreover, cobalt complexes of Schiff's base have been discovered as agent with antimicrobial properties. Moreover, cobalt based salen complexes have revealed an excellent selectivity towards poly (propylene carbonate) development. These observations pave the way for putting significant efforts towards the synthesis of complexes as well as studing the reaction mechanism. Recently there has been an observation that Co(III) salen metal complexes have been employed in the reduction of CO_2 to CO, functioning as a multi-electron reducing agent. In the presence of Quinolinium chlorochromate (QCC), Co(III) is xidized to form keto acid complexes of both bound and unbound hydroxyl acids in Ammonium lauryl sulphate (ALS) and dimethyl dioctadecyl ammonium chloride (DMDAC), without involving electron transfer (DDAC)[10-16&17]. It is possible that the transition state may contain a lower number of electrons. Slow C-H bond fission and hydride ion transfer allow sub transition state [17-21].Sync C-C bond fission and electron transfer to Co(III) [22-24&28]. The relationship between the oxidation rate of Co(III) and its concentration is determined by thermodynamic parameters. Unbound lactic acid reacts faster than Glycolic and Mandelic acids. Finding novel mild and anhydrons oxidation reagents has long been a challenge in organic chemistry. Water, Acetic acid, N-N dimethyl formamide and other oxidants are non-photosensitive and non-hydroscopic QCC [25-26&29].PCC is an oxidant in micellar media.

α-hydroxy acids (AHA) generally known as 2-hydroxy acids which consist of a carboxylic acid group with a hydroxyl group substituent on the adjacent (alpha) carbon. Some familiar examples are citric acid, lactic acid, mandelic acid and glycolic acid. AHAs are a group of acids that occur naturally in fruits, sugar cane, and milk. The increase in acidity is ascribed to the hydrogen bonding present within the molecule. They have found many applications in different fields of sciences. Mainly, they are applicable industry as feed additives and as basis for polymers. Additionally, they are frequently employed in enhancing the appearance of products by providing moisture and performing chemical exfoliation. Moreover, they contribute to refining skin texture and minimizing acne [30]. Nevertheless, they are not always fit for darker skin tones. QCC known as Quinolinium chlorochromate, is a gentle and specific oxidizing agent. Originally recognized for its ability to oxideze alcohols, QCC finds its application in the field of synthetic organic chemistry.

In general, the dispersion of micelles in a suitable solvent lead to the formation of aggregated amphiphiles and are in existence with equilibrium with free and unaggregatedamphiphile. Micellar solutions arise when the concentration of amphiphil surpasses the critical micellar concentration (CMC) or critical aggregation concentration (CAC). The micelle concentration continues to increase until the amphiphil concentration reaches a slightly higher level, resulting in the formation of a lyotropic liquid crystal phase. Micelles are depicted as spherical in shape, but based on the amphiphile chemical structure they could be cylindrical or oblate. Micellar solutions are typically present in isotropic phases[31-34]. Intially, these micellar particles demonistrated significant efficacies in skincare products back in 1931 in France. When a French pharmaceutical company Bioderma released a product called Sensibio H_2O micellar water in 1991 it has bagged the international recognition and this product is so popular that is sold for every 2 seconds worldwide even today. Micelles possess all the physical properties of colloid and are ultramicroscopic units observed in colloids. During their formation, three or more molecules aggregated in a particular medium under thermodynamic equilibrium state to create anisotropic interfacial lining between polar and nonpolar regions impacts the chemical and physical properties of the system. The determination of reaction rates in micellar phases is a typically carried out using pseudo phase approach, which allows for the distinction of aqueous, organic and surfactant components between the reactant and the product[35]. The distribution follows the traditional principles of kinetic and mass transfer.

Chemical kinetics is a field of study that explains the connection between reactants and products in relation to the nature of the reaction. The rate constant, or specific rate constant, serves as a measure of the relationship between the rate of a chemical reaction and the concentrations of the reactants.

2. Materials and Methods

Analytical-grade reagents were acquired and utilized in their original state without further purification. The rate of reaction was calculated by monitoring the reaction mixture with a UV-Vis spectrophotometer with respect to time.

2.1 Preparation of Quinolinium Chlorochromate (QCC)

100 grams of chromium trioxide in solid form and kept it in a beaker. To this 6 molar HCl (184 mL) is addad. The mixture is stirred well for about 5-10 minutes. Then a homogeneous solution is formed. Then it was cooled up to 15°C. To this mixture, 1molar (129.5g) of quinolin was added dropwise for about half an hour. After adding the quinoline, the mixture was cooled to 15°C again. The result is a precipitated form of quinolinium chlorochromate in the form of a yellow powder is obtained. The powder was dried under a vacuum for 3 hours. The yield obtained is about 96%, and its melting point is 239°C.

2.2 Characteristics of Quinolinium chlorochromate (QCC)

Molecular formula of QCC is $C_9H_7N\ ClCrO_3H$.

It is a good oxidizing agent used for the oxidation process in the reactions of complex compounds. It is economical and cost effective. It is a non-hygroscopic yellow coloured solid with melting point value as 130^{0}C. It is soluble in water and in acetic acid. A little work was done with QCC as oxidant so far. Kinetics and oxidation mechanism of diphenylsulphide and α-hydroxy acids by QCC in $HClO_4$ was studied with the help of an absorption spectra at 360 nm.

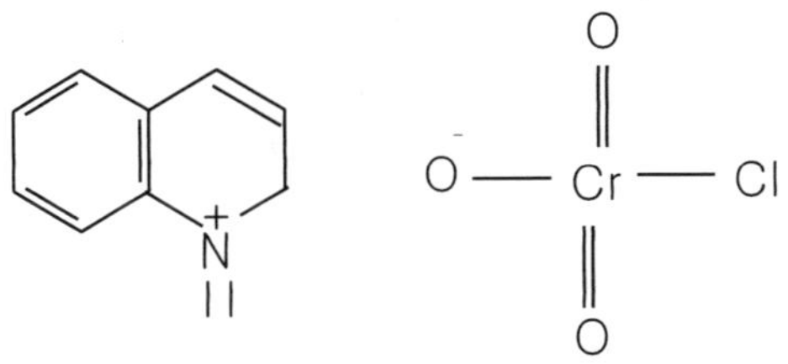

Molecular structure of QCC

2.3 Mechanism of pentaamine cobalt (III) complexes of α-hydroxy acids by oxidation with Quinolinium Chlorochromate (QCC) in micellar medium.

The Kinetics and stoichiometric result obtained in the oxidation of pentaamine cobalt (III) complexes with QCC in both bound and unbound α-hydroxy acids in ALS, DDAC and Polysorbate 80 medium are presented in Table 1, Table 2, Table 3 and the results are summarized as follows.

- Firstly, QCC oxidation reaction rate relies on the Cobalt (III) complex concentration.
- The reaction follows first-order kinetics on the concentration of Cr (IV) in the system.
- For bimolecular reactions, Arrhenius dependence is evident and thermodynamic parameters are in the predicted order.
- The lactato-Co (III) lactato complex reacts fastest of the three complexes.
- The α-hydroxy acids bound to Cobalt (III) are oxidised at a greater rate than unbound ligands, and the Cobalt (III) centre is completely reduced.
- It was observed that electron transfer mechanism in cobalt (III) complexes of α-hydroxy acids in the presence of QCC is significantly higher than that of other micelles.

- In this reaction, 1.00 mol of Cobalt (III) complex reacts with around 0.35 mol of QCC, giving 100 % Co(III) and 100 % carbonyl compounds. One mol of α-hydroxy acids interacts with roughly 0.65 mol of QCC to produce carbonyl compounds and CO_2 with a yield of 100%.

The following mechanism (Fig.1) has been postulated for the oxidation of pentaamine cobalt (III) complexes of α-hydroxy acids with QCC is in accordance with the findings.

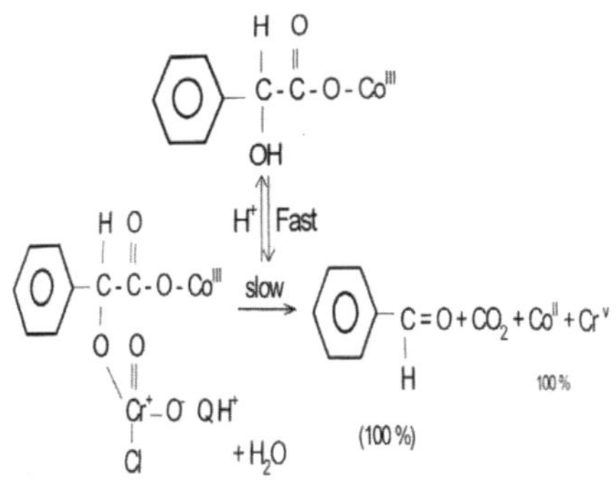

Figure 1: Plausible mechanism of QCC oxidation of pentaamminecobalt (III) complexes of α-hydroxy acids

Table 1: Experimental Data of Stoichiometric data for qunilonium chlorochromate oxidation of Cobalt (III) bound and unbound α- hydroxy acids in the presence ALS at 31°C.

Sulphuric acid = 1.00 mol dm^{-3}

Ammonium lauryl sulphate = 2.00 x 10^{-4}mol dm^{-3}

10^3[Compound] mol dm^{-3}	10^2 Qunilonium chlorochromate Initial	10^2 Qunilonium chlorochromate Final	$\Delta 10^3$ Qunilonium chlorochromate mol dm^{-3}	Compound: Δ Qunilonium chlorochromate
$CH_3CH(OH)COOH$				
4.0	2.0	1.681	2.60	1.00 : 0.63
5.0	2.0	1.652	2.99	1.00 : 0.62
6.0	3.0	2.64	3.39	1.00 : 0.59
$C_6H_5CH(OH)COOH$				
4.0	2.0	1.71	2.51	1.00 : 0.64
5.0	2.0	1.70	2.99	1.00 : 0.62
6.0	3.0	2.65	3.39	1.00 : 0.56
$OHCH_2COOH$				
4.0	2.0	1.69	2.29	1.00 : 0.66
5.0	2.0	1.68	3.12	1.00 : 0.64
6.0	3.0	2.67	3.13	1.00 : 0.54
Cobalt(III) Lactato				
4.0	2.0	1.69	2.35	1.00 : 0.49
5.0	2.0	1.70	2.79	1.00 : 0.46
6.0	3.0	2.65	3.29	1.00 : 0.46

Cobalt (III) Mandelato				
4.0	2.0	1.79	1.89	1.00 : 0.48
5.0	2.0	1.78	2.29	1.00 : 0.46
6.0	3.0	2.71	2.60	1.00 : 0.43
Cobalt (III) Glycolato				
4.0	2.0	1.72	1.69	1.00 : 0.44
5.0	2.0	1.69	2.06	1.00 : 0.42
6.0	3.0	2.74	2.56	1.00 : 0.45

Table 2: Experimental Data of Stoichiometric data for Qunilonium chlorochromate oxidation of Cobalt (III) bound and unbound α-hydroxy acids in the presence Dodecyl ammonium chloride at 31°C.

Sulphuric acid = 1.00 mol dm^{-3}

Dodecyl ammonium chloride = 2.00 x 10^{-4}mol dm^{-3}

10^3 [Compound] mol dm^{-3}	10^2 Qunilonium chlorochromate Initial	10^2Qunilonium chlrochromate Final	$\Delta 10^3$ Qunilonium chlorochromate	Compound: Δ Qunilonium chlorochromate
$CH_3CH(OH)$ COOH				
4.0	2.0	1.73	2.71	1.00 : 0.67
5.0	2.0	1.68	3.21	1.00 : 0.64
6.0	3.0	2.56	3.61	1.00 : 0.61
$C_6H_5CH(OH)CO$ OH				
4.0	2.0	1.74	2.65	1.00 : 0.66
5.0	2.0	1.68	2.99	1.00 : 0.61
6.0	3.0	2.82	3.69	1.00 : 0.62
$OHCH_2COOH$				
4.0	2.0	1.72	2.58	1.00 : 0.64
5.0	2.0	1.68	3.01	1.00 : 0.62
6.0	3.0	2.59	3.32	1.00 : 0.57
Cobalt (III) Lactato				
4.0	2.0	1.74	2.47	1.00 : 0.41
5.0	2.0	1.70	2.85	1.00 : 0.43
6.0	3.0	2.65	3.21	1.00 : 0.42
Cobalt (III) Mandelato				
4.0	2.0	1.78	2.26	1.00 : 0.43
5.0	2.0	1.71	2.75	1.00 : 0.40
6.0	3.0	2.59	3.02	1.00 : 0.41

Cobalt (III) Glycolato				
4.0	2.0	1.81	1.78	1.00 : 0.42
5.0	2.0	1.77	2.24	1.00 : 0.41
6.0	3.0	2.67	2.32	1.00 : 0.43

Table 3: Experimental Data of Stoichiometric data for oxidation of Qunilonium chlorochromate oxidation of Cobalt(III) bound and unbound α- hydroxy acids in the presence polysorbate 80 at 31°c.
Sulphuric acid = 1.00 mol dm^{-3}
[Polysorbate 80] = 2.00 x 10^{-4}mol dm^{-3}

10^3[Compound] mol dm^{-3}	Qunilonium chlorochromate Initial	Qunilonium chlorochromate Final	Δ10^3 Qunilonium chlorochromate	Compound: Δ Qunilonium chlorochromate
$CH_3CH(OH)COOH$				
4.0	2.0	1.73	2.69	1.00 : 0.68
5.0	2.0	1.67	3.31	1.00 : 0.66
6.0	3.0	2.59	3.72	1.00 : 0.64
$C_6H_5CH(OH)COOH$				
4.0	2.0	1.73	2.59	1.00 : 0.65
5.0	2.0	1.67	3.02	1.00 : 0.59
6.0	3.0	2.59	3.54	1.00 : 0.64
$OHCH_2COOH$				
4.0	2.0	1.74	2.43	1.00 : 0.61
5.0	3.0	1.70	2.85	1.00 : 0.60
6.0	3.0	2.62	3.12	1.00 : 0.59
Cobalt(III) Lactato				
4.0	2.0	1.67	2.34	1.00 : 0.42
5.0	2.0	1.71	2.69	1.00 : 0.44
6.0	3.0	2.61	2.99	1.00 : 0.41
Cobalt(III) Mandelato				
4.0	2.0	1.79	1.93	1.00 : 0.45
5.0	2.0	1.74	2.31	1.00 : 0.43
6.0	3.0	2.67	2.54	1.00 : 0.44
Cobalt(III) Glycolato				
4.0	2.0	1.84	1.62	1.00 : 0.41
5.0	2.0	1.79	1.88	1.00 : 0.39
6.0	3.0	2.99	2.12	1.00 : 0.36

3. Results and Discussion

Study of kinetics in oxidation of pentaammine cobalt (III) complexes of lactic, mandelic and glycolic acids by Quinolinium chlorochromate.

Studying the reaction kinetics between Quinoliniumchlorochromate and cobalt (III) bound lactate (lactic, mandelic, and glycolic) acids in the presence of micellar medium at 31°C has been done.

3.1 Dependence on the micelles' content on Quinolinium chlorochromate.

As Quinolinium chlorochromate concentration increases, so does the rate at which lactic, mandelic, and glycolic acid Cobalt (III) complexes. The rate at which Quinolinium chlorochromate concentration changes in any given run is constant, and the graphs between logarithm of the concentration vs time are found linear (Fig.2).

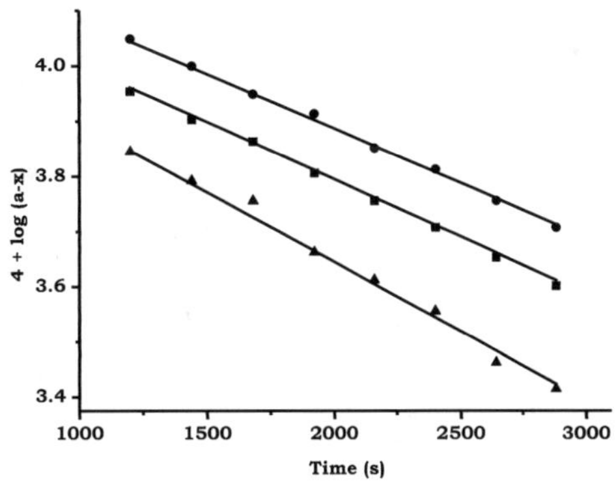

Figure 2: First order dependence plots.

■ Ammonium lauryl sulphate $K1 = 4.202 \times 10^{-4}\ s^{-1}$
● Dimethyl Dioctadecyl Ammonium Chloride $K1 = 4.158 \times 10^{-4}\ s^{-1}$
▲ Polysorbate 80 $K1 = 4.053 \times 10^{-4} s^{-1}$

A first order relationship on Quinolinium chlorochromate concentration may be inferred from the figure, which show that the particular rates determined which are in good agreement with the integrated rate equation. Specific rates remain unchanged even when Quinolinium chlorochromate concentration is increased by a factor of five in micellar medium with an initial concentration of Cobalt (III) complexes at 1.00×10^{-3} versus 5.00×10^{-3}mol dm^{-3} (Table 4.1, 4.2 & 4.3). $-\frac{d[Cr(VI)]}{dt} = K_1[Cr(VI)]$

Table 4: Kinetic studies of penta ammine Cobalt (III) complexes of withlactic, mandelic, and glycolic acidsat a concentration of 2 x10^{-2} moldm^{-3}of was carried out in acid medium(1 moldm^{-3}) and 2x10^{-3}moldm^{-3} concentration of Qunoliniumchlorochromate in different surfactants at 31^{0}C

Table 4.1

Time (s)	Ammonium lauryl sulphate	
	10^3(a-x) mol dm^{-3}	10^4k_1 (s^{-1})
240	1.44	4.1583
480	1.15	4.1584
720	1.12	4.1586
960	1.01	4.1587
1200	0.90	4.1589
1440	0.80	4.1580
1680	0.73	4.1582
1920	0.64	4.1586
2160	0.57	4.1589
2400	0.51	4.1584
2640	0.45	4.1588
2880	0.40	4.1581

Table 4.2

Time (s)	Dimethyl dodecyl ammonium chloride	
	10^3(a-x) mol dm^{-3}	10^4k_1 (s^{-1})
240	1.59	4.2013
480	1.49	4.2015
720	1.26	4.2018
960	1.16	4.2019
1200	1.12	4.2021
1440	1.00	4.2013
1680	0.89	4.2016
1920	0.82	4.2014
2160	0.71	4.2017
2400	0.65	4.2022
2640	0.57	4.2019
2880	0.51	4.2020

Table 4.3

Time (s)	Polysorbate 80	
	10^3(a-x) mol dm^{-3}	10^4k_1 (s^{-1})
240	1.13	4.0530
480	0.96	4.0532
720	0.84	4.0534
960	0.77	4.0533
1200	0.70	4.0539
1440	0.62	4.0535
1680	0.57	4.0531
1920	0.46	4.0536
2160	0.41	4.0537
2400	0.36	4.0539
2640	0.29	4.0532
2880	0.26	4.0534

The rate of Quinolinium chlorochromate oxidation on Cobalt (III) bound lactic, mandelic and glycolic acids in Ammonium lauryl sulphate medium are found to be:
Cobalt (III) Lactato(1.92)>Cobalt(III) Mandelato (1.56)> Cobalt(III) Glycolato(0.86).
Similarly, the rate of Cr (VI) oxidation in Dimethyl dioctadecyl ammonium chloride medium are
Cobalt (III) Lactato(3.03)> Cobalt (III) Mandelato(2.46)> Cobalt (III) Glycolato(1.36).
Similarly, the in Polysorbate 80 medium are Cobalt (III) Lactato(1.18)> Cobalt (III) Mandelato(0.99)> Cobalt (III) Glycolato (0.54).
An accelerated rate of C-H decomposition results in an increase in oxidation rate for the complexes Cobalt (III) lactato and Glycolic acid (Cobalt (III) mandelato and Glycolic acid) if the reaction is carried out via chromate ester. Mandelato complex can also sterically inhibit the production of chromate ester. The lactato complex's higher reactivity may be due to the methyl proton's increased acidity and lack of steric retardation.

3.2 Dependence of rate on lactic, mandelic and glycolic acids concentration

The effect of adjusting the concentration of unbound lactic, mandelic, and glycolic acids in micellar medium has been examined in the range of 1.00 x 10^{-3} to 8.00 x 10^{-3}mol dm-3. Unbound lactic, mandelic, and glycolic acids increase the rate of Quinolinium chlorochromate oxidation with increasing substrate concentrations (Table 5.1, 5.2 & 5.3).

Table 5: Kinetic studies of penta ammine Cobalt (III) complexes of lactic, mandelic and glycolic acids with concentration 2 x10^{-2} moldm^{-3} concentration was carried out in acid medium (1 moldm^{-3}) and 2x10^{-3}moldm^{-3} concentration of Quniloniumchlorochromate in different surfactants at 31^{0}C.

Table 5.1						
10^2[$(NH_3)_5Co^{III}L$] mol dm-3	Ammonium lauryl sulphate		Dimethyl Dioctdecyl Ammonium Chloride		Polysorbate 80	
	10^4K_1 (s^{-1})	10^2k_2 dm^3 mol^{-1} s^{-1}	10^4 k_1 (s^{-1})	10^2k_2 dm^3 mol^{-1} s^{-1}	10^4 k_1 (s^{-1})	10^2k_2 dm^3 mol^{-1} s^{-1}
L = Lactic acid						
1.00	1.59	1.59	2.02	2.02	0.80	0.80
2.00	2.24	1.59	2.70	2.02	1.13	0.80
3.00	3.10	1.59	3.91	2.02	1.39	0.80
4.00	3.90	1.59	5.02	2.02	2.05	0.80
5.00	5.50	1.59	6.50	2.02	2.58	0.80
Table 5.2						
10^2[$(NH_3)_5Co^{III}L$] mol dm-3	Ammonium lauryl sulphate		DimethylDioctdecylAmmoniumChloride		Polysorbate 80	
	10^4K_1 (s^{-1})	10^2k_2 dm^3 mol^{-1} s^{-1}	10^4K_1 (s^{-1})	10^2k_2 dm^3 mol^{-1} s^{-1}	10^4K_1 (s^{-1})	10^2k_2 dm^3 mol^{-1} s^{-1}
L = Mandelic acid						
1.00	1.13	1.13	1.55	1.55	0.64	0.64
2.00	1.60	1.13	2.05	1.55	0.82	0.64
3.00	2.05	1.13	2.76	1.55	1.05	0.64
4.00	2.70	1.13	3.64	1.55	1.55	0.64

5.00	3.72	1.13	5.02	1.55	2.05	0.64
Table 5.3						
$10^2[(NH_3)_5Co^{III}L]$ mol dm-3	Ammonium lauryl sulphate		DimethylDioctdecylAmmoniumChloride		Polysorbate 80	
	10^4K_1 (s^{-1})	10^2k_2 dm^3 mol^{-1} s^{-1}	10^4K_1 (s^{-1})	10^2k_2 dm^3 mol^{-1} s^{-1}	10^4K_1 (s^{-1})	10^2k_2 dm^3 mol^{-1} s^{-1}
L = Glycolic acid						
1.00	0.80	0.80	1.33	1.33	0.40	0.40
2.00	1.05	0.80	1.66	1.33	0.52	0.40
3.00	1.33	0.80	2.24	1.33	0.65	0.40
4.00	1.87	0.80	3.02	1.33	0.92	0.40
5.00	2.52	0.80	4.03	1.33	1.30	0.40

A graph of the specific rate k2 vs concentration of lactic, mandelic and glycolic acids shows that the rate law for the Quinolinium chlorochromate oxidation of unbound lactic, mandelic and glycolic acids is linear with a slope roughly equal to unity (Fig. 2 to 4). $-\frac{d[Cr(VI)]}{dt} = K_1[Cr(VI)] \times [\alpha\text{-Hydroxy acid}]$

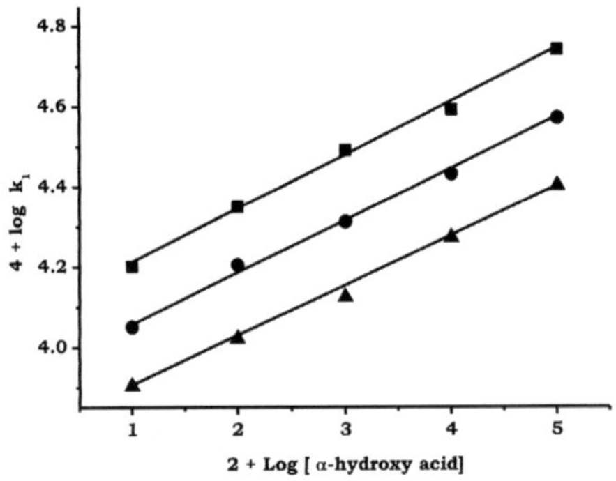

Figure 3: Dependence of rate on [α-hydroxy acid] in Ammonium lauryl sulphate

■ Ammonium lauryl sulphate Slope = 0.7982
● Dimethyl Dioctadecyl Ammonium Chloride Slope = 0.9575
▲ Polysorbate 80 Slope = 1.1857

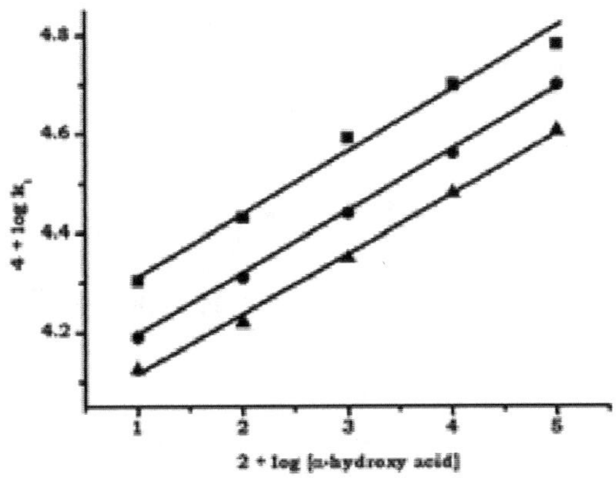

Figure 4: Dependence of rate on [α-hydroxy acid] in Dimethyl Dioctadecyl Ammonium Chloride

■ Ammonium lauryl sulphate Slope = 0.8694
● Dimethyl Dioctadecyl Ammonium Chloride Slope = 0.9623
▲ Polysorbate 80 Slope = 1.0392

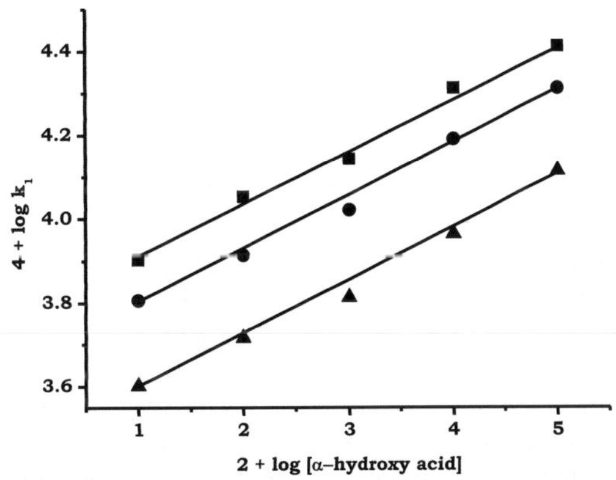

Figure 5: Dependence of rate on [α-hydroxy acid] in Polysorbate 80

■ Ammonium lauryl sulphate Slope = 1.3506
● Dimethyl Dioctadecyl Ammonium Chloride Slope = 1.3546
▲ Polysorbate 80 Slope = 1.2073

Conclusions

An analogous oxidant was previously used to induce electron transfer processes, mostly in Ce (IV) and pentaammine Cobalt (III) complexes. Because reaction routes are divided, the percentage of Co (III) produced varies. In the presence of micelles, QCC is discovered in the pentaammine Cobalt (III) complex of α-hydroxy acid. QCC showed an inducible electron transfer process. ALS, DDAC, Polysorbate 80. The α-hydroxy acid bound to Cobalt (III) is oxidised faster than the unbound ligand,

and the Cobalt (III) centre is reduced by 100%, which is worthy of noting and scope for further investigations as it is useful for controlling such oxidations. Co (III) complexes on micellar medium undergo QCC oxidation via C-C bond fission. The oxidation rate of oxidised QCC is higher in micellar DDAC than in the other two micelles.

Acknowledgements

The authors thank the authorities of Islamiah College, Vaniyambadi, Tamil Nadu, India-655752 to facilitate for all research work. The support of principal and management DNR College, Bhimavaram and colleagues of chemistry department for their support and encouragement in the tenure of research work is commendable

References

1. A. Padurangan, S. Kavitha and I. Alphonse, (2005)*Indian J Chem.*, **44A**, 715
2. S. Udhayavani and K. Subramani 2012Mn(III)acetate-Induced Electron Transfer in Pentaammine Cobalt(III)Complexes of α-Amino Acids in Micellar Medium *E-Journal of Chemistry*, **9(2),** 912.
3. R. Gurumurthy, M. Gopalakrishnan, B. Karthikeyan and M. Selvaraju1998 Kinetics and mechanism of oxidation of some para-, meta-, and ortho- substituted ethyl S-phenylmercaptoacetates by chloramine- B(CAB)*Asian J Chem.,* **10(3),** 476.
4. N. K. Mohanty and P. K. Nanda 1982Kinetics of copper ion catalyzed peroxydisulphate oxidation of glyoxylate*Indian J Chem.,* **21A**, 52.
5. N. K. Saran and R.C. Acharya 1985 Synthesis and characterization of some new mixed ligand complexes of ruthenium(II)*Indian Chem Soc.*, **62**, 747.
6. K. R. Sankaran and C. Anbuselvan 1998 Kinetic Study Of the Oxidation of Pentaamminecobalt(III) Complexes of NLGУ-hydroxy acids by Imidazolium Dichromate.*Asian J Chem.*, **10**, 806.
7. B. Ramkumar 2003Synthesis and Characterization of Some New Mixed Ligand Complexes of Ruthenium(II) with 2,2'-Bipyridine and 1,10-Phenanthroline *Afinidad*, **60**, 257
8. A. MohmedFarook 2007 Kinetics of Oxidation of 4-Oxoacids by N-Chlorosaccharin in aqueous acetic acid medium *J Sol. Chem.*, **36**, 345.
9. K. Subramian and S. Udhayavani 2012 Kinetic Study Of Oxidation Of Pentaamminecobalt (III) Complexes Of A-Amino Acids By Mn (III) Acetate, Mn (IV) Heteropolyanion And Mn (III) Perchlorate In Micellar Medium*Journal of Chem. Bio and Physciences.*, **2**, 551.
10. B. Anandaratchagan, Md. Nawaz and K. Subramani 2011Induced Electron Transfer Reactions in Pentaammine Cobalt (III) Complexes of Alpha – Hydroxy Acids by Quinolinium Dichromate in Acetic Acid (70%) Medium: A Kinetic Study*Acta. Chim. Pharm India.*, **1**, 44.
11. A. Thaminum Ansari 2009 Pyridiniumchloro chromate – Induced electon transfer in Pentaammine Cobalt (III) complexes of ∝ -hydroxy acids in micellar medium*Int. Journal of Chem Tech Research.*, **1**, 308.
12. Z. Abdul Vaheith and Mansur Ahmed 2015 Nanocellulose: From Fundamentals to Advanced Applications*Int. J of Frontiers in Science and Technology.,* **3**, 49.
13. K.G. Sekar and R.V. Sakthivel2013d-Glucose as green ligand for selective copper-catalyzed phenol synthesis *J Solution Chem.,* **42**, 1748.

14. P. Kannan, S. Gopalakrishnana and B. Befekadu 2015Synthesis, single crystal growth, characterization and comparison of two new enone shifted chalcones and their NLO behaviour*Int. J Chem. Sci.,* **13**, 759.
15. K. Anji Reddy, V. SeethaRamaiahand A Mansur, *IJARST*, **6**, 707 (2017).
16. A. Palanivel, P. Rajkumarand and K. Subramani 2016 Phytochemical Screening, GC-MS Analysis and Antioxidant Activity of Marine Algae Chlorococcumhumicola*International Journal of Chemistry and Pharmaceutical Sciences*, **4**, 534.
17. A. Palanivel, P. Rajkumarand and K. Subramani 2017 Phytochemical Screening, GC-MS Analysis and Antioxidant Activity of Marine Algae Chlorococcumhumicola*International Journal of Chemical Studies*, **5**, 83.
18. A. Palanivel, P. Rajkumar, T. Palanisamyand and R. Dayanandhan, *International Journal of Frontiers in Science and Technology*, **4**, 22 (2016).
19. C. Anbuselvanand and K. R. Sankaran 1998 2,2′-Bipyridinium chromate oxidation of cobalt(III)-bound and unbound α-hydroxy acids*Oxidation Communications*, **21**, 257.
20. K. Subramian and S. Udhayavani 2012 Kinetic Study of Oxidation of Pentaamminecobalt (III) Complexes of α-Amino Acids by Mn (III) Acetate, Mn (IV) Heteropolyanion and Mn (III) Perchlorate in Micellar Medium *Journal of Chem. Bio and Phy sciences.*, **2**, 551.
21. Z. Abdul Vaheith and M. Ahmed 2015Oxidation of cobalt (III) complexes of α-hydroxy acids by piperazinium dichromate in surfactants*Int. J. of Frontiers in Science and Technology.*, **3**, 49.
22. K. Pownsamy, S. Gopalakrishnana and B. Befekadu 2015 Adsorption of Methylene Blue Dye from Aqueous Solution Using Water Hyacinth Root Powder as a Low Cost Adsorbent *Int. J Chem. Sci.,* **13**, 759.
23. N. K. Mohantyand and R.K. Nanda 1982 Kinetics of Acid and Base Catalysed Hydrolysis of cis-a- and cis-β 2-[Co(trien)(aniline)Cl] 2+ Cations*Transtion Met. Chem.,* **7**, 202.
24. R. Gurumurthy, B, Karthikeyan and M. Selvaraju 1998 Reactivity OfCyclanols Towards QuinoliniumChlorochromate Oxidation*Oriental J. of Chem.*, **14**.
25. Udhayavani and K. Subramani 2012 Kinetic Study of Oxidation of Pentaamminecobalt (III) Complexes of α-Amino Acids by Mn (III) Acetate, Mn (IV) Heteropolyanion and Mn (III) Perchlorate in Micellar Medium*Acta.Chem Pharma. Indica*, **2**, 213.
26. B. MohammedNawaz, K.Subramani and A. NazimAhmed 2012 Synthesis, Characterization and Antimicrobial Activity of Some Novel Surfactant-Copper(II) Complexes*J. Chem. Bio. Phy. Sci., Sec'A'*, **2**, 668.
27. F. Basalo, and R.G. Pearson 1967 *Mechanisms of Inorganic Reactions*(2nd Ed., Wiley Eastern Ltd.).
28. A. G. Sykes 1966 *Kinetics of Inorganic Reactions*(Pergarmon, Newyork).
29. H. Taube 1970 "*Electron Transfer Reactions of complex Ions in Solution*(Academic Process New York).
30. Tang SC, Yang JH 2018 The title of the article or work is not included in the information you provided*Molecules***10**;23(4):863.
31. Elsevier Hand book, 1996 Comprehensive Analytical Chemistry, Chapter 1 Behaviour and properties of amphiphilic molecules and their assemblies,**31,** 1-91.
32. S.J. Holder, N. A. J. M. Sommerdijk 2011 The title of the article is not included in the information provided*Polym. Chem.,***2**, 1018.

33. Y. Zhang, X. Zeng; S. Cheng, X. Yu, A. Tian1999 Micellar Catalysis Of Composite Reactions I Micellar Effect On The Consecutive First Order Reaction *Journal of Dispersion Science and Technology.***20** (3): 1009–1024.
34. Davidson P, Penisson C, Constantin D, Gabriel JP 2018.Isotropic, Nematic, And Lamellar Phases In Colloidal Suspensions Of Nanosheets*Proc Natl AcadSci U S A*. **26**;115(26):6662-6667.
35. B.Samiey, C.H. Cheng,J. Wu 2014 Effects of Surfactants on the Rate of Chemical Reactions*Journal of Chemistry,***2014**, Article ID 908476.

Investigations on Microstructure and Mechanical Properties of HAp Nanocomposites reinforced with ZrO_2 and TiO_2

Dr.V **Ajay Kumar**[1, 3,5], Dr.P. Rama Murty Raju[2,4] **Dr.R Bhargavi**[1, 3]Dr.S **Madhusudan**[2, 3], **and Dr. S Rajesh**[2, 4]

[1]Associate Professor, Department of Mechanical Engineering.
[2]Professor, Department of Mechanical Engineering.
[3]Usharama College of Engineering and Technology, Telaprolu, Andhra Pradesh, India.
[4]SRKR College of Engineering, Bhimavaram, Andhra Pradesh, India.
[5]Email: ajayvijaya04@gmail.com.

Abstract. In the contemporary context, approximately 15 million individuals in India grapple with bone-related issues stemming from accidents, fractures, bone mass depletion, and the natural aging process. To address this challenge, a diverse array of biomaterials has been innovated to rectify or enhance bone deficiencies within the human body. Nevertheless, over the past decade, calcium phosphate-based bio-ceramic materials have gained significant prominence as synthetic alternatives for bone and dental applications in the fields of orthopaedics and dentistry. One such bioactive ceramic, Hydroxyapatite (HAp), boasts exceptional versatility and can find utility in both dental and orthopaedic contexts. In the current research, exhaustive research was conducted to explore the impact of HAp/ (ZrO_2+TiO_2) hybrid nano-composites synthesized through a mechanical synthesis process employing high-energy ball milling (HEBM), incorporating varying weight proportions. The resulting "HAp/ (ZrO_2+TiO_2) nano-composites" were subjected to characterization through scanning electron microscopy combined with energy-dispersive X-ray spectroscopy (SEM-EDX) and X-ray diffraction (XRD). Furthermore, the study assessed the influence of these nano-composites regarding the HAp nano-composites' mechanical characteristics.

1. Introduction

Hydroxyapatite ($Ca_{10}(PO_4)_6(OH)_2$, HAp) emerges as a pivotal material in the realm of medical applications, particularly in the context of human health. Its chemical composition closely mirrors that of the solid tissues found in humans, such as bones and teeth. Existing literature underscores the outstanding biocompatibility and bioactivity exhibited by HAp. HAp actively contributes to the generation of fresh bone tissues. Nevertheless, despite its commendable biocompatibility traits, HAp faces constraints due to its inherent brittleness and limited strength. The incorporation of robust reinforcing agents has been instrumental in significantly enhancing the mechanical attributes of ceramics. These reinforcing agents have been found to possess higher strength and elastic modulus when compared to the matrix material employed in ceramic-matrix composite constructions. One approach for augmenting the mechanical properties of HAp is to reinforce it with other ceramics. The strength of the secondary phase in a ceramic-matrix composite must be greater than the matrix material's for such a reinforcement to be effective. One promising avenue involves the creation of bio-ceramic composites, wherein a combination of HAp, TiO_2, and ZrO_2 is employed to yield enhancements in both mechanical properties and bioactivity. Zyman et al [1] have noted that the inclusion of robust reinforcing agents significantly improves mechanical properties. Xihua et al [2] have produced cost-effective, high-performance HAp matrix ceramic composites and investigated the impact of alumina additions to HAp. Lee et al [3] have identified that ZrO_2/Al_2O_3, in combination, exhibit elevated biocompatibility, oxidation resistance, and wear resistance, serving as both matrix and reinforcement phases.Shi et al [4] have harnessed ZrO_2/La_2O_3 binders to enhance the bending and tensile strength of HAp, while Aminzare et al [5] have explored nano-composites to achieve highly dense structures with the requisite mechanical properties. These authors have also delved into the impact of titania and alumina nano particles on mechanical properties, microstructure, and phase breakdown. Viswanath and Ravishankar [6] have conducted a detailed investigation into interfacial reactions within HAp/alumina nanocomposites, revealing complete alumina interaction with HAp.Mobasherpour et al [7] have chosen to create nanocrystalline HAp by precipitation and scrutinized the effects of ZrO_2/Al_2O_3 on the mechanical characteristics of HAp powder. Lee et al [8] have elucidated a multi-pass extrusion process aimed at enhancing the mechanical attributes of HAp-(t-ZrO_2/Al_2O_3-(m-ZrO_2) composites. Khalil and Almajid [9] have explored Mg/HAp nano-composites reinforced with grains that are around 40 nm in size and sintering under high-frequency induction heat conditions. This research has revealed micro hardness and compressive strength values of about 60 HV and 194.5 MPa, respectively. Gain et al. [10] have manufactured HAp-ZrO_2 nanocomposites via powder metallurgy, incorporating polymethyl methacrylate powders as pore-forming agents, and subsequently examined their characteristics. The study has highlighted the substantial influence of nanocomposite microstructure on mechanical strength. Rapacz-Kmita et al. [11] have probed the mechanical properties of Zirconia combined with HAp, discerning superior mechanical properties within needle-grained structures across a range of temperatures.Choi et al. [12] have generated nanoparticles of Al_2O_3 and Ni_3Al and reinforced them with HAp composites., revealing marginal enhancements in fracture toughness and fracture strength with the incorporation of reinforced elements. Morks et al. [13] have executed HA/SiO2/Ti plasma spray coatings on SUS 304 substrates using a Gas-Tunnel Plasma Torch. Their findings have highlighted the notable impact of SiO_2 and TiO_2 reinforcements on microstructure, mechanical properties, and corrosion resistance.Against this backdrop, the present study focuses on the utilization of Hydroxyapatite (HAp) nano-powders with an average particle size ranging from 20 to 40 nm. The chemical composition and structure of Hydroxyapatite (HAp) align closely with that of human hard tissues, specifically bones and teeth, underscoring its biocompatibility and bioactivity. Despite HAp's remarkable biocompatibility, its utility is limited due to its inherent weakness and brittleness. Therefore, it is imperative to reinforce HAp mechanically for long-term performance by combining it with other ceramics and metals, such as Zirconia(ZrO_2)andTitania(TiO_2). Furthermore, AZ31's enhanced ductility enables it to sustain impact and deformations throughout the forming process, hence preventing any disastrous failure. Reinforcement can be incorporated into AZ31 to strengthen its mechanical properties and increase the range of potential uses. The use of reinforcements, like as fibers and nanoparticles, has the potential to significantly enhance certain alloy characteristics. According to Zhang et al. [14]; Ye et al. [15], this enhancement may lead to increased wear characteristics, high tensile strength, and higher fatigue resilience. By carefully adjusting the composition and microstructure of AZ31, its performance can be tailored to meet the requirements of any given application (Singh Arora et al., [16]).

Here, TiO2 and HAp represent viable options for reinforcement in bio-medical applications due to their biocompatible characteristics. TiO2, a ceramic, is well known for its exceptional strength, wear resistance, and potential to improve corrosion resistance. Because of its similar qualities to bone material, HAp, a biocompatible material represented by the formula Ca10(PO4)6(OH)2, has found use as a replacement for natural bone (Mehta and Singh, [17]; Singh et al., [18]).

The study employs Zirconia particles with an average size of 80-100 nm, benefiting from their superior chemical and thermal stability, wear resistance, high strength, fracture toughness, and bio-inertness. Additionally, Titania particles with an average size of 80-100 nm are utilized, as Titanium and its alloys have gained tungsten carbide widespread adoption as load-bearing implants due to their superior mechanical properties, biocompatibility, and corrosion resistance. In this research, HAp particles serve as the material, while ZrO_2 and TiO_2 particles are introduced as reinforcements to craft hybrid nano-composites with varying weight percentages (5, 10, 15, 20, and 25 wt%).

2. Experimentation

2.1 Sample Preparation

Hydroxyapatite (HAp) and its composite powders with varying weight percentages of ZrO_2+TiO_2 (0%, 5%, 10%, 15%, 20%, and 25% by weight) were amalgamated to fabricate HAp composites. These composites were crafted by reinforcing HAp particles with ZrO_2, TiO_2, or a combination of both using the High Energy Ball Milling (HEBM) technique. In employing tungsten carbide-hardened steel balls of specified size for duration of 1 hour at a rotational speed of 300 revolutions per minute (rpm). The ratio of balls to powders was maintained at 20:1 during the milling process. Subsequently, the milled powders were compacted at 100 bar pressure, with a settling time of 150 seconds, followed by at a sintering temperature of 1200°C for 1 hour. From these compacted powders, cylindrical specimens were fashioned for the assessment of compression strength and hardness. These cylindrical specimens measured 20 mm in diameter and 5 mm in height. Additionally, plain specimens were produced to evaluate flexural strength and fracture toughness. These plain specimens had dimensions of 40mm in length, 10mm width and 3mm thickness. After that, the specimens were heated inside the furnace until the desired temperature was attained. The instant the desired temperature was reached was used to calculate the heating time. After that, the furnace was left to cool naturally for all of the sintered samples.

2.2 Experimental procedure

To differentiate between HAp and its composites, an X-ray diffractometer (XRD) outfitted with Cu-Kα radiation (wave length λ=1.5406 Å0) is utilized.This test is run in the 6-90° range with a step size of 0.010, following the width (β) of their diffraction pattern. Scherer's formula [19] was used to estimate the average size (D). SEM (Scanning Electron Microscopy) provided the microstructural details.The elements Ca, P, O, and Zr are analyzed using FE-SEM-attached EDAX (Energy Dispersive Analysis of X-rays). On a UTM, the compressive strength of nano composite ceramics with dimensions of 20 mm in diameter and 5 mm in height was measured. Using distilled water and the Archimedes technique, the density of sintered samples was determined (relative density).. The machine crosshead speed ($\vartheta = \varepsilon^*.h_o$) is 0.005 mm/s, and the test was conducted under quasi-static loading conditions with a fixed initial strain rate (ε^*) of 0.001 s^{-1}. Additionally, h_o is the specimen's initial height, at which point specimens that are loaded to up to fracture [19].
The polished HAp composites that were examined using the Shimadzu HMV-2000 micro hardness tester are the foundation for evaluating the hardness test (Knoop's hardness indenter). As per ASTM E384, the specimen is subjected to a load of 4.905 N and is held for 15 seconds during the measurement process. The indenter's width to length ratio is 1:7.11. The face angle of the concavity or indention is 130 degrees for the breadth and 172 degrees for the rectilinear edge, respectively, and it is approximately one-third the length of the longer dimension...

By Eq. (1) is possible to ascertain the Knoop's hardness ('H') as below [20]

$$H = 14229\frac{L}{d^2} \qquad (1)$$

The longer diagonal of the indentation relates to "d" (mm), while the load is denoted by "L" (Newton). The following Eq. (2), which is provided below, was used to determine the Young's modulus along the composites. [20]:

$$\frac{B'}{A'} = \frac{B}{A} - \alpha\frac{E}{H} \qquad (2)$$

The indent diagonal ratio is (B^1/A^1), while the indenter dimensions ratio is (B/A). The constant denoted by α has a value of 0.45.

The flexural strength of the nano composite is ascertained by means of the three-point bending test. In order to smooth out surface imperfections, the HAp induced composite specimens (40 mm length, 10 mm width, and 3 mm thickness) were first manufactured and refined. They were then placed on a three-point bend test setup. The load (frame cross head speed of 0.05 mm/min) was put from the top at middle of the bar. Through the indentation technique, the Fracture Toughness K_c(Eq: 3) was calculated, and the values were confirmed through the equation below [21]:

$$K_C = \xi\left[\frac{E}{H}\right]^{2/5}\frac{P}{d\times l^{1/2}} \qquad (3)$$

'K_c' gives Fracture Toughness; 'E' refers to Young's Modulus (GPa), and 'H' gives the Hardness (GPa). 'P' gives the applied load (N), 'ξ' shows the constant which is based on the geometry of the indenter (ξ=0.008).

3. Results &Discussion

3.1 X-Ray Diffraction Analysis of HAp/(ZrO_2+TiO_2) (1:1) Powders

Figure 1shows the XRD patterns of HAp/(ZrO_2+TiO_2) nanocomposite powders prepared using high energy ball milling. XRD phase analysis of HAp composites reveals the presences of ZrO_2 and TiO_2 particles. HAp particles with higher intensity peaks can be observed at 2θ=32.6°. HAp intensity and the bandwidth is increased with the increase of (ZrO_2+TiO_2) reinforcement percentage at 2θ=34.8°. It may be due to the (ZrO_2+TiO_2) which act as an effective grain growth inhibitor to the HAp grains. Additional reflections were not detected from impurities. It can further infer that HAp peak intensities are much stronger and improve the crystalinity compared to pure HAp.

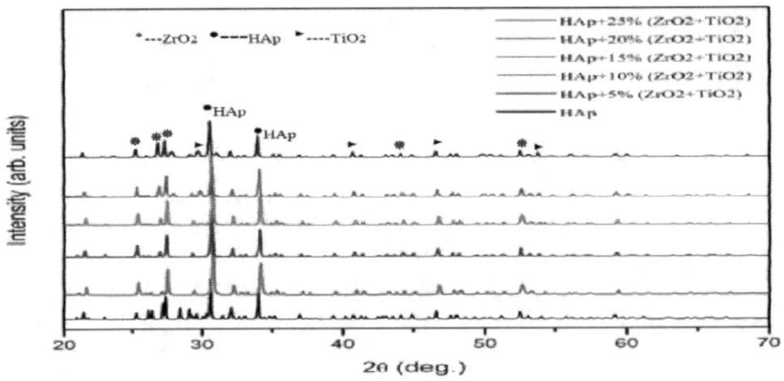

Figure.1 X-ray diffraction patterns of HAp/(ZrO_2+TiO_2)(1:1) Nano-composite powders

3.2 Micro structure analysis of HAp/(ZrO_2+TiO_2)(1:1) powders

The SEM micrographs of the HAp composites containing 5, 10, 15, 20, and 25 weight percent (ZrO_2+TiO_2) are displayed in Figure. 2(a)–(f), respectively. The addition of (ZrO_2+TiO_2) to the pure HAp powders seems to have reduced the size of the grains. The homogeneity of mixing is determined by the sizes of the unbroken clusters during mixing as well as the particle sizes since the two types of particles are blended. The HAp with addition of (ZrO_2+TiO_2) powder particles, results in reduction of the agglomeration of the particles. It may be due to continuous cold welding due to synthesis process of HAp/ (ZrO_2+TiO_2) composites using high energy ball milling. This reduction of agglomeration adversely affects the packing efficiency of the mixed powders and in turn affecting the densification and the homogeneity of composites HAp / (ZrO_2+TiO_2).

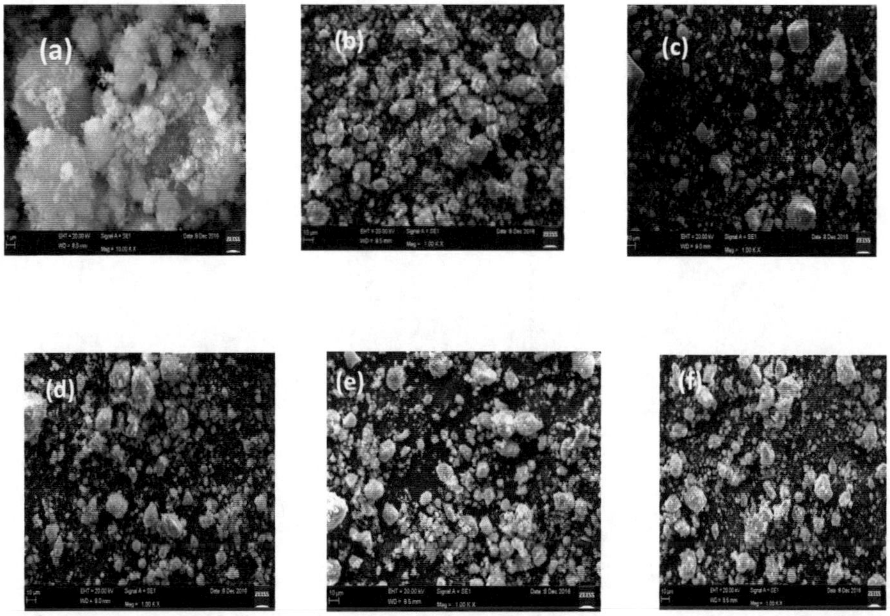

Figure. 2 SEM micrographs after ball milling (a) HAp (b) HAp /5 wt% (ZrO_2+ TiO_2) (c) HAp/10 wt% (ZrO_2+ TiO_2) (d) HAp/15 wt% (ZrO_2+ TiO_2) (e) HAp/20 wt% (ZrO_2+ TiO_2) (f) HAp/25 wt% (ZrO_2+ TiO_2)

3.3. X-Ray Diffraction Analysis of HAp/(ZrO_2 + TiO_2)(1:1) after Sintering

Figure. 3 presents XRD patterns of HAp and its composites (ZrO_2+TiO_2). The reflections are matched with the HAp phase and its composites (ZrO_2+TiO_2). No additional reflections are observed due to the presence of impurities. It is observed that the combination of (ZrO_2+TiO_2) is found to be more structural stability with an increase in the diffraction pattern intensities due to the increase of (ZrO_2+TiO_2) content of composites. Zirconia and HAp are shown with strong tricalcium phosphate (TCP) peaks. The increased weight percentage of ZrO_2 particles along with the excessive interaction between HAp and ZrO_2 may be the cause of TCP production. When TCP structure biomaterials are inserted into the body, the body produces new bone tissue more quickly because of their exceptional osteoconductivity and biocompatibility. (Colon et al [22]). It has been proven that ZrO_2 is dominant on the addition of (ZrO_2+TiO_2) (Miao et al [23]) using EDX and XRD, identification of the protruding crystals is carried out. High Zr and Ti intensities were confirmed by EDX in the projecting crystals

(Figure. 3), while strong Zr but weak Ti intensities were seen in the background grains (Figure. 5). Therefore, it is possible to classify the background grains as zirconia phase and the projecting crystals as TCP phase. But EDX can't pinpoint the exact stages that are now in place. For this reason, phases are analyzed using XRD. Figure 3 shows the XRD patterns of the 25 wt% (ZrO_2+TiO_2) composites sintered at 1200° C, only rutile TiO_2 and zirconia phases are observed. The addition of TiO_2 and the sintering temperature also affected the grain size. It is clear that the addition of TiO_2 and ZrO_2 up to 20 weight percent resulted in a decrease in particle size for the background grains (mostly ZrO_2). Nevertheless, additional addition of ZrO_2 and TiO_2 (25 weight percent) led to an increase in matrix particle sizes, which might be because the matrix grains grew more quickly. The enhanced rate of grain growth of TiO_2 particles and the limited solubility of TiO_2 in ZrO_2 may be the cause of the increase in grain size with TiO_2 and ZrO_2 addition. Additionally, ZrO_2 is highly doped with HAp, which may have increased the diffusion rate in view of TCP formation and, consequently, the rate of grain growth. Using the Debye Scherer equation, the average crystalline size (D) of HAp/ (ZrO_2+TiO_2) nanocomposites was calculated.[19]

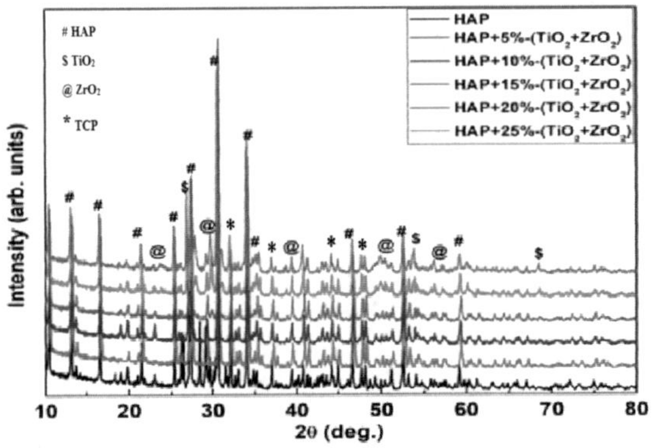

Figure 3 XRD pattern of HAp/(ZrO_2+TiO_2)

3.4. Micro Structure Analysis of HAp/(ZrO_2+TiO_2) Nanocomposite after Sintering

Figure 4(a)-(e) shows that SEM micrographs of (ZrO_2+TiO_2) composites with 5, 10, 15, 20, 25 wt% respectively. High densification or low residual porosity is evident in the SEM micrographs of the HAp sintered composites. In addition to variations in densification, there has been a notable distinction in the compound formation between ZrO_2 and TiO_2 in hybrid composites (HAp/ (ZrO_2+TiO_2).Protruding crystals structure (TCP) compound is observed in Figure. 4(e).

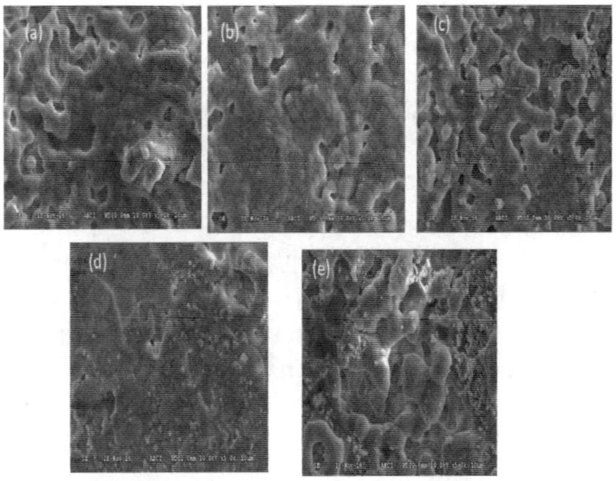

Figure 4 SEM micrographs of HAp/(ZrO_2+TiO_2) composites (a) 5 wt% (b) 10 wt% (c) 15 wt % (d) 20 wt% and (e) 25 wt%

SEM and XRD analysis confirms the above statement. The changes in the micro structural properties due to Zr can be dominating the Ti beyond the 20 wt% of reinforcement content. The composition of Ca, P, Zr and Ti is studied using EDX as shown in Figure 5 shows the HAp specimens after testing. The tested specimens for grain size determination of HAp composites materials is as shown in Figure 6. Using the ASTM line method, the grain size of the composites are determined [23]. The ASTM line method was utilized to evaluate the grain size of HAp composites. It is noticed that the grain size reduces gradually with increasing the reinforcement content in HAp/ (ZrO_2 + TiO_2) composites up to 20wt% (15.9nm), after which an increase in grain size is observed.

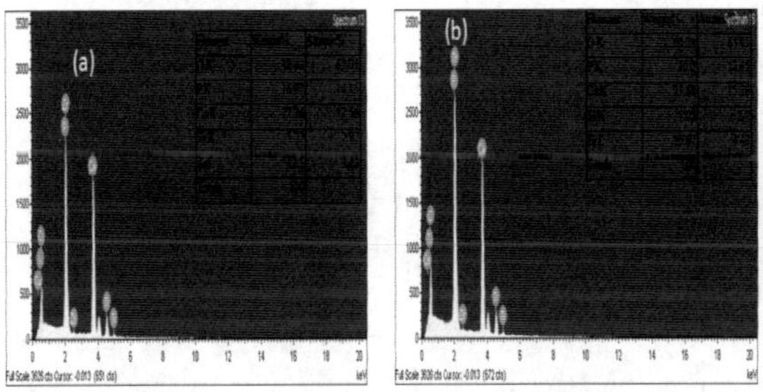

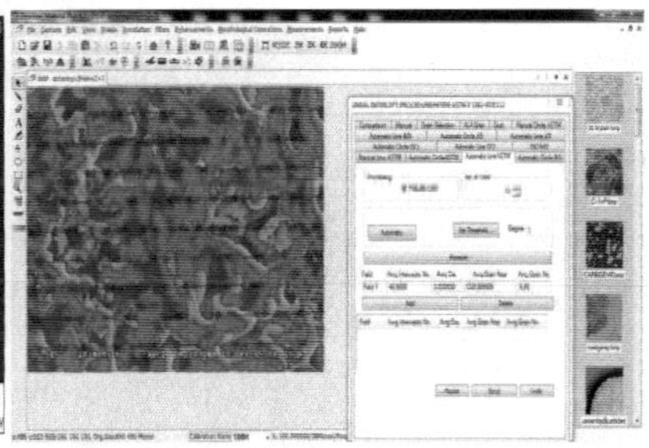

Figure 5.EDX(a)10wt%(b)20wt% HAp/(ZrO_2+TiO_2) Figure6Grain size at 10 wt% HAp/(ZrO_2+TiO_2)

3.5MECHANICAL PROPERTIES

3.5.1EFFECT OF (ZrO_2+ TiO_2) (1:1) CONTENT ON HAp NANOCOMPOSITES

The apatite crystallites seen in living bone tissues are similar in crystallographic structure to hydroxyapatite (HAp). It is being given further consideration as a material for bone implants to encourage orthopedic prosthesis attachment more quickly. The applied load in different clinical applications was acutely restricted due to the observed susceptibility of sintered HAp to fatigue failure. Reinforcing the HAp with metal substrates is one way to address the issue. This configuration ensures that implants are biocompatible because HAp is present, and the metal substrate adds to the mechanical characteristics.

3.5.1.1 Porosity and Density

In Figure 7, the measured porosities and density of HAp/(ZrO_2+TiO_2) composites with (ZrO_2+TiO_2) content are displayed. The samples with the highest densities within comparable sets of composition are represented by each sample that is listed. HAp's densification occurred after sintered for one hour at 1200°C and (ZrO_2+TiO_2) (5, 10, 15, 20, and 25 wt%) samples matched with the corresponding theoretical density. Up to 20 weight percent, the density was observed to rise linearly with the (ZrO_2+TiO_2) concentration; after that, it began to trend downward. As the (ZrO_2+TiO_2) concentration increases, this may be caused by an increase in the smaller (ZrO_2+TiO_2) phase particles, as Fig. 7 illustrates. This demonstrates that 20 weight percent of (ZrO_2+TiO_2) can yield the maximum density. As seen from Figure 7, it is clear that with increasing the (ZrO_2+TiO_2) content, the porosity decreases; the minimum porosity value is found at 20 wt% (ZrO_2+TiO_2) content.

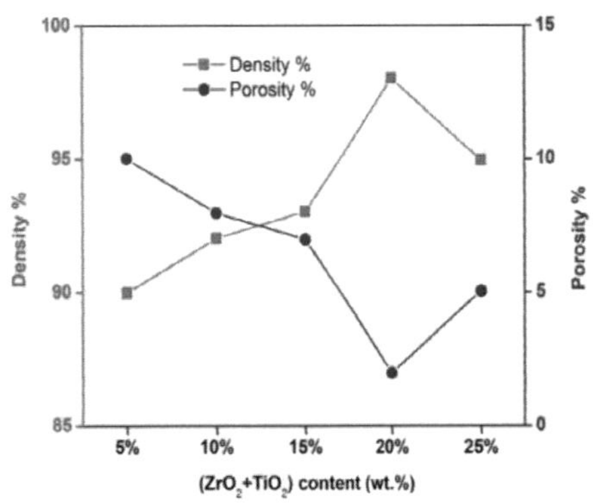

Figure 7. Variation of density and porosity

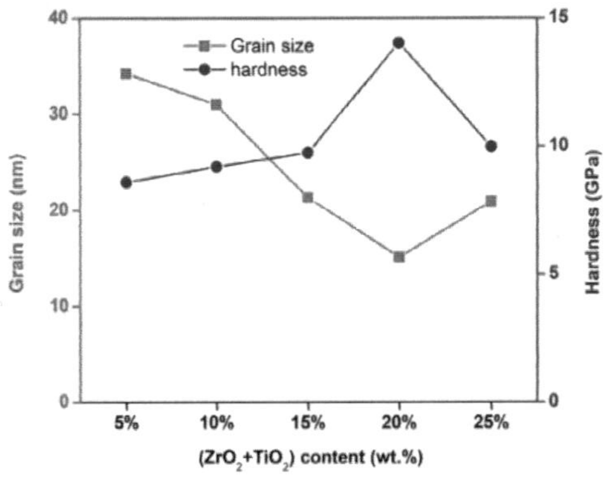

Figure 8. Variation of Hardness and Grain size

3.5.1.2 Effect of Grain size on Micro Hardness

The variation in grain size and Hardness of HAp/(ZrO_2+TiO_2) nano-composites with the increasing of (ZrO_2+TiO_2) content is shown in Fig.8. The hardness of the HAp and (ZrO_2+TiO_2) composite increased from 8.6 to 14.01 GPa, respectively, with (ZrO_2+TiO_2) content. The gradual decrease in grain size and increase of (ZrO_2+TiO_2) wt% content resulted in increase of the hardness. At 25 wt% of (ZrO_2+TiO_2), the increase in grain size alters the decrease in density and hardness. Based on Figure 8, the grain expansion may be caused by HAp up to 20 weight percent (ZrO_2+TiO_2) reinforcing agents. Figure 8 illustrates how the (ZrO_2+TiO_2) content affects density, hardness, particle size, and hardness fluctuations. The HAp/(ZrO_2+TiO_2) composites with a weight percentage of 5, 10, 15, 20, and 25 wt% have the maximum hardness of 14.01 GPa. It is discovered that the hardness values show a progressive increase up to 20 weight percent (ZrO_2+TiO_2), after which there is a sharp decline at 25 weight percent (Hardness value 9.98GPa). This might be explained by a decrease in grain size with (ZrO_2+TiO_2) content as well as an increase in density and dispersion of smaller (ZrO_2+TiO_2) particles in the matrix phase. The mechanical characteristics of the HAp/(ZrO_2+TiO_2) composites were also altered as a result.

3.5.1.3 Young's modulus and Fracture Toughness

The HAp and (ZrO_2+TiO_2) composites' Young's modulus are determined to be between 126.2 and 236.8GPa. Figure 9 illustrates the correlation between the Young's modulus and fracture toughness of sintered samples of HAp and (ZrO_2+TiO_2) composites. The values of fracture toughness

IOP Conf. Series: Journal of Physics: Conf. Series **2765** (2024) 012009 doi:10.1088/1742-6596/2765/1/012009

(Kc), obtained by applying Equations 3.3 and 3.4, range from 2.4 MPa.m$^{1/2}$ for HAp to 3.73 MPa.m$^{1/2}$ for 20 weight percent of (ZrO_2+TiO_2) composite material. Figure 9 illustrates the hardness and fracture toughness of (ZrO_2+TiO_2) and HAp composites in sintered samples as the (ZrO_2+TiO_2) content increases. As a result, these two parameters—fracture toughness 3.08 MPa.m1/2 and Young's modulus 217 GPa maximize at 20 weight percent ZrO_2+TiO_2 and then slightly decline at 25 weight percent ZrO_2+TiO_2. (Young's modulus 217GPa and fracture toughness 3.08 MPa.m$^{1/2}$).

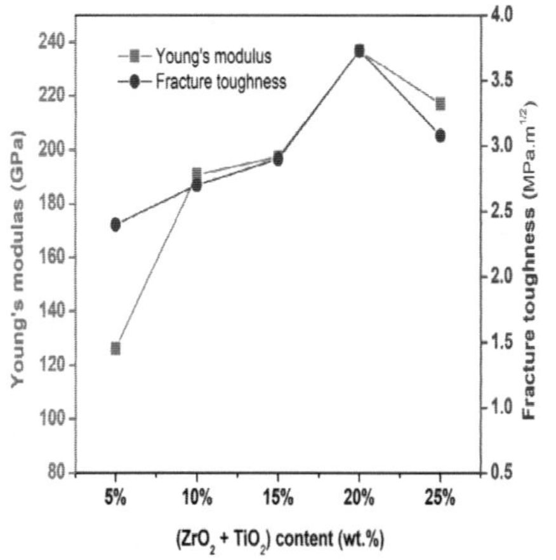

Figure 9 Variation of Young's modulus and Fracture toughness

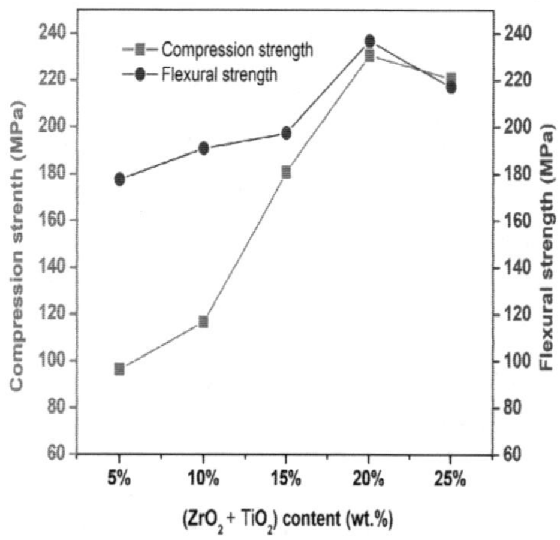

Figure 10 Variation of compression strength and flexural strength

3.5.1.4 Compression Strength and Flexural Strength

Figure 10 shows the compression strength and flexural strength for HAp/(ZrO_2+TiO_2) composites at different (ZrO_2+TiO_2) percentages. With a compressive strength of 54.72 MPa upon sintering, HAp had the lowest strength. With a rise in (ZrO_2+TiO_2) composition, the compression strength rose significantly. Twenty weight percent (ZrO_2+TiO_2) yields higher compression strength of 230.6 MPa. For HAp/(ZrO_2+TiO_2) composites, flexural strengths ranging from 177.63 to 236.8 MPa were achieved. As illustrated in Fig. 10, the highest flexural strength is noted at 20 weight percent of (ZrO_2+TiO_2) content, after which it exhibits a declining trend. This might be because, as seen by the micro structural characteristics, there is a greater dispersion of (ZrO_2+TiO_2) particles in the TCP structure.

4. CONCLUSIONS

The HAp/(ZrO_2+TiO_2) composites using HEBM were prepared successfully to determine the mechanical properties. It is observed that the improved mechanical properties are due to the increase in density and decrease in grain size. With the increase in wt% reinforcement content, HAp/(ZrO_2+TiO_2) composites, an increase in the dominant smaller particles of ZrO_2 in the matrix phase and for the content greater than 20 wt%, a rapid increase of grain size is observed. HAp/(ZrO_2+TiO_2) composites exhibited better mechanical properties compared to HAp/ TiO_2 and HAp/ZrO_2. From this study, the following observations are made:

- The hardness values of HAp/(ZrO_2+TiO_2) found to be in the range of 8.6-14.01GPa. The fracture toughness is found to be in the range of 2.4-3.73 MPa $m^{1/2}$. The composites of HAp/(ZrO_2+TiO_2) with 20wt. % of (ZrO_2+TiO_2) is found to exhibit better hardness and toughness(14.01GPa and 3.73 MPa $m^{1/2}$) compared to that of pure HAp (5.56 GPa and 0.742 MPa $m^{1/2}$).
- With increase in (ZrO_2+TiO_2) content, the Young's modulus of the sintered HAp and (ZrO_2+TiO_2) composites increased from 126.2 to 217.1GPa, and are superior compared to that of the pure HAp (89.09 GPa).
- As compared to pure HAp, composites with 20wt% (ZrO_2 +TiO_2) addition yielded maximum compression strength of 230.6 MPa. Upon further increase in (ZrO_2+TiO_2) content i.e at 25 wt% the value is reduced (220.07MPa).
- The Maximum flexural strength is found to be 236.8MPa for 20wt% (ZrO_2+TiO_2) composite and dropped at 25 wt. % (ZrO_2 +TiO_2)(217.1MPa).

It is discovered that when the concentration of (ZrO_2 +TiO_2) increases, the mechanical properties get better. Significant improvements in mechanical characteristics are observed up to a 20 weight percent addition of reinforcing content. The mechanical characteristics of HAp are observed to be reduced when reinforcement consisting of 25 weight percent (ZrO_2 +TiO_2) is added.

References:

1. Z. Zyman, I. Ivanov, D. Rochmistrov, V. Glushko, N. Tkachenko, and S. Kijko, "Sintering peculiarities for hydroxyapatite with different degrees of crystallinity," J. Biomed. Mater. Res., vol. 54, no. 2, pp. 256–263, 2001
2. Z. Xihua, L. Changxia, L. Musen, B. Yunqiang, and S. Junlong, "Fabrication of hydroxyapatite/diopside/alumina composites by hot-press sintering process," Ceram. Int., vol. 35, no. 5, pp. 1969–1973, 2009.
3. B. T. Lee, C. W. Lee, A. K. Gain, and H. Y. Song, "Microstructures and material properties of fibrous HAp/Al_2O_3-ZrO_2 composites fabricated by multi-pass extrusion process," J. Eur. Ceram. Soc., vol. 27, no. 1, pp. 157–163, 2007.
4. S. L. Shi, W. Pan, M. H. Fang, and Z. Y. Fang, "Reinforcement of hydroxyapatite bioceramic by addition of Ti_3SiC_2," J. Am. Ceram. Soc., vol. 89, no. 2, pp. 743–745, 2006
5. M.Aminzare, A.Eskandari, M.H.Baroonian, A.Berenov, Z.RazaviHesabi, M.Taheri, S.K.Sadrnezhaad, "Hydroxyapatite nanocomposites: Synthesis, sintering and mechanical properties," Ceram. Int., vol. 39, no. 3, pp. 2197–2206, 2013.
6. B. Viswanath and N. Ravishankar, "Interfacial reactions in hydroxyapatite/alumina nanocomposites," Scr. Mater., vol. 55, no. 10, pp. 863–866, 2006.
7. Mobasherpour, M. SolatiHashjin, S. S. RazaviToosi, and R. DarvishiKamachali, "Effect of the addition ZrO_2-Al_2O_3 on nanocrystalline hydroxyapatite bending strength and fracture toughness," Ceram. Int., vol. 35, no. 4, pp. 1569–1574, 2009.
8. B. T. Lee, C. W. Lee, M. H. Youn, and H. Y. Song, "Relationship between microstructure and mechanical properties of fibrous HAp-(t-ZrO_2) /Al_2O_3-(m-ZrO_2) composites," Mater. Sci. Eng. A, vol. 458, no. 1–2, pp. 11–16, 2007.
9. A. Khalil and A. A. Almajid, "Effect of high-frequency induction heat sintering conditions on the microstructure and mechanical properties of nanostructured magnesium/hydroxyapatite nanocomposites," Mater. Des., vol. 36, pp. 58–68, 2012.

10. K. Gain, L. Zhang, and W. Liu, “Microstructure and material properties of porous hydroxyapatite-zirconia nanocomposites using polymethyl methacrylate powders,” Mater. Des., vol. 67, pp. 136–144, 2015.
11. Rapacz-Kmita, A. Ślósarczyk, and Z. Paszkiewicz, “Mechanical properties of HAp-ZrO_2 composites,” J. Eur. Ceram. Soc., vol. 26, no. 8, pp. 1481–1488, 2006.
12. J. W. Choi, Y.M. Kong and H.E.Kim, “Reinforcement of hydroxyapatite bioceramic by addition of Ni_3Al and Al_2O_3,” J. Am. Ceram. Soc., vol. 81, no. 7, pp. 1743–48, 1998.
13. M. F. Morks, N. F. Fahim, and A. Kobayashi, “Structure, mechanical performance and electrochemical characterization of plasma sprayed SiO_2 /Ti-reinforced hydroxyapatite biomedical coatings,” Appl. Surf. Sci., vol. 255, no. 5 PART 2, pp. 3426–3433, 2008.
14. Hongbin Zhang , Jiawen Sun , Kang Chen , Haiping Zhou , Jianbo Jia , Zequn Wang, Yue Lu , Kuidong Gao , Wenhao Ma”Microstructure and thermal stability of nanocrystalline AZ31 Mg alloys reinforced by ultrafine vanadium particles”. J.Mater.Res.Technol., 23 (2023), pp. 943953, 10.1016/J.JMRT.2023.01.082.
15. Junliu Ye , Xianhua Chen , Huan Luo , Jie Zhao , Jianbo Li , Jun Tan , Hong Yang , Bo Feng , Kaihong Zheng , Fusheng Pan “Microstructure, mechanical properties and wear resistance of Ti particles reinforced AZ31 magnesium matrix composites,J. Magnesium Alloys”, 10 (8) (2022), pp. 2266-2279, 10.1016/J.JMA.2022.06.012
16. Gurmeet Singh Arora , Sathish Rao U. , Kuldeep Kumar Saxena “Critical review of Mg matrix composite for bio-implants through powder metallurgy”Mater. Today Proc., 57 (2022), pp. 902-907, 10.1016/J.MATPR.2022.03.036.
17. Mehta, A.,Singh, G.“Consequences of hydroxyapatite doping using plasma spray to implant biomaterials: review paper J. Electrochem. Sci. Eng., 13 (1) (2023), pp. 5-23, 10.5599/JESE.1614
18. Gurbhej Singh, Amrinder Mehta, Amit Bansal,”Electrochemical behaviour and biocompatibility of claddings developed using microwave route: review paper J. Electrochem. Sci. Eng., 13 (1) (2023), pp. 173-192, 10.5599/JESE.1604
19. Ahmed E. Hannora, Sabbah. AtayaStructure and compression strength of hydroxyapatite/titaniananocomposites formed by high energy ball milling. Journal of Alloys and Compounds 658, 222-233 (2016).
20. HongboGuo, KhiamAikKhor, Yin ChiangBoey, XigengMiao, Laminated and functionally graded hydroxyapatite/yttria stabilized tetragonal zirconia composites fabricated by spark plasma sintering. Biomaterials 24(4), 667-675 (2003).
21. G. Colón, M. A. Avilés, J. A. Navío, and P. J. Sánchez-Soto, “Thermal behaviour of a TiO_2-ZrO_2microcomposite prepared by chemical coating,” J. Therm. Anal. Calorim., vol. 67, no. 1, pp. 229–238, 2002.
22. X. Miao, D. Sun, P. W. Hoo, J. Liu, Y. Hu, and Y. Chen, “Effect of titania addition on yttria-stabilised tetragonal zirconia ceramics sintered at high temperatures,” Ceram. Int., vol. 30, no. 6, pp. 1041–1047, 2004.
23. “Standard Test Methods for Determining Average Grain Size,” ASTM E112-10, vol. 96, no. 2004, pp. 1–27, 2010.

Evaluation of tensile properties of ABS material specimen using commercial and manufactured filaments in FDM process

Dr.P.NANDA KUMAR[1], P.S RAVI KUMAR[2], C.LAKSHMI NARAYANA[3] and P. UDAY KUMAR[4]

Professor, Département of Mechanical Engineering, N.B.K.R Institute of Science & Technology - Vidyanagar, Tirupati (Dist.),INDIA

Tel : + 91 9885385064, Fax : 08624228257, email : podaralla.nandakumar@gmail.com

ABSTRACT: Acrylonitrile butadiene styrene (ABS) is the most common of engineering plastic materials; it is copolymer of Styrene-Acrylonitrile and butadiene styrene in special form. It is widely used in substantive and Additive Manufacturing (AM) applications. AM is a promising manufacturing technology that can fulfill future needs as products and their configurations are changing day by day. Fused Deposition Modeling FDM is an AM technique that fabricates parts layer upon layer. This paper analyses the effect of different process variables and orientations on the tensile properties of the FDM manufactured specimens. Standard specimens are printed using commercial ABS and manufactured ABS filaments for analysis of tensile properties. The ABS filament is fabricated using a mini filament extruder. After modeling, the specimens are printed using an FDM printer. The tensile testing of specimens is carried out on the Instron 3360 table frame UTM machine. Tensile properties obtained for different specimens are presented.

Keywords: ABS, FDM, Tensile properties, specimen orientation,

1. Introduction

Manufacturing scenario is changing due to advances in technology. The customer requirements are also changing rapidly and to meet his requirements. Hence existing products are either to be modified or new versions are to be introduced. Rapid prototyping methods shall be very much handy in this changing product scenario. Additive manufacturing emerged from prototype manufacturing processes to produce good quality parts for products. Chou et al.[1] presented basics of various rapid prototyping processes – liquid, solid and powder-based AM processes, STL file issues like file formats, different STL file problems such as gaps, degeneracy, redundancy, overlapping, STL file repair applications of AM processes and pioneers in these technologies. The work of Al-Makky and Mahmoud [2] emphasizes that the industrial objective of AM is to develop parts with good quality, reduction in lead time, low operational cost and light weight parts. They mention that the labour intensive& near contour shapes tooling, medical applications which require medical image info conversion to human body parts& porous structures for bones, automotive industry which required low quantity products & prototypes, aerospace applications where weight reduction is the top requirement are the places where AM gives solution. The work of Rahim et al. [3] expressed that (FDM) is the most extensive of methods used 3D printing which uses polymers as build materials in creating models, prototypes and many times end products also. The usage of FDM has been increasing at a drastic rate from 2009, year by year and there is an expert belief the FDM process possesses the potential to can transform manufacturing in many industrial segments. The principal advantages of FDM are simplicity in fabrication, lower cost than other competing 3D printing techniques and possibility to making intricate geometries& cavities with sufficient accuracy in dimensions. Gordeliera et al.[4] made research review on additive manufacturing components on process parameters and tensile test specimens. Selection of materials, orientation of print, angle of raster, air gap, and height of layer were studied in their research while looking at the result of taking different parameters on the tensile strength of samples printed. The conclusion of this study report offers suggestions for improving tensile strength in the FDM samples printed. Basavaraj and Vishwas [5] in their work discussed about the FDM process variables such as thickness of layer, angle of orientation and thickness of shell that is affecting ultimate value of tensile strength, dimensional accuracy, and production cycle time. Their experimental results were analysed using Taguchi method. They conclude that smaller thickness of layer gives better bonding strength and good capability for carrying axial loads. They also said that as the change in angle of orientation changed the strength of bond between layers. Daniyar et al. [6] focussed on the influence of infill density & pattern, extrusion temperature, thickness of each layer, diameter of nozzle, angle of raster, and orientation of the build on mechanical properties. They felt that variations in the said process variables makes substantial differences in the quality of printed product as well their mechanical properties. A good combination of the said process variables is of utmost importance. Their analysis also considered the influence of part pre-processing, pre-deposition heating, use of IR lamp and use of vacuum environment as means of improving mechanical strength of parts. The work of Pernica et al. [7] focussed on the effect of different nozzle temperatures in fused filament fabrication of PLA, PETG and ABS. They performed experiments with three different temperatures within the limits set by the material supplier. They concluded that high temperatures are recommended for printing to get better mechanical strength while lower temperatures can offer lower mechanical strength offset by printing costs. Gonabadi et al. [8] also investigated upon the effect of process parameters on PLA samples produced by FFF method. They prepared and tested samples with different infill density and different direction of deposition. Further they followed gyroid, concentric, square and triangle patterns for filling up. Their results showed that the ultimate tensile strength as well as Young's modulus increased in quadratic order with density of infill. They also discovered that when the direction of filament deposit coincides with direction of loading, properties obtained are nearly equal to properties of filaments. On contrary, with angle between loading and deposition directions, strength was lesser. They also found that the infill patterns have no influence on properties. Daniel Farbman and Chris Mccoy [9] performed tensile tests on 3D printed plastic specimens made up of ABS and PLA. In fill percentage, geometry, orientation of load and rate of strain were the variables. They considered specific tensile strength as a measure for effect upon by the different parameters. They mention that the specific strength was found to decrease with infill percentage. Hexagonal in fill was found to give more strength. However, their FEA analysis conducted concluded that rectilinear infill gives stiffer component. Ala'aldin Alafaghani et al. [10] discussed in their work the design considerations to optimize part quality. Their experiments include performance of PLA FDM parts at elevated service temperatures, capability in printing overhangs and unsupported bridges and influence of process parameters in accuracy of dimensions. Their paper also discussed about surface roughness limitations, clearances for assemblies, unfused features and islands. The work of Ranvijay Kumar et al. [11] focussed on multi-materials such as ABS, PLA and HIPS for 3D printing. Structural properties such as breaking load, breaking strength, percentage elongation at break and Young's modulus have been analysed for such multi material combinations. Their work included MFI characterisation for determination of flowability, DSC tool for finding glass transition temperature, crystallinity degree and thermal capacity. Pull out testing and 3-point bending tests were also performed. They concluded that few combinations gave better of the said properties. Mohamed Daly et al. [12]

investigated on the effect of parameters in 3D printing speed, on the residual stresses and deflection of ABS specimens using finite element analysis procedure. ABAQUS and DIGIMAT software were used to simulate the thermomechanical behaviour. Their numerical analysis indicated that the residual stresses and deflections were reduced by increasing the speed of printing. Further the temperature difference between platform and part layers produced high residual stresses leading to cracks and delamination. The present work focused on fabricating ABS filaments and use them to print tensile testing specimens using fused Deposition modelling machine. The specimens are then evaluated for its tensile characteristics and are compared with the results of tensile tests on specimens using commercial filaments.

2. FDM Process:

The Fused deposition modeling (FDM) is sometimes called Fused Filament Fabrication (FFF) process. A schematic of the working model of this process is presented in Fig. 1. This FDM machine has a table and a movable extrusion head. The table moves in X and Y directions to print on a horizontal surface area. The movable head extrudes a hot semisolid thread of filament attached to the Z-axis slide, which moves up and down to accommodate the print in a vertical direction. The heated soft material is deposited on the substrate a little above the molten temperature to have better bonding between the layers in all directions and to obtain the solid part.

The CAD software, such as CATIA, Solid works, etc., are used create the object to print. The developed CAD model is saved in stereo lithography format i.e., STL format to export to 3D slicing software like Ultimaker CURA. The slicing software uses CAD model .stl file as input file and generates G-Code to perform printing layer after layer by giving input parameters in order with respect to the desired quality of the component to be obtained.

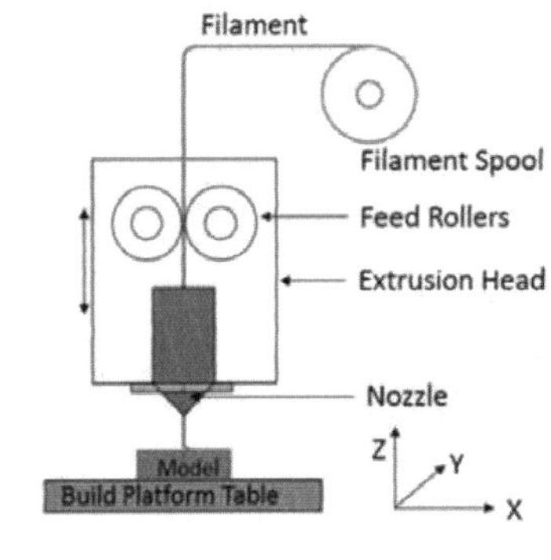

Fig.1 FDM machine working model

3. Materials:

The specimens are fabricated by using two types of ABS filaments: commercial ABS filaments and manufactured ABS filaments. The commercial ABS filament is purchased directly from the market. The manufactured filaments are fabricated from the polymer-granular ABS material using a mini filament extrusion machine, as shown in Fig. 2. The ABS granules are preheated before pouring into the hopper of the extrusion machine to remove any absorbed moisture. The nozzle is fixed to the extrusion end of the barrel, and the barrel with the rotating screw conveyor of the extruding machine is heated above the melting temperature of the ABS. The softened hot filament coming from the nozzle was cooled immediately using a cooling water bath and wound using an adjustable (electronically controlled) speed winding machine that has simultaneous rotation and traversing spool.

Fig.2 Mini filament extrusion machine

4. Methodology to evaluate the tensile properties of specimens:

Tensile properties give valuable information about the components to analyze their quality. The properties are evaluated for components produced by FDM process. The methodology adopted is in the sequence as discussed in Fig.3.

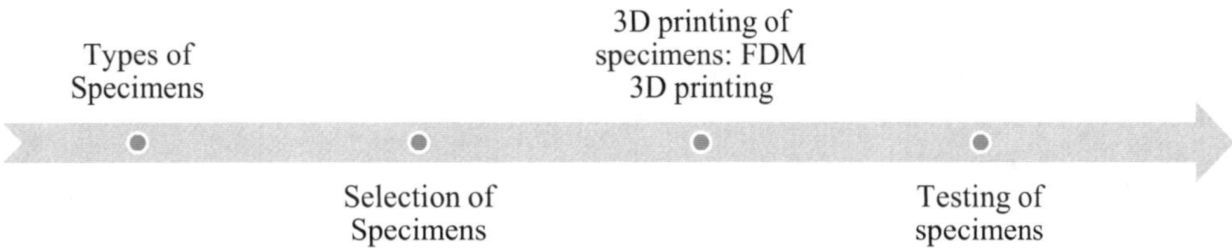

Fig.3 Methodology adopted

4.1. Specimen Selection:

Tensile characteristics can be used to examine a component's response to tension-based loads applied to it. It is necessary to find the tensile properties because doing so gives details about the ultimate load, breaking load, ultimate tensile stress, Young's modulus, % elongation strain, and other many more properties. Tensile testing results in a load - elongation curve which is subsequently converted ınto a stress - strain curve that is used to determine the above said tensile properties. Fig.4 shows a standard tensile test specimen ASTM D638, that is popular for testing tensile properties of polymeric materials. Hence this specimen geometry has been used in the present work also to evaluate tensile properties.

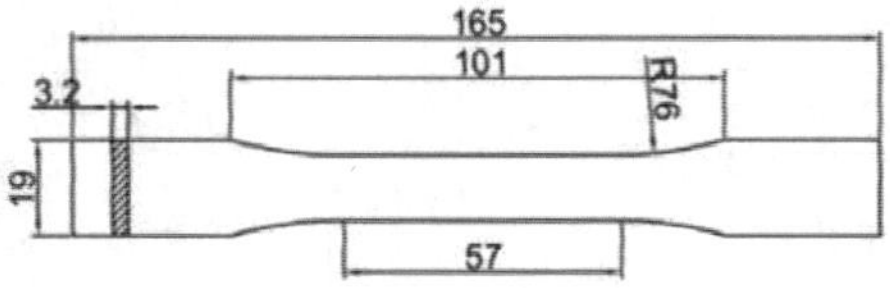

All Dimensions are in mm

Fig.4 ASTM D-638 Type I

4.2. 3D printing of Specimens: FDM 3D printing

The three dimensional model of ASTM D638 specimens was modeled using Catia V5 modeling software and saved in .STL format to export to slicing software. This file is imported into the Ultimaker Cura software. Before generating the G codes the setting parameters like thickness of Layer Height, number of Top/bottom layers, type of infill pattern, bottom base build plate temperature, temperature of printing nozzle, the speed of printing and travelling, etc. have been selected based the information available in literature, machine capabilities with trial-and-error process. These parameters are presented in Table 1. The G-code pattern is then generated by selecting the FDM machine and setting those process parameters. The G-codes are generated separately for the orientations viz., horizontal, vertical, and inclined, shown in Fig. 5a Parts were fabricated using an Arjun 3D Printer developed by M/S Vignan Technology, Tirupati. The printer head has a nozzle diameter of 0.4 microns as shown in Fig. 5b The process of printing is initiated by printing base of the component, followed by printing of outer walls and then printing inner portion with triangular infill pattern. For uniformity in testing, same process parameters have been selected for printing both commercial and manufactured specimens. Layer height of 0.15 mm has been set up as prints were observed to be smooth. For higher layer height than 0.15mm, pattern was found to be rough in appearance. Hence 0.15mm layer height was followed. The top and bottom layers selected were 4 in number because ironing was observed to be good. Triangular infill pattern has been followed because of strong infill observed. The material used for part fabrication was ABS commercial and ABS fabricated filament. After the well-deserving slicing of the specimen, the file has been stored on the SD card to make it machine readable. The specimens are made to print as per the orientation, i.e., horizontal, vertical, or inclined, following ASTM D638.

Table 1. . Parameters used in FDM printing process

Parameters	Values
Layer Height	0.15mm
Wall Count	4
Top/bottom layers	5
Infill Density	100%
Infill pattern	Triangles
Build plate temperature	60^0C
Nozzle temperature	200^0C
Print Speed	60mm/sec
Travel Speed	130mm/sec
Jerk speed	8mm/sec

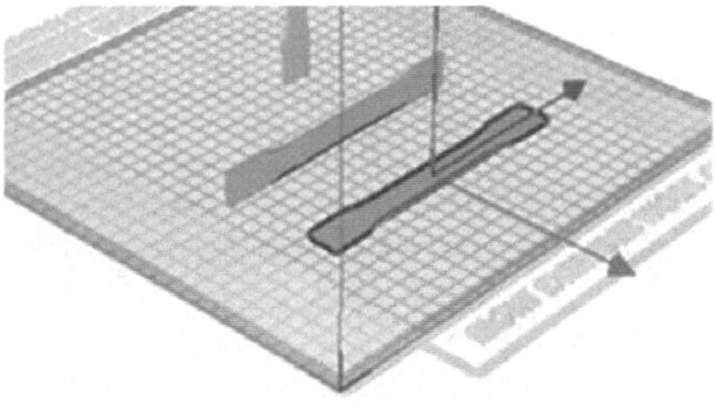

Fig.5a Orientations of the specimen

Fig.5b 3D printing of the specimen

4.3. Testing of specimens:

The specimens cut as per ASTM D638 are tested for tensile properties on an Instron 3360 universal testing machine (UTM). Tensile property measurement is carried out on a Shimadzu, Japan-made UTM machine whose specifications are given in Table 2. The width and thickness of each sample are measured for individual specimens. Specimens are held in the machine with help of composite rubber grippers. Load is applied gradually at a prescribed strain rate until the specimens fail. Figures 6a and 6b show the gripping of the specimens with ABS commercial and manufactured filaments, respectively, in the jaws of the UTM.

Table 2. Specifications of Universal Testing Machine

Parameter	Details
Model no.	3360 Series Dual Column Table Frame
Range	5–50 kN
Accuracy	0.01 N
Make	Shimadzu Japan
Applicable test	Tensile, compressive, flexural, tear, shear, elongation and modulus

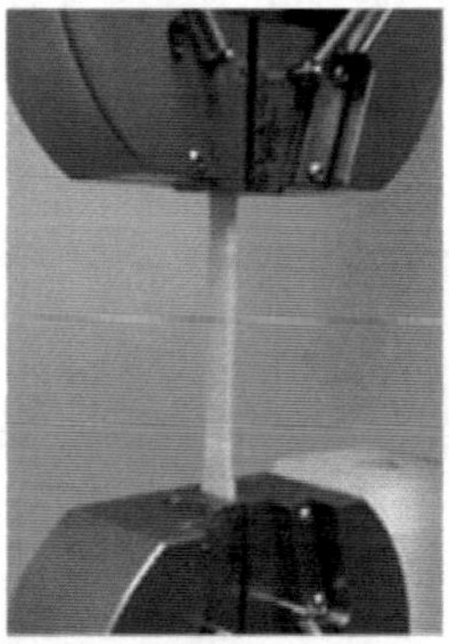

Fig.6a Testing of specimen with commercial filament

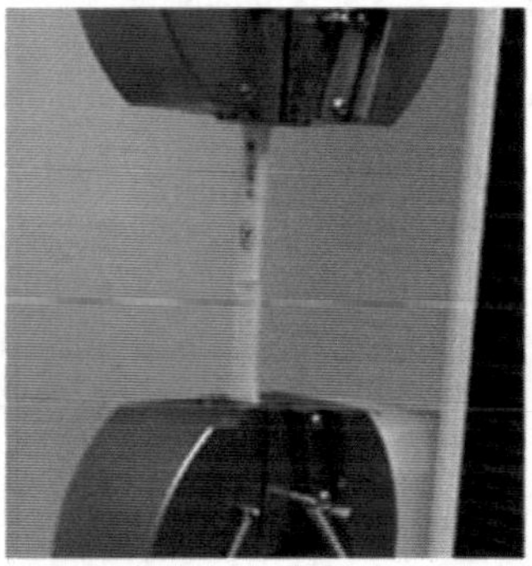

Fig.6b Testing of specimen with manufactured filament

The specimens are gripped on both ends of the UTM machine, loading is applied on both ends, and various readings like ultimate load, breaking load and associated elongations are noted. Graphs are plotted. After testing, the fractured specimens are presented in Fig. 7a and Fig. 7b with commercial and manufactured ABS filaments, respectively.

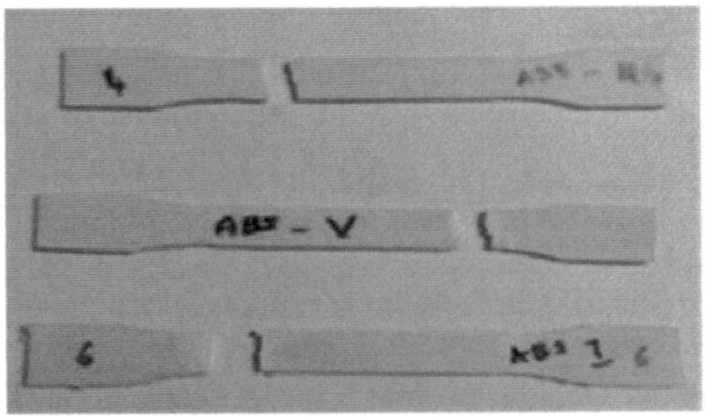

Fig.7a Specimens fabricated by commercial ABS filament after testing

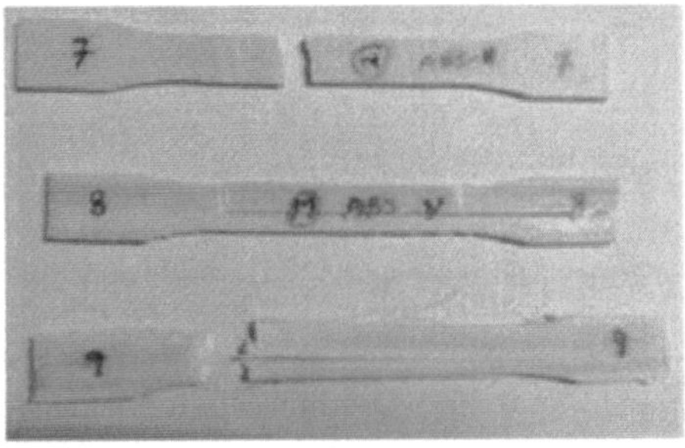

Fig.7b Specimens fabricated by Manufactured ABS filament after testing

5.Results& Discussions

After performing tensile tests on the Instron 3360 series, the obtained results are compared to study the effectiveness. The 3-D printed specimens, fabricated using commercial ABS as well as manufactured ABS filaments, are tested. The specimens after testing are presented in Figs.7a and 7b. The results obtained are discussed as case I, case II and case III. The comparison of tensile properties between commercial ABS and manufactured ABS for horizontal orientation is referred as case I. Case II discusses the comparison of obtained tensile properties for commercial ABS and manufactured ABS for vertical orientation. Case III discusses the comparison of tensile properties obtained for commercial ABS and manufactured ABS for inclined orientation.

Case(I): Comparison of tensile properties obtained for commercial ABS and Manufactured ABS specimens (horizontal orientation)

The results of the tensile test conducted on specimens obtained with horizontal orientation of commercial ABS and manufactured ABS filaments are presented in Table 3.The results showing that the propties obtained by manufacted filament are nearer to commercial filament as presented in the Figs.8a-d.

Table 3. Comparison of tensile properties obtained for commercial ABS and Manufactured ABS specimens (horizontal orientation)

Property (Horizontal Orientation)	Commercial ABS	Manufactured ABS	% Difference
Maximum load (N)	1160.41	905.63	21.9
Load at break (N)	1151.53	876.45	23.8
Tensile stress (MPa)	38.68	30.19	21.9
Young’s modulus (MPa)	1264.30	1011.53	19.9
Tensile strain	0.04583	0.05083	-10.9

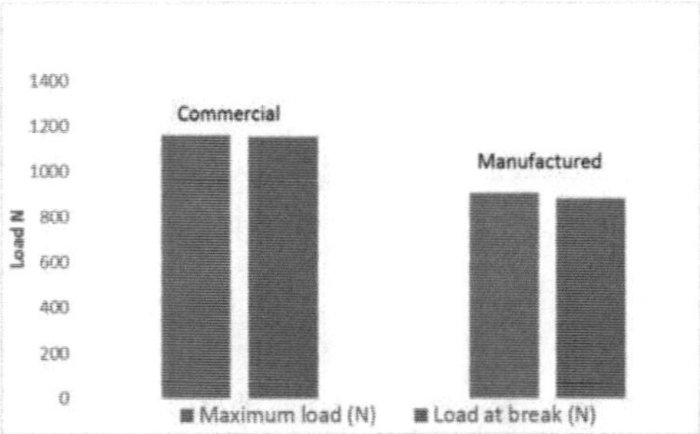

Fig.8a Comparison of Maximum load & Load at Break in Horizontal orientation

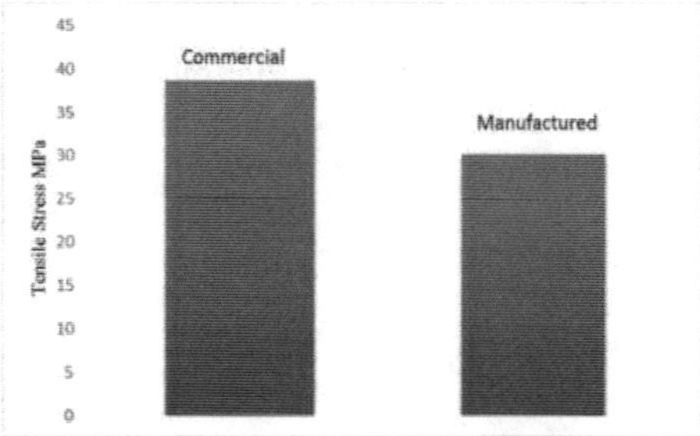

Fig.8b Comparison of Tensile stress in Horizontal orientation

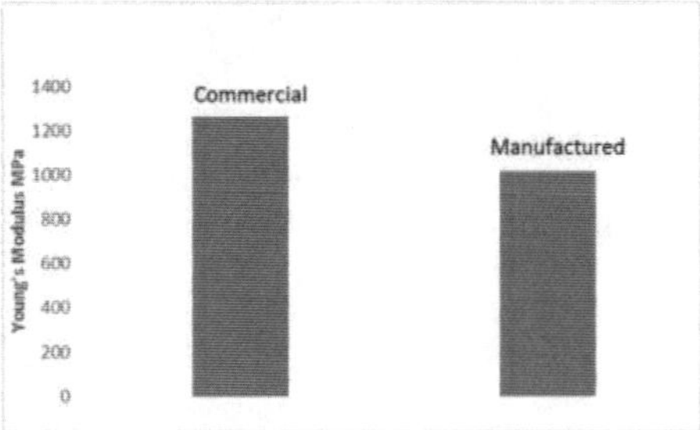

Fig.8c Comparison of Young's modulus in Horizontal orientation

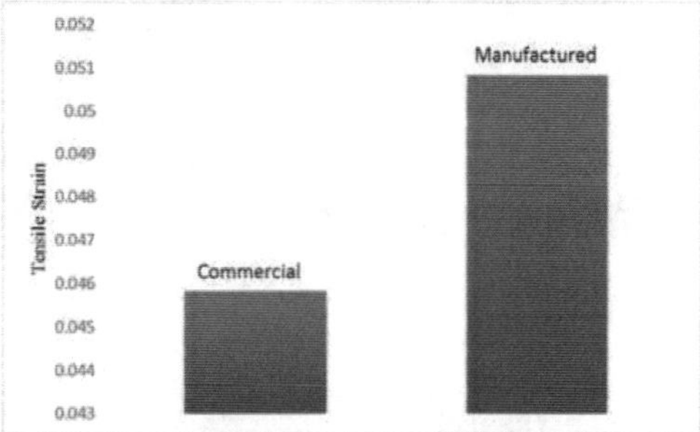

Fig.8d Comparison of Tensile strain in Horizontal orientation

Case (II) : Comparison of tensile properties obtained for commercial ABS and Manufactured ABS specimens (vertical orientation)

The tensile characterstic results of vertical orientation compared in the same way as horizontal orientation. From Table 4, we can see that commercial ABS exhibits higher tensile properties than manufactured ABS, except for tensile strain. The third column in the table shows the percentage difference between their properties. The change in percentage of all the properties looks similar except tensile strain. Tensile strain has the least percentage change.

A histogram is plotted for the tensile properties of specimens of commercial ABS and manufactured ABS filaments and presented in Fig.7. The maximum load, load at breaking point, and tensile stress are showing lower values than those for specimens printed using commercial ABS filament.

Table 4. Comparison of tensile properties obtained for commercial ABS and Manufactured ABS specimens (vertical orientation)

Property (Vertical orientation)	Commercial ABS	Manufactured ABS	% Difference
Maximum load (N)	1640.27	945.55	42.35
Load at break (N)	1029.52	879.90	14.53
Tensile stress (MPa)	53.84	31.52	41.44
Young's modulus (MPa)	1313.34	1108.21	15.6
Tensile strain	0.06583	0.0615	6.57

The maximum % difference has been seen in maximum load and tensile stress, and the least % difference is seen in tensile strain. A histogram is plotted between the specimens of commercial and ABS-manufactured filaments and presented in Fig.9. The maximum load, load at breaking point, and tensile stress showed less than the values of specimens printed using ABS commercial filament. The maximum load is high for commercial ABS filament when compared to manufactured ABS filament. The percentage tensile strain is higher for manufactured ABS when compared to commercial ABS.

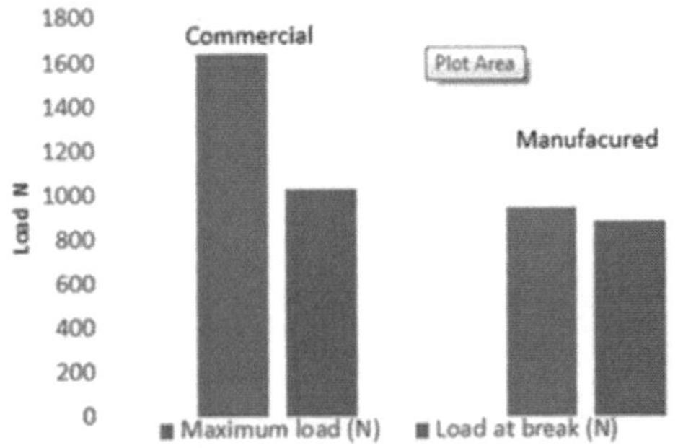

***Fig.9a** Comparison of maximum load and Load at break in vertical orientation*

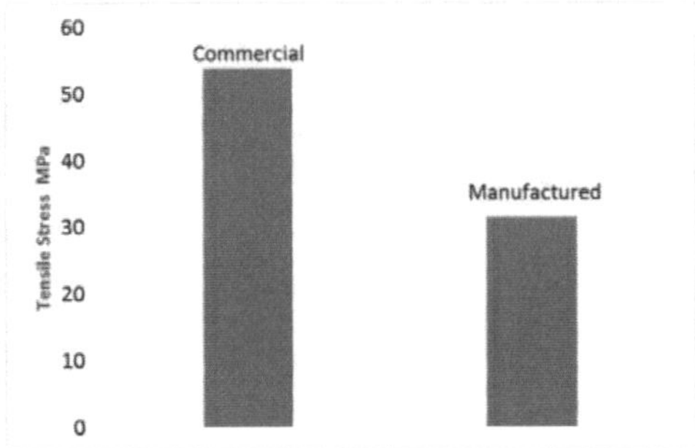

Fig.9b Comparison of Tensile stress in vertical orientation

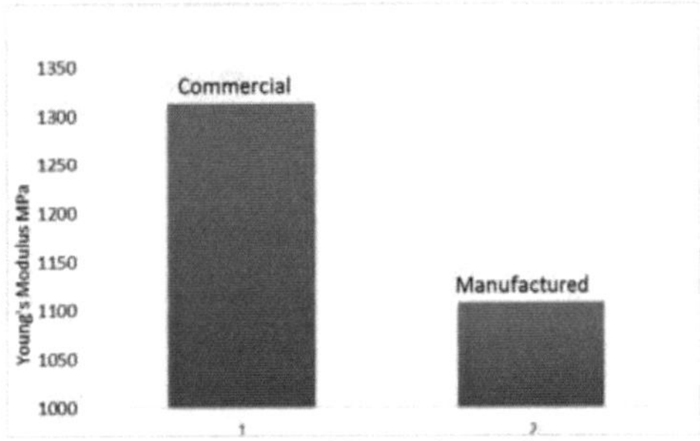

***Fig.9c** Comparison of Young's modulus in Vertical orientation*

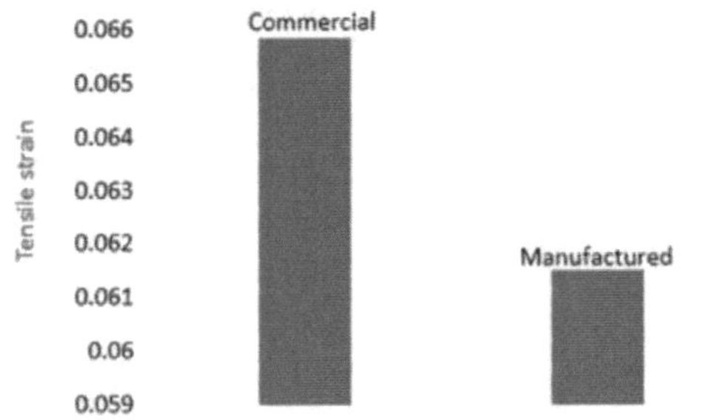

Fig.9d Comparison of Tensile strain in Vertical orientation

As we can see from Figs.9a-d, all the properties are higher for commercial ABS than for manufactured ABS.

Case (III): Comparison of tensile properties obtained for commercial ABS and Manufactured ABS specimens (Inclined orientation)

Table 5. Comparison of tensile properties obtained for commercial ABS and Manufactured ABS specimens (Inclined orientation)

Property (Inclined Orientation)	Commercial ABS	Manufactured ABS	% Difference
Maximum load (N)	1551.20	1272.24	17.98
Load at break (N)	1082.29	771.43	28.72
Tensile stress (MPa)	51.71	42.41	17.98
Young's modulus (MPa)	1299.91	1061.82	18.31
Tensile strain %	0.09017	0.08866	1.67

The Table 5 shows the comparison between the tensile properties of commercial ABS and manufactured ABS with respect to inclined orientation. The maximum % difference can be seen in load at break point and least percentage change is seen in tensile strain and the remaining properties shows similar % changes.

A histogram is plotted between the specimens of ABS commercial and ABS manufactured filaments and presented in the Figs.10a-d. The maximum load, load at breaking point and tensile stress are showing lesser values than the values of specimen printed using ABS commercial filament.

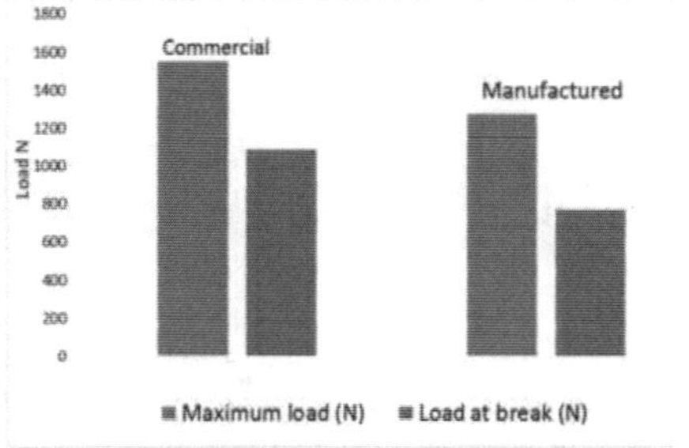

Fig.10a Comparison of maximum load and Load at break in Inclined orientation

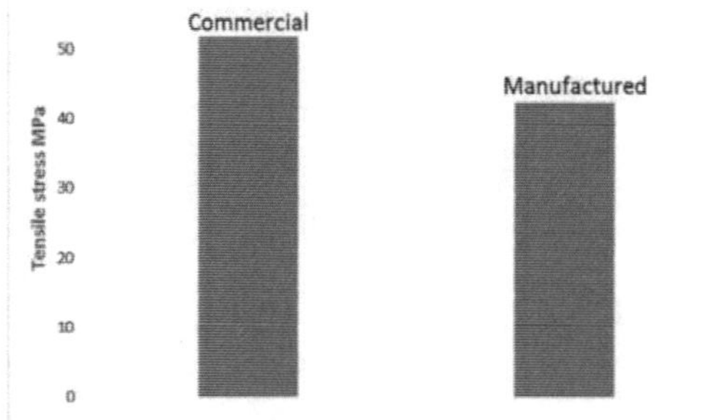

Fig.10b Comparison of Tensile stress in Inclined orientation

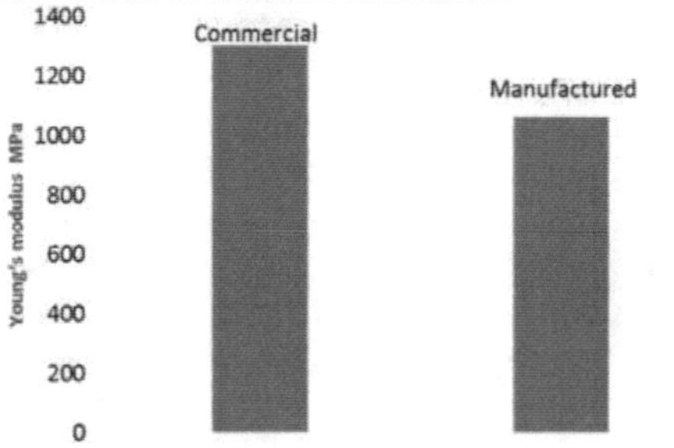

Fig.10c Comparison of Young's modulus in Inclined orientation

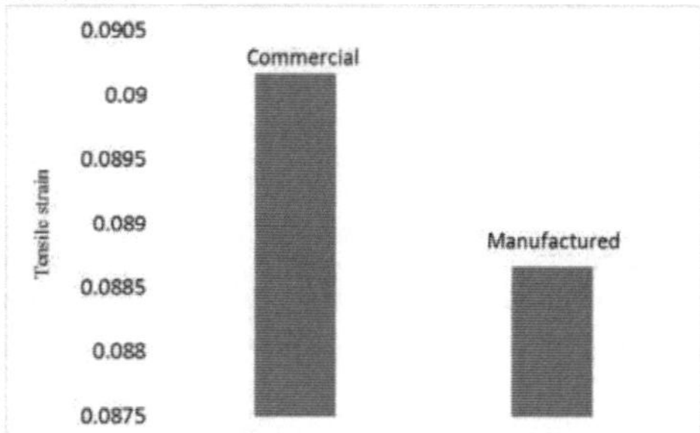

Fig.10d Comparison of Tensile strain in Inclined orientation

The maximum load is the load at the ultimate strength. So, this point separates the strain hardening region and the necking region. At this point, the maximum amount of strain hardening has occurred. The specimens with inclined orientation are observed with higher tensile property values than other orientations, hence components produced with this orientation gives higher bearing loads in the plastic deformation range.

In both of the inclined orientation specimens, tensile strength is more, which indicates, resistance of an object to a force that could tear it apart is good in this orientation. The inclined orientation specimens are better to avoid plastic deformation of component with better stability in comparison with other orientations.

Young's modulus is a constant that describes how much stress occurs can occur for under for given amount of strain with in the elastic limit. The stress-strain curve's linear region's slope is displayed. As seen in the commercial filament specimens, the values are greater and do not vary excessively in any of the orientations. In horizontal orientation is exhibiting highest values in specimens obtained by commercial and manufactured filaments.

The higher strain that has been measured coincides with the inclined orientation components. High tensile strength components can withstand a substantial amount of tensile strain before breaking, whereas low tensile strength components are susceptible to breaking at smaller amounts of strain.

The stress and strain diagram with ABS specimens obtained using both Commercial and manufactured filament for horizontal, vertical and inclined orientation presented in the Fig 11.a. &11b.

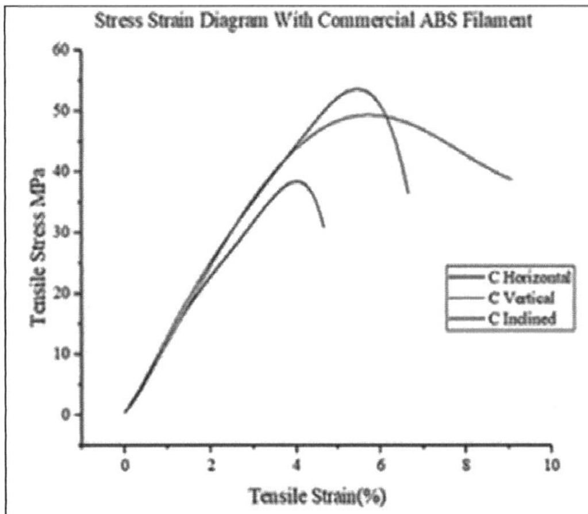

Fig.11a stress and strain diagram with ABS specimens using Commercial filament

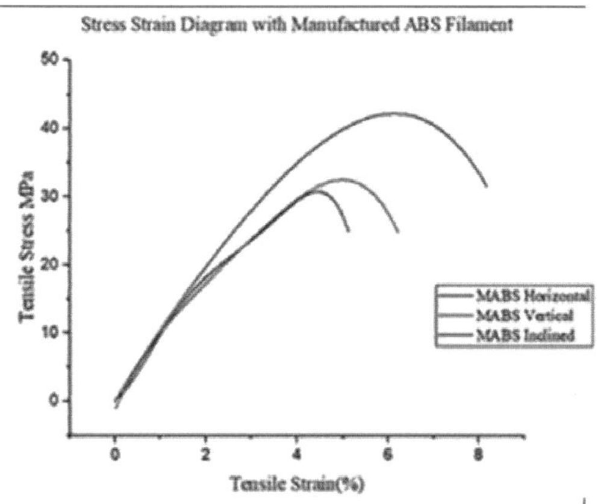

Fig.11b stress and strain diagram with ABS specimens using manufactured filament

The specimens obtained by use of commercial ABS filaments showed higher tensile stress for inclined orientation followed by vertical and horizontal orientations. However, the strain at failure did not follow the same order. Vertically oriented specimen showed higher failure strain followed by inclined and horizontal orientations. Hence the vertical orientation in this case showed a larger failure strain. Further this strain was much larger after ultimate load.

On other hand the manufactured filaments showed a different pattern compared with commercial filaments. The highest ultimate stress and breaking stress were observed for horizontal orientation followed by vertical and inclined orientations. The breaking strain also followed the same pattern. The strain after ultimate load was larger in case of horizontal orientation compared with vertical case. In case of inclined orientation, it was still smaller.

6.Conclusions:

This study focused on the tensile properties of components obtained by FDM additive manufacturing. The specimens are printed using Commercial ABS as well as Manufactured ABS. Such tensile specimens were 3-D printed using FDM-based Arjun 300 pro 3D printer. The following are the properties and outcomes:

When specimens are printed in horizontal orientation using Manufactured ABS, Commercial ABS, maximum tensile properties are observed in Commercial ABS than in Manufactured ABS. Maximum tensile stress of 38.68MPa is observed for a maximum load of 1160.41N for ABS whereas maximum stress of 30.19MPa is observed for a maximum load of 905.63N for Manufactured ABS. Maximum tensile strain of 0.05083 is observed in Manufactured ABS compared with 0.04583 in Commercial ABS. When specimens are printed in vertical orientation using Manufactured ABS, Commercial ABS, Maximum tensile properties are observed in Commercial ABS than Manufactured ABS. Maximum tensile stress of 53.84MPa is observed for a maximum load of 1640.27N for Commercial ABS whereas maximum stress of 31.52MPa is observed for a maximum load of 945.55N for Manufactured ABS. Maximum tensile strain of 0.06582 is observed in Commercial ABS than 0.0615 in Manufactured ABS. The similar tensile properties are observed even when the specimen is printed in inclined orientation. Maximum tensile properties are observed in Commercial ABS. Maximum tensile stress of 51.71MPa is

observed for a maximum load of 1551.20N for Commercial ABS whereas maximum stress of 42.41MPa is observed for a maximum load of 1272.24N for Manufactured ABS. Maximum tensile strain of 0.09017 is observed in Commercial ABS than 0.08866 Manufactured ABS.
3D printing of specimens using FDM process with horizontal orientation giving good quality in comparison with other orientations.
The specimen with manufactured ABS filament showed properties slightly lagging behind the properties of specimen using commercial ABS filament. It is due to further improvement expected in the filament extrusion process which used laboratory setup in the present work. However even at this level there exists a scope for adding up reinforcements and developing novel composite filaments.

Acknowledgments
The authors thank the AICTE-Faculty Development Cell (FDC). This work has been supported by a grant from the AICTE of Research Project Scheme, Project Number: F. No. 8-80/FDC/RPS (Policy-1)/2019-20.

References
[1] Chua C.K., Leong K.F. and Lim C.S., 2003, *Rapid Prototyping: Principles and Applications*, World Scientific Publishing Co. Pte. Ltd.
[2] Al-Makky M. and Mahmoud D., 2016, *Proceedings of the International Conference on Applied Mechanics and Mechanical Engineering*, **17**, 1-14.
[3] Tuan Noraihan, Azila Tuan Rahim, Abdul Manaf Abdullah and Hazizan Akil Md., 2019, *Polymer Reviews*, **59,** 4, 589.
[4] Gordeliera T, Thiesa P.R, Johanninga L and Turner L, 2019, *Rapid prototyping journal*, 1-35.
[5] Basava Raj C K and Viswas M, 2016, *Proceedings of international conference series – Materials Science and Engineering IConAMMA-2016,* **149**, 1-11.
[6] Daniyar Syrlybayev, Beibit Zharylkassyn, Aidana Seisekulova, Mustakhim Akhmetov, Asma Perveen and Didier Talamona, 2021, *Polymers* **13**, 1-35.
[7] Jakub Pernica, Michal Sustr, Petr Dostal, Martin Brabec and David Dobrocky, 2021, *Manufacturing Technology, Engineering Science & Research Journal* (Elsevier) **21**, 3.
[8] Gonabadi H, Yadav A, Bull S J, 2020, *The International Journal of Advanced Manufacturing Technology,* **111**, 695-709.
[9] Daniel Farbman and Chris Mccoy, 2016, *Proceedings of the ASME 2016 International Manufacturing Science and Engineering Conference MSEC2016*, 1-12.
[10] Alaaldin Alafaghani, Ala Qattawi, Muhammad Ali Ablat, 2017, *Open journal of applied sciences*, (Scientific research publishing) **7**, 291-318.
[11] Ranvijay Kumar, Rupinder Singh and Ilenia Farina, 2018, *PSU Research review* (Emerald publishing limited), **2**, 2, 115-137.
[12] Mohamed Daly, Mostapha Tarfaoui, Manel Chihi, Chokri Bouraoui, 2023, *International journal of advanced manufacturing technology*, **126**, 5307-23.

Investigating the Laser engraving on Polymethyl methacrylate sheets and process parameters optimization using GRA technique.

Dr. P.Nanda Kumar[1], V.Muni Varma[2], J.Venkata Jaswanth[3]

1. Professor of Mech. Engg. Department, N.B.K.R Institute of Science & Technology, Vidyanagar-524413, Tirupati (Dist),India

2&3 UG students N.B.K.R Institute of Science & Technology, Vidyanagar-524413,Tirupati (Dist),India

Tel : + 91 9885385064, Fax : 08624228257, email : podaralla.nandakumar@gmail.com

ABSTRACT

Polymethyl methacrylate (PMMA) is sound-resistant because it can lessen the transmission of outside sound waves. PMMA is widely used in many applications, such as automotive lighting, electronics and energy, health, sanitation, etc. Laser engraving for PMMA sheet panels with process parameters of good quality is a major problem in the process industry. There is a demand for surface roughness that is affected by the volume of material removal. Therefore, in this work, multi-value response optimization was performed on MRR and surface roughness during laser engraving of PMMA sheets using an optical fiber diode in a dry environment through gray relational analysis. Also, regression models are developed, and their responses are compared for their correlation. The GRA combined with the Taguchi method has proved efficient in the decision-making of multi-responses in a laser engraving environment.

Key words: PMMA, Surface roughness, MRR, Multi objective optimization, Regression analysis.

1. INTRODUCTION

Manufacturing is crucial to a nation's or regions economic progress in the global economy. Nobuya Haraguchi et.al focused on growth rate of manufacturing is crucial to the sustainability of developing nations. If manufacturing remains the primary engine of growth and if its proportion in developing nations has not changed, we may legitimately conclude that its significance for economic development has not been affected [1]. Avanish Kumar Dubey et.al studied on Laser Beam machining as thermal energy and non- contact type method. Due to the complex shapes and uncommon sizes of the work materials, the demanding design specifications, and the development of innovative engineering materials, the use of traditional machining has been limited. [2]. Torben Lenau et.al worked on selection of suitable manufacturing method and the process of selecting the right materials and manufacturing techniques for the product's individual components based on the product's capabilities and constraints in order to make the product easily producible [3]. Obilor et.al reviewed the Micro-texturing of polymer surfaces using lasers. Collective surface alteration techniques for polymers are reviewed with an emphasis on the mechanisms causing changes to the surface morphology, texture geometry, and surface characteristics that have an impact on functionality. According to studies, many texturing approaches have been investigated in order to develop functional surfaces, compare texturing methods, and understand how polymers respond to process parameters. [4]. Cakir et.al suggested that the DFM approach make equilibrium between production and variables affecting. The design engineer and process engineer must work well together to integrate these DFM elements. Typically, the design engineer lacks the significant process engineering skills needed for a prompt evaluation of efficient and inexpensive manufacturing [5]. Santo et.al. gave an overview of current research on the subject is given, along with descriptions of various processes, including laser welding, cutting and marking, micro molding assisted by lasers, curing thermosetting powder used for painting, and serigraphic ink curing. The outcomes in novel fields are also covered, along with the benefits and drawbacks of high lighting compared to traditional methods. [6] Deepa e.al studied optimizing the machining parameters, such as speed and power, for polypropylene (PP), using grey relational analysis. Tensile test and flexure test results were taken into consideration while determining laser power and machining speed. The Taguchi method is employed in research. [7] Pawade et.al. Made the Grey relational theory research on multi-objective optimization is used to gauge how well ceramic composites are ultrasonically machined. SR, TWR, and MRR were selected as the performance metrics. [8]. Forte et.al reviewed on PMMA as Suitable Substrate for ALD. Chemically, PMMA is an ester of poly (methacrylic acid) and a polymer of methyl methacrylate (MMA). In addition to anionic and coordination polymerizations, PMMA can also be produced through the radical polymerization of MMA. PMMA is a type of methacrylic resin, also called acrylic resin. PMMA is a thermoplastic that is transparent and is commonly used in place of inorganic glass because of its resistance to impact, weather, and chemicals. [9]. Frazer et.al found as PMMA as useful material The PMMA material is not a replacement for glass. PMMA has been used in cementing bones, contact lenses, intraocular lenses, screw fixation in bones, and filling defects in the skull and cavities of bones. The material is helpful in stabilizing the vertebrae of patients suffering from osteoporosis [10]. Nanda Kumar et.al worked with design-of-experiments approach is used for experimentation, and later the experimental data is used to develop the regression model [11]. The new processing methods are essential to adopt economically for the upcoming materials. Khan et al. carried out the laser engraving process for optimization of energy consumption, CO2 emissions, and removal rates [12]. The new processing methods are essential to adopt economically for the upcoming materials. It is proposed to work on laser engraving on a polymer PMMA sheet and the selection of its process parameters based on a multi-objective optimization function using GRA and the analysis of the variance of the Grey Relational Grade.

2. Experimentation

The foremost goal of this paper is to engrave small rectangle shapes on PMMA sheets with a view to investigating the influence of input parameters on the response variables and to note the surface roughness with a minimum value and removal rates with a higher volume. With an input of 12 V and 3 A, the laser engraving machine's laser source is transformed into Optical output power: 4.5–5 W, frequency: 1 KHz, wave length: 455 ± 5, plus duration: 1 ms. A laser engraving machine has a fiber optic laser source. Pump light from what are known as laser diodes is used in fiber lasers. Light from those diodes is sent through the fiber-optic cable. Then, a specific wavelength is generated and amplified using optical components that are positioned within the cable. Figure 1 illustrates the laser's structure. The electrical energy input of the laser diodes is converted into photons, or light, to be pumped into the fiber-optic cable. Which is known as a pump source. In the core, light travels. The core is made of silica glass and partially includes rare earth elements. The core is encircled by cladding. The light traveling out of direction is reflected back to the core as soon as it touches the cladding. The electrons of the doped fiber particles become more energetic as a result of their interactions with light. When the light is amplified, it penetrates through the lens system used. The well-collimated laser beam exits from the cavity because of the guiding property of the fiber optic.

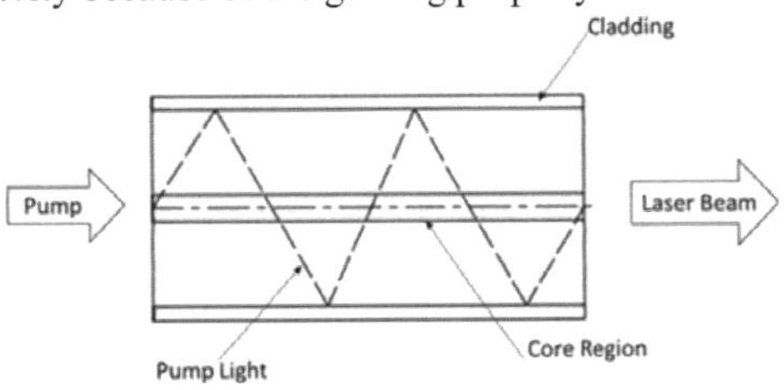

Figure 1. Structure of laser

Figure 2. Pockets engraved on PMMA sheet

Only three major factors were taken into consideration when designing the experiment in order to facilitate the collection of experimental data. A full-three-factorial design was used to plan the experimental program. [11]. The laser's power, speed, and distance are taken into consideration. Table 1 illustrates the three levels: low, medium, and high, and their ranging values for each of them were set. Focusing on this set-up, a total of 27 experiments were conducted, each of which included a combination of various factor levels as indicated in Table 2. The experiments are conducted on a laser engraver. The PMMA sheet is considered for making engraving pockets. The surface roughness and material removal rate are measured with a digital surface roughness measuring instrument, and the weight of material removed is measured with respect to time. Under the investigation, dependent variables such as surface roughness and MRR are considered, while speed, power, and distance are the independent variables.

Table.1 Values of test variables

Variable designation	Description	Values of different levels		
		Low(L)	Medium(M)	High(H)
S	Speed m/min	500	800	1200
P	Power in percentage	400	700	1000
D	Distance in mm	25	30	35

The experiments are conducted with respect to the process parameter combinations presented in Table 2 for rectangular pocket engraving. Figure 2 represents pockets engraved on a PMMA sheet. The Atomstack 20W laser engraving machine used the engraving process. The measured surface roughness and MRR are presented in Table 2. The MRR is computed by the following equation: 1.

$$MRR = \frac{(W_1 - W_2)}{\rho X g} \tag{1}$$

W_1= weight of the work piece before engraving in gm.
W_2= weight of the work piece after engraving in gm.
ρ = Density of the work piece material 0.00105 gm. /mm^3
g= gravity force in 0.00981mm/s^2

Table.2 Laser Engraving outcome of surface roughness and Material removal rate

Exp. No	Process parameters			Experimental results	
	Speed (s)	Power (p)	Distance (d)	R_a µm	MRR mm^3

1	500	40	25	19.554	0.5
2	500	40	30	22.019	0.45
3	500	40	35	24.484	0.45
4	500	70	25	23.502	1.36
5	500	70	30	25.967	1.17
6	500	70	35	28.432	1.15
7	500	100	25	20.45	1.9
8	500	100	30	29.915	1.76
9	500	100	35	32.38	1.73
10	800	40	25	18.633	0.3
11	800	40	30	21.098	0.26
12	800	40	35	23.563	0.19
13	800	70	25	22.581	0.83
14	800	70	30	25.046	0.72
15	800	70	35	27.511	0.71
16	800	100	25	26.529	1.26
17	800	100	30	28.994	1.16
18	800	100	35	31.459	1.13
19	1200	40	25	17.405	0.12
20	1200	40	30	19.87	0.07
21	1200	40	35	22.335	0.03
22	1200	70	25	21.353	0.53
23	1200	70	30	23.818	0.49
24	1200	70	35	26.283	0.46
25	1200	100	25	25.301	0.83
26	1200	100	30	27.766	0.76
27	1200	100	35	30.231	0.74

3. Optimization methodology using grey relational analysis

The influence and relationship of various factors in a multi-response system are intricate and vague. This is called "grey," denoting inaccurate and imprecise information. With the help of grey relational grade, this suggested procedure (grey relational analysis) considers this ambiguity in the multiple responses, such as surface roughness and MRR, in a laser engraving system and optimizes it. Consequently, an optimization problem with a single response, called single relational grade, is derived from a multi-response optimization problem.

Step.1. The obtained data from experiments needs to be normalized in order to prevent various units and reduce inconsistency. It is basically necessary because each data variation is different from the variation of other data. To construct the array from 0 to 1, a proper value is obtained from the actual number. It is, in general, a process of transforming the actual data into equivalent data. If the desired outcome is to be minimized, then smaller-the-better and, for maximization, larger-the-better features are meant for normalization using the formula that follows to scale it into a suitable range.

Smaller-the-better

$$Y_i^*(k) = (\, MaxY_i(k) \, - Y_i(k))/(MaxY_i(k) - MinY_i(k)) \tag{2}$$

Larger-the-better

$$Y_i^*(k) = (\, Y_i(k) \, - MinY_i(k)/(MaxY_i(k) - MinY_i(k)) \tag{3}$$

In the above equations i = 1,..., m; k = 1,...,n., in which n is the number of responses and m is the number of experimental data. $Y_i(k)$ gives the actual sequence, $Y_i^*(k)$ gives the order of the response data, max

$Y_i(k)$ indicates the highest value of $Y_i(k)$, min $Y_i(k)$ indicates the lowest value. of $Y_i(k)$, and Y is the expected value.
The data from experiments is normalized for surface roughness using Smaller-the-better equation 2 and Larger-the-better for MRR using equation 3 are shown in Table 3 known as grey relational generations.

Table.3 Normalized values response variables.

Exp. No	R_a Smaller the better	MRR Larger the better
1	0.856	0.251
2	0.692	0.225
3	0.527	0.225
4	0.593	0.711
5	0.428	0.610
6	0.264	0.599
7	0.797	1.000
8	0.165	0.925
9	0.000	0.909
10	0.918	0.144
11	0.753	0.123
12	0.589	0.086
13	0.654	0.428
14	0.490	0.369
15	0.325	0.364
16	0.391	0.658
17	0.226	0.604
18	0.062	0.588
19	1.000	0.048
20	0.835	0.021
21	0.671	0.000
22	0.736	0.267
23	0.572	0.246
24	0.407	0.230
25	0.473	0.428
26	0.308	0.390
27	0.144	0.380

Step.2. In this step compute Grey relational coefficient ξ_i (k) with normalized data of Table 3.

$$\xi_i(k) = (\Delta_{min} + \xi\Delta_{max})/(\Delta_{0i}(k) + \xi\Delta_{max}) \quad (4)$$

Δ_{0i} is the reference sequence's deviation sequence as well as the comparability sequence.

$$\Delta_{0i} = \|Y_0(k) - Y_i(k)\|$$

Where $Y_0(K)$ is normalized reference sequence and $Y_i(k)$ comparability sequence. Δ_{max} and Δ_{min} are the largest and smallest values of the comparing sequences. Equation 4 has been used to find the grey relational coefficients of the normalized data of Table 3. Given that both response characteristics have been given equal weighting, the value of the Grey relational coefficient ξi(k) is assumed to be 0.5. Table 4 presents the findings.

Table.4 Grey relational coefficient and grey relational grade values

	Evaluation of Δ_{0i}		Grey relational coefficient		GRG	Rank
	R_a	MRR	R_a	MRR		
1	0.144	0.749	0.777	0.400	0.589	7
2	0.308	0.775	0.619	0.392	0.505	14
3	0.473	0.775	0.514	0.392	0.453	21
4	0.407	0.289	0.551	0.634	0.593	5
5	0.572	0.390	0.467	0.562	0.514	13
6	0.736	0.401	0.404	0.555	0.480	15
7	0.203	0.000	0.711	1.000	0.855	1
8	0.835	0.075	0.374	0.870	0.622	3
9	1.000	0.091	0.333	0.846	0.590	6
10	0.082	0.856	0.859	0.369	0.614	4
11	0.247	0.877	0.670	0.363	0.516	12
12	0.411	0.914	0.549	0.353	0.451	22
13	0.346	0.572	0.591	0.466	0.529	10
14	0.510	0.631	0.495	0.442	0.469	19
15	0.675	0.636	0.426	0.440	0.433	25
16	0.609	0.342	0.451	0.594	0.522	11
17	0.774	0.396	0.392	0.558	0.475	17
18	0.938	0.412	0.348	0.548	0.448	23
19	0.000	0.952	1.000	0.344	0.672	2
20	0.165	0.979	0.752	0.338	0.545	8
21	0.329	1.000	0.603	0.333	0.468	20
22	0.264	0.733	0.655	0.406	0.530	9
23	0.428	0.754	0.539	0.399	0.469	18
24	0.593	0.770	0.458	0.394	0.426	26
25	0.527	0.572	0.487	0.466	0.477	16
26	0.692	0.610	0.420	0.451	0.435	24
27	0.856	0.620	0.369	0.446	0.407	27

Step.3. To calculate the grey relational grade (GRG)

$$\gamma_i = 1/n \sum_{k=1}^{n} \xi i\ (k) \qquad (5)$$

' n' is the number of response characteristics and is the necessary gray relational grade for the ith experiment. The interrelationship sandwiched between the sequence used as a reference and the comparability sequence is represented by the grey relational grade, which is an overall representation of all the quality attributes.

Equation 5 is then used to determine the grey relational grade (GRG) based on the grey relational coefficient results. In Table 5, the GRG result is displayed. When converting the multiple responses into a single grade, this outcome is utilized to optimize the responses.

IOP Conf. Series: Journal of Physics: Conf. Series **2765** (2024) 012011 doi:10.1088/1742-6596/2765/1/012011

Step .4. Then, using a greater grey relational grade, which denotes a higher level of product quality, an appropriate level of process varaiables is established. This is achieved by calculating the mean grade values for each process parameter level that is subsequently be figured as a mean response table. The mean response table's larger average grade values are taken as the ideal combination of parameters for multiple responses.

The main effects of all process parameter at various levels are displayed in Fig. 5 based on the value of GRG, and Table 6 presents the mean grey relational grade. On the basis of higher mean grey relational grade values from Table 5, the best parametric combination is selected. Better performance and a better correlation to the reference sequence are implied by a higher grey relational grade value. Subsequently, the best configurations for multiple responses become S1-P3-D1 i.e. speed of 500mm/min, power of 1000 and distance of 25mm respectively. The least value of surface roughness and the largest value of MRR are indicated by the higher mean grey relational grade values. For the laser machining parameters, the difference between the maximum and smallest mean GRG values was 0.136 for distance, 0.086 for power, and 0.043 for speed.

Table. 5 Main effects on grey relational grade.

Factors	Mean Grey relational Grade			Max-Min	Rank
	Level 1	Level 2	Level 3		
Speed	0.578	0.495	0.492	0.086	2
Power	0.535	0.493	0.537	0.043	3
Distance	0.598	0.506	0.462	0.136	1
Total mean Grey relational grade :0.522					

Step.5. The analysis of variance (ANOVA) is then conducted to identify the relevant parameters influencing the multi-responses at a 95% confidence level, which provides crucial information about the data from the experiments. The Taguchi method will not be accustomed to determining the impact of every variable on the multiple responses; therefore, the ANOVA study will be useful to determine the contribution percentages to determine the impacts. If the P-value is less than 0.05, the significance of the chosen response is determined using the F-ratio.

4. Analysis of Variance of Grey Relational Grade (GRG)

Figure 3 and Table 6 display the main effects plots and the analysis of variance (ANOVA) for the GRG, respectively. From the ANOVA table, it is found that speed and distance have a noteworthy contribution to the inconsistency of the performance features in laser machining of PMMA sheets. Given that the p-value for speed and distance is equal to 0.05, it is determined to be statistically significant at the 95% confidence interval. It is the likelihood that the GRG's 95% of observations fall within the normal curve's 95% spread, or the z-distribution. On the multi-response parameters, however, the power is statistically insignificant; this is a smaller amount of significance with a 66% (C.I./100 = 1-p value) confidence level. This finding suggests that, in a laser machining operation, distance has a greater influencing effect on multi-responses than speed and power. Distance > speed > power is the order in which process parameters are important for multiple responses.

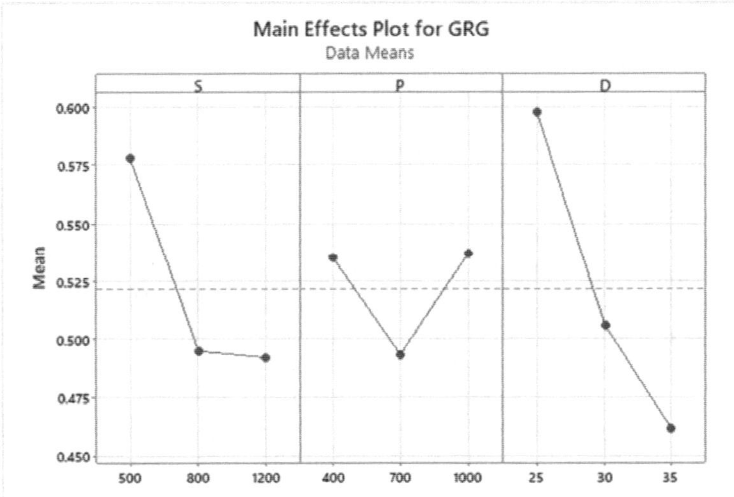

Figure 3. Main effects plots for grey relational grade

Table. 6 Results of ANOVA on GRA

Source	DF	Adj SS	Adj MS	F-Value	P-Value	Remarks
Speed (S)	2	0.04255	0.021276	4.49	0.025	Significant
Power (P)	2	0.01084	0.005419	1.14	0.339	Insignificant
Distance (D)	2	0.08687	0.043437	9.16	0.001	Significant
Error	20	0.09483	0.004741			
Total	26	0.23509				

As was previously mentioned, within the chosen range of speed and distance, the contributions of both are greater to the variability of GRG. In this work, the material removal rate and surface roughness of the PMMA sheet are two multi-performance characteristics for which the laser engraving parameters are optimized with Taguchi Grey Relational Analysis. It has been discovered that the following parameter combination yields the biggest value of the GRG, which is indicative of better process performance: Speed: 500, Power: 1000, and Distance: 25 mm.

It is acknowledged after the main effects plots, as shown in figure 3, that as the speed accelerates from 500 to 1000, there is a significant decline in the GRG assessment from 0.578 to 0.495, and on increasing further from 1000 to 1200, the GRG decreased minutely to 0.492. When the distance increases from 25 to 30, the GRG value declines from 0.598 to 0.506; when the distance increases from 30 to 35, the GRG value declines from 0.506 to 0.462. In the case of power rises from 400 to 700, a decline in the GRG value from 0.535 to 0.493 is witnessed. On further increase in distance from 700 to 1000, an increase in the GRG assessment from 0.493 to 0.537 is witnessed.
Step.6.

The next stage involves conducting confirmatory experiments to improve the grey relational grade once the ideal combination of process variables has been found. The following is how to obtain the predicted value of the grey relational grade for the optimal level:

$$\gamma^{\wedge} = \gamma_m + \sum_{i=1}^{0} (\gamma_i^{-} - \gamma_m) \qquad (6)$$

Where 0 is the number of significant control factors, γ_m is the total mean grey relational grade, and γ_i is the mean grey relational grade at the optimal level of each parameter.

The predicted or approximated grey relational grade (optimum) is equal to the mean grey relational grade and the sum of the differences between the overall mean grey relational grade and the mean grey relational grade for individual response parameters with that of the actual parameters by using the optimal engraving combinations. Equation 6 is used to calculate the predicted grey relational grade (optimum), which is displayed in Table 7. It has been discovered that the grey relational grade has improved by 18.63%.

Table 7. Results of confirmation experiment

	Predicted experiment	Confirmation experiment	Improvement of GRG
Level combination	S1 P3 D1	S1 P3 D1	
Speed		500	
Power		700	
Distance		25	
GRG	0.669094	0.8554	0.186306

5. Mathematical model

For both responses, a multiple regression model has been created, including surface roughness and material removal rate with a 95% degree of confidence, taking laser engraving speed (S), power (P), and distance (D) as input parameters. The mathematical model's acceptability has been tested by estimating its regression coefficients. From the experimental results, regression models are developed, and regression equations are presented in equations 7 and 8 below. These equations are used to compute the predicted values and are compared with the experimental results presented in figures 4.a and 4.b.

Regression Equation for R_a

$$R_a = 0.94 - 0.00201S + 0.01186\,P + 0.5708\,D \qquad (7)$$

Regression Equation for MRR

$$MRR = 0.810 - 0.001005\ \text{Speed}\ S + 0.001648\ \text{Power}\ P - 0.01156\ \text{Distance} \qquad (8)$$

IOP Conf. Series: Journal of Physics: Conf. Series **2765** (2024) 012011 doi:10.1088/1742-6596/2765/1/012011

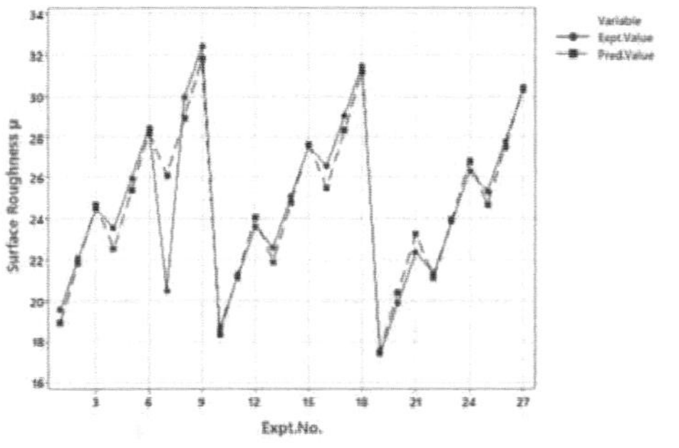

Figure 4.a.Experimental vs. predicted values of Surface roughness

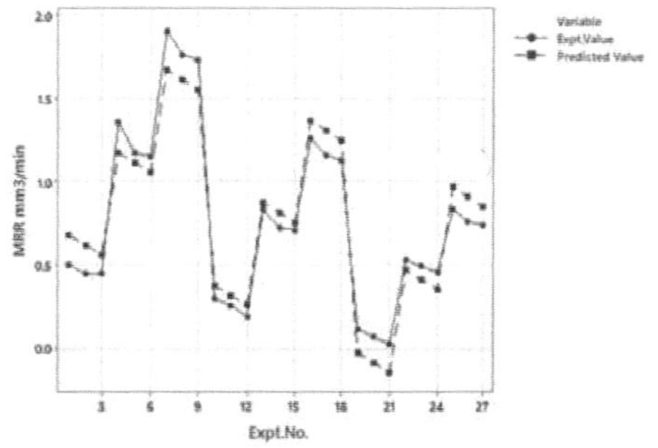

Figure4.b.Experimental vs. predicted values of Material removal rate

6.Conclusions

This work reports the summary of a multi-criteria optimization of process parameters for laser engraving of PMMA sheets in a dry environment using a 5W optical fiber diode. The full factorial approach and grey relational analysis were combined to make this decision.Furthermore, a regression model was created in advance of experimentation in order to choose the process parameters and their outcomes. The analysis leads to the following conclusions:

- In all experimental runs, laser engraving produces a material removal rate and surface roughness values that are less than the criterion limit of 1.9 mm^3 and 32.38 µm, respectively. This means that laser engraving can be successfully applied to the laser engraving of PMMA sheets.
- The optimal parameter sets for multi-responses are S1-P3-D1, i.e., speed of 500 mm/min, power of 40%, and a distance of 25 mm, respectively. The yield of multi-responses is a surface roughness Ra of 20.45 µm and an MRR of 1.9 mm^3.
- The multi-objective GRA optimization technique yielded the results observed for experimental response variables. The larger the better is MRR i.e.1.9 mm^3 showing the largest MRR of the experimental response. The smaller the better, Surface Roughness Ra 20.45 µm is nearer to the lowest value (17.45 µm) of the experimental response.
- The ANOVA table indicates that, with a 95% confidence level, distance is the most significant controlled process parameter that affects multiple responses in laser engraving of pockets, followed by speed.in terms of enhancing the rate of material removal and reducing surface roughness. The responses are independent of other parameters, such as power.
- By selecting the best possible parametric combination, the grey relation grade (0.186306) is greatly improved. As a result, manufacturing industries can use the outcomes of parametric optimization to lower production costs and improve engraving efficiency.
- When compared to a first-order model, the presented regression model shows the best fit. The model shows strong correlations between the expected and experimental outcomes.

- Using a case study of laser engraving a pocket on a PMMA sheet, the suggested Grey-based full factorial methodology has been shown to be effective in resolving multi attribute decision-making problems.
- It is observed that the multi-objective optimization made in this paper is in agreement with References 7 and 8. However, the approach followed is slightly different as regards the process of optimization.

Acknowledgments
The authors thank the AICTE- Faculty development cell (FDC). This work was supported by a grant from the AICTE of Research Project Scheme, project number F. No. 8-80/FDC/RPS (Policy-1)/2019-20 and thanks Institute for facilitating the equipment and supporting to work beyond working hours.

References;

[1] Nobuya Haraguchi, Charles Fang Chin Cheng and Eveline Smeets., 2017, The Importance of Manufacturing Econ. Development: Has This Changed? Elsevier World Development, **93**, 293–315. ISSN 0305-750X
[2] Avanish Kumar Dubey, Vinod Yadava, 2008, , Laser beam machining—A review, International Journal of Machine Tools and Manufacture, **48**, 6, 609-628
[3] Torben Lenau, Jan Haudrum, 1994,Cost evaluation of alternative production methods, Materials & Design **15**, 4,235-247.
[4] Obilor, A.F., Pacella, M., Wilson, A,2022, Micro-texturing of polymer surfaces using lasers: a review, *Int J Adv Manuf Technol* **120**, 103–135 . https://doi.org/10.1007/s00170-022-08731-1.
[5] M. C. Cakir &O. O. Cilsal,2008,Implementation of a contradiction-based approach to DFM International Journal of Computer Integrated Manufacturing , **21**, 7, 839-847.
[6] L. Santo, F. Trovalusci, J.P. Davim,2014 Laser Applications in the Field of Plastics Elsevier Comprehensive Materials Processing **9**, 243-260
[7] A Deepa, K Padmanabhan and P Kuppan,2017,Analysis and optimization of machining parameters of laser cutting for polypropylene composite,India IOP Conf. Series: Materials Science and Engineering **263** ,062043.
[8] R. S. Pawade, S. B. Bhosale,2013,Grey Relational Parameter Optimization in Ultrasonic Machining of Ceramic Composite (Al2O3/Zro2) , Journal of the Association of Engineers India,**83**, 2 (2013), 63-77 .
[9] Forte, M.A.; Silva, R.M.; Tavares, C.J.; Silva, R.F.e, 2021,Is Poly(methyl methacrylate) (PMMA) a Suitable Substrate for ALD?: AReview. Polymers, **13**, 1346
[10] Frazer RQ, Byron RT, Osborne PB, West KP,2005, PMMA: an essential material in medicine and dentistry,J Long Term Eff Med Implants,**15**,629-39.
[11] Dr. P. Nanda Kumar, Dr. G. Ranga Janardhana,2007, Development of empirical model for prediction of surface roughness in cnc end milling, Journal on Future Engineering & Technology, **3**,65-70.
[12] Khan, M.M.A.; Saha, S.; Romoli, L.; Kibria, M.H,2021, Optimization of Laser Engraving of Acrylic Plastics from the Perspective of Energy Consumption, CO2 Emission and Removal Rate, J. Manuf. Mater. Process,**5**,78

Investigation of acoustical excess parameters and FTIR studies for binary liquid mixtures of tetrachloroethylene and cyclopentanone at different temperatures.

M Bonjubabu[1], P Mahesh[2], P V S L Aparna[3], K Poornima[4], K Jayarao[4] and K. Samatha[2]

1 Department of HBS(Physics), Welfare Engineering College, Pinagadi, Visakhapatnam-531173 A.P, India.
2 Department of Physics, Andhra University, Visakhapatnam-53003, A.P, India
3 Department of Physics, S.R.K.R. Engineering College (A), China Amiram, Bhimavaram-534204, West Godavari (AP), India.
4 Department of HBS, Welfare Engineering College, Pinagadi, Visakhapatnam- 531173 A.P, India

Mobile: + 917671985262 Email: bonjumphd86@gmail.com

Abstract: Speed of sound (U), Density (ρ), and viscosity (η) values of liquid mixtures of Tetrachloroethylene with Cyclopentanone and pure liquids were determined at regular intervals of molar range at different temperatures (303.15, 308.15, 313.15, 318.15) K. From the experimentally determined values, thermo-acoustic parameters such as excess molar volume (V^E) and excess free length (L_f^E), excess Gibb's free energy (ΔG^E) and excess enthalpy (H^E) have been estimated. The variations of the above parameters discussed based on intermolecular interactions existing in this binary mixture. The computed values of sound speed in the mixture from various theories are compared with experimentally calculated values and have a clear check to validate the theories for the present system, and FTIR studies provide detailed analysis of molecular interactions.

Keywords: viscosity, density, ultrasonic velocity, excess parameters, molecular interaction.

1. Introduction

The volumetric and viscometry survey together with Speed of sound of liquid mixtures are of significant parameters and play a vital role in realizing the intermolecular interactions obtained in and around the molecules of finding large usage in various manufacturing technological industries[1,2]. Diverse scholarly articles found with density, viscosity, and speed of sound calculations for huge range of binary liquid mixtures contain ketones as one of the components[3-8] and these tracts were construed in terms of specific and nonspecific interactions.

Ketones are tinted for dipole – dipole interactions and hydrogen bonding. Ketones have a key role for their applications in chemistry, biology and interaction studies of hydrogen and other liquid bindings. This study reveals that ketone molecules play a vital role in forming hydrogen bonding among molecules. In depth investigation of molecular association in liquids related to aromatic functional group with various functional groups is carried out in this study. The study of thermodynamic properties of multi component mixtures and other different computational models are beneficiary for pharmaceutical and other industries applications[9].

The functional parameters of thermodynamics were sensitively depending on various intermolecular forces and on size and shape[10]. From the determined values of speed of sound and viscosity and density, we will extract the excess thermo-acoustic parameters like excess ultrasonic velocity (U^E), excess adiabatic compressibility (β^E_{ad}), excess molar volume (V^E), excess free length (L_f^E), excess Gibb's free energy (ΔG^E) and excess enthalpy (H^E). Now this current epilogue shows the speed of sound and viscosity values had been a huge empirical bondage in the liquid. In this study the results were discussed for liquid mixtures of tetrachloroethylene with cyclopentanone at temperatures of (303.15, 308.15, 313.15 and 318.15) K. FT-IR spectra of binary mixtures were obtained at room temperature for different compositions and pure liquids. The FT-IR spectroscopy is one of the tools to identify the strength of interactions, type of bonding and existence of hydrogen bond with shape of the absorption peaks and frequency shift of the band.

2. Experimental

2.1 Materials

Tetrachloroethylene (TCE) with cyclopentanone is used in this study which was AR grade obtained from Merck Co. Inc., Germany, by 99% purity with a standard purified method[11].

2.2 Measurements

By amalgamation of all these chemical mixtures were kept in observation in air tight bottles for 72 hours' for viscosity, velocity and density measurements. Viscosity was measured by using Ostwald viscometer. The uncertainty in viscosity measurement is up to 0.001mPa-s. For measurement of density was accurately using 10 ml specific gravity bottles. The average uncertainty in the measured density is ±0.001 kg/m^3. The speed of sound is measured with a single crystal variable path interferometer operating at a frequency of 2MHz that had been calibrated tetrachloroethylene with and cyclopentanone used for collecting the different temperatures. The FT/IR-6000 FT-IR spectrometer was used to get the data of binary mixtures.

2.3 Computational Details

The values of experimentally determined density and speed of sound for the binary mixtures of tetrachloroethylene with cyclopentanone at 303.15 K, 308.15 K, 313.15 K, and 318.15 K for different compositions.

In the present work under study by using formulas excess values were calculated by using the expressions given in literature[12] as follows,

Excess free length $$L_f^E = L_f - KT\,(Ks)^{1/2} \quad ...1$$

Excess Gibbs free energy $$\Delta G^E = RT\left[\ln\left(\frac{\eta V}{\eta_2 V_2}\right) - x_1 \ln\left(\frac{\eta_1 V_1}{\eta_2 V_2}\right)\right] \quad ...2$$

Excess enthalpy $$H^E = H - (x_1 H_1 + x_2 H_2) \quad ...3$$

Excess adiabatic compressibility $$\beta^E_{ad} = \frac{1}{\rho U^2} \quad ....4$$

Excess volume $$V^E = \frac{x_1 M_1 + x_2 M_2}{\rho} - \left(\frac{x_1 M_1}{\rho_1} + \frac{x_2 M_2}{\rho_2}\right) \quad ...5$$

3. Results and discussion

The comparison experimental values and literatures of speed of sound(u), density(ρ) and viscosity (η) in case of the binary liquid mixture (Tetrachloroethylene + Cyclopentanone) under study at temperatures, T =303.15 K are given in Table 1.

Liquid	Ultrasonic Velocity (u) $m.s^{-1}$		Density(ρ) $x10^{-3}$ Kgm^{-3}		Viscosity(η) x 10^{-3} $Ns.m^2$	
	EXP	**LIT**	**EXP**	**LIT**	**EXP**	**LIT**
Tetrachloroethylene	1028.80	1028.80[13]	1606.06	1606.34[14]	0.797	0.798[15,16]
Cyclopentanone	1378.05	1374.00[17]	939.11	939.40[18]	0.997	0.9346[19]

Table 2:- From the data of density(ρ), speed of sound(U) and viscosity(η), adiabatic compressibility, the values of molar volume (V_m), Intermolecular free length (L_f) and molar volume (V_m) were calculated.

Mole Fraction of TCE	ρ x 10^{-3} Kgm^{-3}	U $m.s^{-1}$	η x 10^{-3} $Ns.m^2$	β_{ad} x 10^{-10} $kg^{-1}.m\ s^2$	L_f x10^{12} m	V_m x 10^{-5} m^3/mol^{-1}
Tetrachloroethylene +Cyclopentanone at 303.15 K						
0.0000	939.11	1378.05	0.997	5.6072	4.9135	89.5741
0.1478	1051.21	1329.34	0.972	5.5988	4.9098	92.0011
0.3025	1145.26	1277.25	0.947	5.6664	4.9394	94.6668
0.4645	1250.90	1220.62	0.919	5.7822	4.9896	97.3122
0.6343	1358.78	1159.64	0.887	5.8656	5.0254	100.0520
0.8126	1462.89	1094.32	0.847	5.9214	5.0493	102.8883
1.0000	1606.06	1028.80	0.797	5.9415	5.0521	105.9826
Tetrachloroethylene +Cyclopentanone at 308.15 K						
0.0000	933.80	1358.42	0.932	5.8033	5.0468	90.0835
0.1478	1045.75	1307.65	0.917	5.8005	5.0456	92.5532
0.3025	1139.22	1255.32	0.899	5.8664	5.0742	95.2252
0.4645	1244.30	1198.25	0.869	5.9927	5.1285	97.8628
0.6343	1351.58	1137.65	0.845	6.0777	5.1648	100.5928
0.8126	1455.16	1075.42	0.811	6.1332	5.1883	103.4385
1.0000	1597.72	1010.80	0.769	6.1624	5.2006	106.5155
Tetrachloroethylene +Cyclopentanone at 313.15 K						
0.0000	929.71	1337.93	0.862	6.0087	5.1844	90.4798
0.1478	1040.27	1285.47	0.849	6.0114	5.1856	93.0954
0.3025	1133.17	1232.85	0.837	6.0811	5.2155	95.8066
0.4645	1237.60	1175.12	0.817	6.2122	5.2715	98.4465
0.6343	1344.34	1114.24	0.798	6.3002	5.3087	101.1216
0.8126	1447.41	1054.34	0.768	6.3581	5.3330	103.9855
1.0000	1589.35	993.00	0.736	6.3809	5.3425	107.0081
Tetrachloroethylene + Cyclopentanone at 318.15 K						

0.0000	924.42	1320.10	0.810	6.2075	5.3193	90.9975
0.1478	1034.78	1263.25	0.796	6.2323	5.3299	93.6556
0.3025	1127.10	1211.54	0.781	6.3076	5.3620	96.3684
0.4645	1230.90	1152.32	0.765	6.4431	5.4193	98.9806
0.6343	1337.07	1095.39	0.751	6.5332	5.4570	101.6769
0.8126	1456.21	1036.84	0.729	6.5792	5.4741	104.4712
1.0000	1580.90	978.26	0.709	6.6097	5.4889	107.4835

Table 3: Values ofexcess ultrasonic velocity (U^E), excess adiabatic compressibility (β_{ad}^E), excess enthalpy (H^E), Intermolecular free length (L_f^E), Molar volume (V_m^E), excess Gibb's free energy (ΔG^E) and for the binary liquid mixtures of tetrachloroethylene with cyclopentanone at 303.15to 318.15K.

Mole Fraction of TCE	**U^E $m.s^{-1}$**	**β_{ad}^E x 10^{-10} $N^{-1}.m^2$**	**HE $KJmol^{-1}$**	**L_f^E x10^{12} M**	**V_m^Ex 10^{-5} m^3/mol^{-1}**	**ΔGE x 10^{-20} $kJ.mol^{-1}$**
Tetrachloroethylene + Cyclopentanone at 303.15 K						
0.0000	0.0000	0.0000	0.0000	0.0000	0.0000	0.0000
0.1478	-22.1255	-0.0480	0.2151	-0.0212	-0.3424	0.0012
0.3025	-35.2556	-0.0796	0.3815	-0.0327	-0.7226	0.0017
0.4645	-40.3565	-0.0917	0.4578	-0.0362	-1.0404	0.0018
0.6343	-36.1745	-0.0874	0.3922	-0.0322	-0.8445	0.0015
0.8126	-21.9871	-0.0609	0.2221	-0.0202	-0.4265	0.0009
1.0000	0.0000	0.0000	0.0000	0.0000	0.0000	0.0000
Tetrachloroethylene + Cyclopentanone at 308.15 K						
0.0000	0.0000	0.0000	0.0000	0.0000	0.0000	0.0000
0.1478	-24.1521	-0.0417	0.2612	-0.0185	-0.4265	0.0011
0.3025	-38.2155	-0.0694	0.4454	-0.0291	-0.8522	0.0016
0.4645	-44.1221	-0.0804	0.5172	-0.0322	-1.1645	0.0017
0.6343	-39.2548	-0.0772	0.4440	-0.0287	-0.9545	0.0014
0.8126	-23.5859	-0.0531	0.2714	-0.0181	-0.5258	0.0008
1.0000	0.0000	0.0000	0.0000	0.0000	0.0000	0.0000
Tetrachloroethylene + Cyclopentanone at 313.15 K						
0.0000	0.0000	0.0000	0.0000	0.0000	0.0000	0.0000
0.1478	-26.5248	-0.0343	0.3127	-0.0157	-0.5117	0.0011
0.3025	-42.1976	-0.0580	0.5041	-0.0252	-0.9922	0.0016
0.4645	-47.9256	-0.0687	0.5714	-0.0282	-1.2955	0.0017
0.6343	-42.3132	-0.0668	0.5064	-0.0252	-1.0895	0.0014
0.8126	-25.8856	-0.0468	0.3215	-0.0160	-0.6334	0.0008
1.0000	0.0000	0.0000	0.0000	0.0000	0.0000	0.0000
Tetrachloroethylene + Cyclopentanone at 318.15 K						
0.0000	0.0000	0.0000	0.0000	0.0000	0.0000	0.0000
0.1478	-29.3215	-0.0272	0.3637	-0.0131	-0.6152	0.0009

0.3025	-46.1405	-0.0471	0.5601	-0.0209	-1.1560	0.0014
0.4645	-51.5425	-0.0565	0.6192	-0.024	-1.4241	0.0015
0.6343	-45.5214	-0.0563	0.5608	-0.0213	-1.2166	0.0011
0.8126	-28.3798	-0.0402	0.3727	-0.0136	-0.7345	0.0007
1.0000	0.0000	0.0000	0.0000	0.0000	0.0000	0.0000

Table 4: - The Figures of excess values of ultrasonic velocity (U^E), adiabatic compressibility (β_{ad}^E), enthalpy (H^E), Intermolecular free length (L_f^E), Molar volume (V_m^E), Gibb's free energy (ΔG^E) and for the binary liquid mixtures of tetrachloroethylene with cyclopentanone at 303.15 to 318.15K.

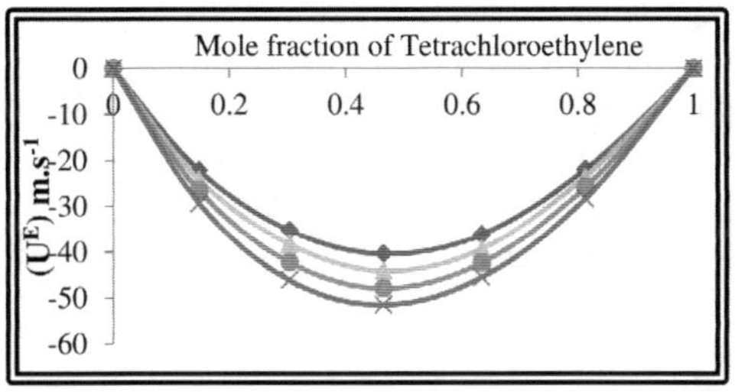

Figure 1. Variation of excess Ultrasonic velocity

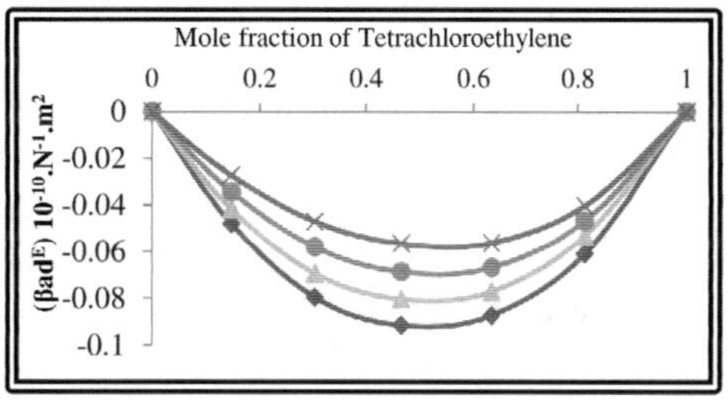

Figure 2. Variation of excess adiabatic compressibility

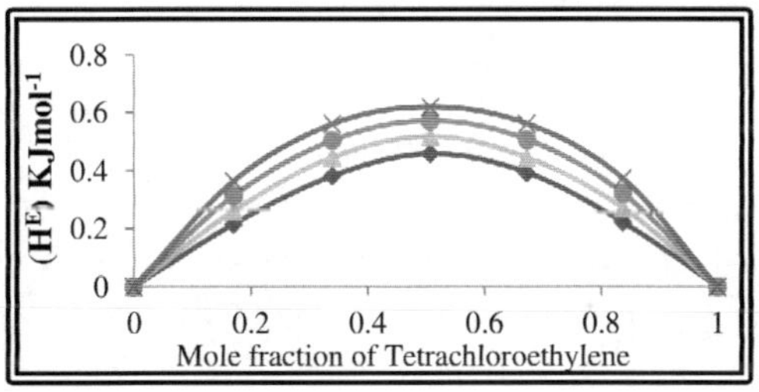

Figure 3. Variation of excess enthalpy

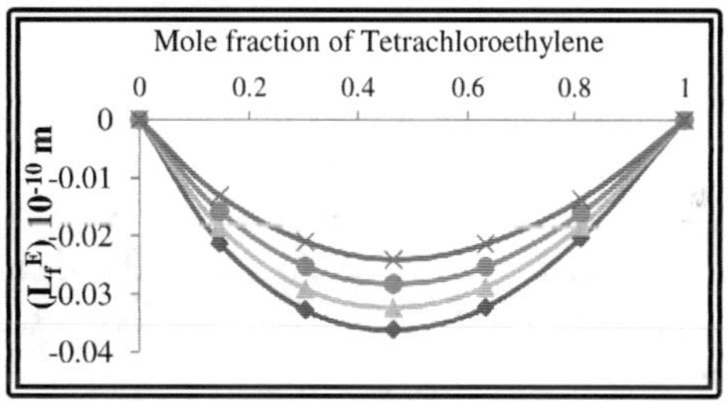

Figure 4. Variation of excess free-length

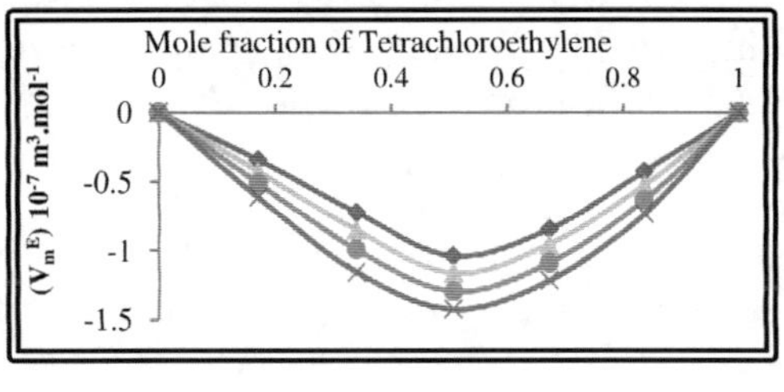

Figure 5. Variation of excess Molar volume

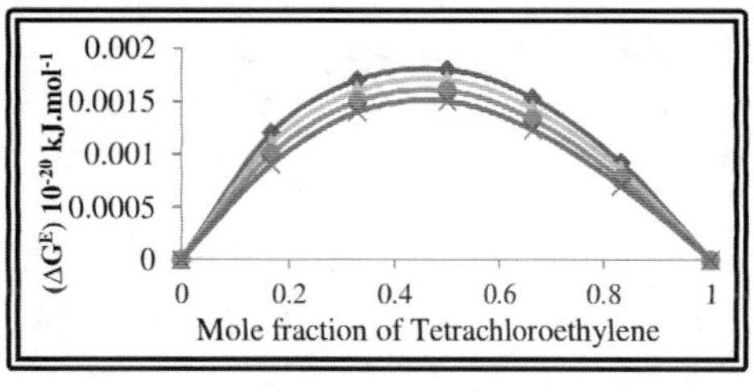

Figure 6. Variation of excess Gibb's free energy

From the above Table1 stated the literature survey and from Table 2 and Table 3 we get the value of parameters and also excess parameters of binary mixtures contain tetrachloroethylene with cyclopentanone were computed against mole fractions at different temperatures, these are mentioned Fig. 1-6.

From Figure-1, the presence of strong interactions in these mixtures is confirmed through the negative values. This sign of the excess ultrasonic velocity (U^E) plays a significant role in the compactness of association among molecules in liquid mixtures through charge transfer, dipole – dipole interaction between different molecules of different shapes and size stated by fort et al[20].

Figure-2 indicates the corresponding graphs against the binary liquids of tetrachloroethylene with cyclopentanone can be visualised the action of binary mixtures at various temperatures. As it was stated that more negative variation of excess adiabatic compressibility(β_{ad}^E)[21] is observed. Jacobson observed the alteration in free length by the values adiabatic compressibility assigned to strong interaction. The dispersion forces indicate the vigour interaction had stated by fort and more[22].

Figure-3 shows the excess enthalpy (H^E) which designates the behaviour of fluids in clarifying the internal structure. The complex information of H^E is important in chemistry, and other biological branches. It also reveals the depth and type of the molecular collaborations and signifies the energy of the hydrogen bonding. Positive values of H^E indicates vital interactions among the components in liquids[23,24]. Increasing content of TCE marks that dispersion compels in the liquids and molecules[25] and confirms the significance of feeble dipole-dipole interactions in this system[26].

From figure-4, in a keen observation it reveals that L_f^E values had negative pattern corresponding to what observed in the study U^E. The negative values of L_f^E indicate particular interactions among dissimilar molecules[27]. The variation of excess molar volume (V_m^E) with the mole fraction X_1 is mentioned in figure 5. This intermolecular association will be explained by using the magnitude and sign of V_m^E. In this case study the V_m^E values are highly negative and might be attributed to the establishment of hydrogen bond and breaking of alcohol self-association. It is a clear result of geometrical fitting of molecules of different size and structure of different components within the mixture. These intermolecular forces of attraction in a molar volume can be affected by temperature increase is connected to the presence of weakening of forces[28] by Garcia et all[29,30] and similar corresponding works were observed by previously observed due to thermal agitation.

From Figure-6, the excess Gibb's free energy of activation (ΔG^E) shows positive changes with changes in mole fraction and increasing values with temperature. ΔG^E variation indicates the presence of hydrogen bonding among constituent component molecules under study[31]. This property was used in desulfurization of sulphur coals, since the removal rate of organic sulphur become higher in Tetrachloroethylene solvents[32].

Computed values of speed of sound in the liquid mixtures with various relations like Impedance dependence relation, Nomoto's relation, Junjie's theory, Ideal mixing relation, Rao's specific velocity relation and free length theory over the whole mole fraction range of tetrachloroethylene at certain temperatures are given in Table 5.

Table 5: - Ultrasonic velocity experimental and theoretical values at temperatures (303.15, 308.15, 313.15, and 318.15) K for the Tetrachloroethylene mole fraction with Cyclopentanone.

Mole Fraction(X)	U_{EXP} ms^{-1}	U_J ms^{-1}	U_{NOM} ms^{-1}	U_{IMX} ms^{-1}	U_{RAO} ms^{-1}	U_{FLT} ms^{-1}	U_{IDR} ms^{-1}
303.15 K							
0.0000	1378.05	1378.05	1378.05	1378.05	1378.05	1378.05	1378.05
0.1478	1303.48	1297.78	1315.07	1297.26	1557.52	1298.66	1298.12
0.3025	1241.34	1229.13	1254.04	1228.30	1578.32	1254.19	1229.30
0.4645	1175.82	1169.53	1194.93	1168.57	1556.64	1176.93	1169.41
0.6343	1120.13	1117.11	1137.71	1116.25	1444.77	1104.74	1116.82
0.8126	1074.43	1070.54	1082.34	1069.99	1231.94	1090.86	1070.28
1.0000	1028.80	1028.80	1028.80	1028.80	1028.80	1028.80	1028.80
308.15 K							
0.0000	1358.42	1358.42	1358.42	1358.42	1358.42	1358.42	1358.42
0.1478	1283.96	1277.87	1295.09	1277.66	1536.34	1279.62	1278.12
0.3025	1223.23	1208.97	1233.76	1208.6	1555.14	1208.94	1209.01
0.4645	1158.04	1149.13	1174.40	1148.69	1532.19	1185.54	1148.88
0.6343	1103.34	1096.50	1169.70	1096.09	1420.18	1127.23	1096.11
0.8126	1058.52	1049.73	1061.45	1049.46	1209.07	1056.14	1049.41
1.0000	1007.80	1007.80	1007.80	1007.80	1007.80	1007.80	1007.80
313.15 K							
0.0000	1337.93	1337.93	1337.93	1337.93	1337.93	1337.93	1337.93
0.1478	1264.55	1258.73	1275.64	1258.51	1509.87	1256.73	1259.04
0.3025	1204.65	1190.96	1215.31	1190.58	1528.71	1213.54	1191.09
0.4645	1140.47	1132.08	1156.91	1131.64	1506.52	1166.07	1131.94
0.6343	1086.59	1080.30	1100.41	1079.88	1397.21	1098.54	1079.99
0.8126	1042.41	1034.27	1045.79	1033.99	1190.28	1045.64	1034.00
1.0000	993.00	993.00	993.00	993.00	993.00	993.00	993.00
318.15 K							
0.0000	1320.10	1320.10	1320.10	1320.10	1320.10	1320.10	1320.11
0.1478	1245.23	1241.60	1258.34	1241.45	1491.16	1271.25	1241.87
0.3025	1186.00	1174.43	1198.54	1174.17	1508.9	1211.40	1174.51
0.4645	1122.90	1116.09	1140.66	1115.77	1486.18	1183.52	1115.89
0.6343	1069.94	1064.77	1084.67	1064.46	1377.76	1110.42	1064.42
0.8126	1026.68	1019.15	1030.55	1018.95	1317.01	1057.71	1018.86
1.0000	978.26	978.26	978.26	978.26	978.26	978.26	978.26

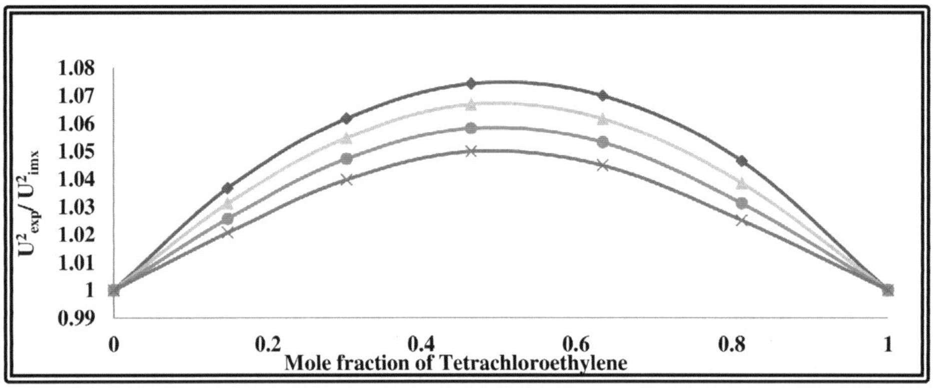

Figure 7. Variaton of U^2_{exp}/U^2_{imx} with mole fraction of Tetrachloroethylene + Cyclopentanone

It is observed that the positive percentage deviation for Rao's theory and in other theoretical values it is negative. Thus, experimental data of free length theory is well suited for this mixture. In all the theories, the deviations of ultrasonic velocity represent strong interactions. The occurrence of negative deviations in five models confirms the presence of strong hydrogen bonding between the components. From Fig. 7 the interaction parameter also positive denotes the strong possible interactions. The value of U^2_{exp}/U^2_{imx} disperses with temperature and the interaction between dissimilar molecules reduces with increasing temperature.

3.1 FT-IR Studies on Tetrachloroethylene with Cyclopentanone binary mixtures

The solution structure and physical information about intermolecular interactions can be investigated using the spectroscopic technique like IR which also gives information about the functional groups that are influenced by the surrounding condensed medium[33]. The infrared spectra are recorded for the one binary mixture Cyclopentanone at room temperature of 298.15 K as a function of the composition of TCE to study of interactions of the molecules between the binary liquid mixtures.

Figures 8, 9 and 10 show the recorded IR-spectra of pure Tetrachloroethylene, its binary mixture with Cyclopentanone (TCE + CYP = 1:1 ratio) and pure Cyclopentanone respectively. The compounds like Cyclopentanone of the ketone functional group gives C–H alkyl stretching 2800 – 3000 cm-1, C = O stretching 1670 – 1820 cm^{-1} and C-C stretching about 1310 cm^{-1}.

It is observed from the recorded infrared spectra of TCE that the C=C Bending band is at 902.77 cm^{-1} with the C-Cl stretching at 775.10 cm^{-1}. Continuous change is observed in the frequency of IR bands varies on mixing Tetrachloroethylene + Cyclopentanone binary liquid mixture for different concentrations.

The observed band frequency value of C=O stretching is changed from 1728 cm^{-1} and maximum frequency deviation is observed at equal composition of constituent liquids. It is observed that strong interactions exist between tetrachlorethylene and Cyclopentanone binary liquid mixture. The variation of excess thermodynamic factors discussed on focussing the interactions is confirmed by the IR data[32-35].

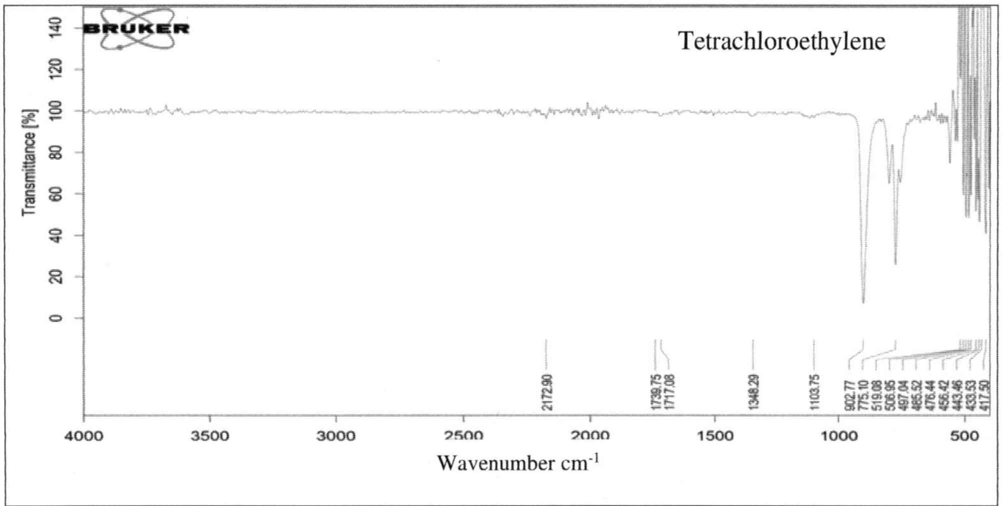

Fig 8 FT-IR Spectrum of Tetrachloroethylene

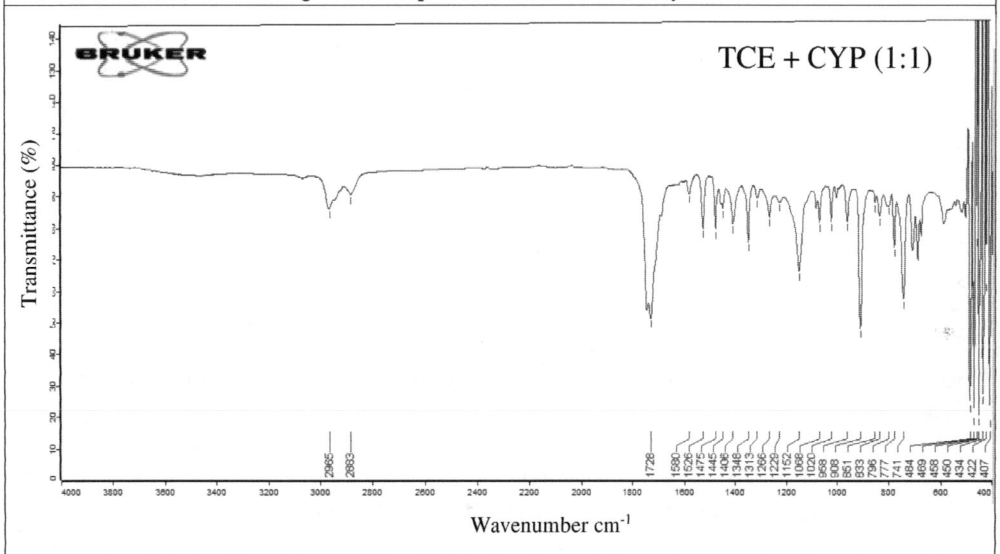

Fig 9 FT-IR Spectrum of Tetrachloroethylene + Cyclopentanone (1:1)

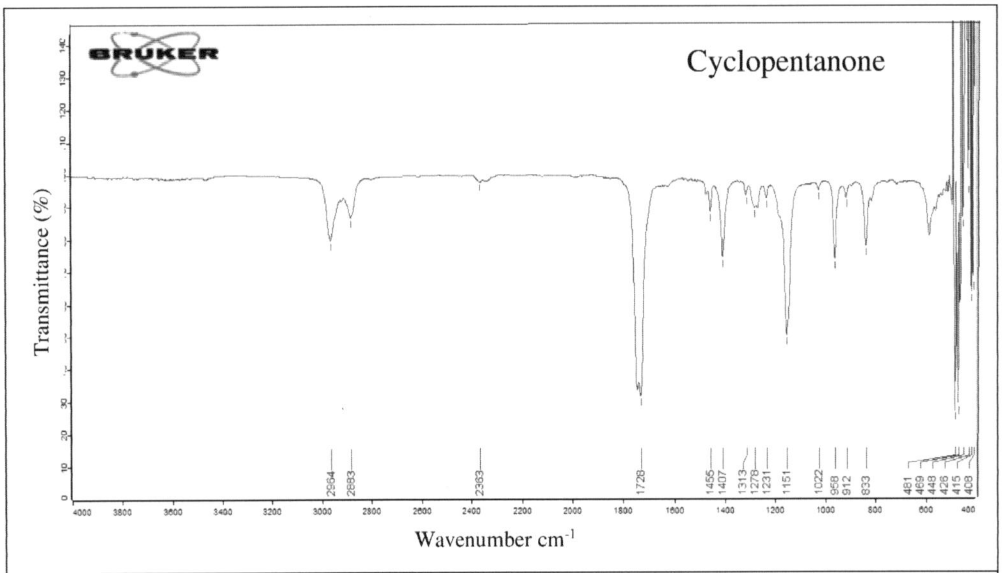

Fig 10 FT-IR Spectrum of Cyclopentanone

4. Conclusions

In the present study, it reveals that all the excess parameters like U^E, V_m^E, β_{ad}^E, L_f^E. and H^E, ΔG^E is computed from the determined speed of sound(u), density(ρ) and viscosity(η) values and the creation of hydrogen bonds among the molecules of all mixtures is identified. The excess molar volume had slightly changed from positive to negative for the binary system. This is an apt sign for the proximity of hydrogen bonding with that of other, various theories that were correlated with the experimental values. Speed of sound values was obtained with free length theory relation is in good agreement with the experimental value. As a common conclusion, it can be said that the FTIR spectral analysis in the liquid mixtures of Tetrachlorethylene studied are supported well by the experimental results from excess acoustical and thermodynamic techniques. This is confirmed by the presence of strong interactions.

5. Acknowledgements

The author expresses his gratitude to Prof. K. Samatha, for continuous support in carrying out experimental work and conveys special thanks to HOD, Department of Physics for providing Lab facilities.

6. Reference Journals

1. Oswal S L, Oswal P and Shalak R P1998 *J. Sol. Chem.,* **27**.
2. Kumar H, Kaur M, Gaba R and Kaur K 2011 *J. Therm. Anal. Calorim.,* **105.**
3. Zorebski E and Waligora E 2008 *J. Chem. Eng. Data.,* **53**.
4. Boruń A, Żurada M and Bald A 2010 *J. Therm. Anal. Calorim.,* **100**.
5. Sastry S S, Babu S, Vishwam T, Parvateesam K and Tiong H. S. 2013. *Phys B.,* **420**.
6. Checoni R F 2010 *J.Therm. Anal. Calorim.,* **101**.

7. Sastry S S, ShaikBabu and Vishwam T 2014. *J. Chem. Thermodynamics.,**68***.
8. Sastry S S, Ibrahim S M, Tanuj Kumar L and Babu S (2015) *Int. J. Eng. Res. & Tech.,***4**.
9. Resa J M, Gonzalez C and Goenaga J M 2007 *J Therm. Anal Calorim.,* **45**.
10. Sharma S Jasmin B Ramani J Patel R 2011 *Phys. Chem. Liq.,* **49**.
11. Vogel A I 1989 *John Wiley.,* **5**.
12. Narendra K, Srinivasu Ch and Kalpana Ch 2012 *J. Therm. Anal. Calorim.,***107**.
13. Venkatramana L, Gardas R L and Reddy R L 2015 *J. Sol. Chem.,* **44,** 359.
14. Singh M and Agarwal D 2004 *J. Chem. Eng. Data.,* **49,** 1224.
15. Iloukhani H and Reddy K D Rao M V P 2007 *J. Sol. Chem.,* **36,** 714.
16. Venkatesulu D, Venkatesu P and Rao M V P 1997 *J. Chem. Eng. Data.,* **42**, 367.
17. Sankar M G, Venkateswarlu P and Kumar K S Ramabu S S 2013 *J. Sol. Chem.,* **42,** 935.
18. Reddick J A, Bunger W B and Sakano T K 1986 *Int. Sci.,* **45**, 2.
19. Swain N, Singh S K, Panda D and Chakravortt V 2000 *J. Mol. Liq.,* **85,** 32.
20. Fort R J and Moore W R 1965 *Trans. Faraday Soc.,* **61**.
21. Uvarani R and Sivapragasam J 2009 *J. Chem.,* **6(4),** 1150.
22. Mehra R and Gaur A K 2009 *J. Ind. Council Chem.,* **26(1)**, 85.
23. Wankhede D S, Wankhede N N, Lande M K and Arbad B R 2008 *J. Mol. Liq.,* **138**, 124.
24. Vesely F, Uchytil P, Zabransky and M Pick J 1979 *Collect Czech, Chem. Eng. Data.,* **44**, 5869.
25. Bhuiyan M M H and Uddin M H 2008 *J. Mol. Liq.,* **138,** 139.
26. Ali A and Nain A K 2002. *Bull. Chem. Soc. Japan.,* **74,** 681.
27. Oswal S L, Pandiyan V, Kumar B and K Vasantharani P 2010 *Thermo. Acta.,* **507**.
28. Bhatia S C, Rani R Bhatia R and Anand H 2011 *J. Chem. Thermodyn.,* **43**.
29. García B, Aparicio S, Navarro A M and Alcalde R. 2004 *J. Phys. Chem. B.,* **108.**
30. Sreeharisastry S, Babu S, Vishwam T and Tiong H S 2014 *J. Therm. Analysis and Calorimetry.,* **116**.
31. Ashwin Kumar, 2020, *J. Chem. Thermodynamics*, **19**.
32. Raju R, Ravikumar S, Arokiaraj R G, Karlapudi S, Sivakumar K and Pandian, 2020, *Chem. Data Collections,* **29**.
33. Dubey G P and Dhingra L, 2021, *J. Chem. Thermodynamics,* **157**.
34. Christina Almeida, Sofia S, Claudia A, Ana P, Sandra N and Celia A R, 2013 *J. Braz. Chem. Soc.*, **24**, 1403.
35. Hendricson J B, Cram G S and Hammond *Org. Chem, McGraw-Hill Koga kusha Ltd.,* Tokyo.

EXPERIMENTAL STUDIES ON POLYESTER - TITANIUM FUNCTIONALLY GRADED MATERIALS

S **Madhusudan**[1], **Rajesh Mudunuri**[2*], **R Bhargavi**[1], **V Ajay Kumar**[1]

[1]Department of Mechanical Engineering, Usharama College of Engineering and Technology, Telaprolu, Andhra Pradesh, India.
[2]Department of Mechanical Engineering, Sagi Rama Krishnam Raju Engineering College, Bhimavaram, Andhra Pradesh, India.
*Email: rajesh_m@srkrec.ac.in

Abstract. In the present work an attempt has been made to fabricate Polyester based Titanium Functionally Graded Materials. Titanium particles are produced through conventional filing operation, by rotating Titanium rod of 50 mm diameter on lathe machine. Filing operation has been carried out at 650 rpm using a course file. The obtained powder was demagnetized to remove the foreign material related iron. Powders are subjected to sieving operation to get desired range of particle size i.e 300 microns. FGMs with 2.5 gm, 5 gm and 10 gm titanium by weight are fabricated through sedimentation technique. The quality of the FGMs is examined through liquid penetrant non-destructive testing. FGMs are tested and analyzed from density, hardness and optical microscopy point of view. Obtained results are compared and analyzed.

1.Introduction

The scientific community has created a great deal of novel materials in today's inventive environment. High strength/weight and stiffness ratio composite materials that are the lightest in weight have found effective uses in the aircraft industry and other engineering fields. Functionally graded materials (FGMs) are a novel class of composite materials that have garnered a lot of attention lately. By layered mixing two materials with distinct thermo-mechanical properties and volume ratios, one can create Functionally Graded Materials (FGM) by progressively altering each layer until the final layer has the highest volume/weight ratio of the first phase and only a small number of particles from the second phase. In order to maximize heat resistance and mechanical properties, functionally graded materials (FGMs) distribute the material functions throughout the material body. This makes them perfect for spacecraft, where one side may be exposed to extremely high temperatures and the other side to extremely low temperatures. Many of the analytical techniques created for conventional composites with distinct phases may not be directly applicable to FGMs due to the constantly changing material properties that are a characteristic of FGMs. Functionally graded materials have gained a lot of attention lately due to their many uses in industries including energy, defense, aerospace, electronics, and medicine. The type of processing method employed determines the final characteristics of FGM. The two primary processes in the creation of FGMs are consolidation, which entails placing the graded structure on a substrate, and nonhomogeneous spatial gradation of the component parts. FGMs with compositional gradients are created using a variety of techniques. The segregation phenomenon, which is brought on by the different specific gravities of the alloy liquids and reinforcement particles, makes it possible to produce Functionally Graded Materials (FGMs) with gradient distributed reinforcement particles using centrifugal casting. By managing the solidification process, the gradient can be preserved.

Metal and ceramic-based systems have been the subject of the majority of research. Sedimentation is a specific process used in the production of particle reinforced FGMs. The difference in density between the particle and the matrix is the basis for the sedimentation process. If gravitational force is used properly, a graded structure can be created. Hassanin and Jiang [1] developed Multi-layer FGMs using highly stable alumina/zirconia suspensions. Suspension stability was achieved by controlling both pH and the concentration of dispersant with respect to the powder weight.

Materials with varying functional grades are useful in a wide range of applications. FGMs made of Al_2O_3/Y and ZrO_2 are utilized in medical applications like knee or hip prostheses [2]. The energy industry [3], high current connectors, sensors, and capacitors [4-6] all use graded dielectrics. When Youngming et al. [7] studied the Al_2O_3/Ti_3SiC_2 FGM system, they found that as the weight percentage of Al_2O_3/Ti_3SiC_2 content increased, the FGM's electrical conductivity increased as well. Additionally, it was discovered that the Al_2O_3/Ti_3SiC_2 content of the FGM directly correlated with its increased hardness. Afzal et al. synthesized a bio FGM zirconia, hydroxyapatite, and alumina matrix and examined its structural integrity [8]. Das et al. [9] summarized current advances in engineering and the processing of materials have led to a new class of graded multilayered materials called functionally graded materials (FGMs). This article looks at the best processing technologies and applications of the advanced, high quality products generated in FGMs. It also highlights about the future research scope in FGMs.

According to Jha et al. [10], the tailored structural properties, including morphological properties, are the primary advantage of FGMs over other composites. When comparing FGMs to non-graded FGMs, Pan E found that the residual stresses were lower in FGMs [11]. Thermal barrier coatings are used in gas turbine engine components and tungsten-coated dampers, according to research by Kim, J.H. et al. The thermal mismatch between the coating material and the base, however, presents one of the main obstacles in the development of thermal barrier coatings [12]. Andrew et al. [13] demonstrated that thixotropic casting successfully produced regular gradients in hydroxyapatite-316L stainless steel biomaterial FGMs. An impeller dry blending (IDB) case study in this chapter showed the potential of IDB for producing excellent linear gradients in metal-ceramic FGMs. A hydrostatic shock forming case study showed its potential as a densification method for continuous bulk FGMs for which the metal and ceramic components have greatly differing melting points. Ti-TiB FGMs were fabricated using pressure-assisted densification for ballistic applications. The disadvantages of residual stresses were addressed by the developed Ti-TiB FGMs. It has been reported that mechanical characteristics like the modulus of elasticity, thermal coefficient of expansion, and ballistic performance are significantly impacted [14]. Functionally graded polyster was created, and its mechanical properties were examined by Lucignano and Quadrini [15].

Spark plasma sintering (SPS) is one of the most popular solid phase methods for fabricating FGM [16–18]. The SPS technique is useful for producing nano products, metal matrix composites, fiber reinforced metal matrix composites, intermetallic compounds, FGMs, and a wide range of other materials with high wear resistance, including sandblasting nozzles [19]. The SPS process is appropriate for the creation of FGMs because it can produce varying phase melting temperatures. Compared to other conventional techniques, the SPS technique offers comparatively easy activation and purification of the powders. Current activated tip based sintering was explained by EI - Desounky et al., who also demonstrated a direct correlation between green density and FGM size [20]. Using phenolic resin and nanofibers and a powder metallurgy process, Bafekrpour et al. [21-23] created polymeric functionally graded nano composites and studied their viscoelastic and thermo mechanical properties. Furthermore, their findings showed that the composites with the highest percentage of graphite in the top and bottom layers and the lowest percentage in the center showed improvements of about 97% in thermo mechanical properties and 34.7% in creep resistance. Ashish et al. [24] presented an overview of FGM is presented regarding the concept, need, and manufacturing methods used to manufacture FGM to date and mathematical models representing the FGM effective properties are discussed. Current trends regarding the application of AM methods on manufacturing FGM are discussed from both the academic and industrial points of view.

Watanebe et al. [25] investigated the viability of using centrifugal casting to create Ni-aluminide steel clad pipe. Jedamark et al. [26] used the electrochemical gradation technique to fabricate the W/Cu FGM system. The manufactured FGM produced the necessary chemical gradient for the FGM and revealed that certain process parameters, including electrolytic porosity, resistivity, density, and geometry, had an impact on the morphological structure and shape. Ankush Mehta et al. [27] proposed an FGM using two separate materials in the form of adhesive and silicone-based gels (A and B). The 3D bio-printer was used for developing an FGM in the form of tensile specimens and mechanical, and electrical results for the developed FGM were highlighted. Also, two possible combinations of FGM (A + B, B + A) in the same weight proportion were explored in the 3D bioprinter, and the variation in the thickness or non-uniformity in the shape was observed which led to the change in dielectric properties of FGM.

Wilson and Shin [28] used the laser deposition technique to create Ni-based FGMs. It was stated that adding more TiC powder to Inconel 690 powder would result in a more refined structure. A liquid precursor dissolved in a liquid solvent is used in the straightforward chemical solution deposition (CSD) thin film processing method. The mechanical behavior of ZrO_2/Al_2O_3 and WC/Co, which were created using the electrophoretic deposition technique, was investigated by Put et al. [29]. For the ZrO_2/Al_2O_3 system, the hardness value variation was found to range from 13.5 GPa at the core to 19.5 GPa at the edges; in contrast, for the WC/Co FGM system, the hardness value variation ranged from 21 GPa to 9 GPa from the core to edges. The Al_2O_3-W FGM system was created by Katayama et al. [30] by employing the slip casting process and then sintering it at a temperature of 18,000 C. A higher percentage of dispersant was found to increase the raw materials' dispersibility. Additionally, a gradual change in W's grain size was noted along with the composition of the layer. The gel casting technique is one of the most popular methods for creating intricately shaped ceramics with a dense and porous structure. One method for processing colloidal materials that has a high yield at a low cost, a high green capacity, and a short formation time is gel casting. Al_2O_3/ZrO_2 FGM system was created by Park et al. [31] by dip coating substrates that had been prepared using the gel casting technique. Mechanical and microstructural characteristics of the manufactured FGM system were described. There was a smooth transition in hardness from the substrate to the outer layer. Ion plating, chemical vapour deposition, and plasma spraying are a few examples of gas-based techniques. In the case of vapour deposition systems, the gradient composition of the deposited products is determined by the control of the production system and the ratio of the phase's reaction. For coating applications, compositional gradient composites made using liquid phase techniques are widely used. The type of processing technique used determines the final property of the FGM. The primary challenges in the production of FGMs are the non-uniform spatial gradation of the component parts and consolidation, which entails placing the graded structure on a substrate [32]. Metal and ceramic-based systems have been the subject of the majority of research. Consequently, there aren't many research studies on the processing methods for polymeric-based FGMs.

Metal and ceramic-based systems have been the subject of the majority of study. As a result, there are few research reports on the processing methods for FGMs based on polymers accessible. Hence, in the present work an attempt has been made to fabricate polyester based titanium FGMs through sedimentation technique and evaluating density, microstructure and hardness properties.

2. Experimental Procedure

2.1 Materials and Methods

In the present work, polyester resin used as matrix material and titanium powder particles as reinforcement material. Table 1 and 2 shows the properties of pure polyester and titanium.

Table 1. Properties of Polyester

Property	Value
Viscosity at $25^0C(ns/m^2)$	250-350
Density (g/cm^3)	1.09
Modulus of elasticity E(Gpa)	3.3
Flexual strength (Mpa)	4.5
Tensile strength(Mpa)	40
Maximum elongation	1

Table 2. Properties of Titanium

Property	Value
Density (g/cm^3)	4.5
Melting point(*c)	1670
Thermal conductivity(w/m-k)	15-22
Elastic modulus (Pa)	115
Reactivity with oxygen	High+
Corrosion resistance	High+

2.2 *Production of Reinforcements (particles)*

Titanium particles are produced through conventional filing technique. Figure 1 shows the production of powder particles. Titanium rod of diameter 50 mm was subjected to filing operation on a lathe machine using course file with an operating speed of 650 RPM. Figure 2 shows the obtained powder particles.

Figure 1. Filing operation on lathe machine.

Figure 2. Obtained particles after filing.

Figure 3. Sieving of Ti particles.

The produced powder is subjected to demagnetization in order to remove ferrous particles. In order to select certain range of particles, the obtained powder was subjected to sieving for 10 minutes. Figure 3 shows the sieving operation. The Ti particles are sieved up to 300 microns as per IS 460:1982 standards.

2.3 Fabrication of FGMs

2.3.1 Sedimentation Method

PVC pipe of diameter 2.4 cm and length 4.2 cm is taken for manufacturing of polyester - Ti FGM's by sedimentation technique. Sieved Ti particles are taken with varying percentages of 2.5, 5 and 10 gms by weight for fabricating of FGM's. Measured Ti (2.5gm) particles are mixed with polyester and poured in the prepositioned pipe which is placed in vertical direction. After pouring, settling time of 1hr to 2 hrs is allowed for sedimentation.

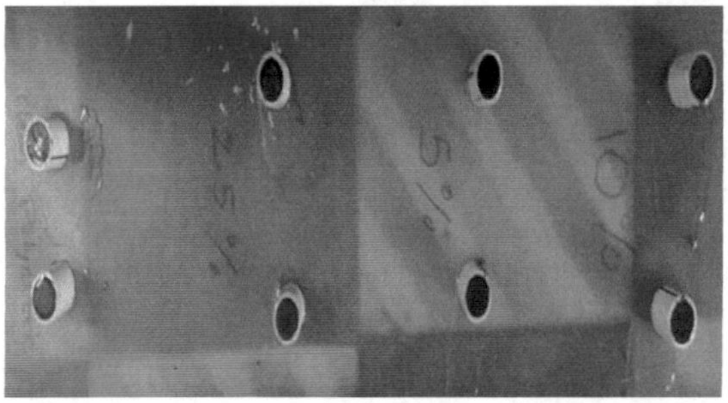

Figure 4. Fabrication of FGMs through Sedimentation process.

Figure 5. Obtained samples from sedimentation process.

Figure 4 shows the process of sedimentation, similar procedure was adopted for the production of polyester-5Ti and polyester-10Ti FGM's. For the purpose of comparison pure polyester (without titanium powder) was also fabricated under similar processing conditions. Figure 5 shows the obtained specimens with varying percentages of Ti, i.e., pure polyester, 2.5gm Ti particles, 5gm Ti particles, 10 gm Ti particles.

2.4 Testing

The fabricated FGMs were subjected to the following testing for quality estimation and properties evaluation.

2.4.1Dye penetrant test

Any homogenous substance can be inspected using the penetration approach. The current test makes use of both red penetrant and white developer. The test section is now treated with a dry powder after the penetrant has seeped in. This powder expands the region of penetrant indication by pulling the penetrant from the defect and acting as a sponge.

2.4.2 Density

After the fabrication, each sample was weighed and their densities were estimated using the following formula

$$P = Mass/Volume$$

2.4.3 Hardness Test

All the samples were subjected Vickers hardness test both on top and bottom surface at a constant load of 0.5 kg.Three readings were taken on each sample and average values are reported.

2.4.4 Microscopic examination

The fabricated samples are subjected microscopic examination after following the sequential polishing techniques. Both top and bottom surfaces were subjected microscopic examination inorder to witness the presence of titanium particles and as well as polyester base.

3. Results &Discussion

3.1Liquid penetrant NDT

Figure 6 shows the sample obtained from liquid penetrant test. The red colour indicates the cracks or flaws on the specimen. The defects in the specimen finds out after the developer (white colour) has been applied to the specimen. It is evident from the figure that except at the tip there are no significant defects that are present in the specimen.

Figure 6. Defect identification through liquid penetrant test.

3.2 Density

Table 3 shows the summary of densities obtained for Pure polyster, 2.5gm Titanium, 5gm Titanium and 10gm Titanium FGMs. Figure 7 shows the variation of densities for pure polyester, 2.5Ti, 5Ti and 10Ti FGMs. The obtained values are 1.33 gm/cm^3 for 2.5Ti FGM and 1.40 gm/cm^3 for 5Ti FGM and 1.59 gm/cm^3 for 10Ti FGM. Increase in density can be observed with respect to Titanium increment. This is true because the Titanium density is higher as compared to pure polyester as a result FGMs are with high density. Sedimentation process aided in more densification, hence the incremental values. Rate of increment in density is also proportional to the addition of Ti content in polyester resin. For a given volume, the increment in density depends on how effectively the Ti powder particles have been doped. From the results it is evident that the doping was effective and sedimentation technique is successful.

Table 3. Summary of densities for FGMs

S.NO	Ti by weight (gm)	Density (g/cm^3)
1	0	1.21
2	2.5	1.33
3	5	1.4
4	10	1.59

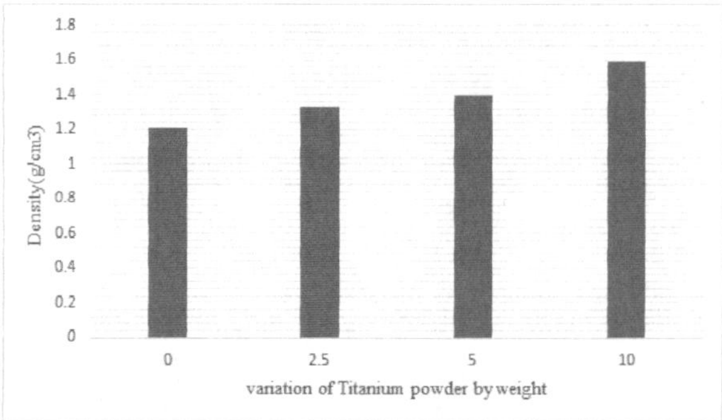

Figure 7. Comparison of densities for various FGMs.

3.3 Hardness

Figure 8 & Figure 9 shows the Vickers hardness values for 2.5 gm FGM. Figure 8 represents the pure polyester side, where as Figure 9 reveals the hardness on titanium side. It is understood that pure polyester side is soft, depth of indentation is more and hardness is less. However, hardness value is 613HV at Ti side. For a given load, small impression of indenter indicates hard material, and the opposite is true for large impressions.

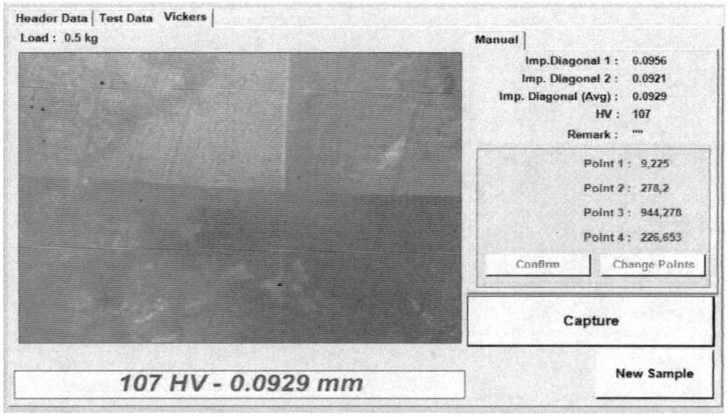

Figure 8. Hardness values of Pure Polyester Side (2.5 gm).

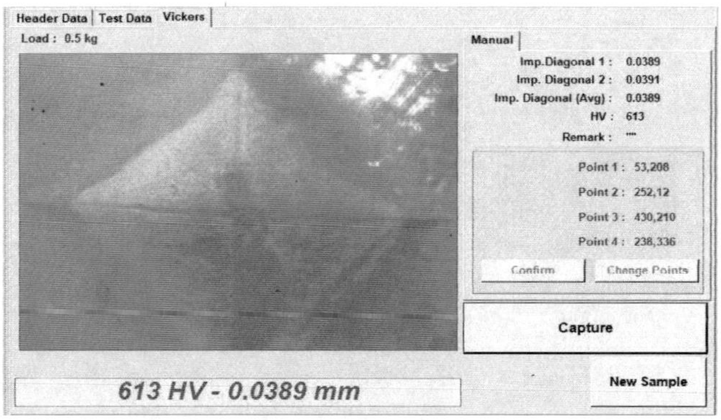

Figure 9. Hardness values of Titanium rich side (2.5gms).

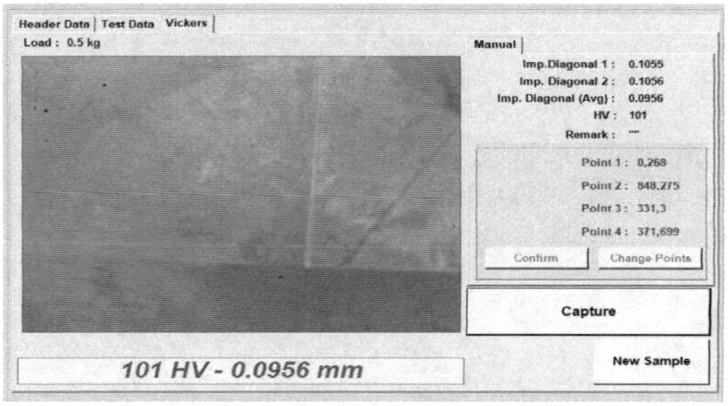

Figure 10. Hardness values of Titanium (5 gm) pure Polyester Side.

Figure 10 and Figure 11 shows the Vickers hardness values for 5Ti FGM. Figure 10 shows 101HV on polyester rich side, where as Fig 11 shows 771HV on titanium rich side. The hardness is higher on Ti side as compared 2.5 Ti FGM. Enhanced Ti content made the FGM harder. Harder the material, higher is the height to which the indenter ball rebounds. Greater and deeper indentation and a lower hardness index number are indicative of softer materials. The same can be observed from indentation dimensions i.e., 0.0956 mm and 0.0347 mm.

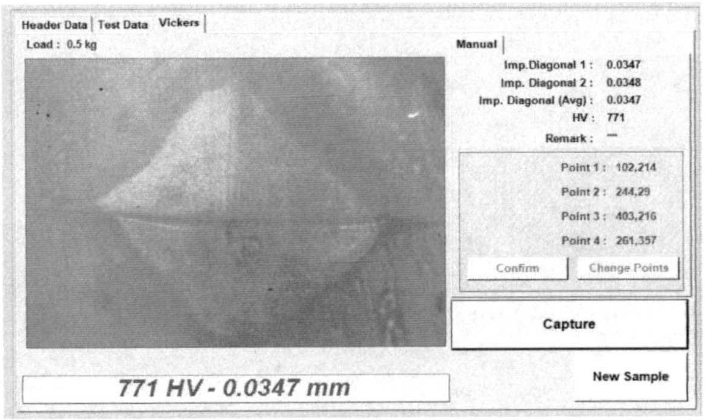

Figure 11. Hardness values of Titanium rich side (5gms).

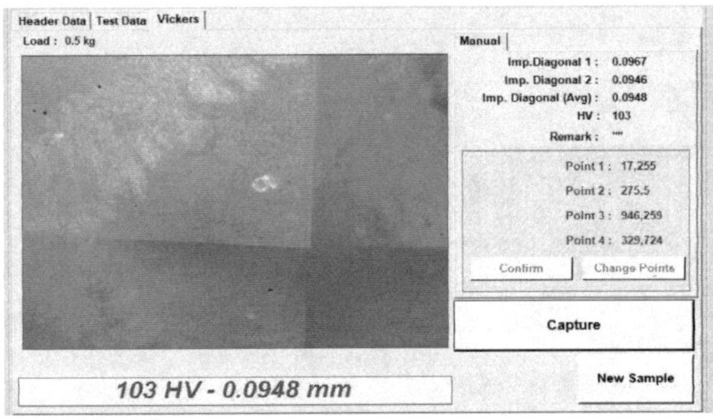

Figure 12. Hardness values of Pure Polyester Side.

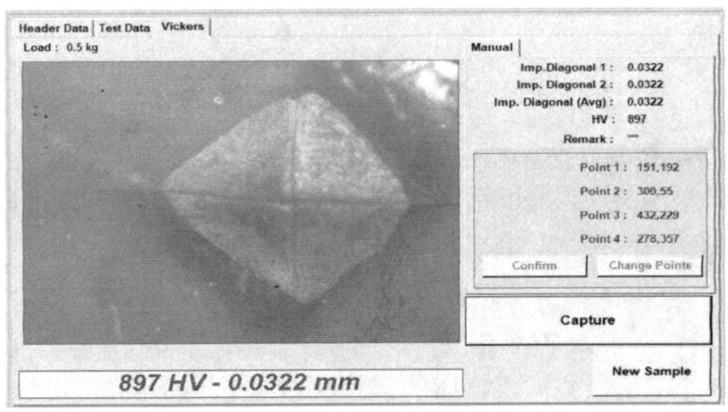

Figure 13. Hardness values of Titanium rich side (10gms).

Figure 12 and Figure 13 shows the hardness images for 10Ti FGM. Figure 12 shows the value corresponding to polyester side and Figure 13 shows the value corresponding to Ti rich side. Ti rich side the hardness value is 897HV. This is highest as compared to 2.5 Ti and 5 Ti FGMs. The material's elastic limit, or yield strength, determines hardness to a greater extent than the work hardening coefficient. The modulus of elasticity also has an impact on hardness. However, the resin side the value is much low i.e. 103 HV.

3.4Optical Microscopy

Figure 14 shows the optical microscope image of pure polyester resin. Since it is the image taken only on polyester there are no traces of metallic inclusions. i.e. titanium powder. Figure 15 shows the optical microscope image of 10 Ti FGM on rich side. It can be seen that dominant metallic presence of titanium powder.

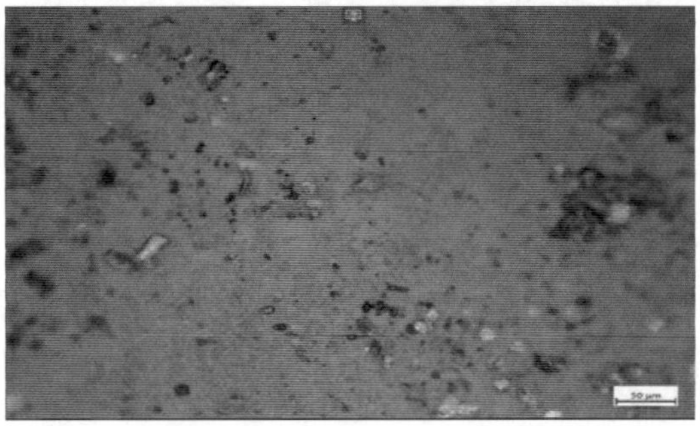

Figure 14. Optical microscope image of pure polyester resin.

Figure 15. Optical microscope image of 10 Ti FGM

Figure 16. Optical microscope image of 5 Ti FGM on rich side.

However, rare presence of polyester that was impregnated in to metallic zones can also be witnessed. Figure 16 shows the optical microscope image of 5 Ti FGM on rich side. The presence of titanium powder evidence is less as compared to 10 Ti content FGM. Increased content of polyester which appears to be deposited/ covered the presence of Ti powder. This can be attributed to good bonding between powder particles makes the FGM more strong and sustainable. Figure 17 shows the optical microscope image of 2.5 Ti FGM on rich side. It can be seen that poor evidence related to metallic titanium powder presence. However, presence of polyester can also be seen which appears to be in smoke colour deposited on metallic surfaces.

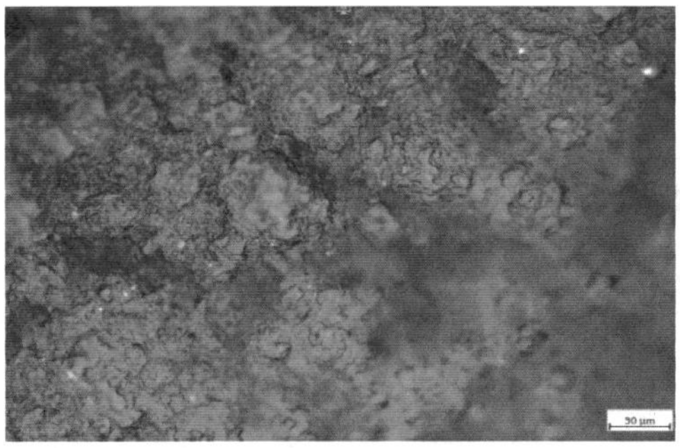

Figure 17 Optical microscope image of 2.5 Ti FGM.

4. Conclusions

1) Polyester based Titanium with varying weight fractions are successfully fabricated using sedimentation technique.
2) Liquid penetrant NDT method revealed that the samples are of good quality without any cracks and porosity.
3) With increasing titanium content, increased values of hardness has been observed from polyester rich surface to titanium rich surface.
4) Improved densities are observed for 2.5Ti, 5Ti, 10Ti FGMs with increasing titanium content.
5) Optical microscope images corroborate the concept of functionally graded materials.

References:

[1] Hassanin, H., & Jiang, K. (2014). Fabricating functionally graded ceramic micro-components using soft lithography. In Advances in Ceramic Matrix Composites (pp. 611-623). Woodhead Publishing.

[2] Petr Lukas, MiroslavVrana, JefVleugels, Guy Anne, Omer Van der Biest, Materials Science Forum (Volume 571-572), Pages: 309-314.

[3] Benjamin Hatton, Patrick S. Nicholson, Journal of the american ceramic society, Volume84, Issue3, Pages 571-576(2001).

[4] Koyel Banerjee Ghosh, JayantaMukhopadhyay, Rajendra N. Basu, Journal of Power Sources, Volume 328, 1 Pages 15-27 (2016).

[5] E Müller, C Drasar, J Schilz, W.A Kaysser, Materials Science and Engineering: A, Volume 362, Issues 1–2, Pages 17-395 (2003).

[6] R. Slowak, S. Hoffmann, R. Liedtke & R. Waser,Integrated ferroelectrics, Vol 24, issue 1-4, Pages 169-179 (1999).

[7] LuoYongming, Pan Wei, Li Shuqin, Wang Ruigang, Li Jianqiang, Materials Science and Engineering: A, Volume 345, Issues 1–2, Pages 99-105.(2003).

[8] Mohammad AtifFaiz Afzal, PallaviKesarwani, K. Madhav Reddy, SushmaKalmodia, Basu B and KanteshBalani, Materials Science and Engineering: C, Volume 32, Issue 5, 1 Pages 1164-1173 (2012).

[9] Das, S., Das, S., Nampi, T., & Roy, K. (2021). Functionally grade composite material production.
[10] D.K. Jha, Tarun Kant, R.K. Singh, Composite Structures, Volume 96, PP 833 – 849 (2013).
[11] Pan, E. Journal of composite materials, 37 (21), 1903-1920.(2003).
[12] J.H. Kim, M.C. Kim, C.G. Park, Surface and coatings Technology, Volume 168, Issues 2, Pages 275-280 (2003).
[13] Andrew J. Ruys and Brett A. Sutton 'Metal-ceramic functionally graded materials (FGMs)', Metal-Reinforced Ceramics, 2021, Pages 327-359
[14] Ruys, A. J. (2020). Metal-reinforced ceramics. Woodhead Publishing.
[15] Carmine Lucignano and FabrizioQuadrini, Measurement, Volume 42, Issue 6, Pages 894-902. (2009).
[16] Huiping Yuan, Junguo Li, QiangShen, Lianmeng Zhang, International Journal of Refractory Metals and Hard Materials, Volume 34, Pages 3-7 (2012).
[17] A. Teber, F. Schoenstein, F. Tetard, M. Abdellaoui, N. Jouini, International Journal of Refractory Metals and Hard MaterialsVolume 30, Issue 1, Pages 64-70 (2012).
[18]ZhuhuiQiao, Jan Rathel, Lutz-Michael Berger, Mathias Herrmann, International Journal of Refractory Metals and Hard Materials., Volume 38, Pages 7-14(2013).
[19] Jerzy J. Sobczak , Ludmil Drenchev , Journal of Materials Science & Technology, Volume 29, Issue 4, Pages 297-316 (2013).
[20] Ahmed El-Desouky, Samuel K. Kassegne, Kee S. Moon, J. McKittrick, K. Morsi, Journal of Materials Processing Technology, Volume 213, Issue 8, Pages 1251-1257 (August 2013).
[21] EhsanBafekrpour, George P. Simon, Jana Habsuda, MinooNaebe, Yang C and Bronwyn Fox, Materials Science and Engineering: A, 2012, Volume 545, Pages 123-131 (2012).
[22] EhsanBafekrpour, George P. Simon, Bronwyn Fox, *Journal of Thermal Analysis and Calorimetry,*109, pages 1169 – 1176 (2012).
[23] Bafekrpour, E., Simon, G. P., Naebe, M., Habsuda, J., Yang, C., & Fox, B. Annual Technical Conference Proceedings, April 2-4, Orlando (2012).
[24] Mishra, A. K., Yadav, K., & Kumar, A. (2023). Selective laser melting of functionally graded material: Current trends and future prospects. In Advances in Additive Manufacturing (pp. 281-297). Elsevier.
[25] Yoshimi Watanabe, Yoshimi Iwasa, Hisashi Sato, Akira Teramoto, Abe k and Miura-Fujiwara, Journal of materials processing Technology, Volume 211, Issue 12, Pages 1919-1926 (2011).
[26] R. Jedamzik, A. Neubrand, J. Rodel, Journal of Materials Science , 35, pages 477–486 (2000).
[27] Ankush Mehta , Rupinder Singh , B.S. Pabla , Kamaljit Singh Boparai 'Bio-printing on the thermoplastic substrate for functionally graded materials with improved dielectric properties, Reference Module in Materials Science and Materials Engineering, 2023.
[28] J. Michael Wilson, Yung C. Shin, Surface and Coatings Technology, Volume 207, 25 Pages 517-522 (2012).
[29] S Put, J Vleugels, G Anne, O Van der Biest, Colloids and Surfaces A: *Physicochemical and Engineering Aspects,*Volume 222, Issues 1–3, Pages 223-232.(2003).
[30] Katayama, T, Sukenaga, S, Saito, N, Kataga. H, & Nakashima. K. Inn IOP Conference series: Materials Science and Engineering (Vol. 18, No. 20, p 202023) IOP Publishing.(2011).
[31] Park, S.H, Kang, J.H, Jung, Y.G, & Paik, U. In AIP Conference Proceedings (Vol. 973, No.1, pp 273 – 278) AIP (2008).
[32] Pravin Malik, RavikiranKadoli, Journal of Vibration and Control, Volume 24, Issue 6, 1077546316659422 (2016).

Structural and Quantum chemical insights into the atomic orbitals of 2-benzoylbenzoic acid: Density functional theory approach

P.V.Prasanna Kumar[1], G.V.Gaurav [2], D. Jagadeeswara Rao[3*], S.Sreedhar [4]

[1] Department of Engineering Physics, S.R.K.R Engineering College, Bhimavaram, A.P, India
[2] Department of Chemistry (FED), Anil Neerukonda Institute of Technology and Sciences (A), Sangivalasa, Bheemunipatnam (M), Visakhapatnam.
[3] Department of Physics, S.V Govt. Degree College, Parvathipuram , A.P, India.
[4] Department of Mathematics, GITAM University, Andhra Pradesh, India.
*Corresponding author: dasarijagadeesh@gmail.com

Abstract

The present work aims to explore the structural insights and the molecular orbitals associated with 2-benzoylbenzoic acid through density functional theory approach. The geometrical parameters of the optimized structure of 2-benzoylbenzoic acid were obtained at B3LYP/6-311++G basis set. The atomic sites which are prone to electrophilic/nucleophilic attack were identified from molecular electrostatic potential (MEP) surface analysis. The molecular orbitals (lone pair, bonding and anti-bonding) which are favorable to the charge delocalization within the system and stabilize the molecule were studied. Besides, the vibrational assignments of the characteristic group frequencies and coupled modes vibrations were made using Cartesian coordinate displacement analysis. Moreover, the electronic transitions between the filled and unfilled molecular orbitals of 2-benzoylbenzoic acid were investigated.

Keywords: 2-benzoylbenzoic acid, Electronic transitions, Electrophilic/Nucleophilic attack, Vibrational assignments, Charge delocalization.

IOP Conf. Series: Journal of Physics: Conf. Series **2765** (2024) 012014 doi:10.1088/1742-6596/2765/1/012014

1. Introduction

Many Aromatic compounds such as benzoylbenzoic acid derivative compounds are commonly used in industry and medicinal fields. Several benzoylbenzoic acid derivatives are recognized as photosensitizing compound. Investigations indicate that thiols can be protected in a similar manner as primary and secondary alcohols using 2-benzoylbenzoic acid, which is an efficient photolabile protective group. [1]. The photophysical properties like ultraviolet absorption, fluorescence excitation and emission spectra of the zinc complex of 2-benzoylbenzoic acid has been reported in a novel work [2]. In recent years, benzoylbenzoic acid have been the frequent subjects of experimental and theoretical research because of their significant applications in the industry [3-10]. Literature survey discloses that no theoretical investigation of 2-benzoylbenzoic acid have been reported so far. Therefore, it is worthwhile to make a study on 2-benzoylbenzoic acid.

It has been effectively shown that density functional theory (DFT) can describe the structural and electronic properties of a vast variety of substances, ranging atoms to complex systems. As a result, DFT is now frequently used to explore molecules theoretically. Using vibrational spectroscopy, materials can be identified and details on their structure, composition, functional groups, and intramolecular interactions can be obtained. Additionally, it provides unique spectra for every individual molecule. Using DFT and the B3LYP functionality along with an expanded basis set of 6-311++G, that we have determined the equilibrium geometry, harmonic vibrational wavenumbers, absolute Raman scattering activities, and infrared absorption intensities. In order to understand the molecular wave function in terms of donor-acceptor interactions, hybridization, resonance, Lewis structure, charge, bond order, and bond type, the NBO analysis was conducted. The title compound's polarization has also been investigated by the plotting of the molecular electrostatic potential (MEP) surface.

2. Computational methods

Gaussian 03 software [11] and GaussView molecule imaging application [12] were employed to perform quantum chemical computations on the molecule mentioned in the article's title. For the purpose of modeling the structure, optimizing the geometry was done for the title compound through B3LYP (Becke's three-parameter hybrid model using the Lee-Yang Parr correlation functional) methods [13], which combine the hybrid exchange functional initially doi:10.1088/1742-6596/2765/1/012014

proposed by Becke along with the Lee-Yang–Parr correlation functional [14]. The optimized structural parameters are used in the vibrational frequency calculations at DFT level. Applying the B3LYP/6-311++G basis set, the donor and acceptor occupation as well as the stabilization energy of the various overlaps for the title molecule were investigated. Furthermore, the molecular electrostatic potential were evaluated using B3LYP method to investigate the reactive sites of the title compound.

3. Results and discussion

3.1. Molecular geometry

Geometry optimization in the ground state was applied to the chemical 2-benzoylbenzoic acid, which is the name of the substance. The optimised geometry is part of the groupsymmetry of C1 points. Figure 1 displays the molecular structure using the study's chosen atomic numbering approach. The following is a list of the optimal geometrical characteristics of 2-benzoylbenzoic acid as determined by B3LYP using the 6-311++G basis set. The geometry determined belong to a true minimum proven by real wavenumbers in the vibrational analysis. The optimized geometry reveals that COOH group substituted in benzene ring is laying slightly out-of-plane. Due to the substitution, the torsional angle is -68.53° between C3, C2, C1 and O8. The benzene ring appears to be little distorted with (C2–C3) and (C2–C7) bond lengths exactly at the substitution place bylonger than the remaining bonds(C3–C4), (C4–C5), (C5–C6) and (C6–C7). The bond length between the atom C1 and C2 is 1.4752 Å. The torsional angle 27.4° (C2, C3, C10 and C11) shows that both benzene rings in the title compound are not in the same plane. In the ring at the substitute their position, the CCC angle is noticeably smaller than the remaining positions. The CCC angles that stand were nearly 120°. The computation indicates that, at the C3 point, the angle C2-C3-C4 is 0.9° shorter than its normal value of 120°.The other parameters for the title compound can be computed using the computed geometric variables, which are a good approximations.

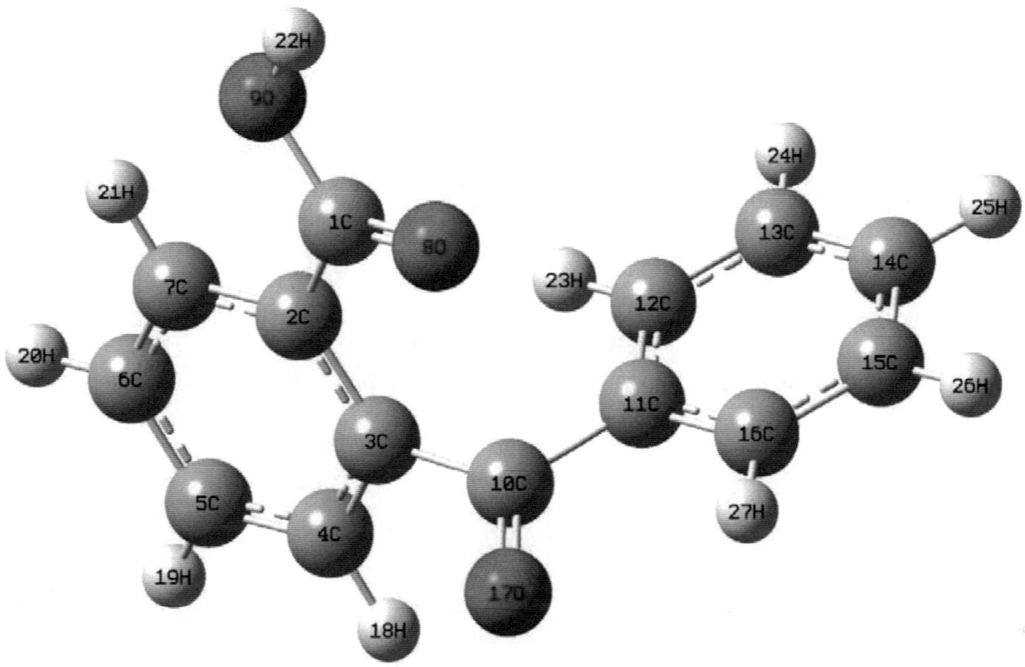

Fig. 1. Optimized geometrical structure of 2-benzoylbenzoic acid.

3.2. NBO analysis

NBO analysis has been performed on the title molecule at the by B3LYP with the 6-311++G basis arranged to clarify the rehybridization, delocalization, and the intramolecular electron density inside the molecule. In addition to offering a practical foundation for researching charge transfer or conjugative interaction in molecular structures, naturally occurring bond orbital analysis is an effective technique for examining intra- and intermolecular bonding and interaction between bonds. There have already been reports on certain electron donor orbitals, acceptor orbitals, and the interaction stabilisation potential arising from the second-order perturbation theory [15, 16]. The results of the NBO analysis are shown in Table 1. The coupling causes a vacant non-Lewis orbital to replace the concentration of electron NBO in the assumed Lewis structure. The stabilizing energy E (2) linked to the delocalization i→j is computed for every donor (i) and acceptor (j).

$$E^{(2)} = -n_\sigma \frac{\langle\sigma|F|\sigma\rangle^2}{\varepsilon_\sigma{}^* - \varepsilon_\sigma} = -n_\sigma \frac{F_{ij}{}^2}{\Delta E}$$

Where

$\langle\sigma|F|\sigma\rangle^2$ or ${F_{ij}}^2$ is the Fock matrix element i and j NBO orbital's, ${\varepsilon_\sigma}^*$ and ε_σ are the energies of σ and σ * NBOs and n_σ is the population of the donor σ orbital.

The larger the E(2) value, the more intensive is the interaction between donors and acceptors. The extent of conjugation of the system depends on the donating tendency from donors to electron acceptors. The stabilized state of the donor-acceptor association is indicated by the delocalization of electron density between occupied Lewis-type (bond or lone pair) NBO orbitals and formally empty (antibond or Rydberg) non-Lewis NBO orbitals.

The NBO computations revealed a wide range of energy levels. The primary interactions energies (π – π*) associated with molecular coherence involve electron donation by the electron donating groups' π(C–C) atoms to the anti-bonding acceptor π*(C–O) atoms. This corresponding interactions (π (C_{11}–C_{12}) → π*(C_{10}–O_{17}) of the molecule has theE(2) value of 26.23 kJ mol^{-1}. This larger energy shows the hyperconjugation between the electron donating groups and acceptor. The orbital overlap among the bonding (C–C) and anti-bonding (C–C) orbitals forms the within the molecule interaction. This intramolecular transfer of charge results in the stability of the entire system. The electron density (ED) in the C–C anti-bonding orbital increases as a result of all these interactions, weakening the corresponding bonds. The strong intramolecular hyper conjugative interaction of the π electron of (C_3– C_4) distribute to σ* C_{10}– C_{11} andC_{10}–O_{11}in the substitution place. On the other hand, the π (C_3–C_4) in the ring conjugate to the anti-bonding orbital of π*(C_5–C_6) which leads to strong delocalization of 46.08 kJ/mol. The π (C_1–O_8) bond is contributing energy by 4.92 kcal/mol with π*(C_2–C_7). In reverse trend, the π (C2–C7) bond is contributing energy by 33.53 kcal/mol with π * (C_1– O_8).

Table 1: Analysis of the Fock matrix in NBO basis corresponding to the intramolecular bonds of 2-benzoylbenzoic acid using second-order perturbation theory.

Donor (i)	ED (i) (e)	Acceptor (j)	ED (j) (e)	E(2) (kJ / mol)	E(j) - E(i) (a.u.)	F(i,j) (a.u.)
π(C5-C6)	1.63585	π*(C2-C7)	0.37175	53.24	0.46	0.14
π(C13-C14)	1.65721	π*(C11-C12)	0.36835	47.9	0.47	0.135
π(C15-C16)	1.64690	π*(C13-C14)	0.32019	47.15	0.47	0.134
π(C3-C4)	1.64309	π*(C5-C6)	0.31339	46.08	0.48	0.134
π(C2-C7)	1.64725	π*(C3-C4)	0.31030	45.17	0.49	0.134
π(C11-C12)	1.65521	π*(C15-C16)	0.28147	40.01	0.49	0.127
π(C15-C16)	1.64690	π*(C11-C12)	0.36835	39.47	0.46	0.121
π(C3-C4)	1.64309	π*(C2-C7)	0.37175	38.32	0.46	0.12
π(C5-C6)	1.63585	π*(C3-C4)	0.31030	36.27	0.48	0.119
π(C11-C12)	1.65521	π*(C13-C14)	0.32019	36.06	0.48	0.118
π(C13 -C14)	1.65721	π*(C15-C16)	0.28147	34.12	0.49	0.117
π(C2-C7)	1.64725	π*(C1-O8)	0.18301	33.53	0.47	0.117
π(C2-C7)	1.64725	π*(C5-C6)	0.31339	33.48	0.48	0.115
π(C11-C12)	1.65521	π*(C10-O17)	0.10156	26.96	0.49	0.11
π(C1 -O8)	1.98610	π*(C2-C7)	0.37175	4.92	0.68	0.057
σ(C12-H23)	1.98067	σ*(C11-C16)	0.02109	5.36	1.52	0.081
σ(C3-C4)	1.97209	σ*(C2-C3)	0.02697	5.11	1.73	0.084

$E^{(2)}$ means energy of hyper conjugative interaction (stabilization energy).
E(j) - E(i) is the energy difference between donor and acceptor i and j NBO orbitals.
F(i,j) is the Fock matrix element between i and j NBO orbitals.

3.3. Electronic absorption spectra

In UV–visible spectrum region All molecules allow strong σ–σ* and π–π* transitions with high extinction coefficients [17]. Using fully optimized ground-state structure of the title molecule, Low-lying excited states of benzoyl benzoic acid are being calculated using the B3LYP/6-311++ G technique.. The theoretical UV spectrum was calculated by keeping methanol as solvent. Figure 2 shows the UV-vis absorbance spectrum of the molecule. The Calculated vertical excitation energies (E), oscillator strength (f) and wavelength (λ) have been reported in the Table2. According to the Frank-Condon principle, the molecule's vertical activation transitions is what causes the greatest absorption peak in a UV-visible spectrum.. The theoretical calculation in methanol predicts two intense electronic transitions at 222.77 and 229.85 nm. The excitation energy and oscillator strength of the peak at 222.77 nm are 5.5657eV and 0.0391, respectively. The excitation energy and oscillator strength of the second (intense) peak at 229.85 nm are 5.3941 eV and 0.2499, respectively. The UV absorption calculation supports the absorption peaks are due to electron transition from the highest occupied molecular orbit (HOMO) to lowest unoccupied molecular orbital (LUMO).

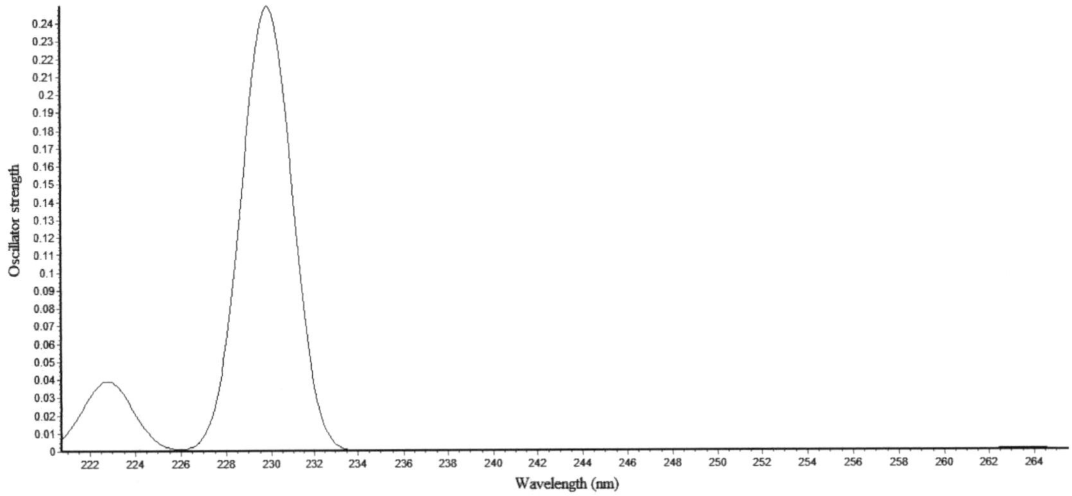

Fig. 2. UV–vis spectrum of the 2-benzoylbenzoic acid.

Table 2: Experimental value of absorption wavelength (nm), excitation energies E (eV) and oscillator strengths (f) of 2-benzoylbenzoic acid.

Wavelength (nm)	Excitation energy (eV)	Oscillator strength
263.47	4.7058	0.0014
229.85	5.3941	0.2499
222.77	5.5657	0.0391

3.4. Vibrational spectral analysis

Molecular vibrational spectrometry has proven to become the most efficient approach of resolving a variety of biological and chemical issues. This is being widely applied to scientific research. Vibrational wavenumbers for title molecule were calculated based on the optimized geometries by using B3LYP level with the 6-311++G basis set. It assumes that the title's molecule has C1 point group symmetry. A non-linear compound with N atoms can have a maximum amount of 3N-6 potentially active normal mode vibrations [18]. The total number of atoms in the molecule is 27, hence there are 75 normal modes. For N-atomic molecule, 2N-3 of its vibrations are planar and N-3 are non-planar. Therefore, 51 of these modes should be planar and 24 should be non-planar in the title molecule [19]. Table 3 summarizes the computed vibrational frequency ranges, determined IR intensities, computed Raman scattering activities, and vibrational frequency assignments for the title chemical. Additionally, Figures 3 and 4 provide the molecule's computed Raman and IR spectra at the B3LYP/6-311++G level, respectively. It is generally acknowledged that systematic errors in DFT level computations cause the frequencies of vibration to be overestimated. Vibrational harmonics were thus normalized by 0.967.

The OH stretching vibration, which is located in the spectrum between 3000 and 3500 cm-1, is extremely sensitive to intra- and intermolecular hydrogen bonding. It produces three types of vibrations: out-of-plane bending, in-plane bending, and stretching vibrations. Strong intramolecular hydrogen bonding in the structure is responsible for the OH stretching vibration that is responsible for the absorption band in the 3558 cm-1 area. At 75 cm^{-1}, the OH stretching vibration is also collected.. Generally, the OH rocking vibrations lie in the region 1300–1400 cm^{-1}. The absorptions at 1434 cm^{-1} is attributed to OH rocking vibrations in the molecule. The OH out-plane bending vibrations are absorbed in the region 750 – 800 cm^{-1}. The band observed at 1707 cm^{-1} was assigned to OH in-plane bending vibration of the title compound.

The vibrations of aromatic hydrocarbons (CH) bending, bending in plane, and out of plane occur in frequency ranges of 3100–3200 cm^{-1}, 1100–1400 cm^{-1}, and 800–1000 cm^{-1}, respectively. The aliphatic CH stretching vibrations of the molecule under description represent the absorption bands between 3114 and 3177 cm^{-1}. Furthermore, vibrations related to in-plane bending and out-of-plane CH have been observed at 1115–1385 cm^{-1} and 818–1025 cm^{-1}, correspondingly. The frequency regions of 1300–1100 cm^{-1} are anticipated to exhibit the greatest band of CO stretching vibrations. The title compound's absorption band, which was seen at 1094 cm^{-1}, was attributed to the CO stretching mode of vibration. The aromatic compounds' CC stretching vibrations are responsible for the absorption bands at 1600–1400 cm^{-1}. The aromatic rings of the molecule under consideration exhibited CC stretching modes between 1437 and 1594 cm^{-1}. The C=O stretching vibrations of the molecule are absorbed at 1616 and 1679 cm^{-1}. The bands at 167 cm^{-1} and, 273 and 377 cm^{-1} are attribute to the C=O out-plane and in-plane bending vibrations of the molecule. At 415 cm^{-1}, the COOH bending vibration is captured. The

IOP Conf. Series: Journal of Physics: Conf. Series **2765** (2024) 012014 doi:10.1088/1742-6596/2765/1/012014

group of carboxyl in the molecule is responsible for the twisting and rocking vibrations shown by the bands at 161 and 554 cm^{-1}.

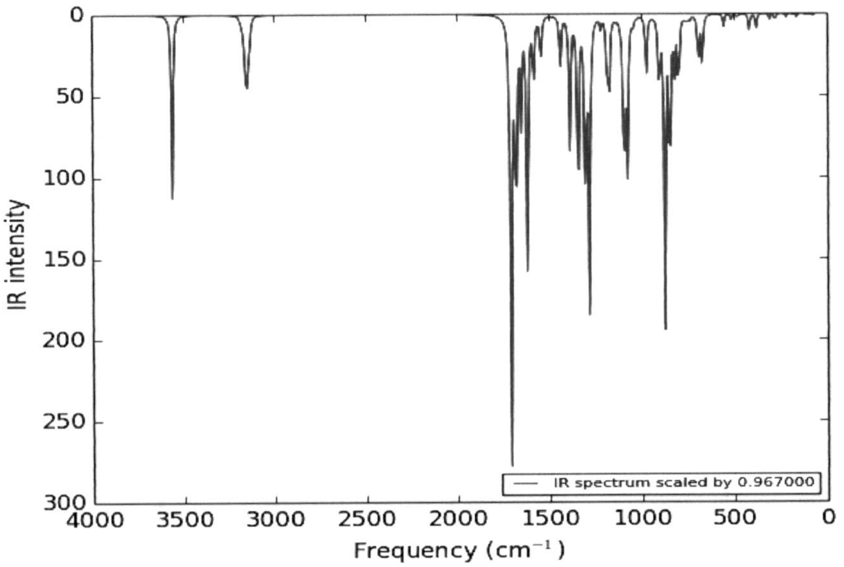

Figure 3. FT-IR spectrum of 2-benzoylbenzoic acid

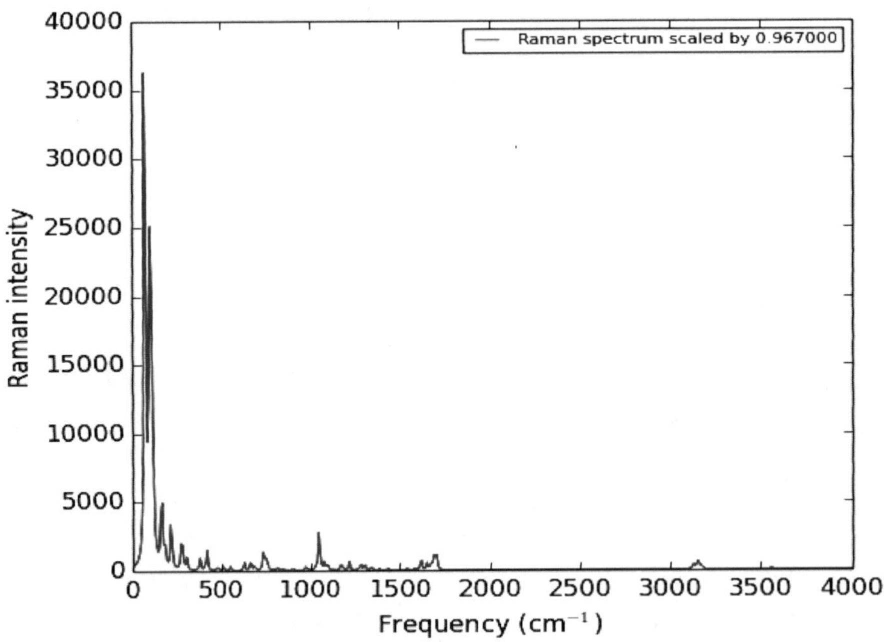

Figure 4. FT-Raman spectrum of 2-benzoylbenzoic acid

IOP Conf. Series: Journal of Physics: Conf. Series **2765** (2024) 012014 doi:10.1088/1742-6596/2765/1/012014

Table 3: The calculated values of IR intensities and Raman scattering activity of 2-benzoylbenzoic acid

Frequency		IR intensity	Raman Activity	
Unscaled	scaled	(km/mol)	(Å4/amu)	assignment
78	75	1.8	5.0	τ OH
167	162	2.3	1.2	τ COOH
173	168	0.3	2.0	β_{out} C=O
283	274	2.3	2.5	β_{in} C=O
290	280	1.9	1.8	β_{in} OH
315	305	3.5	1.6	ρ COOH
390	377	8.8	2.6	β_{in} C=O
429	415	11.0	5.9	β_{in} COOH
573	554	8.3	2.0	ρ COOH
691	669	0.3	6.6	β_{in} CC
693	671	36.2	1.5	β CO
778	753	2.2	8.8	β_{out} OH
789	763	2.3	4.7	β_{out} OH
825	798	49.5	1.5	β_{out} OH
847	819	34.2	3.7	β_{out} CH
874	846	116.5	3.1	β_{out} CH
902	873	197.2	0.5	β_{out} CH
931	901	20.6	1.9	β_{out} CH
935	904	27.8	1.7	β_{out} CH in R1 + R2
1005	972	39.6	6.9	β_{out} CH
1016	982	0.8	0.4	β_{out} CH in R2
1060	1025	0.8	3.7	β_{out} CH in R1
1080	1044	5.8	89.3	β_{in} CC in R1
1110	1074	86.7	15.3	β_{in} CC in R1
1116	1079	10.7	0.4	β_{in} CC in R1
1131	1094	114.3	14.3	υ CO
1154	1115	2.0	0.3	β_{in} CH in R2
1165	1127	0.3	0.1	β_{in} CH in R1+R2
1172	1133	0.1	0.7	β_{in} CH in R1+R2
1173	1135	0.1	0.2	β_{in} CH in R1
1177	1139	1.7	0.2	β_{in} CH in R2
1211	1171	37.7	11.4	β_{in} CH
1217	1177	17.7	7.0	β_{in} CH
1225	1185	30.4	3.9	β_{in} CH

1259	1218	7.1	26.0	β_{in} CH
1312	1268	12.2	3.7	β_{in} CH
1325	1282	182.8	18.8	β_{in} CH
1346	1302	27.1	4.6	β_{in} CH
1349	1305	94.6	16.0	β_{in} CH
1388	1342	133.7	15.1	β_{in} CH
1433	1386	84.0	6.3	β_{in} CH
1484	1435	14.7	3.5	ρ OH
1487	1438	20.3	6.1	υ CC
1592	1539	13.3	8.9	υ CC
1596	1543	18.4	1.0	υ CC
1636	1582	43.2	7.6	υ CC
1649	1595	1.4	1.8	υ CC
1672	1617	189.5	70.0	υ C=O in carbonyl group
1707	1651	61.6	41.5	υ CH in R1+R2
1727	1670	15.1	15.9	υ CC in R2
1737	1679	137.3	40.2	υ C=O in carboxyl group
1753	1695	29.5	124.9	υ CC in R2
1766	1708	269.5	82.4	β_{in} OH
3221	3115	0.3	44.8	υ asymmetric CH in R2
3232	3125	2.0	56.6	υ asymmetric CH in R1
3233	3127	10.6	115.4	υ asymmetric CH in R2
3246	3139	38.0	39.6	υ asymmetric CH in R2
3248	3140	10.8	86.4	υ asymmetric CH in R1
3257	3150	18.3	162.2	υ asymmetric CH in R2
3261	3153	16.7	172.7	υ asymmetric CH in R1
3271	3163	8.1	139.8	υ symmetric CH in R2
3286	3177	3.8	124.0	υ symmetric CH in R1
3679	3558	117.9	115.0	υ OH

υ – stretching; β – bending ; ρ – rocking; τ – twisting ; R1 – (C2-C7); and R2 – (C11 – C16).

3.5. Molecular electrostatic potential

The molecular electrostatic potential is an exceptionally powerful technique for recognizing the sites of intermolecular and intramolecular interactions [20]. The reactive site of the electrophilic and nucleophilic regions for the title molecule was studied by calculating molecular electrostatic potential of molecule at B3LYP methods with the 6-311++G basis set. The plot of the molecular electrostatic potential map of the title molecule is shown in the figure 5.The negative (red and yellow) regions of molecular electrostatic potential map were related to electrophilic reactivity and the positive (blue) region is to nucleophilic reactivity. The figure shows that the molecule has several possible sites for electrophilic attack. Negative regions in the title molecule were found around the carbonyl group (O17) and carboxyl group (O8 and O9) in the molecule. The calculated electrostatic charges are tabulated in Table 4. The electrostatic charges of the O17, O8 and O9 are −0.5325 a.u., −0.54 a.u. and −0.5901 a.u. respectively. The atom O9 of the carboxyl group has more negative charges due to presence of positive region of atom H22. Nevertheless, the most positive regions is found around the atom H22 of carboxyl group in the molecule. This positive region atom has the electrostatic charge value of 0.3972 a.u. According to these electrostatic potential calculation, the negative potential sites are on electronegative atoms (oxygen) as well as the positive potential sites are around thehydrogen atoms in the molecule. These sites give information about the regionfrom where the compound can have non-covalent interactions. The molecular electrostatic potential surface shows that the more negative region is associated with O9 atom and more positive region is with H22 atom. This study reveals that an intramolecular interaction may occur on this site. When the intramolecular interaction happens the electrostatic potential of the negative atom changes to less positive one in the molecule [21].

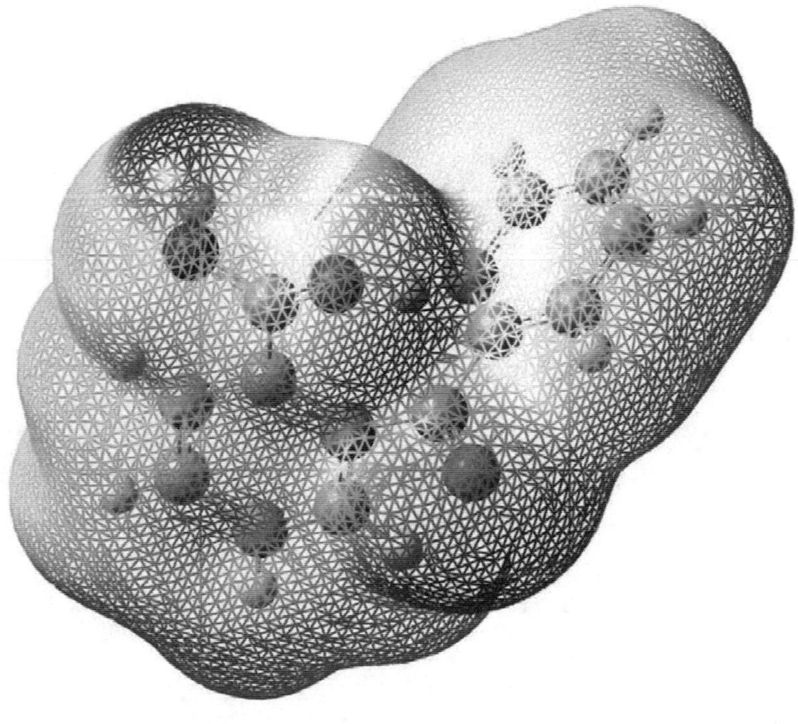

Figure 5 molecular electrostatic potential map of 2-benzoylbenzoic acid

Table 4: Calculated electrostatic charges of 2-benzoylbenzoic acid

Atoms	Electrostatic point charges (Positive)	Atoms	Electrostatic point charges (Negative)
C1	0.7089	C2	-0.2561
C3	0.2511	C4	-0.2203
C5	0.0547	C6	-0.1925
C7	0.0499	O8	-0.5400
C10	0.5772	O9	-0.5901
C12	0.0659	C11	-0.2480
C14	0.1024	C13	-0.0577
C16	0.2687	C15	-0.2570
H18	0.0679	O17	-0.5325
H19	0.0858	H23	-0.0199
H20	0.1001	H27	-0.0161
H21	0.0922		
H22	0.3972		
H24	0.0164		
H25	0.0155		
H26	0.0765		

4. Conclusion

Since the molecule 2-benzoylbenzoic acid is capable of photosensitizing, the key parameters and structural insights which are responsible for its characteristic behavior have been studied in this work to understand the molecule at an atomic level. Quantum chemical computational method implemented in this work provided an amble opportunity to retrieve the geometrical parameters, favorable sites for electrophilic/nucleophilic reactions, lone pair/anti-bonding species for charge delocalization, electronic transition wavelengths. In addition, the IR and Raman spectra of the molecule were simulated and the vibrational assignments of characteristic frequencies and couple modes were made on the basis of Cartesian coordinate displacement analysis. The absorption wavelength found at 263 nm is due to electronic transition from HOMO to LUMO orbital. The predicted values of second-order perturbation energies show that charge transfer from π to π* orbitals stabilizes the molecule.

Reference:

[1] Paul B. Jones, Michael P. Pollastri, and Ned A. Porter, 2-Benzoylbenzoic Acid: A Photolabile Mask for Alcohols and Thiols, *J. Org. Chem.,* 61 (1996) 9455–9461.

[2] Yishan Song, Bing Yan & Zhenxia Chen, Synthesis and crystal structure of a novel luminescent zinc complex of 2-benzoylbenzoic acid, *J. Coord. Chem.,* 58 (2005) 1417-1421.

[3] R. Tesser, M. Di Serio, M. Ambrosio, E. Santacesaria,Heterogeneous catalysts for the production of anthraquinone from 2-benzoylbenzoic acid,Chem. Eng. J.,90 (2002) 195-201.

[4] Naiqun SUN, Xijin ZHAO, Yamin YANG, Liping LI, Aiqin ZHANG, Husheng JIA, Xuguang LIU,Synthesis and luminescent properties of terbium complex containing 4-benzoylbenzoic acid for application in NUV-based LED,*J. Rare Earth.*,34 (2016) 130-136.

[5] Kun Huang, Defang Han, Xianglin Li, MengniPeng, Xianshun Zeng, Linhai Jing, Dabin Qin,A new Cu2+-selective fluorescent probe with six-membered spirocyclichydrazide and its application in cell imaging,Dyes Pigm.,171:107701 (2019) .

[6] ShunsukeTakenaka, Norio Matsuura, NiichiroTokura,Induced circular dichroism of benzoylbenzoic acids in β-cyclodextrin,*Tetrahedron Lett.,*15 (1974) 2325-2328,

[7] V. Balachandran, A. Nataraj, T. Karthick,Molecular structure, spectroscopic (FT-IR, FT-Raman) studies and first-order molecular hyperpolarizabilities, HOMO–LUMO, NBO analysis of 2-hydroxy-p-toluic acid,*Spectrochim.Acta A,* 104(2013) 114-129.

[8] A. Nataraj, V. Balachandran, T. Karthick, M. Karabacak, A. Atac,FT-Raman, FT-IR, UV spectra and DFT and ab initio calculations on monomeric and dimeric structures of 3,5-pyridinedicarboxylic acid,*J. Mol.Struct.,*1027(2012) 1-14.

[9] A. Nataraj, V. Balachandran, T. Karthick,FT-IR and Raman spectra, DFT and SQMFF calculations for geometrical interpretation and vibrational analysis of 3-nitro-p-toluic acid,*J. Mol. Struct.,* 1022 (2012) 94-108.

[10] T. Karthick, V. Balachandran, S. Perumal, A. Nataraj,Spectroscopic studies, HOMO–LUMO and NBO calculations on monomer and dimer conformer of 5-nitrosalicylic acid, *J. Mol. Struct.,* 1005 (2011) 192-201.

[11] M.J. Frisch et. al., Gaussian 03, Revision D.01, Gaussian Inc., Wallingford, CT, 2004.

[12] A. Frisch, A.B. Nielson, A.J. Holder, GAUSSVIEW User Manual, Gaussian Inc., Wallingford, CT, 2003.

[13] C.T. Lee, W.T. Yang, R.G. Parr, *Phys. Rev. B* 37 (1988) 785–789

[14] A.D. Becke, *J. Chem. Phys.* 98 (1993) 5648–5652.

[15] L. J. Bellamy, *The Infrared Spectra of Complex Molecules*, John Wiley & Sons, NewYork, NY,USA, 3rd edition, 1975.

[16] E. Koglin, E.G.Witte, and R. J. Meier, The vibrational spectra of metabolites of methabenzthiazuron: 2-amino-benzothiazole and 2-(methylamino)benzothiazole, *Vib. Spectrosc.*,33 (2003) 49–61.

[17] F.A. Cotton, C.W. Wilkinson, *Advanced Inorganic Chemistry,* third ed., Interscience Publisher, New York, 1972.

[18] M. Silverstein, G. Clayton Bassler, C. Morril, *Spectroscopic Identification of Organic Compounds,* John Wiley, New York, 1981.

[19] E.B. Wilson, J.C. Decius, P.C. Cross, *Molecular Vibrations*, Dover Pupl. Inc., New York, 1980.

[20] P. Politzer, D.G. Truhlar, *Chemical Applications of Atomic and Molecular Electrostatic Potentials,* Plenum, New York, 1981.

[21] W. Hussein, C.G. Walker, Z. Peralta-Inga, J.S. Murray, Computed electrostatic potentials and average local ionization energies on the molecular surfaces of some tetracyclines, *Int. J. Quantum Chem.* 82 (2001) 160-169.

A Comprehensive Review of the Effect of Additives on Microalgae Biodiesel Performance and Emissions

Adina Srinivasa Vara Prasad[1,2,3], K V Murali Krishnam Raju[3 *], Krishna Bhaskar K[1], and Tarun Kumar Kotteda[3]

[1] Department of Mechanical Engineering, University College of Engineering Kakinada, JNTUK Kakinada (AP), India.
[2] Department of Mechanical Engineering, Shri Vishnu Engineering College for Women(A), Bhimavaram, West Godavari (AP), India.
[3] Department of Mechanical Engineering, Sagi Rama Krishnam Raju Engineering College, Bhimavaram, West Godavari (AP), India.

*Corresponding author e-mail: dr.kvmkraju@gmail.com

Abstract. The production and utilization of biodiesel derived from microalgae have gained significant attention as a sustainable alternative to fossil fuels. However, the intrinsic properties of microalgae biodiesel can limit its performance and lead to undesirable emissions. This paper summarizes the impact of additives on microalgae biodiesel, focusing on their influence on performance and emissions. Additives play a crucial role in improving the overall quality and efficiency of microalgae biodiesel. They encompass a range of compounds, including antioxidants, stabilizers, and combustion enhancers. These additives have been studied extensively to elucidate their effects on fuel properties, engine performance, and exhaust emissions. By modifying the physicochemical properties of microalgae biodiesel, additives can enhance its stability, cetane number, and oxidative resistance, thereby promoting efficient combustion and reducing emissions of harmful pollutants. This review concludes that the usage of additives to the algae biodiesel leads to improved combustion performance and emission reduction.

Keywords: Biodiesel, Microalgae, Additives, CI Engine, Performance, Emissions.

1. Introduction

In recent decades, the search for sustainable and environmentally friendly alternatives to fossil fuels has become a paramount concern in addressing the challenges posed by climate change and diminishing global energy resources. Biodiesel, a renewable and biodegradable fuel derived from natural sources such as vegetable oils and animal fats, has emerged as a promising candidate to mitigate the negative impacts of conventional diesel fuels [1–6]. However, despite its advantages, biodiesel presents certain challenges related to its properties [7,8], [9,10]performance, and emissions that demand thorough investigation [11–13].

The potential of biodiesel as a viable substitute for diesel fuel lies in its renewable nature and its ability to reduce greenhouse gas emissions and reliance on fossil fuels [14]. To fully realize its benefits, researchers and engineers have turned their attention to the incorporation of additives into biodiesel formulations. These additives, often chemical compounds or substances, are introduced to enhance the fuel's overall characteristics, address its limitations, and optimize its compatibility with existing diesel engines [15–17].

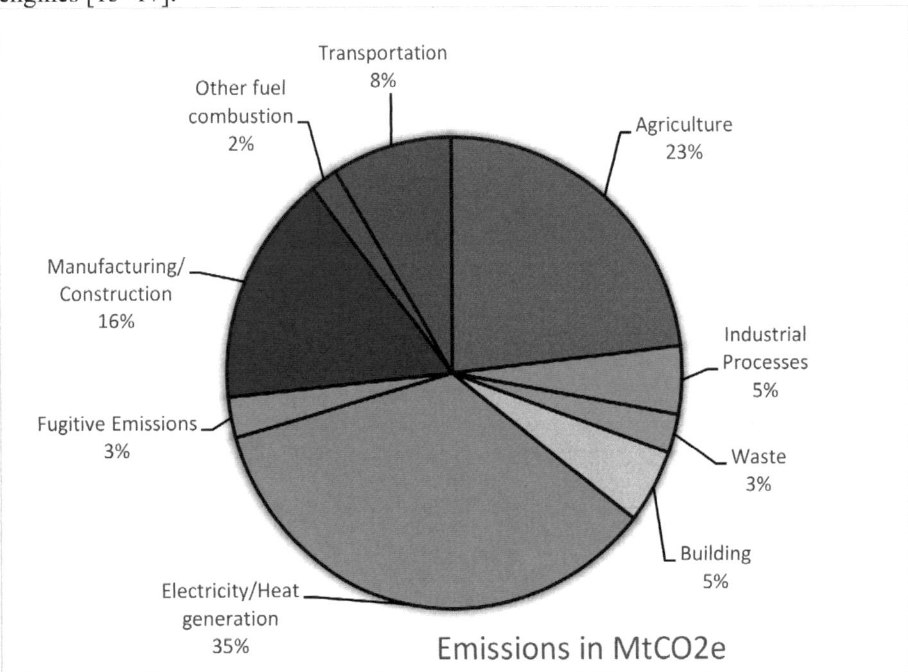

Figure 1. Sector-wise Greenhouse gas emissions (million metric tons of CO_2 equivalent) in India in 2020 [18].

This comprehensive review aims to delve into the intricate relationship between biodiesel additives and their effects on key aspects: properties, performance, and emissions. By methodically examining the existing body of knowledge, this review seeks to provide a comprehensive understanding of how different additives influence the physical and chemical properties of biodiesel. Additionally, it will explore their impact on engine performance parameters, such as combustion efficiency, power output, and fuel consumption, as well as their role in shaping the emission profiles of biodiesel-powered engines.

2. Microalgae Biodiesel

Microalgae biodiesel is a sustainable and promising alternative to conventional fossil fuels due to its numerous benefits. Microalgae are tiny, photosynthetic organisms that can efficiently convert sunlight and carbon dioxide into lipids (oils) through photosynthesis [19]. These lipids can be extracted and processed into biodiesel, a renewable and environmentally friendly fuel.

Microalgae biodiesel offers several advantages. Firstly, it has a high oil yield per unit area of cultivation compared to traditional oil crops, minimizing land use [20]. Additionally, microalgae can be grown in diverse environments, including non-arable land and wastewater, reducing competition with food production [21]. The cultivation of microalgae can also serve as a carbon capture and wastewater treatment solution, mitigating environmental issues [22,23].

However, challenges remain. The production process requires optimization to enhance lipid productivity and reduce costs. Efficient harvesting, lipid extraction, and conversion methods are crucial for commercial viability [24,25]. Economic viability hinges on achieving cost-effective large-scale cultivation and refining processes.

3. Biodiesel Additives

Additives play a vital role in diesel-biodiesel blends when used in Compression Ignition (C.I.) engines, commonly known as diesel engines. These engines are widely used in various applications, such as transportation, industrial machinery, and power generation. The integration of additives into diesel-biodiesel blends brings about several benefits that are crucial for optimizing engine performance, enhancing fuel efficiency, and mitigating operational challenges.

Diesel-biodiesel blends can sometimes lead to the formation of injector deposits due to differences in their chemical properties [26,27]. Differences in chemical properties such as higher oxygen content and lower volatility in biodiesel compared to diesel can lead to incomplete combustion, resulting in the formation of carbonaceous deposits on injector nozzles. These deposits can impair fuel atomization, reduce combustion efficiency, and negatively affect engine performance. Additives with detergency properties help prevent and remove injector deposits, promoting consistent and efficient combustion, enhancing fuel efficiency, and reducing exhaust emissions [28].

Biodiesel has a higher cloud point and pour point compared to conventional diesel, making it prone to gelling in cold temperatures which can hinder fuel flow and lead to engine startup issues. Additives known as cold flow improvers modify the crystallization behavior of the fuel, preventing gelling and ensuring that the blend can flow smoothly even in low temperatures, crucial for reliable engine startup and operation [29,30]. Biodiesel is more susceptible to oxidative degradation over time, leading to the formation of gums, varnishes, and other deposits that can clog fuel filters and affect engine performance. Antioxidant additives are used to enhance the oxidative stability of biodiesel, preventing degradation and maintaining fuel quality during storage and usage [29,31].

Diesel-biodiesel blends are known to influence exhaust emissions, with potential reductions in certain pollutants like particulate matter and CO_2. However, the combustion of biodiesel can also lead to increased nitrogen oxide (NOx) emissions [32–34]. Additives designed for emissions control, such as NOx reduction catalysts, can help mitigate these emissions and maintain compliance with emission standards. Biodiesel's higher oxygen content and more acidic nature can impact fuel system components and storage tanks. Corrosion inhibitors and stabilizers minimize corrosion and ensure compatibility with engine materials and infrastructure, safeguarding the longevity of the engine and the fuel delivery system [35].

Biodiesel has a relatively higher cetane number compared to diesel, which can influence ignition quality. By adjusting the cetane number through additives, combustion efficiency is improved, resulting in better engine start-up, reduced noise, and smoother operation [36,37].

IOP Conf. Series: Journal of Physics: Conf. Series **2765** (2024) 012015 doi:10.1088/1742-6596/2765/1/012015

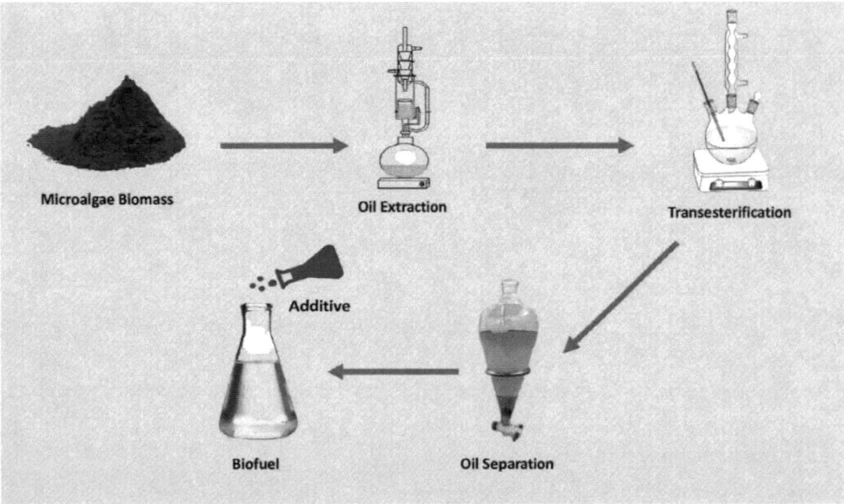

Figure 2. Process flow of conversion of Algae into Biofuel.

In terms of their application, additives are categorized as follows.

3.1. Metal-based additives

Metal-based nano additives enhance diesel engine performance by improving combustion efficiency, lubrication, and emissions reduction. Nano-sized metal particles facilitate better fuel atomization, leading to more complete combustion and increased power output [38,39]. These additives also reduce friction and wear between engine components by forming a protective layer of nanoparticles on the surfaces of moving parts. This layer acts as a lubricant, reducing direct contact between metal surfaces and minimizing friction and wear during engine operation.

Moreover, metal nanoparticles act as catalysts, promoting cleaner combustion and lowering nitrogen oxide (NO_x) and particulate matter emissions, aligning with environmental standards [40].

3.2. Oxygenated additives

Oxygenated additives are compounds containing oxygen that can be added to diesel fuel to improve engine performance and reduce emissions in diesel engines. These additives can include substances like alcohol, ethers, and esters. When added to diesel fuel in small quantities, oxygenated additives enhance combustion efficiency by promoting more complete fuel oxidation. This leads to reduced carbon monoxide (CO) emissions and unburned hydrocarbons, contributing to cleaner exhaust gases [41]. Additionally, oxygenated additives can improve the cetane number of diesel fuel, enhancing ignition quality and reducing ignition delay, which results in smoother engine operation [42]. However, the selection of appropriate oxygenated additives requires consideration of factors such as fuel compatibility, engine design, and regulatory standards. Proper use of oxygenated additives can lead to better engine efficiency, lower emissions, and improved overall engine performance in diesel engines otherwise may lead to compatibility issues with engine components and potential impacts on fuel stability. [43].

3.3. Cetane number improver additives

Cetane number improver additives are utilized in diesel engines to enhance the ignition quality of diesel fuel. The cetane number indicates the fuel's ignition delay period – a lower number signifies quicker ignition [44]. These additives accelerate combustion by reducing ignition delay, resulting in smoother engine start-up, reduced noise, and improved combustion efficiency. This leads to enhanced power output and lower emissions, including reduced nitrogen oxide (NOx) and particulate matter [45],[46]. Cetane improvers are particularly valuable when using lower-quality diesel fuel or biodiesel blends with varying cetane numbers. However, the appropriate dosage of these additives should be carefully determined to prevent overstimulation of combustion processes.

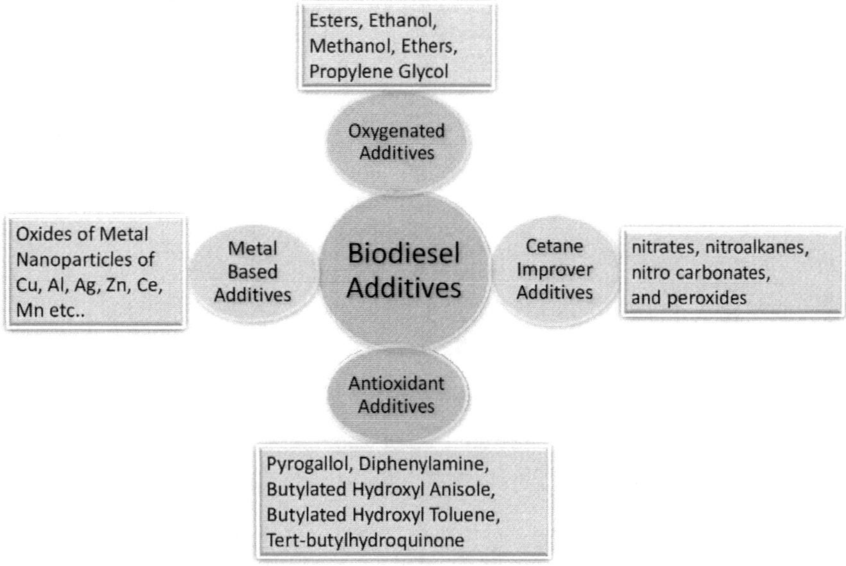

Figure 3. Different types of Biodiesel additives.

3.4. Antioxidant additives

Antioxidant additives play a crucial role in diesel engines by preventing the degradation of the fuel and enhancing its stability. Diesel fuel is susceptible to oxidation, especially during storage and exposure to air. This oxidation can result in the formation of deposits, gums, and varnishes that clog fuel filters and injectors, leading to reduced engine efficiency and performance [47,48]. Antioxidant additives are introduced to diesel fuel to inhibit the chain reaction of oxidation, preserving fuel quality and preventing the development of harmful byproducts. By maintaining the integrity of the fuel, these additives ensure that the engine receives a consistent and clean fuel supply, which contributes to optimal combustion efficiency, reduced emissions, and extended engine life [49,50]. The usage of antioxidant additives is essential in maintaining fuel stability and preventing the detrimental effects of oxidation in diesel engines, ensuring reliable and efficient operation.

The variation in the properties of different strains of microalgae biodiesel with the additives are shown in the table 1.

Table 1: Effect of additives on microalgae biodiesel properties.

Micro Algae	% of Additives	Density at 15 °C (Kg/m³)	Viscosity at 40 °C (mm²/S)	Flash point °C	Pour point °C	Cetane number	Calorific value (MJ/kg)	Oxygen content % mass	Ref.
Diesel		860-900	1.9–6.0	101 min	6-18 max	47 min	43.2	-	[51,52]
C. protothecoides (B85)		865	4.5	164.5	-	49.5	39.1	11.5	[53]
C. protothecoides (B85)	15% DEE & 1000ppm DPPD	845	3.85	133.8	-	61.5	38.3	13.2	[53]
Botryococcus braunii (B20)		830	4.73	176	-4	58	43.54	-	[54]
Botryococcus braunii (B20)	25 ppm CuO_2	831	5.1875	176.5	-4	54.5	44.512	-	[54]
Botryococcus braunii (B20)	50 ppm CuO_2	831.7	5.6449	177	-4	51.5	45.483	-	[54]
Botryococcus braunii (B20)	75 ppm CuO_2	831.8	5.6671	177.5	-4	51.75	45.501	-	[54]

Botryococcus braunii (B20)	100 ppm CuO_2	831.9	5.6894	178	-4	52	45.519	-	[54]
Kappaphycus Alvarezil (B5)		844	2.5	-	-	48.2	-	0.0052	[55]
Kappaphycus Alvarezil (B5)	10 ppm A.oleic + Al_2O_3	845	2.7	-	-	48.8	-	0.0109	[55]
Kappaphycus Alvarezil (B5	20 ppm A.oleic + Al_2O_3	846	2.8	-	-	49.4	-	0.0166	[55]
Kappaphycus Alvarezil (B5	50 ppm A.oleic + Al_2O_3	846	2.8	-	-	49.9	-	0.017	[55]
Naviculla Sp. (B20)		878	3.912	109	-4	50	37	-	[56]
Naviculla Sp. (B20)	50 ppm MgO	879	3.913	114	-4	56	38.9	-	[56]
Naviculla Sp. (B20)	100 ppm MgO	881	3.916	116	-4	61	40.8	-	[56]
Neochloris oleoabundans (B20)		832	4.71	175	-4	58	43.59	-	[57]
Neochloris oleoabundans (B20)	25 ppm CeO_2	831.9	5.1776	175.5	-4	54.75	44.6065	-	[57]
Neochloris oleoabundans (B20)	50 ppm CeO_2	831.8	5.6451	176	-4	51.5	45.623	-	[57]
Neochloris oleoabundans (B20)	75 ppm CeO_2	831.7	5.6669	176.5	-4	51.75	45.921	-	[57]
Neochloris oleoabundans (B20)	100 ppm CeO_2	831.6	5.6888	177	-4	52	46.219	-	[57]
Naviculla Sp.(B20)		878	3.912	109	-4	52	37.2	-	[58]
Naviculla Sp.(B20)	50 ppm La_2O_3	880	3.9	110	-4	55	38.4	-	[58]
Naviculla Sp.(B20)	100 ppm La_2O_3	882	3.912	111	-4	58	39	-	[58]
Chlorella vulgaris (B25)		846.5	4.30	76	-10	52	43.051	5.15	[59]
Chlorella vulgaris (B25)	5 ml MTBE	854	3.98	81	-9	52.4	43.222	7.13	[59]
Chlorella vulgaris (B25)	10 ml MTBE	864.3	3.77	97	-8	52.8	43.590	9.08	[59]
Chlorella vulgaris (B25)	15 ml MTBE	878.2	3.56	159	-7	53	43.863	10.21	[59]
- (B20But30)		840.3	4.166	66.4	-	40.5	39.90	-	[60]
- (B20But30)	25 ppm TiO_2	838.3	4.10	168	-	41	39.922	-	[60]
-(B20But30)	50 ppm TiO_2	836.5	4	164	-	41.5	39.94	-	[60]
-(B20But30)	75 ppm TiO_2	833	3.90	163	-	41.59	39.96	-	[60]
-(B20But30)	100 ppm TiO_2	830	3.77	161	-	43	39.999	-	[60]

4. Effects of additives on the performance, and emissions of diesel engines

Based on the preceding discussion, various categories of additives demonstrate encouraging outcomes in terms of their application to both biodiesel and diesel fuels. The researcher's findings in numerous articles highlight promising results. However, the choice of additives must be grounded in the specific characteristics of diesel and biodiesel, as they pertain to their use within diesel engines. Researchers analyzed the effect of various additives including antioxidants, oxygenating agents, metal-based compounds, and cetane number enhancers on the performance and the emissions from the engines. The literature reviews in this article demonstrate the effects of using various additives with biodiesel fuels on engine performance, and emission.

Chlorella protothecoides microalgae biodiesel added with DEE in 10%, 15%, 20%, and DPPD in 1000ppm, 2000ppm was tested on a thermal barrier-coated diesel engine [53]. The outcomes revealed that the additives improved the biodiesel's resistance to oxidation and its cetane number, simultaneously leading to a reduction in viscosity. Furthermore, there is a decrease in smoke, NOx, and CO emissions, but an increase in unburned hydrocarbons is detected. The MB85DEE15DPPD1000 ppm blend, when combined with EGR, demonstrated a noteworthy decrease of 18.5% in NOx emissions in comparison to regular diesel fuel. According to research on the combustion analysis of the addition of CuO_2 nanoparticles to a blend of Botryococcus braunii algae biodiesel in a CI engine [54], the addition of CuO_2 nanoparticles to the B20 fuel blends increased cylinder gas pressure and temperature, which improved combustion and decreased CO, HC, and smoke emissions. However, because B20 has a lower calorific value than other fuels, the test fuels B20 + 75 ppm and B20 + 100 ppm saw a significant drop in BTE. CuO_2 Nano additives were added to neochloris oleoabundans algae B20 biodiesel in 25, 50, 75, and 100 ppm dosage [61]. B20 biodiesel blended with CuO_2 nanoparticles results in higher cylinder pressure than pure B20 due to the presence of oxygen, resulting in lower cetane numbers and longer ignition delay periods. Initial heat release during ignition delay is low due to fuel vaporization. Combustion leads to rapid heat release, peaking in the premixed fuel-air stage. CuO_2 nanoparticles in B20 improve premixed combustion, enhancing net heat release. Butylated hydroxytoluene (BHT) antioxidant additive was mixed with B20 microalgae biodiesel and tested on a 4.4 kW stationary engine coupled to an eddy current dynamometer [62]. The study revealed that B20 biodiesel exhibited 38% higher NOx emissions than regular diesel due to its elevated oxygen content. Introducing BHT to B20 led to a 25% average reduction in nitrogen oxide emissions, slightly surpassing diesel levels. This reduction might be attributed to BHT's cooling impact, caused by its high latent heat of vaporization. On average, B20 biodiesel showed a 21% reduction in smoke opacity compared to diesel, likely because of higher oxygen levels and oxygen availability in fuel. Incorporating BHT into B20 increased smoke opacity by around 36%. On average, B20 biodiesel exhibited a 15% reduction in hydrocarbon emissions compared to diesel. However, introducing BHT to B20 led to an average 50% increase in unburned hydrocarbon (UBHC) emissions, possibly due to incomplete combustion and inadequate atomization. B20 biodiesel showed slightly lower efficiency than diesel, possibly due to the need for extra energy to break down larger hydrocarbon chains and aromatic compounds. Introducing antioxidants to B20 led to a small reduction in its brake thermal efficiency.

Kappaphycus Alvarezil-Brown algae biodiesel in B5 blend ratio added with Al_2O_3 and $C_{18}H_{34}O_2$ additives 10, 20, and 50 ppm was tested on a CI engine [55]. The additive improves the quality of the fuel mixtures without compromising the standards. It lowers the sulfur and increases the oxygen, which can lead to cleaner and more efficient combustion. However, it also raises the risk of low lubricity and nitrous oxide emissions. The SFC was reduced by 0.21% with a B5+20ppm additive blend. When the same additives Al_2O_3 and $C_{18}H_{34}O_2$ were mixed with Alvarezil-Brown algae biodiesel, the viscosity increased, of sulfur dioxide and particulate matter was reduced. But NOx emissions increased [63]. A blend of magnesium oxide nanoparticles in 0, 25, 50, and 100 ppm and biodiesel from Naviculla Sp. algae oil (B20) was tested on a single-cylinder diesel engine and the results indicated a reduction in BSFC and an improved BTE with B20+100ppm MgO blend. Also, the Co, HC, and Smoke emissions were reduced with this blend whereas the NOx emissions were increased [56]. A mixture of 20% neochloris oleoabundans algae oil (B20) and regular diesel fuel was used, combining cerium oxide (CeO2) nanoparticles at different concentrations: 25 ppm, 50 ppm, 75 ppm, and 100 ppm into a single-cylinder air-cooled engine, connected to an electrical dynamometer [57]. The results showed an enhancement in brake thermal efficiency and a reduction in exhaust emissions from the engine. Additionally, the specific fuel consumption decreased, except when using 25 ppm of CeO2 nanoparticles, which caused an increase in NOX emissions. The research investigated the impact of blending Naviculla Sp. algae biodiesel with La_2O_3 nanoparticles on the performance and emissions of a four-stroke single-cylinder diesel engine [58]revealed that La_2O_3 nanoparticles with biodiesel led to higher calorific value compared to the B20 blend. The reduction in BSFC resulted from the nanoparticles' positive effects on fuel properties and shorter ignition delay time. The presence of nanoparticles improved combustion due to their higher surface-to-volume ratio, facilitating more efficient fuel-air interaction, thus enhancing Brake Thermal Efficiency (BTE). Also. La_2O_3 blend fuels exhibited lower CO, HC, and smoke emissions compared to B20, while they produced higher NOx emissions. Chlorella vulgaris biodiesel blend (B25) with Methyl tert butyl ether additive in 5ml, 10ml, and 15ml was used on the Kirloskar TV-1 engine . The performance analysis demonstrated that incorporating 15ml of additives into B25 biodiesel led to a minor enhancement in brake thermal efficiency while significantly lowering BSFC compared to regular diesel. The study also observed a significant reduction in exhaust gas temperature alongside the improvements in BSFC and BTE. In terms of emissions, the B25 blend with 15ml additives exhibited substantial reductions in HC, CO, and smoke emissions. Furthermore, the biodiesel blends with additives emit less CO_2 than pure diesel. However, an increase in O_2 levels due to the additive concentration resulted in a slight rise in NO_x emissions [59]. As an additive, a B50 blend of Scenedesmus obliquus algae was incorporated with n-pentane in varying amounts (5, 10, and 15 ml per liter). Optimal results were achieved with 15 ml of n-pentane, leading to increased BTE (7.1%) and decreased BSFC (6.4%) compared to B50. While hydrocarbon emissions dropped by 7.2%, nitrogen oxides increased due to elevated exhaust temperature and oxygen content in B50. Carbon dioxide rose by 22.3%, while carbon monoxide and oxygen concentrations decreased by 17.35% and 9.5%, respectively, with B50 [64]. To investigate the combustion, emissions, and performance of a 4-stroke diesel engine (Kirloskar AV1, 5.2 kW) running on Botryococcus braunii algal biodiesel, acetylene was introduced with varying mass flow rates (100, 150, 200, and 300 g/hr). The acetylated biodiesel notably improved brake thermal efficiency (BTE) by 3% to 5%. While NOx emissions increased with acetylated biodiesel, CO2, HC, and CO emissions decreased due to acetylene addition [65]. Five different biodiesel blends were prepared from B20But30 (consisting of 50% biodiesel, 20% butanol, and 30% diesel) with varying concentrations (0, 25, 50, 75, and 100 ppm) of TiO_2. The results showed that compared to pure diesel at full load, BTE increased by 0.48% for B20But30, and by 4.87%, 9.48%, 11.06%, and 12% for B20But30TiO_2 (25 ppm), B20But30TiO_2 (50 ppm), B20But30TiO_2 (75 ppm), and B20But30TiO_2 (100 ppm) respectively. The exhaust gas temperature also varied, with the highest value at 220°C for B20But30TiO_2 (100 ppm). The experimental study demonstrated that adding TiO_2 to the blends led to reduced levels of NOx, CO_2, and SOx emissions, while hydrocarbon (HC) emissions increased [60].

An oxygenated additive called triacetin is used with a blend of spirogyra algae oil biodiesel and diesel fuel (ABD40) to investigate its impact on combustion, engine performance, and exhaust emissions in a single-cylinder 3.5 kW diesel engine[66]. Researchers tested different concentrations of triacetin (2%, 4%, 6%, and 8% by volume) mixed with ABD40 while maintaining a constant engine speed and varying engine loads. The results of the combustion tests at full load (100%) showed that ABD40 with 4% triacetin achieved higher combustion pressure compared to both pure diesel and other biodiesel blends. Additionally, it demonstrated a higher BTE of 30.39% compared to other biodiesel blends, although it was 7.23% lower than mineral diesel fuel. It's worth noting that biodiesel blends consumed more fuel due to their higher density and kinematic viscosity compared to DF. Moreover, the emissions from the engine running on ABD40 with 4% triacetin were significantly lower than those from pure diesel fuel. Specifically, there was a reduction of 25.51% in CO emissions, 12.81% in NOx emissions, 70.53% in unburned hydrocarbons, and 14.85% in smoke emissions at 100% load compared to diesel [66]. Researchers examined the use of algae biodiesel blended with methanol in three different proportions: B10 (10% diesel + 90% algae oil + 2.5% methanol), B20 (20% diesel + 80% algae oil + 5% methanol), and B30 (30% diesel + 70% algae oil + 7.5% methanol). The results showed that B20 biodiesel performed very similarly to pure diesel fuel, with only slight increases in HC and CO emissions. However, there was a significant reduction in NOx emissions for all three blends (B10, B20, and B30) when compared to pure diesel fuel [67]. B25 algae biodiesel was mixed with metallic oxide nanoparticles like CeO_2, SiO_2, and TiO_2 at a concentration of 150 mg/liter to study their influence on diesel engines. The study's findings indicated that when compared to traditional diesel fuel, the use of the B25+TiO_2 blend resulted in notable enhancements, including a 23.4% increase in in-cylinder pressure, a 16.6% rise in heat release rate, a 32.4% improvement in BTE, a 65.09% increase in CO_2

emissions, and a 25.62% reduction in BSFC. Additionally, this blend reduced CO emissions by 33.4%, NOx emissions by 68.3%, unburned hydrocarbons by 22.75%, and smoke opacity by 30.99% [68]. The research was conducted using a single-cylinder, direct injection, liquid-cooled diesel engine, where cerium oxide (CeO2) and copper peroxide (CuO2) nanoparticles were added to a B20 blend of Neochloris oleoabundans green microalgae biodiesel. The findings revealed that the addition of nanoparticles led to improvements in several performance indicators, including BTE by 0.79% to 3.70%, cylinder pressure by 0.75% to 1.67%, and Heat Release Rate by 2.35% to 4.98%. Additionally, there were reductions in BSFC by 5.19% to 9.5% and CO_2 emissions by 5.18% to 9.5%. However, this improvement came at the expense of increased NOx emissions, which saw a substantial rise of 22.67% to 70.69% when nanoparticles were introduced into the Neochloris oleoabundans biodiesel blend[69]. The study involved blending spirulina microalgae biodiesel with diesel at different ratios (10% and 20%) and adding varying amounts of Fe_3O_4 nanoparticles (25, 50, and 75 ppm) to the 20% MAB and 80% diesel (20MAB80D) mixture. The most significant enhancements were observed when using the 20MAB80D75IO fuel blend, with average increases of 28.64% in BTHE and 13.18% in BMEP, compared to using pure diesel. The 20MAB80D75IO fuel blend exhibited an average BSFC reduction of 19.02% when compared to D100, while the 20MAB80D blend showed an average increase of 11.31% in BSFC. The 20MAB80D75IO fuel achieved average reductions of 24.21% for CO and 57.36% for HC compared to D100. Similarly, the 20MAB80D blend resulted in average reductions of 20.14% for CO and 22.62% for HC. However, the 20MAB80D75IO fuel exhibited an 11.63% higher NOx production compared to D100, while the 20MAB80D blend showed a 3.83% increase in NOx emissions[70]. A study compares Scenedesmus sp. and Pongamia biodiesels and the effects of cerium oxide (CeO2) nanoparticles with these biodiesels in an indirect injection CI engine. Nanoparticles improved engine efficiency by 4.2% for Scenedesmus sp. and 3.8% for Pongamia biodiesel blends. Peak pressures were highest in nanoparticle-added blends, with peak heat release rates slightly shifted. NOx emissions were lower than diesel for A30 and P30 but increased by 2.17% and 6.17% with nanoparticles. Smoke emissions were significantly lower than diesel for all biodiesel variants. A30C100 achieved diesel-like efficiency at higher loads, making it a promising alternative for CI engines[71].

5. Conclusions

In the quest for sustainable and environmentally responsible energy sources, microalgae biodiesel has emerged as a promising contender. Its potential to provide a renewable and cleaner alternative to conventional diesel fuels has garnered substantial attention from researchers, policymakers, and industries worldwide. However, challenges exist in optimizing its performance, stability, and environmental impact as with any emerging technology. This review has delved into a critical aspect of microalgae biodiesel production – the effect of additives. Over the course of this comprehensive analysis, we have explored the diverse array of additives employed to enhance microalgae biodiesel, aiming to boost its overall performance and mitigate emissions. As we conclude this exploration, we summarize the key findings and implications that shed light on the role of additives in shaping the future of microalgae biodiesel as a sustainable energy solution.

- The comprehensive analysis of various additives in microalgae biodiesel production has revealed their potential to significantly enhance the performance and stability of biodiesel. Additives can improve oxidative stability, cetane number, and cold flow properties, making microalgae biodiesel a more attractive alternative to conventional diesel fuels.
- The utilization of additives in microalgae biodiesel has shown a remarkable reduction in harmful emissions, particularly in terms of HC, CO, CO_2, and particulate matter. The NO_x emissions were also reduced significantly with some additives. This emphasizes the environmental benefits of microalgae biodiesel as a cleaner and greener energy source.
- While additives offer promising benefits, it is crucial to recognize that their effectiveness can vary depending on the specific microalgae feedstock and processing methods. There may also be trade-offs in terms of increased cost and potential impacts on other fuel properties. A careful balance between additives and their concentrations needs to be established to optimize performance and cost-effectiveness.

- To transition additives from the laboratory to commercial-scale microalgae biodiesel production, economic considerations, scalability, and regulatory approvals must be addressed. Collaboration between researchers, industry stakeholders, and policymakers is essential to bring these innovations to the market successfully.

The findings underscore the importance of microalgae biodiesel as a sustainable and renewable energy source. Additives can play a pivotal role in making microalgae biodiesel a viable and environmentally friendly option for reducing our reliance on fossil fuels and mitigating climate change.

References:

[1] Atabani A E, Silitonga A S, Ong H C, Mahlia T M I, Masjuki H H, Badruddin I A and Fayaz H 2013 Non-edible vegetable oils: A critical evaluation of oil extraction, fatty acid compositions, biodiesel production, characteristics, engine performance and emissions production *Renewable and Sustainable Energy Reviews* **18** 211–45

[2] Demirbas A H and Demirbas I 2007 Importance of rural bioenergy for developing countries *Energy Convers Manag* **48** 2386–98

[3] Altarazi Y S M, Abu Talib A R, Yu J, Gires E, Abdul Ghafir M F, Lucas J and Yusaf T 2022 Effects of biofuel on engines performance and emission characteristics: A review *Energy* **238**

[4] Kumar N, Varun and Chauhan S R 2013 Performance and emission characteristics of biodiesel from different origins: A review *Renewable and Sustainable Energy Reviews* **21** 633–58

[5] Hurajama T, Oh Y-T, Miyamoto N, Chikahisa T and Takagi N 2018 *Mdiwancing Mobility 400 COMMONWEALTH DRIVE WARRENDALE, PA 15096*

[6] Kumar K T, Prasad A S V and Raju B 2020 Experimental Investigation on The Performance and Emissions of A CI Engine Tested with The Blends of Diesel-Rapeseed Methyl Ester and N-Butanol as an Additive *International Journal of Advance Science and Technology* **29** 8417–27

[7] Ashraful A M, Masjuki H H, Kalam M A, Rizwanul Fattah I M, Imtenan S, Shahir S A and Mobarak H M 2014 Production and comparison of fuel properties, engine performance, and emission characteristics of biodiesel from various non-edible vegetable oils: A review *Energy Convers Manag* **80** 202–28

[8] Hoekman S K, Broch A, Robbins C, Ceniceros E and Natarajan M 2012 Review of biodiesel composition, properties, and specifications *Renewable and Sustainable Energy Reviews* **16** 143–69

[9] Kotteda T K, Raju Chekuri R B, Raju B N, Kantheti P R and Balakumar S 2021 Analysis on emissions and performance of ceramic coated diesel engine fueled with novel blends using artificial intelligence *Advances in Materials Science and Engineering* **2021**

[10] Prakash Varma P S and Subbaiah K V 2022 CFD Analysis of piston bowls geometry for CI direct injection engine using finite element analysis *Int. J. Nonlinear Anal. Appl* **13** 2008–6822

[11] Silitonga A S, Masjuki H H, Mahlia T M I, Ong H C, Chong W T and Boosroh M H 2013 Overview properties of biodiesel diesel blends from edible and non-edible feedstock *Renewable and Sustainable Energy Reviews* **22** 346–60

[12] Mathew G M, Raina D, Narisetty V, Kumar V, Saran S, Pugazhendi A, Sindhu R, Pandey A and Binod P 2021 Recent advances in biodiesel production: Challenges and solutions *Science of the Total Environment* **794**

[13] Meira M, Quintella C M, Ribeiro E M O, Silva H R G and Guimarães A K 2015 Overview of the challenges in the production of biodiesel *Biomass Convers Biorefin* **5** 321–9

[14] Manzanera M, Molina-Muñoz M L and González-López J 2008 *Biodiesel: An Alternative Fuel* vol 2

[15] Rashedul H K, Masjuki H H, Kalam M A, Ashraful A M, Ashrafur Rahman S M and Shahir S A 2014 The effect of additives on properties, performance and emission of biodiesel fuelled compression ignition engine *Energy Convers Manag* **88** 348–64

[16] Madiwale S, Karthikeyan A and Bhojwani V 2017 A Comprehensive Review of Effect of Biodiesel Additives on Properties, Performance, and Emission *IOP Conference Series: Materials Science and Engineering* vol 197 (Institute of Physics Publishing)
[17] Hosseinzadeh-Bandbafha H, Tabatabaei M, Aghbashlo M, Khanali M and Demirbas A 2018 A comprehensive review on the environmental impacts of diesel/biodiesel additives *Energy Convers Manag* **174** 579–614
[18] Anon Climate Watch Historical Country Greenhouse Gas Emissions Data
[19] Alaswad A, Dassisti M, Prescott T and Olabi A G 2015 Technologies and developments of third generation biofuel production *Renewable and Sustainable Energy Reviews* **51** 1446–60
[20] Tan T, Yu J and Shang F 2011 Biorefinery Engineering *Comprehensive Biotechnology, Second Edition* vol 2 (Elsevier Inc.) pp 815–28
[21] Mat Aron N S, Khoo K S, Chew K W, Veeramuthu A, Chang J S and Show P L 2021 Microalgae cultivation in wastewater and potential processing strategies using solvent and membrane separation technologies *Journal of Water Process Engineering* **39**
[22] Tan J Sen, Lee S Y, Chew K W, Lam M K, Lim J W, Ho S H and Show P L 2020 A review on microalgae cultivation and harvesting, and their biomass extraction processing using ionic liquids *Bioengineered* **11** 116–29
[23] Srimongkol P, Sangtanoo P, Songserm P, Watsuntorn W and Karnchanatat A 2022 Microalgae-based wastewater treatment for developing economic and environmental sustainability: Current status and future prospects *Front Bioeng Biotechnol* **10**
[24] Vasistha S, Khanra A, Clifford M and Rai M P 2021 Current advances in microalgae harvesting and lipid extraction processes for improved biodiesel production: A review *Renewable and Sustainable Energy Reviews* **137**
[25] Jee Young Kim J-M J S J Y-K P Y F T K-Y A L Y-E C E E K 2022 Biodiesel from microalgae: Recent progress and key challenges *Prog Energy Combust Sci* **93**
[26] Liaquat A M, Masjuki H H, Kalam M A and Rizwanul Fattah I M 2014 Impact of biodiesel blend on injector deposit formation *Energy* **72** 813–23
[27] Desantes J M, Payri R, García A and Manin J 2009 Experimental study of biodiesel blends' effects on diesel injection processes *Energy and Fuels* **23** 3227–35
[28] Zak G, Ziemiański L, Stępień Z and Wojtasik M 2014 Engine testing of novel diesel fuel detergent-dispersant additives *Fuel* **122** 12–20
[29] Sorate K A and Bhale P V. 2015 Biodiesel properties and automotive system compatibility issues *Renewable and Sustainable Energy Reviews* **41** 777–98
[30] Chiu C W, Schumacher L G and Suppes G J 2004 Impact of cold flow improvers on soybean biodiesel blend *Biomass Bioenergy* **27** 485–91
[31] Khurana D and Agarwal A K 2011 Oxidation Stability, Engine Performance and Emissions Investigations of Karanja, Neem and Jatropha Biodiesel and Blends *SAE Int J Fuels Lubr* **4** 76–83
[32] Hoekman S K and Robbins C 2012 Review of the effects of biodiesel on NOx emissions *Fuel Processing Technology* **96** 237–49
[33] Palash S M, Kalam M A, Masjuki H H, Masum B M, Rizwanul Fattah I M and Mofijur M 2013 Impacts of biodiesel combustion on NOx emissions and their reduction approaches *Renewable and Sustainable Energy Reviews* **23** 473–90
[34] Chen H, Xie B, Ma J and Chen Y 2018 NOx emission of biodiesel compared to diesel: Higher or lower? *Appl Therm Eng* **137** 584–93
[35] Ma I A W A Sh , K S S A , R K , R S 2022 A concise review on corrosion inhibitors: types, mechanisms and electrochemical evaluation studies *J Coat Technol Res* **19** 241–68
[36] Mahmudul H M, Hagos F Y, Mamat R, Adam A A, Ishak W F W and Alenezi R 2017 Production, characterization and performance of biodiesel as an alternative fuel in diesel engines – A review *Renewable and Sustainable Energy Reviews* **72** 497–509

[37] Ileri E, Karaoglan A D and Akpinar S 2020 Optimizing cetane improver concentration in biodiesel-diesel blend via grey wolf optimizer algorithm *Fuel* **273**

[38] M.S. Gad Ü A A A H P S J N A A Q A S E-S 2023 A comprehensive review on the usage of the nano-sized particles along with diesel/biofuel blends and their impacts on engine behaviors *Fuel* **339**

[39] A. Devaraj M N D Y O P Y A R A S 2022 Influence of nano-additives on engine behaviour using diesel-biodiesel blend *Mater Today Proc* **62** 2266–70

[40] Ndolomingo M J, Bingwa N and Meijboom R 2020 Review of supported metal nanoparticles: synthesis methodologies, advantages and application as catalysts *J Mater Sci* **55** 6195–241

[41] Kashyap D, Das S and Kalita P 2022 Influence of Oxygenated Fuel and Additives in Biofuel Run Compression Ignition Engine *Potential and Challenges of Low Carbon Fuels for Sustainable Transport* ed A K Agarwal and H Valera (Singapore: Springer Singapore) pp 183–243

[42] Adem Siraj Mohammed S M A A V R G A 2023 A comprehensive review on the effect of ethers, antioxidants, and cetane improver additives on biodiesel-diesel blend in CI engine performance and emission characteristics *Journal of the Energy Institute* **108**

[43] Vijayashree and Ganesan V 2020 A Comprehensive Review on Oxygenated Fuel Additive Options for Unregulated Emission Reduction from Diesel Engines *Alternative Fuels and Their Utilization Strategies in Internal Combustion Engines* ed A P Singh, Y C Sharma, N N Mustafi and A K Agarwal (Singapore: Springer Singapore) pp 141–65

[44] Keunsoo Kim W L P W E M J T C-B M K T L 2023 Effects of the cetane number on chemical ignition delay *Energy* **264**

[45] Sivasubramanian H 2018 Effect of Ignition Delay (ID) on performance, emission and combustion characteristics of 2-Methyl Furan-Unleaded gasoline blends in a MPFI SI engine *Alexandria Engineering Journal* **57** 499–507

[46] S. Rabl T J D A P M R F C 2015 Understanding the relationship between ignition delay and burn duration in a constant volume vessel at diesel engine conditions *Proceedings of the Combustion Institute* **35** 2967–74

[47] Usman A and Ahmad Rufai I 2021 Effect of Oxidation Stability on the Fuel and Storage Properties of Balanites Aegyptiaca Biodiesel *Jurnal Kejuruteraan* **33** 883–90

[48] Zahira Yaakob B N N S P S U K , M A P 2014 A review on the oxidation stability of biodiesel *Renewable and Sustainable Energy Reviews* **35** 136–53

[49] K. Varatharajan D S P 2018 Screening of antioxidant additives for biodiesel fuels *Renewable and Sustainable Energy Reviews* **82** 2017–28

[50] Amran N A, Bello U and Hazwan Ruslan M S 2022 The role of antioxidants in improving biodiesel's oxidative stability, poor cold flow properties, and the effects of the duo on engine performance: A review *Heliyon* **8**

[51] Bureau of Indian Standards 2016 Biodiesel (B100)- Fatty Acid Methyl Esters(Fame)-Specification

[52] Anon Standard Specification for Biodiesel Fuel Blend Stock (B100) for Middle Distillate Fuels 1

[53] Krishna Kolli V, Gadepalli S, Deb Barma J, Krishna Maddali M, Barathula S and Kumar reddy Siddavatam N 2019 Establishment of lower exhaust emissions by using EGR coupled low heat loss diesel engine with fuel blends of microalgae biodiesel-oxygenated additive DEE-antioxidant DPPD *Thermal Science and Engineering Progress* **13**

[54] Dharmaprabhakaran T, Karthikeyan S, Periyasamy M and Mahendran G 2020 Combustion analysis of CuO2 nanoparticle addition with blend of botryococcus braunii algae biodiesel on CI engine *Materials Today: Proceedings* vol 33 (Elsevier Ltd) pp 2874–6

[55] Karthikeyan S, Prathima A, Periyasamy M and Mahendran G 2020 Performance analysis of Al2O3 and C18H34O2 with Kappaphycus Alvarezil-Brown algae biodiesel in CI engine *Materials Today: Proceedings* vol 33 (Elsevier Ltd) pp 4180–4

[56] Arunprasad J and Elango T 2020 Performance and emission characteristics of engine using Naviculla Sp. Algae oil methyl ester with MgO nanoparticles *Materials Today: Proceedings* vol 33 (Elsevier Ltd) pp 3164–8

[57] Kalaimurugan K, Karthikeyan S, Periyasamy M and Mahendran G 2020 Emission analysis of CI engine with CeO2 nanoparticles added neochloris oleoabundans biodiesel-diesel fuel blends *Materials Today: Proceedings* vol 33 (Elsevier Ltd) pp 2877–81

[58] Arunprasad J and Elango T 2020 Performance and emission analysis of engine using Naviculla Sp. Algae oil methyl ester with La2O3 nanoparticles *Materials Today: Proceedings* vol 33 (Elsevier Ltd) pp 3177–81

[59] Adityan K, Soundararajan R, AshokaVarthanan P, Raja A and Prem Anand B 2021 Experimental investigation on a diesel engine fuelled by algae methyl ester with additives *International Journal of Ambient Energy* **42** 1258–68

[60] Verma S, Upadhyay R, Shankar R and Pandey S P 2023 Performance and emission characteristics of micro-algae biodiesel with butanol and TiO2 nano-additive over diesel engine *Sustainable Energy Technologies and Assessments* **55**

[61] Kalaimurugan K, Karthikeyan S, Periyasamy M and Dharmaprabhakaran T 2020 Combustion analysis of CuO2 nanoparticles addition with neochloris oleoabundans algae biodiesel on CI engine *Materials Today: Proceedings* vol 33 (Elsevier Ltd) pp 2573–6

[62] Saravanan S and Krishnamoorthy N 2020 Investigation on reduction in consequences of adding antioxidants into the algae biodiesel blend as a CI engine fuel *Fuel* **276**

[63] Karthikeyan S, Prathima A, Periyasamy M and Mahendran G 2020 Emission assessment of Kappaphycus Alvarezil-Brown algae biodiesel Al2O3 and C18H34O2 *Materials Today: Proceedings* vol 33 (Elsevier Ltd) pp 4192–7

[64] Elkelawy M, Alm-Eldin Bastawissi H, El Shenawy E A, Taha M, Panchal H and Sadasivuni K K 2021 Study of performance, combustion, and emissions parameters of DI-diesel engine fueled with algae biodiesel/diesel/n-pentane blends *Energy Conversion and Management: X* **10**

[65] Aisosa Oni B, Eshorame Sanni S, Daramola M and Victoria Olawepo A 2021 Effects of oxy-acetylation on performance, combustion and emission characteristics of Botryococcus braunii microalgae biodiesel-fuelled CI engines *Fuel* **296**

[66] Kalyani T, Prasad L S V and Kolakoti A 2023 Effect of triacetin as an oxygenated additive in algae biodiesel fuelled CI engine combustion, performance, and exhaust emission analysis *Fuel* **338** 127366

[67] Boopathi M, Sathiskumar S, Manideep B, Jayakrishnan S, Praveen S B, Gokul V, Sakthi Ganesh P and Gokulraj V 2023 Experimental investigation on performance and emission characteristics of algae oil biodiesel with methanol additive in CI engine *Mater Today Proc*

[68] Jegan C D, Selvakumaran T, Karthe M, Hemachandu P, Gopinathan R, Sathish T and Ağbulut Ü 2023 Influences of various metal oxide-based nanosized particles-added algae biodiesel on engine characteristics *Energy* **284** 128633

[69] Singh Rajpoot A, Choudhary T, Chelladurai H, Rajak U and Kumar Sahu M 2023 Comparison of the effect of CeO2 and CuO2 nanoparticles on performance and emission of a diesel engine fueled with Neochloris oleoabundans algae biodiesel *Mater Today Proc*

[70] USLU S, MAKI D F and AL-GBURI A S K 2023 Synthesis of spirulina microalgae biodiesel, and experimental research of its effects on compression ignition engine responses with iron II-III oxide (Fe3O4) nanoparticle supplementation *Energy Convers Manag* **293** 117457

[71] Deshmukh S, Kumar R, Bala K and Chandra R 2023 Comparison and testing of cerium oxide nanoparticle added Pongamia and, Scenedesmus microalgae biodiesel blends *Chemical Engineering Research and Design* **197** 871–83

Analysis Of Delamination During Drilling Process Of Aramid Fibre Reinforced Polymers

V Durga Prasada Rao [a], Sk R S Mahaboob Ali [a], Sk M Z M S Ali [a,*], B S Santhoshi [a]
[a] Dept. of Mechanical Engineering ,S.R.K.R. Engineering. College, Bhimavaram – 534204, India
*Corresponding Author email: saqheebali@gmail.com

Abstract. Aramid Fibre Reinforced Polymer (AFRP) is a modern engineering material which is highly versatile in nature. These materials are usually used to produce impact resistant structures. Aramids have high modulus, tensile strength, stiffness, and low weight density. AFRP, under cyclic loading, shows good performance. As far as textiles and concrete structures are considered, AFRP is one of the best fiber-reinforced composites. Delamination is a failure mode that causes fracture of a variety of materials including concrete and laminate composites. A material, in general, fractures into layers due to delamination effect, i.e., separation into its constituent layers. In the present study, the AFRP polymer is prepared by hand layup method, drilling operation is done on the composite under different machining conditions, measurements regarding delamination are taken using a tool room microscope, and delamination factor is calculated by using a suitable equation. Rate of feed, spindle speed and diameter of drill are the drilling parameters under consideration. Each of these parameters are considered at three different levels. A fully crossed design consisting of 27 tests is considered to conduct experiments. The results of different experiments are analyzed statistically using Fisher variance analysis (ANOVA) to determine the impact of variables of drilling on the delamination factor. The main effects plots disclose that the lowest delamination factor is obtained for a rate of feed of 0.4 mm/rev, a speed of spindle of 1000 rpm and for a drill diameter of 15 mm. It is also noted that the delamination factor increases as the speed of spindle and diameter decreases. The interaction plot of rate of feed-spindle speed indicates that the interaction has moderate effect on delamination factor, whereas interaction plot of rate of feed -diameter represents that the rate of feed on delamination factor banks on level of drill diameter, and interaction plot of speed of spindle speed-diameter of drill demonstrates that the interaction has moderate impact on delamination factor. Also, regression analysis is done to establish the general linear model relating delamination factor and the drilling parameters under consideration.

1. Introduction

A polymer reinforced with fibre is called Fibre Reinforced Polymer (FRP) composite. Among FRPs, Aramid fibre reinforced polymer (AFRP) composites have a number of merits like high strength, light weight, high modulus, non-conductivity, good impact and corrosion resistances [1]. As a result, it has found a large amount of utilization in industrial projects, and applications in civil construction projects like bridges, harbours, subways and tunnels. As such, the important literature regarding the AFRP is collected and is given here under the next paragraphs.

Kalantar and Drzal [2] studied that the thermal and physical properties of composite can significantly be affected by composite interphase. They found that, low aramid-epoxy adhesion is mainly due to the aramid fibre morphology and combined effect of its physical and chemical properties. Cervenka, et.al [3] demonstrated that the effect of hygroscopicity on the action of aramid/epoxy composites, including the interfacial region can best be followed by Raman spectroscopy. Won and Dharan [4] conducted tests of drilling on fibre reinforced composite laminates of carbon and aramid. Machining parameters along with semi-empirical relationships between cutting parameters and drilling forces are established for damage-free drilling of these materials. Tsao and Hocheng [5] considered drilling operation of carbon FRP (CFRP) composites and gave complete analysis of delamination caused by drill wear. Prediction of critical thrust force acting at beginning of delamination on worn drill is done and contrasted with that of ideal drill. Tsao [6] studied experimentally the step drilling of CFRP laminates for thrust force and found that the drill performance is mostly affected by the geometry of drill. They found that drill thrust is main influencing parameter for delamination to occur in the entrance as well as exit planes.

Palanikumar [7] used response surface methodology to model delamination factor and the surface finish for machining of Graphite FRP (GFRP) composites. This work is carried out by employing a three-factor and five-level central composite design. Kilickap [8] studied the damage caused by delamination at inlet and outlet when drilling GFRP composite, and examined the influence of variables of drilling, viz., point angle, rate of feed and cutting speed on delamination. Gao, et.al [9] analysed the reason for fuzzing damage during drilling operation of aramid/epoxy composites. They presented a method similar to sawing and grinding mechanism which includes developing compatible tool of sawing and drilling, having negative rake angle. As a result, the fuzzing damage is successfully restrained during drilling operation. Wang, et.al [10] focussed on three-dimensional delamination morphology during drilling of CFRP. The variation of delamination is described by establishing a relation between thrust force and delamination. Saheed Olalekan Ojo, et.al [11] proposed a new analytical model according to modified theory of first-order shear deformation relating delamination and critical thrust force during drilling of CFRP laminates. Díaz-Álvarez, et.al [12] focused on the damage induced during drilling of aramid composite, and analysed the influence of drill geometry on damage, torque and thrust. They developed mechanistic models by relating cutting parameters and quality of the final product.

Anarghya, et.al [13] carried out drilling of AFRPs with solid carbide tool and using Taguchi L54 experimental layout optimized machining parameters with an objective of minimum torque and thrust to reduce the delamination. They also proposed a multi-layer perception neural network prediction model as a tool of optimization. Untimanon, et.al [14] investigated structural behaviour and durability of AFRP concrete beams having a fixed fibre volume fraction (1%) in the simulated marine environment. Five hybrid AFRC beams and a control beam without fibre are studied under a constant sustained load of 3kN. Liu, et.al [15] considered three representative tools, viz., burr tool, twist and brad drills to investigate cutting mechanisms during dry drilling of AFRP. They evaluated hole quality and cutting performance by conducting organized experiments, and provided extensive data regarding tool execution for AFRP drilling. Jongvivatsakul, et.al [16] studied how the physical properties of aramid fibre-reinforced concrete is influenced by length and shape of fibre. Sinan Liu, et.al [17] introduced two delamination mechanisms in drilling of AFRP according to actual tool/ material mating conditions, and called them hole wall delamination/ hole exit delamination. Malleswararao et.al [18] investigated the effect of varying process parameters such as tool rotational speed, weld speed, and tool tilt angle on the metallurgical behaviour and particle distribution and their influence on the mechanical properties, viz., ultimate tensile strength and tensile elongation. The results obtained gained a clear perspective of friction stir welding process parameters on AA6061-T6 and AA2017-T6 joints with optimal mechanical properties. Chen, et.al [19] developed aramid fibre (AF)–filled polyamide 12 (PA12) composite and used it to fabricate 3D printed parts with improved performance using a Multi Jet Fusion test bed.

It is observed from the literature that delamination can severely deteriorate the mechanical performance of FRP materials in service. Delamination is a major challenge in the machining of laminate and this explains the extensive trials that have been done to quantify, correlate or to model delamination in relation to the machining parameters. Entry and exit delamination are damage types that take place in the transient sections during drilling and are controlled by various process parameters. Thus, the objective of this study, keeping in view the wide variety of applications of AFRP composites and the importance of delamination problem on their performance, the present study deals with investigation of the effect of variables of drilling, viz., speed of spindle, rate of feed and drill diameter, on the delamination factor using ANOVA analysis.

2. Experimentation

The two important processes by which fibre reinforced polymer composite is produced are vacuum bag laminating and hand lay-up process. In the present study, AFRP sample is fabricated by using hand layup process. Then drilling operation on the AFRP sample is done by using a CNC machine, and a tool maker's microscope is used to complete measurement process as a part of finding the delamination effect on the drilled holes.

The process of drilling FRP composites is frequently encountered with delamination. Separation of layers due to loss of coating adhesion is nothing but delamination. It is quantified in terms of a dimensionless metric called delamination factor. The ratio of maximum diameter (d_{max}) of delamination to nominal diameter of hole (d) is known as delamination factor. The diameter of circle which encloses all the delamination zone with its center coincident with the centre of hole is nothing but the maximum diameter (Figure 1).

The formula for calculating delamination factor (F_d) is given by:

$$F_d = D_{max} / D \quad (1)$$

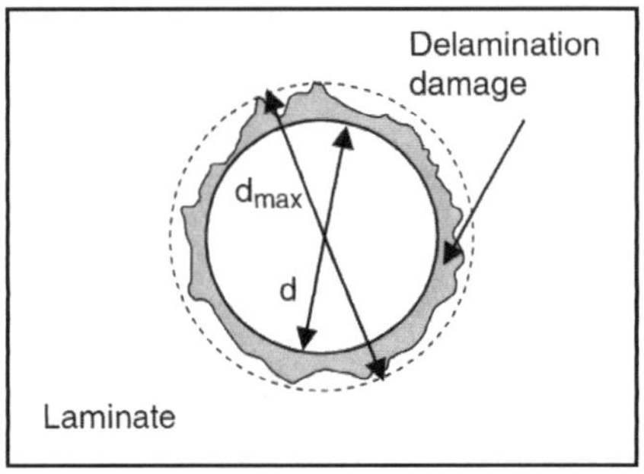

Figure 1: Measurements of Delamination Factor

2.1 Preparation of AFRP sample

In the present work under the hand lay-up process, eight aramid fibres of each 0.3 mm thick (830 gsm) are used to form a final AFRP sheet of 4 mm thick with a size of 300 mm X 300 mm. The epoxy resin of about 150 gms for each layer of AFRP is used so that the ratio of weight of aramid fibre to epoxy resin is 1:2. Also, the ratio of weight of resin to hardener used in the work is about 3:1. Figure 2 shows the preparation of AFRP used in the study.

The first step in the sample preparation is to mix the epoxy resin and hardener in a container, cut the fabrics and core materials to shape and place them in a convenient area on the work table for wet-out or placement in the mould. Later the first layer of fibre reinforcement is laid on the work area. This layer must be wetted with resin and then softly pressed using a brush or a roller. At this stage a second layer of aramid fibre is added, and special care must be taken to eliminate all air bubbles possible. This can be accomplished by either rolling any air bubbles out with a small hand rolling tool or brushing out the air bubbles with a paintbrush. This step is repeated until the desired thickness is obtained.

(a) Before (b) After

Figure 2: Preparation of AFRP sheet

2.2 Experimental design

Experimental design is a promising tool of analysis in designing the modes of an experiment to be carried out. The specifications of the CNC machine used in the study are given in Table 1. In the present work, drilling on the prepared sheet of AFRP is planned by considering three levels of parameters of drilling, viz. speed of spindle, rate of feed, and drill diameter as given in Table 2. The possible combinations of theses parameters will result in fully crossed experimental design, i.e., 27 experiments in total are defined as shown in Table 3.

Table 1: Specifications of CNC machine used for drilling AFRP

Table size (mm)	900*500
Travel (mm)	X-620, Y-420, Z-520
Speed of spindle (rpm)	60-6000
ATC	16U Tools
Taper of spindle	BT-40
Motor of spindle	5.5/7.5 (10-hp) Full Enc

Table 2: Levels of process parameters used

Level	Rate of feed (mm/rev)	Speed of spindle (rpm)	Diameter of drill (mm)
1	0.2	500	5
2	0.4	750	10
3	0.6	1000	15

2.3 Drilling process of AFRP on CNC machine

Owing to the nature and damage prone trend of drilling operation, composites are treated as low machinability materials. For instance, in aerospace applications, the high rate of component rejection during drilling is due to the induced damage, erroneous drill geometries, deficient drilling parameters, and immoderate tool wear. In general, all varieties of composites are subjected to induced damage during drilling. But the damage mode may be different, and is dependent on the type of composite and significant drill parameters. On the whole, the critical damage mechanism in FRPs is reported as delamination, whereas fraying is more dominant in biodegradable composites.

The use of AFRP is growing every day because of its qualities of high tensile strength, low density, high hardness, thermal resistance, and resistance to chemical agents and corrosion. But poor surface finish due to uncut layers at entrance and exit of hole is main characteristic in drilling of AFRP. Delamination is the prevailing dominant damage factor in such cases. Therefore, study of this phenomenon is very essential. In the present study, CNC machining process is performed on aramid fibre reinforced plate with HSS twist drill at distinct rates of feed, speeds of spindle and tool diameters. Each of these variables has three levels of values so that 27 possible holes are drilled (Figure 3) with possible combination of values of these parameters.

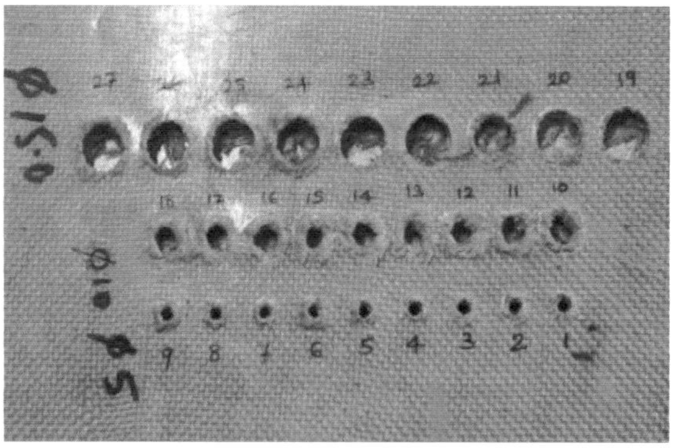

Figure 3: Drilled holes of different diameters

Figure 4: Measurement of hole diameter using microscope

2.4 Measurement of Drilled holes' sizes

The shape, size, and position of small items and features of components are conveniently measured by a multi-functional device, known as the tool maker's microscope. The horizontal table of microscope has an arrangement for translatory motion in X and Y directions. The reticle of the microscope, which is an aid in locating the object or feature of the component placed on the table, provides a datum while viewing the image of the feature through eyepiece. A reference line on the image of the feature of workpiece is aligned with the two crosswires of the reticle. Then, a reading R1 on the microscope scale is noted when a measuring point on the feature (drilled hole) is aligned with one of the crosswires. Another measuring point on the feature is aligned with the crosswire by moving the XY table. Now, the dimension between the two points is obtained by taking the difference between the two readings (i.e., diameter of drilled hole) and is converted into actual reading. The process is done for each drilled hole and the maximum diameter for each is measured at the entrance of drilling operation (Figure 4). The maximum and actual measurements of all the drilled holes are shown in Table 3.

3. Results and Analysis

There are many tools available to find the influence of machining variables on response factor of drilled holes. An important technique for analysing the influence of drilling variables (rate of feed, speed of spindle and diameter of drill) on a response parameter (delamination factor) is to conduct ANOVA.

3.1 Analysis of Variance

The experimental data of different input and output parameters is conveniently analysed by the statistical model called ANOVA. It helps in finding the effect of each input parameter on the variability of response parameter, and/or the amount of variability referable to each input parameter. In the present study, three process parameters viz., drill diameter, speed of spindle and rate of feed with three levels each are taken to study variability in delamination factor of drilled holes. In this case, a multi factor ANOVA is adequate as it involves more than one factor (input parameter). Under ANOVA test, simple effects plots area drawn to show the impact of each of the drilling variables on delamination factor while keeping the other parameters constant. Also, simple-slope plots are drawn for each possible combination of two process parameters to find the influence of their interaction on the delamination factor.

Table 3: Drill parameters and Delamination factors of 27 holes

S. No.	Drill hole Code	Rate of feed (mm/rev)	Speed of spindle (rpm)	Diameter of drill (mm)	Maximum diameter (mm)	Delamination Factor (DF)
1	D1F1S1	0.2	500	5	5.3600	1.07200
2	D1F1S2	0.2	750	5	5.3400	1.06800
3	D1F1S3	0.2	1000	5	5.3200	1.06400
4	D1F2S1	0.4	500	5	5.2500	1.05000
5	D1F2S2	0.4	750	5	5.2085	1.04170
6	D1F2S3	0.4	1000	5	5.2095	1.04190
7	D1F3S1	0.6	500	5	5.2050	1.04100
8	D1F3S2	0.6	750	5	5.1850	1.03700
9	D1F3S3	0.6	1000	5	5.1700	1.03400
10	D2F1S1	0.2	500	10	10.5000	1.05000
11	D2F1S2	0.2	750	10	10.4800	1.04800
12	D2F1S3	0.2	1000	10	10.2000	1.02000
13	D2F2S1	0.4	500	10	10.2800	1.02800
14	D2F2S2	0.4	750	10	10.2000	1.02000
15	D2F2S3	0.4	1000	10	10.1200	1.01200
16	D2F3S1	0.6	500	10	10.5700	1.05700
17	D2F3S2	0.6	750	10	10.5600	1.05600
18	D2F3S3	0.6	1000	10	10.2710	1.02710
19	D3F1S1	0.2	500	15	15.7200	1.02000
20	D3F1S2	0.2	750	15	15.3000	1.01600
21	D3F1S3	0.2	1000	15	15.4200	1.01300
22	D3F2S1	0.4	500	15	15.3000	1.02460
23	D3F2S2	0.4	750	15	15.1800	1.02200
24	D3F2S3	0.4	1000	15	15.8550	1.01860
25	D3F3S1	0.6	500	15	15.8400	1.00660
26	D3F3S2	0.6	750	15	15.4065	1.01373
27	D3F3S3	0.6	1000	15	15.3000	1.00933

3.1.1 Main effects plots

As there are three process parameters and these factors are likely to interact with respect to the response parameter, a multi factor ANOVA is appropriate in this study. Both the simple effects and interactions between the factors can be estimated as a part of this ANOVA test. The simple effects plots, also called main effects plots obtained from ANOVA analysis of the experimental data shown in Table 3 for the aramid fibre reinforced polymer are represented in Figures 5 (a), (b) and (c). MINITAB 20.3 software is used for the purpose of analysis of the experimental data.

Figure 5(a) representing the graph of rate of feed versus mean factor of delamination, affirms that drill hole with a rate of feed of 0.2 mm has obtained the maximum mean delamination factor of 1.04122 and the drill hole with rate of feed of 0.4 mm has obtained lowest mean factor of delamination of 1.02876 and the drill hole with a rate of feed of 0.6 mm has obtained a mean delamination factor below the reference line (1.03376) with a value of 1.03131. From Figure 5(b), i.e., the graph of speed of spindle versus mean delamination factor, it is noted that the drill hole with speed of spindle of 500 rpm has obtained the highest delamination factor of 1.0388 and the drill hole with speed of spindle of 1000 rpm has obtained minimum delamination factor of 1.02666.

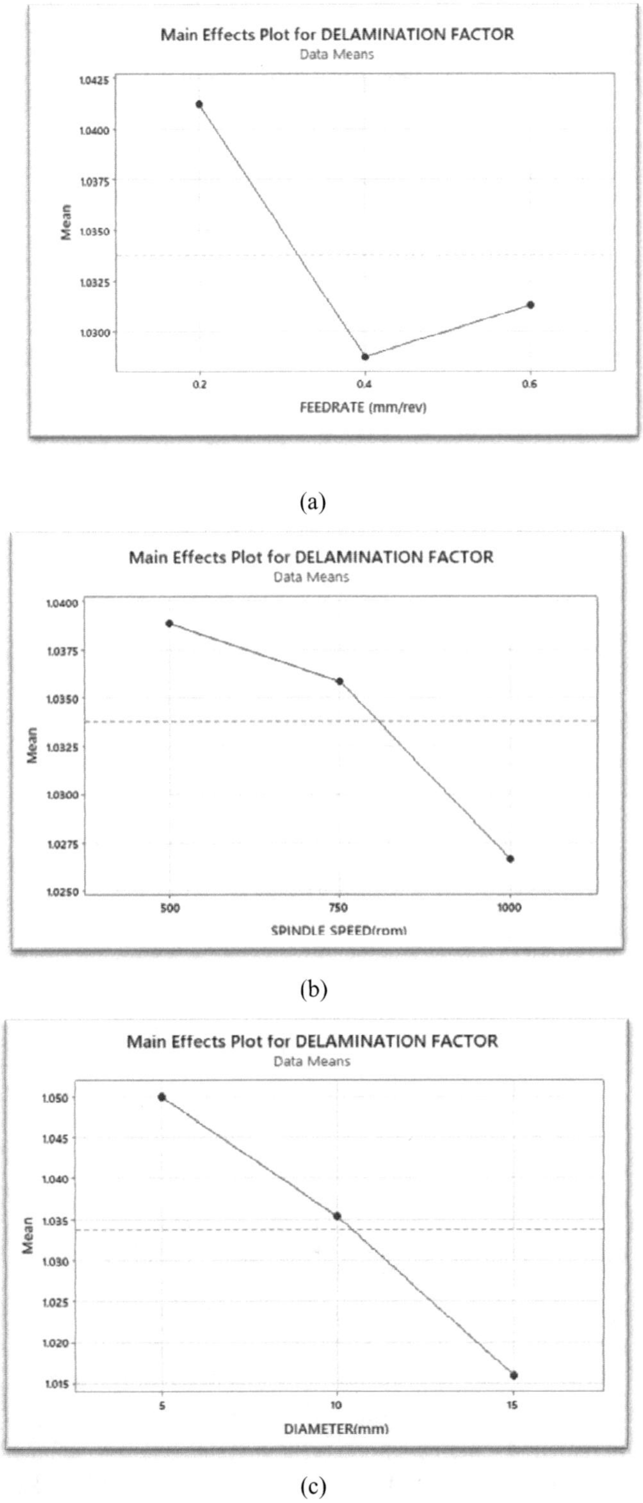

Figure 5: Main Effects plots for delamination factor

From Figure 5(c), i.e., the graph plotted between drill diameter and mean delamination factor, it is noted that the drill hole with 5 mm diameter has obtained maximum mean factor of delamination of 1.04996 and drill hole with a diameter of 15 mm obtained the minimum mean delamination factor of 1.01598. Finally, it can be concluded individually from the simple effects plots that the lowest delamination factor is attained for a rate of feed of 0.4 mm/rev, speed of spindle of 1000 rpm and drill diameter of 15 mm. Also, an increasing trend is observed in delamination factor with the decrease in values of spindle speed and diameter.

3.1.2 Interaction effect plots

Simple slope plots, also called interaction effect plots show the dependence of an independent variable, e.g., feed rate on the level of drill diameter with respect to the dependent variable, i.e., delamination factor. The greater the difference in slope between lines, the higher the degree of interaction. In other words, an interaction plot reveals the impact of changing the settings of one experimental factor with respect to the other factors.

In this work the simple slope plots of three input variables are plotted to find notable ones and their interactions with respect to the output parameter. The graph of spindle speed-feed rate (Figure 6 a) is unparalleled and represents that the influence of speed of spindle on the factor of delamination banks on the level of feed rate. Also, the graph of rate of feed – speed of spindle indicates that their association has moderate effect on delamination factor ranging from 1.024 to 1.0475. The diameter-rate of feed interaction plot (Figure 6 b) represents that the effect of diameter on factor of delamination banks on the level of rate of feed. Also, the graph of rate of feed - drill diameter represents that the effect of rate of feed on factor of delamination (ranging from 1.01 to 1.068) banks on level of drill diameter. The graph of speed of spindle – diameter of drill (Figure 6 c) indicates that the interaction has moderate influence on factor of delamination ranging from 1.014 to 1.054. Also, the drill diameter – speed of spindle interaction plot represents that the effect of diameter on factor of delamination banks on level of spindle speed.

3.2 General Linear Model (GLM)

It is apt to seek for GLM in the present study to apply to the experimental results as all the three input factors are quantitative parameters and have interactions among them. The variance terms for the three factors and the source of error produced under GLM calculations are presented in Tables 4 and 5. ANOVA analysis provides significant rating of the three factors analyzed in this study. Based on this rating, factors, which influence the objective function (delamination factor) significantly, could be identified and proper control measures be adopted.

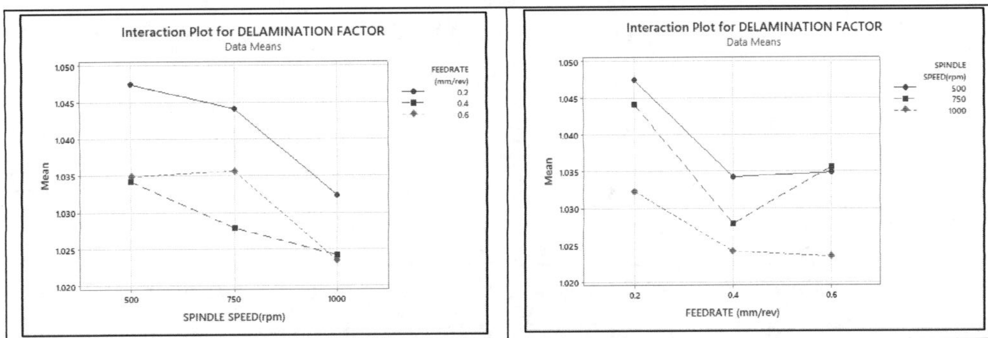

(a) DF vs. rate of feed and speed of spindle

IOP Conf. Series: Journal of Physics: Conf. Series **2765** (2024) 012016 doi:10.1088/1742-6596/2765/1/012016

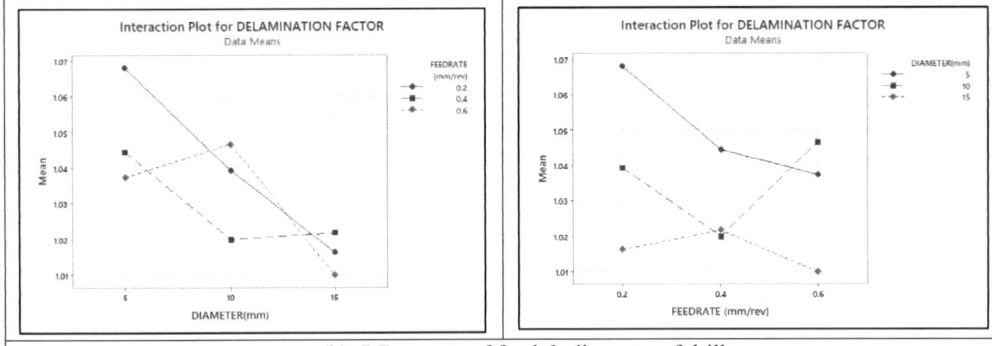

(b) DF vs. rate of feed & diameter of drill

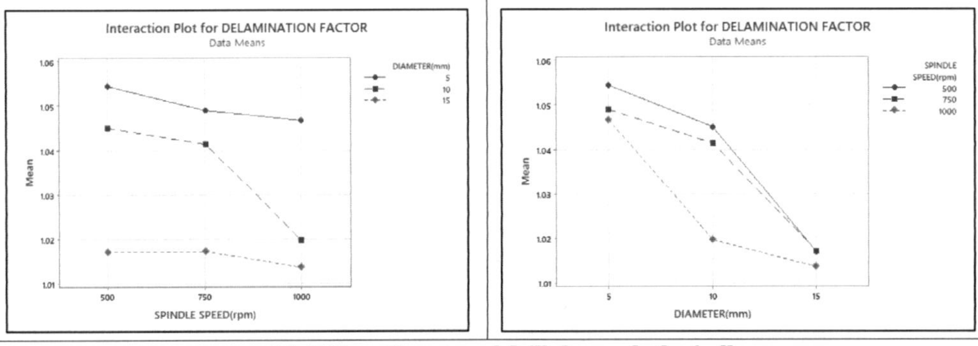

(c) DF vs. diameter of drill & speed of spindle

Figure 6: Interaction Effect plots

Table 4 ANOVA for DF

Source	Degree of Freedom	Adj SS	Adj MS	F-Value	P-Value
X_1	1	0.000442	0.000442	3.12	0.091
X_2	1	0.000663	0.000663	4.68	0.041
X_3	1	0.005193	0.005193	36.63	0.000
Error	23	0.003261	0.000142		
Total	26	0.009560			

X_1 = Feed Rate; X_2 = Spindle speed; X_3 = Diameter

Table 5 Summary of model

S	R-Sq	R-Sq (Adj)	R-Sq (Pred)
0.0119065	65.89%	61.44%	54.46%

Table 4 shows the full linear model fitted for delamination factor, having a regression P-value below 0.05. On seeing the P-values, it can be deduced that the diameter of drill and speed of spindle are major influential factors for delamination factor. Also, on seeing the model summary as given by Table 5, it can be deduced that the r-squared value, which represents the proportion of complete variation in the terms of the model, is 65.89% for delamination factor.

3.3 Regression Analysis

Regression Analysis is a set of statistical processes for estimating the relationships among variables. It models and analyses several variables, when the focus is on the relationship between a dependent variable (delamination factor) and one or more independent variables (predictors). Forecasting and prediction are a part of model development and can conveniently be done by regression analysis. The linear regression model developed by using the experimental data between the input and output parameters is expressed in eq. (2). This equation is obtained by MINITAB 20.3 software and helps one in predicting the value of delamination factor for any combination of drilling parameters selected within the range of their values.

$$\text{Delamination Factor} = 1.0959 - 0.0248\, X_1 - 0.000024\, X_2 - 0.003397\, X_3 \qquad (2)$$

The experimental and predicted data of delamination factor of all 27 drilled holes is compared in a graph (Figure 7). An average proportion of absolute error between predicted and experimental values of 0.9075 % is quite reasonable. Therefore, the equation developed under regression analysis is adequate. The optimum (minimum) value of factor of delamination as regards to the best set of drilling variables as attained by regression equation is 1.011 (for hole or specimen no.24).

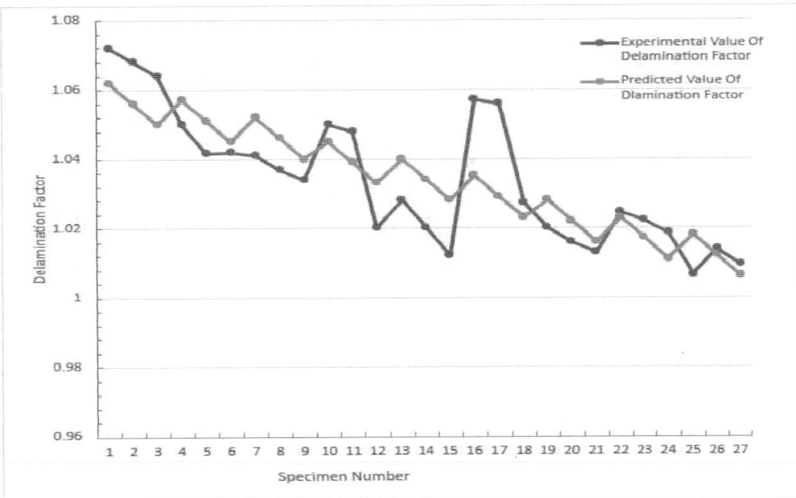

Figure 7: Experimental vs. Predicted values of delamination factor

3.4 Pareto diagram and Normal plot of Delamination Factor

The significant effects of drill parameters and their interactions on delamination factor are identified and shown in a Pareto chart of ANOVA (Figure 8). Considering a significance level of 5%, the absolute values of standardized effects of these parameters and their interactions past the reference line are certainly significant. Figure 9 shows a normal plot of the standardized effects, comparing magnitude of main and interaction effects. The line on the graph conforms to normal distribution with standard deviation equal to 1. There is a significant influence of variables and their interactions which lie distant from the line, and the sources near to the line are insignificant. It can be noted from the two plots that the variables C, B (i.e., drill diameter and spindle speed) are significantly vital and their corresponding influence on delamination factor is in declining order.

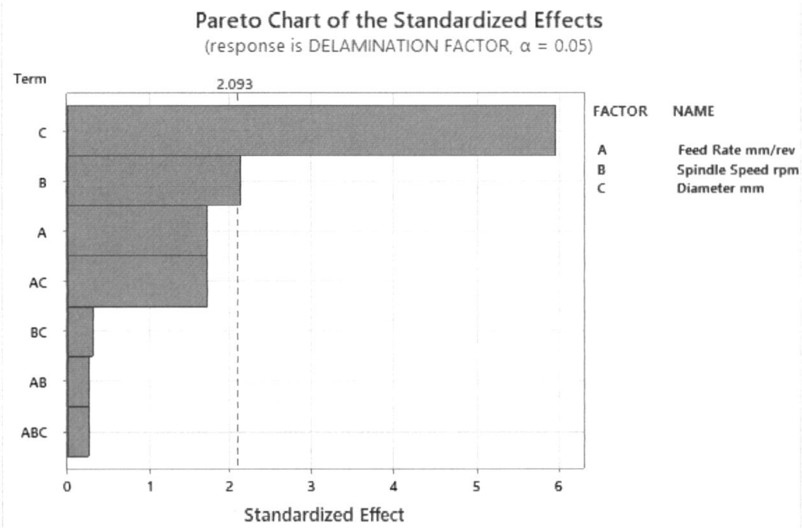

Figure 8: Pareto diagram for delamination factor

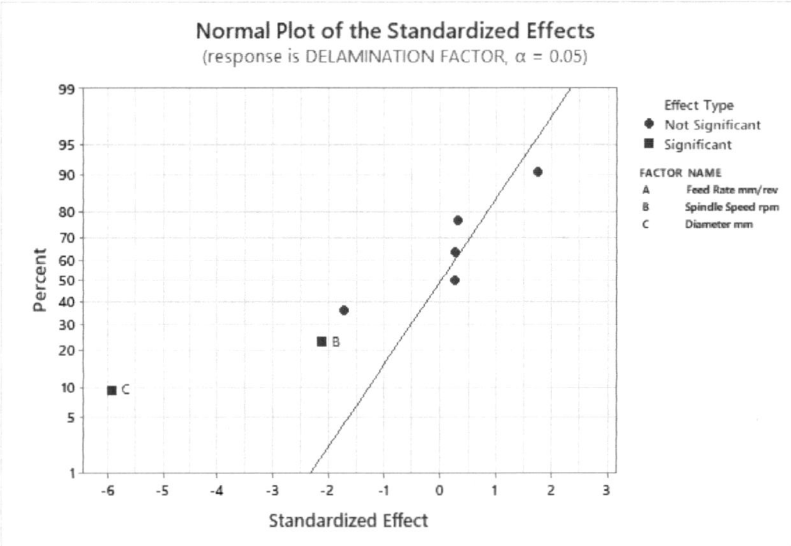

Figure 9: Normal plot for delamination factor

4. Conclusions

The influence of drilling parameters viz., drill diameter, speed of spindle and rate of feed on factor of delamination of Aramid fibre reinforced polymer is presented in this paper. The analysis of results is done by ANOVA. The main offerings and conclusions out of this investigation are:

- It is found from the main effects plots that, the lowest value of factor of delamination of 1.011 is obtained for a rate of feed of 0.4 mm/rev, speed of spindle of 1000 rpm and drill diameter of 15 mm. It is also noted that, there is an increasing trend of delamination factor with respect to the decrease in speed of spindle and drill diameter.

- A remarkable effect of interaction between rate of feed and diameter of drill is observed on factor of delamination, whereas the interaction plot of rate of feed - speed of spindle indicates that the interaction has moderate influence on delamination factor. Also, the interactions between rate of feed – speed of spindle and diameter of drill -spindle speed have moderate influence on delamination factor.
- From the response table of ANOVA, the p-values represent that the drill diameter (0.00) has notable effect followed by speed of spindle (0.041) and rate of feed (0.091) on the factor of delamination of AFRP.
- Also, the Pareto chart and normal plot discloses that diameter of drill and speed of spindle are significantly vital and their corresponding influence on delamination factor is in decreasing order.

Hence on an overall perspective, this study recommends the use of low feed and high spindle speed combo for drilling AFRP composites. Also, while selecting a suitable drill, possibly a moderate diameter of 10 – 15 mm is recommended to reduce the delamination effect while drilling AFRP composites.

5. Scope for future work

In the present work, the fibre plies are placed parallel to each other to form the final sheet of AFRP. The study can be extended to FRPs prepared under different ply orientations. Also, the surface morphology can be further examined on the top and bottom surfaces near the drilled holes to correlate the results and arrive at a better conclusion.

References

[1] Malik Abdul Karim, Mohamad Zaki Abdullah, Ahmed Farouk Deifalla, Marc Azab, Ahsan Waqar, An assessment of the processing parameters and application of fibre-reinforced polymers (FRPs) in the petroleum and natural gas industries: A review, Results in Engineering, 2023, **18** (101091)

[2] J.Kalantar and L.T.Drzal, The bonding mechanism of aramid fibres to epoxy materials, *Journal of Materials Science*, 1990, **25(10)**: 4186-4193

[3] A.J. Cervenka and D.J Bannister and R.J. young, Moisture absorption and interfacial failure in aramid*epoxy composites*, 1998, pp. 2–22

[4] M.S. Won and Hari Dharan, Drilling of Aramid and Carbon Fiber Polymer Composites. *Journal of Manufacturing Science and Engineering*,_ 2002, **124(4)**

[5] C.C. Tsao and H. Hucheng, Effect of tool wear on delamination in drilling composite materials *International Journal of Mechanical Sciences*, 2007,**49(8)**:983-988

[6] C.C.Tsao._Prediction of thrust force of step drill in drilling composite material by Taguchi method and radial basis function network, *The International Journal of Advanced Manufacturing Technology*, 2006,**36(1)**:11-18

[7] K. Palanikumar, Modeling and Analysis of Delamination Factor and Surface Roughness in Drilling GFRP *Composites Materials and Manufacturing Processes*, 2009, **25(10)**:1059-106

[8] Erol Kilickap, Optimization of cutting parameters on delamination based on Taguchi method during drilling of GFRP composite, *Expert Systems with Applications*, 2010 ,**37(8)**: 6116-6122

[9] Hang Gao, Yuan Zhuang, Ben Wang and Jun Liang Huang, Study on the Combined Machining Technology of Sawing and Grinding for Drilling Aramid/Epoxy Composites, *Advanced Materials Research*, 2012, **565**:436-441

[10] Haijin Wang and Jie Sun, Jianfeng Li and Weidong Li, Investigation on delamination morphology during drilling composite laminates, *The International Journal of Advanced Manufacturing Technology* 2014, **74(1-4)**:257-266

[11] Saheed Olalekan Ojo, Sikiru Oluwarotimi Ismail, Marco Paggi and Hom Nath Dhakal, A new analytical critical thrust force model for delamination analysis of laminated composites during drilling operation, *Composites Part B Engineering*, 2017,**124(3–4)**

[12] A. Díaz-Álvarez, M. Rodríguez-Millán, J. Díaz-Álvarez and M.H. Miguélez, Experimental analysis of drilling induced damage in aramid composites, *Composite Structures*, 2018, **202**

[13] A. Anarghya, D.N. Harshith, Nitish Rao, Nagaraj Nayak, Thrust and torque force analysis in the drilling of aramid fiber-reinforced composite laminates using RSM and MLPNN-GA.Heliyon, 2018, **4(7)**: e00703

[14] Suphawit Untimanon, Rajendra Prasad Bohara, Ganchai Tanapornraweekit and Somnuk Tangtermsirikul. Behavior of hybrid aramid fiber reinforced concrete beams under sustained load in marine environment 2019.

[15] Sinan Liu, Tao Yang , Chang Liu, Yu Du, Comprehensive investigation of cutting mechanisms and hole quality in dry drilling woven aramid fibre–reinforced plastic with typical tools Proceedings of the Institution of Mechanical Engineers Part B *Journal of Engineering Manufacture*, 2019, **233(14)**:095440541984485

[16] Pitcha Jongvivatsakul, Chung Nguyen Thi, Ganchai Tanapornraweekit and Linh Van Hong Bui, Mechanical properties of aramid fiber-reinforced composites and performance on repairing concrete beams damaged by corrosion. Songklanakar in *Journal of Science and Technology* 2020, **42(3)**:637-644

[17] Sinan Liu, Yang Tao, Chang Liu, Yan Jin, Dan Sun and Yifei Shen, An analytical delamination model of drilling aramid fiber–reinforced plastics by brad drill, *The International Journal of Advanced Manufacturing Technology*, 2020,**108(14)**:1-12

[18] N Malleswararao Battina, Varaha Siva Prasad Vanthala and Hari Krishna Ch, Effect of friction stir welding process parameters on mechanical and metallurgical behavior of AA6061-T6 and AA2017-T6 tailored blanks, Engineering Research Express, 2021, 3 (2)

[19] Jiayao Chen, Pengfei Tan, Xiaojiang Liu, and Wei Shian Tey, High-strength light-weight aramid fiber/polyamide 12 composites printed by Multi Jet Fusion, *Virtual and Physical Prototyping*, 2022,**17(2)**:1-13

Hall and Ion Slip Current influence on Unsteady MHD Free Convective Boundary Layer Flow Past a Vertical Porous Plate with Thermal Radiation and Chemical Reaction

T R K D Vara Prasad[1] , K V B Rajakumar[2] and T Govinda Rao[3]
[1]Department of E.M & H, S.R.K.R Engineering College,Bhimavaram, India-534204.
[2]Department of Mathematics, Kallam Haranadhareddy Institute of Technology,Guntur, India-522109.
[3] B.S & H Department, Seshadri Rao Gudlavalleru Engineering College, Gudlavalleru, India -521356
Email: trkdvprasad@gmail.com; kvbrajakumar@gmail.com; govind.tangudu@gmail.com

Abstract: The analytical reconnaissance has been effectuated to explicate the influence of Hall and ion-slip current on nonlinear MHD free convective flow near a vertical porous plate in presence of chemical reaction and Dufour effect. In this assessment the governing PDE's are determined analytically. The impact of multifarious physical parameters has been contemplated on velocity, temperature as well as concentration which were exemplified via numerous graphs whereas quantification for the skin fiction, nusselt number and Sherwood number are demonstrated in numeric data form and are reported in table. In this examination mainly observed that velocity declined with the incremental values of Hall current as well as ion-slip current. Meanwhile velocity as well as temperature accelerated by means of the incremental values of Dufour number. However concentration diminished with the accelerated values of chemical reaction along with Schmidt number

Key Words: hall Effect, MHD, heat transfer, Porous plate and chemical reaction

1. Introduction:

The examination of free convective boundary layer flow is crucial in fluid mechanics, especially in various technical applications like spacecraft construction and solar energy collectors designed to replicate oceanic and atmospheric circulations. Despite its significance, the exploration of free convection streams and the effects of pressure work are often overlooked. This analysis sheds light on the often disregarded aspects of difficulty work and its impact on heat transfer features in these systems.

Magnetohydrodynamics (MHD) holds significant importance in both scientific and engineering domains. It is a crucial area of investigation that considers the influence of magnetic fields on the flow of electrically driven fluids. MHD plays a pivotal role in various fields such as astronomy, agriculture, petroleum industries and geophysics. The viscous flow as a result of stretching boundary is distinguished in extrusion phenomenon's from over here sheet material is towed out of an orifice with enhancing velocity. Nehad et.al.[1] addressed influence of MHD on free convective fluid between vertical parallel plates. Kabir et.al.[2] investigated suction/blowing of porous materials on unsteady MHD free convective flow of a incompressible fluid over a vertical plate. Mahamoud Maya et. al.[3] examined influence of MHD heat source on MHD free convection flow of a lid-driven cavity. Alzabut et. al.[4] analysed numerical simulation of magnetohydrodynamics lid-driven free convection flow. Various researchers [5-8] have delved into the examination of MHD by considering different parameters.

The Hall current significantly influences heat transfer mechanisms, and its elevation is attributed to the presence of a magnetic field. The flow and temperature profiles undergo noteworthy changes as a result of the Lorentz force. The magnetic field induces an increase in the Hall current, and the Lorentz force is responsible for substantial alterations in both flow and temperature profiles. Examples include the use of Hall sensors for tasks like measuring linear and angular positions, determining doping levels and assessing electron concentration. Bilal Ali et. al.[9] have been considered Hall and ion impact on mixed convective flow of Casson fluid. Narasimha Rao et. al.[10], Jha et. al.[11], Kanimozi et.al.[12], Yasin et. al.[13] and Hari Babu et.al.[14] have examined the Hall effect within the context of electromagnet hydrodynamic (EMHD) heat transfer flows involving electrically conducting fluids, exploring diverse physical and geometrical scenarios.

A chemical reaction takes place when two substances collide, resulting in the formation of a product with the assistance of catalyst. There are two main types of chemical reactions, reversible and irreversible. Irreversible reactions can be reversed under suitable conditions. In reversible reactions, reactants can convert into products, and products can react to reform the original reactants. Catalysts, substances that speed up reactions without being consumed, play a role in this process. Saikia et. al.[15] studied chemical reaction and radiation on unsteady MHD free convective flow over a infinite oscillatory vertical plate. Mosharrof Hossain et. al.[16] and chandra Shekar et. al.[17] discussed influence of chemical reaction on unsteady MHD convective transport through a vertical porous sheet.

Radiation is defined as propagation of thermal electromagnetic particles through a medium. In thermal radiation, a surface emits electromagnetic waves that transport heat energy. The utilization of thermal radiation is essential for achieving elevated temperatures. Various technological processes, such as the reentry of missiles and spacecraft, take place under high temperature conditions where thermal radiation significantly influences the outcome. Goud et. al.[18] reported effect of thermal radiation on stratified MHD fluid flow through an vertical plate with viscous dissipation. Nagaraju et. al.[19] demonstrated effect of thermal radiation on unsteady MHD free convective flow between vertical plates with hall effect. Rajakumar et. al.[20] concentrated on significance of Casson fluid on unsteady MHD fluid over a hot vertical plate with radiation absorption and chemical reaction. Prasad et.al.[21] and Raghunath et. al.[22] reported Unsteady MHD free convective fluid flow past an inclined porous plate with chemical reaction.

Inspired by the above mentioned researchers and from the literature discussed earlier, we mainly focused on impact of Hall and Ion slip current on unsteady MHD free convective flow over vertical porous plate in terms of chemical reaction and radiation absorption. Real world applications Hall current in heat transfer extend to various fields such as electronics, material science and sensing technologies. Using Perturbation method, a set of nondimensionless equations are solved analytically. Certain important results of the analysis showing the effects of pertinent parameters on the characteristics of the flow, heat and mass transfer are presented and discussed.

2. Problem Formulation:

Examine the time-dependent 2-D magnetohydrodynamics(MHD) free convective movements of a viscous, incompressible and electrically conductive fluid near an endless heated non-parallel porous plate situated within a porous medium. This scenario takes into account the impact of thermal and concentration buoyancy effects.

- Consider the vertical upward direction along the plate as the positive X-axis and the axis normal to plate as the positive Y-axis.

- A magnetic field with uniform strength is applied in a direction that is not parallel to the plate.
- Hall as well as ion-slip current taken into contemplation in the equation of momentum.
- The magnetic Reynolds number and strength of the transverse applied magnetic field are considered to be extremely low, resulting in a negligible produced magnetic field. Additionally, it is assumed that no applied voltage is present, leading to the absence of an electric field.

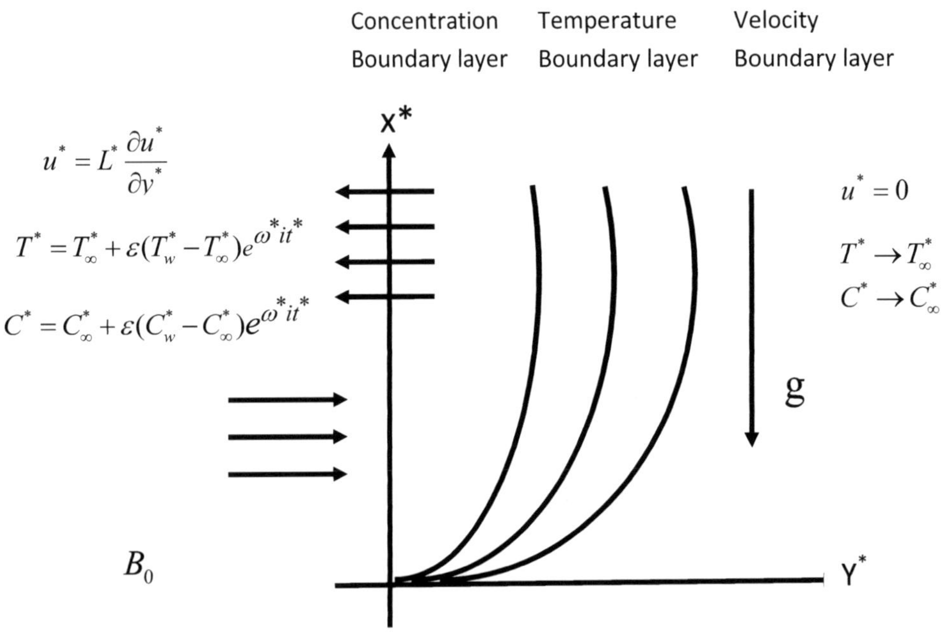

Figure 1: Schematic diagram

The fundamental equations that describe the physical scenario are given by

$$\frac{\partial v^*}{\partial y^*} = 0 \tag{1}$$

$$\frac{\partial u^*}{\partial t^*} + v^* \frac{\partial u^*}{\partial y^*} = \nu \frac{\partial^2 u^*}{\partial y^{*2}} + g\beta(T^* - T^*_\infty) + g\beta^*(C^* - C^*_\infty) - \frac{\nu}{k^*}u^* - \frac{B_0^2 \sigma_e \left[\alpha_e u^* + \beta_e w^*\right]}{\rho\left[\alpha_e^2 + \beta_e^2\right]} \tag{2}$$

$$\frac{\partial w^*}{\partial t^*} + v^* \frac{\partial w^*}{\partial y^*} = \nu \frac{\partial^2 w^*}{\partial y^{*2}} - \frac{\nu}{k^*}w^* - \frac{B_0^2 \sigma_e \left[\alpha_e u^* + \beta_e w^*\right]}{\rho\left[\alpha_e^2 + \beta_e^2\right]} \tag{3}$$

$$\frac{\partial T^*}{\partial t^*}+v^*\frac{\partial T^*}{\partial y^*}=\alpha\frac{\partial^2 T^*}{\partial y^{*2}}-\frac{1}{\rho c_p}\frac{\partial q_r^*}{\partial y^*} \tag{4}$$

$$\frac{\partial C^*}{\partial t^*}+v^*\frac{\partial C^*}{\partial y^*}=D\frac{\partial^2 C^*}{\partial y^{*2}}-k_r^*(C^*-C_\infty^*) \tag{5}$$

The relevant boundary constraints [8] are described as

$$\left.\begin{array}{r} u^*=L^*\frac{\partial u^*}{\partial y^*},w^*=0,T^*=T_\infty^*+\varepsilon(T_w^*-T_\infty^*)e^{\omega^* it^*},C^*=C_\infty^*+\varepsilon(C_w^*-C_\infty^*)e^{\omega^* it^*}\ at\ y^*=0 \\ u^*=0,w^*\to 0,\ T^*\to T_\infty^*,\quad C^*\to C_\infty^*\quad as\quad y^*\to\infty \end{array}\right\} \tag{6}$$

By Rosseland approximation, the vector representing radioactive flux q_r^* can be formulated as:

$$q_r^*=-\frac{4\sigma^*}{3a^*}\frac{\partial T_w^*}{\partial y^*} \tag{7}$$

Under the assumption that the temperature variation within the flow is sufficiently small, we can express T^{*4} as a linear function of the temperature. This is achieved by expanding the expression in a Taylor series around the stream temperature T_∞^* and neglecting higher order terms, thus

$$T_w^{*4}\cong 4T_\infty^{*3}T_w^*-3T_\infty^{*4} \tag{8}$$

From equation (1), the suction velocity at the plate is evidently either constant or dependent on time only. Therefore the suction velocity is perpendicular to the plate is assumed to be in the form

$$v^*=-V_0\left[1+\varepsilon A\exp(iw^*t^*)\right] \tag{9}$$

Now we establish the following non-dimensional factors

$$\left.u=\frac{u^*}{V_0},v=\frac{v^*}{V_0},\eta=\frac{V_0y^*}{v},t=\frac{t^*V_0^2}{4v},w=\frac{4w^*v}{V_0^2},\theta=\frac{T^*-T_\infty^*}{T_w^*-T_\infty^*},\phi=\frac{C^*-C_\infty^*}{C_w^*-C_\infty^*}\right\} \tag{10}$$

After employing Eqs.(7)-(10) in the Eqs.(2)-(5) we obtain modified DE's are

$$\frac{1}{4}\frac{\partial u}{\partial t}-\left(1+\in Ae^{i\omega t}\right)\frac{\partial u}{\partial \eta}=\frac{\partial^2 u}{\partial \eta^2}+Gr\ \theta+Gm\ \phi-\left(\frac{M\left(1+\beta_e\beta_i\right)u+\beta_e w}{\left(1+\beta_e\beta_i\right)^2+\beta_e^2}\right)-\frac{1}{K}u \tag{11}$$

$$\frac{1}{4}\frac{\partial w}{\partial t}-\left(1+\in Ae^{i\omega t}\right)\frac{\partial w}{\partial \eta}=\frac{\partial^2 w}{\partial \eta^2}-\left(\frac{M\left(1+\beta_e\beta_i\right)w-\beta_e u}{\left(1+\beta_e\beta_i\right)^2+\beta_e^2}\right)-\frac{1}{K}u \tag{12}$$

$$\frac{1}{4}\frac{\partial \theta}{\partial t}-\left(1+\in Ae^{i\omega t}\right)\frac{\partial \theta}{\partial \eta}=\lambda\frac{\partial^2 \theta}{\partial \eta^2} \tag{13}$$

$$\frac{1}{4}\frac{\partial C}{\partial t}-\left(1+\in Ae^{i\omega t}\right)\frac{\partial C}{\partial \eta}=\left(Sc\right)^{-1}\frac{\partial^2 C}{\partial \eta^2}-Kr\ C \tag{14}$$

The related boundary constraints are described as

$$\left.\begin{array}{lllll} u=h\dfrac{\partial u}{\partial \eta}, & w=0, & \theta=1+\varepsilon^{\ i\omega t}, & C=1+\varepsilon e^{i\omega t} & At\ \eta=0 \\ u\to 0, & w\to 0, & \theta\to 0, & C\to 0 & As\ \eta\to\infty \end{array}\right\} \tag{15}$$

Introducing the complex variable $\Gamma=u\left(\eta,t\right)+i\ w\left(\eta,t\right)$ (16)

Combine the equations (11) and (12) then we obtained transformed single partial differential equation.

$$\frac{1}{4}\frac{\partial \Gamma}{\partial t}-\left(1+\in Ae^{i\omega t}\right)\frac{\partial \Gamma}{\partial \eta}=\frac{\partial^2 \Gamma}{\partial \eta^2}+Gr\ \theta+Gm\ \phi-N\ \Gamma \tag{17}$$

$$\left.\begin{array}{llll} \Gamma=h\dfrac{\partial \Gamma}{\partial \eta}, & \theta=1+\varepsilon\, e^{i\omega t}, & C=1+\varepsilon e^{i\omega t} & At\ \eta=0 \\ \Gamma\to 0, & \theta\to 0, & C\to 0 & As\ \eta\to\infty \end{array}\right\} \tag{18}$$

Here,

$$\left.\begin{array}{l} R=\dfrac{a^*k}{4\sigma^* T_\infty^{*3} v^2}, K=\dfrac{K^* V_0^2}{v^2},\ h=\dfrac{V_0 L^*}{v},\ Kr=\dfrac{K_r^* v}{V_0^2}, \Pr=\dfrac{\upsilon\rho C_p}{K}, M=\dfrac{\sigma B_0^2\upsilon}{\rho V_0^2} \\ N=\left[\dfrac{1}{K}+\dfrac{M\left[-(1+\beta_e\beta_i)+i\beta_e\right]}{\left[(1+\beta_e\beta_i)^2+\beta_e^2\right]}\right], Gr=\dfrac{\upsilon\beta g(T_w^*-T_\infty^*)}{V_0^3}, Gc=\dfrac{\upsilon\beta^* g(C_w^*-C_\infty^*)}{V_0^3} \\ i=\sqrt{-1}, Sc=\dfrac{\upsilon}{D}, \xi=\lambda\dfrac{iw}{4}Q=N+\dfrac{iw}{4}, \lambda=\dfrac{1}{\Pr}\left(1+\dfrac{4}{3R}\right), \gamma=\left(\dfrac{iw}{4}+Kr\right)Sc \end{array}\right\} \tag{19}$$

3. Method of Solution:

The problems posted in equations (13), (14) and (17) subject to the boundary conditions presented in equation (18) are cannot be solved. Thus, these equations can be simplified form of partial differential equations into a set of ordinary differential equations, the resulting equations can be solved through perturbation method, so that consider:

$$\left.\begin{aligned}\Gamma &= \Gamma_0(y) + \varepsilon e^{i\omega t}\ \Gamma_1(y) + O(\varepsilon^2)\ldots\ldots\ldots\ldots \\ \theta &= \theta_0(y) + \varepsilon e^{i\omega t}\ \theta_1(y) + O(\varepsilon^2)\ldots\ldots\ldots\ldots \\ C &= C_0(y) + \varepsilon e^{i\omega t}\ C_1(y) + O(\varepsilon^2)\ldots\ldots\ldots\ldots\end{aligned}\right\} \tag{20}$$

Substituting Esq. (20) in Equation (13), (14) and (17) and equating the harmonic and non-harmonic terms, and neglecting the higher order terms $o\left(\varepsilon^2\right)$ we obtain the following pairs of equations for ($\Gamma_0, \theta_0, \phi_0$) and ($\Gamma_1, \theta_1, \phi_1$)

$$\Gamma_0'' + \Gamma_0' - N\Gamma_0 = -Gr\ \theta_0 - Gm\ \phi_0 \tag{21}$$

$$\Gamma_1'' + \Gamma_1' - Q\Gamma_1 = -A\ \Gamma_0' - Gr\ \theta_1 - Gm\ \phi_1 \tag{22}$$

$$\theta_0'' + \lambda\theta_0' = 0 \tag{23}$$

$$\theta_1'' + \lambda\theta_1' - \frac{i\omega\lambda}{4}\theta_1 = -2A\lambda\theta_0' \tag{24}$$

$$\phi_0'' + Sc\ \phi_0' - Sc\ Kr\ \phi_0 = 0 \tag{25}$$

$$\phi_1'' + Sc\ \phi_1' - \gamma\ \phi_1 = -Sc\ A\ \phi_0' \tag{26}$$

The corresponding boundary circumstances are

$$\left.\begin{aligned}&\Gamma_0 = h\ \frac{\partial \Gamma_0}{\partial \eta}, \Gamma_1 = h\ \frac{\partial \Gamma_1}{\partial \eta}\quad \theta_0 = 1,\ \theta_1 = 1, \phi_0 = 1,\ \phi_1 = 1 \quad At\ \eta = 0 \\ &\Gamma_0 \to 0,\ \Gamma_1 \to 0\quad \theta_0 \to 0,\ \theta_1 \to 0,\ \phi_0 \to 0,\ \phi_1 \to 0 \quad As\ \ \eta \to \infty\end{aligned}\right\} \tag{27}$$

Solve (21) – (26) subject to boundary Condition (27), then we obtain

$$\Gamma_0 = A_2 e^{-m_5\eta} + N_2 e^{-\lambda\eta} + N_3 e^{-m_1\eta} \tag{28}$$

$$\Gamma_1 = A_3 e^{-m_5\eta} + N_4 e^{-m_3\eta} + N_5 e^{-\lambda\eta} + N_6 e^{-m_2\eta} + N_7 e^{-m_4\eta} + N_8 e^{-m_1\eta} \tag{29}$$

$$\theta_0 = e^{-\lambda\eta} \tag{30}$$

$$\theta_1 = A_0 e^{-m_3\eta} + N_1 e^{-\lambda\eta} \tag{31}$$

$$\phi_0 = e^{-m_1\eta} \tag{32}$$

$$\phi_1 = e^{-m_2\eta} \tag{33}$$

Substituting (28) – (33) in (20), then we obtain;

Fluid Velocity:

$$\Gamma=\left(A_2e^{-m_5\eta}+N_2e^{-\lambda\eta}+N_3e^{-m_1\eta}\right)+\varepsilon e^{i\omega t}\begin{pmatrix}A_3e^{-m_5\eta}+N_4e^{-m_3\eta}+N_5e^{-\lambda\eta}\\+N_6e^{-m_2\eta}+N_7e^{-m_4\eta}+N_8e^{-m_1\eta}\end{pmatrix} \tag{34}$$

Fluid Temperature:

$$\theta=e^{-\lambda\eta}+\varepsilon e^{i\omega t}\left(A_0e^{-m_3\eta}+N_1e^{-\lambda\eta}\right) \tag{35}$$

Fluid Concentration:

$$\phi=e^{-m_1\eta}+\varepsilon e^{i\omega t}\left(e^{-m_2\eta}\right) \tag{36}$$

Skin friction:

$$\tau_w=\left.\frac{\partial\Gamma}{\partial y}\right|_{y=0}=\left(-m_5A_2-\lambda N_2-m_1N_3\right)+\varepsilon e^{i\omega t}\begin{pmatrix}-m_5A_3-m_3N_4-\lambda N_5\\-m_2N_6-m_4N_7-m_1N_8\end{pmatrix} \tag{37}$$

Nusselt number:

$$N_u=\left.\frac{\partial\theta}{\partial y}\right|_{y=0}=-\lambda-\varepsilon e^{i\omega t}\left(m_3A_0+\lambda N_1\right) \tag{38}$$

Sherwood number:

$$S_h=\left.\frac{\partial\phi}{\partial y}\right|_{y=0}=-m_1-m_2\varepsilon e^{i\omega t} \tag{39}$$

4. Results and Discussion:

A numerical analysis is conducted, and the findings are documented to illustrate a comparative study that highlights the influences of multiple parameters on the flow field. During in our analytical computation, assumptions are made regarding the values of different non-dimensional parameters as $Pr = 0.71$, $Sc = 0.60$, $R = 1.0$, $M = 0.1$, A=0.5, $h = 0.2$, $\epsilon = 0.001$, $\omega t = \pi/2$, $Kr = 0.1$, etc., In the analysis, these values remain constant, while the variable parameters are used in the corresponding figures. In the **Figures 2 - 9**, dashed lines specify the influence under slip condition. Influence of Gc on the velocity is illustrated in the **Figure 2:** From the figure the results indicates that for disparate augmentation values of Gc leads to accelerated in the velocity. Impact of Gr on the velocity is shown in the **Figure 3:** From this figure the results indicates that velocity rises with the incremental values of Gr as a result of enhancement of buoyancy force. **Figure 4:** Contemplates that ascend in velocity as a consequence of enhancement of heterogeneous values of Ion-slip parameter β_i owing to the fact that the β_i

lessened the resistive force enforce by magnetic field. **Figure 5:** delineated that the comportment of velocity for disparate values of Hall parameter β_e. From this figure it was establish that increasing the Hall parameter values results in a decrease in velocity, particularly in close proximity to the plate. This is due to the generation of an additional potential difference transverse to the direction of accumulated free charge. The applied magnetic field interacting between opposite surfaces induces as electric current perpendicular to both the magnetic and electric fields. **Figure 6:** elucidates that the reduction in fluid velocity is a result of an increase in the magnetic field parameter (M) in the presence of velocity slip conditions. This behaviour occurs because of resistive Lorentz force; that intensifies due to the strengthening of the magnetic field. **Figure 7:** illustrates the impact of the porous medium parameter K on the velocity, the depicted results indicate that an increase in K results in a decrease in velocity. This is attributed to the presence of the porous medium, which introduces resistance to the flow. **Figure 8: and Figure 13:** Exhibits the influence of chemical reaction parameter Kr on both velocity and concentration is depicted. In this scenario, increasing values of Kr result in a decrease in both velocity and concentration. **Figure 9: & Figure 10:** Established that when the Prandtl number(Pr) varies, showing dissimilar values, there is a corresponding decrease in both velocity and temperature because a fluid with higher Prandtl number tends to have lower thermal conductivity, resulting in reduced conduction and consequently, lower temperature. Certainly, this occurs because fluids with elevated Prandtl numbers exhibit increased viscosity, leading to a reduction in the thickness of the thermal boundary layer.

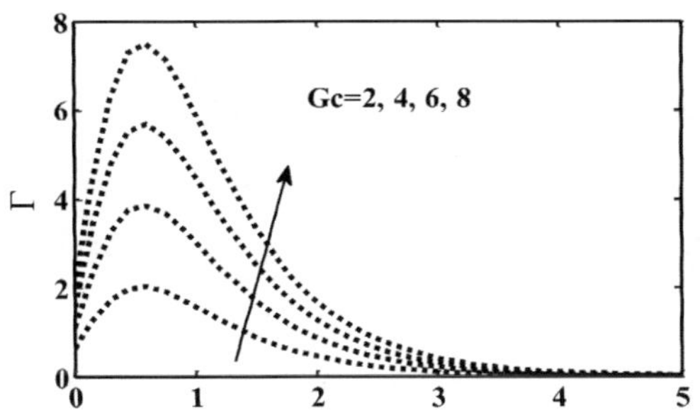

Figure 2: Velocity Profile for Gc

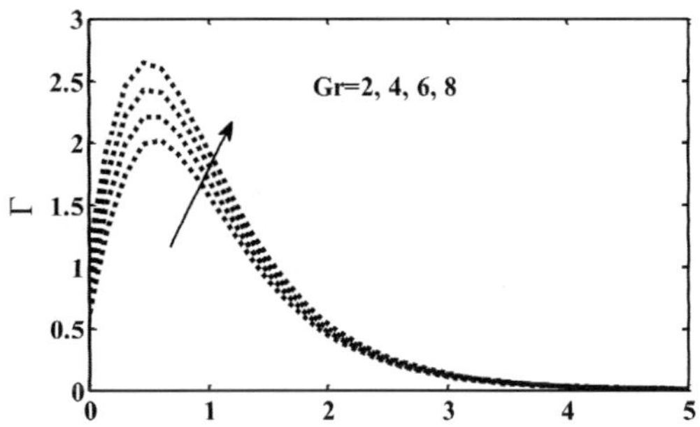

Figure 3: Velocity Profile for Gr

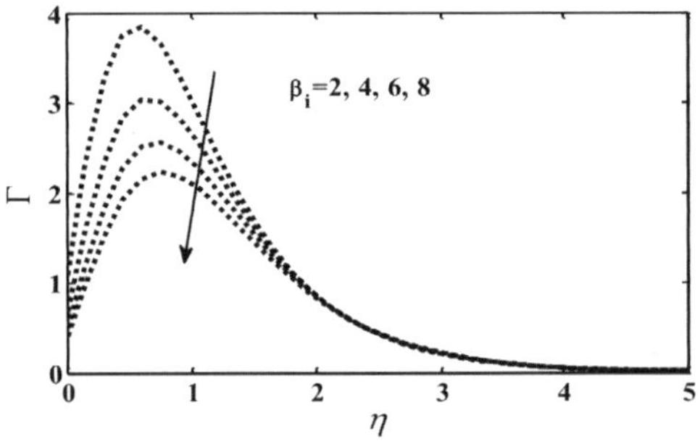

Figure 4: Velocity Profile for β_i

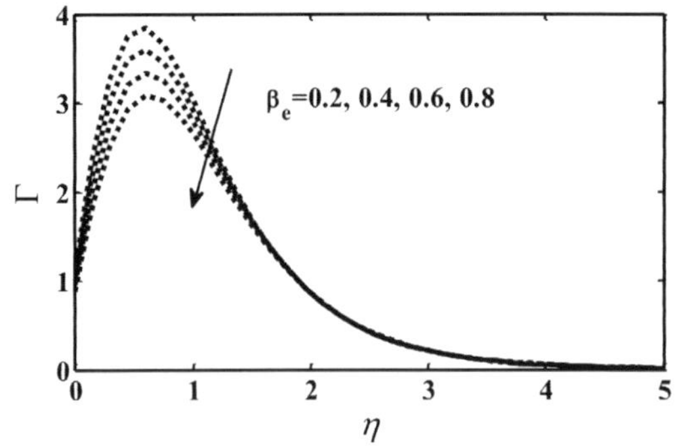

Figure 5: Velocity Profile for β_e

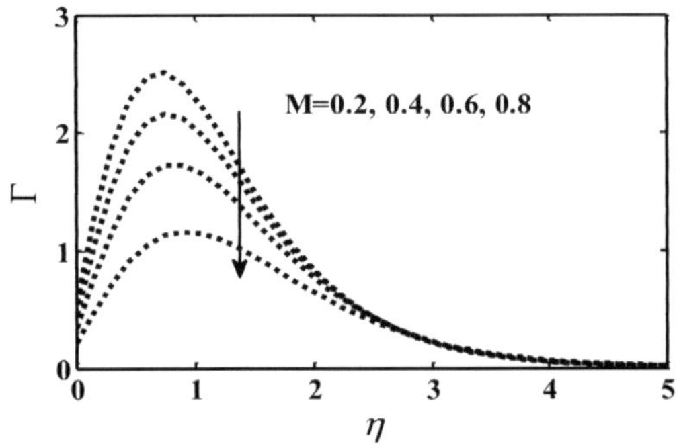

Figure 6: Velocity Profile for M

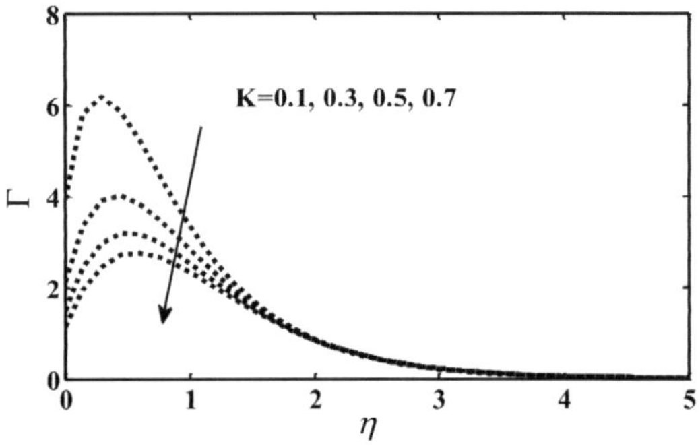

Figure 7: Velocity Profile for K

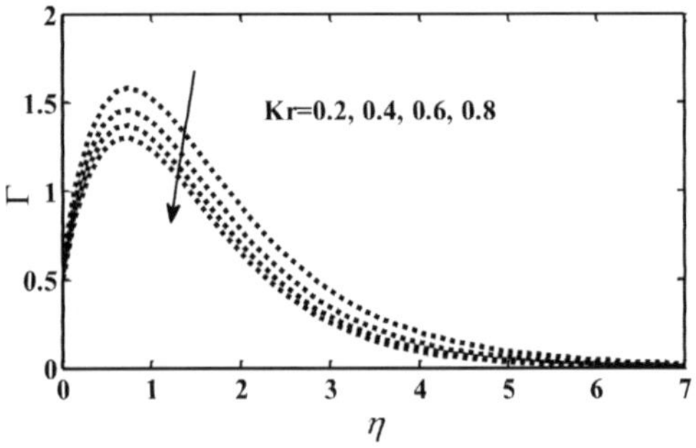

Figre 8: Velocity Profile for *Kr*

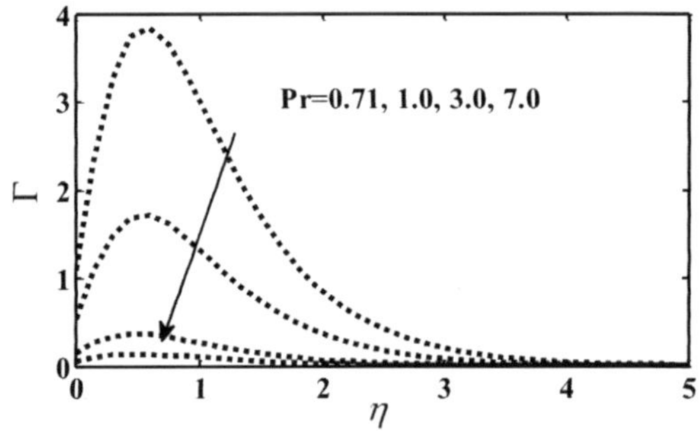

Figure 9: Velocity Profile for *Pr*

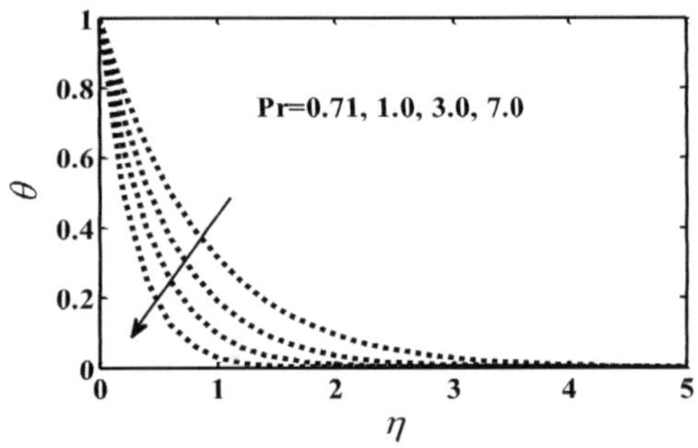

Figure 10: Temperature Profile for *Pr*

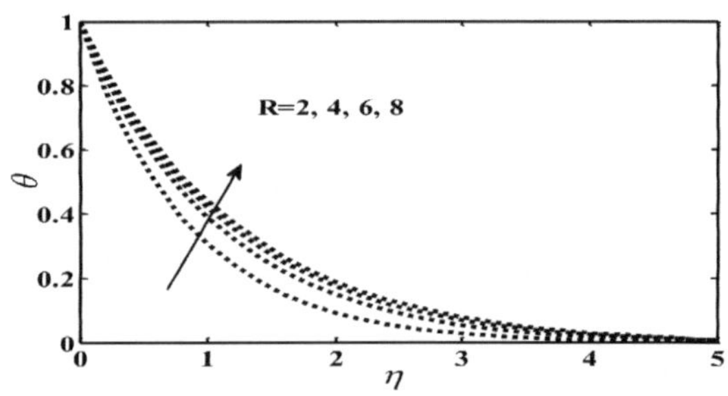

Figure 11: Temperature profile for R

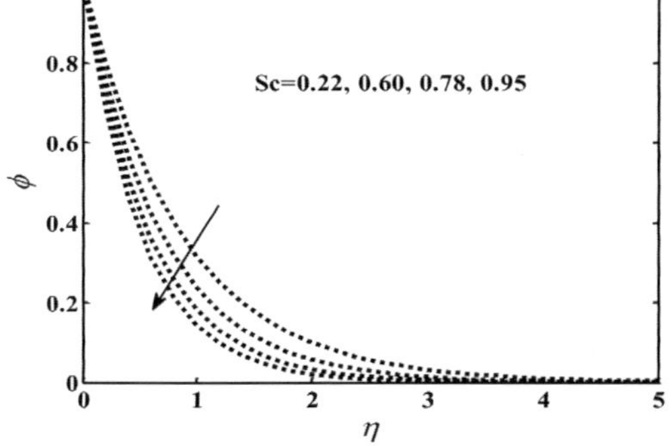

Figure 12: Concentration Profile for Sc

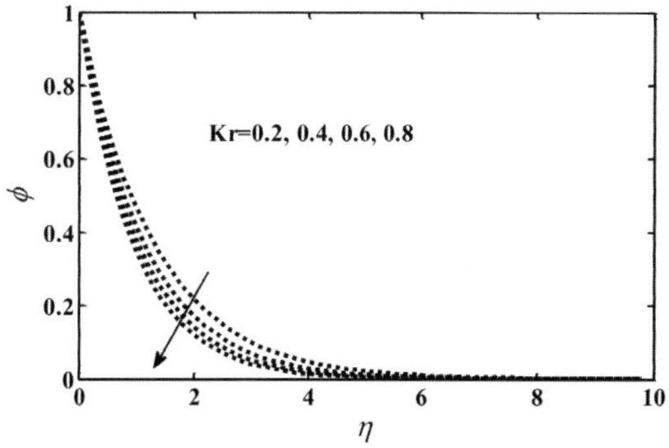

Figure 13: Concentration Profile for Kr

Table 2.1: Influence *Pr, Gr, Gc, R, Sc,* β_i, β_e, *Kr, K* and *M* in Skin friction:

Pr	*Gr*	*Gc*	*R*	*Sc*	β_i	β_e	*Kr*	*K*	*M*	τ
0.71	2	4	1	0.60	0.2	2	0.2	0.1	0.2	5.3724
1.0	2	4	1	0.60	0.2	2	0.2	0.1	0.2	3.9180
3.0	2	4	1	0.60	0.2	2	0.2	0.1	0.2	1.0606
0.71	**2**	4	1	0.60	0.2	2	0.2	0.1	0.2	5.3724
0.71	**4**	4	1	0.60	0.2	2	0.2	0.1	0.2	5.4812
0.71	**6**	4	1	0.60	0.2	2	0.2	0.1	0.2	5.5900
0.71	2	**2**	1	0.60	0.2	2	0.2	0.1	0.2	2.7406
0.71	2	**4**	1	0.60	0.2	2	0.2	0.1	0.2	5.3724
0.71	2	**6**	1	0.60	0.2	2	0.2	0.1	0.2	8.0043
0.71	2	4	**2**	0.60	0.2	2	0.2	0.1	0.2	7.2087
0.71	2	4	**4**	0.60	0.2	2	0.2	0.1	0.2	8.6818
0.71	2	4	**6**	0.60	0.2	2	0.2	0.1	0.2	9.3150
0.71	2	4	1	**0.60**	0.2	2	0.2	0.1	0.2	5.3724
0.71	2	4	1	**0.78**	0.2	2	0.2	0.1	0.2	5.4962
0.71	2	4	1	**0.95**	0.2	2	0.2	0.1	0.2	5.8347
0.71	2	4	1	0.60	**0.4**	2	0.2	0.1	0.2	5.1238
0.71	2	4	1	0.60	**0.6**	2	0.2	0.1	0.2	4.8708
0.71	2	4	1	0.60	**0.8**	2	0.2	0.1	0.2	4.6250
0.71	2	4	1	0.60	0.2	**4**	0.2	0.1	0.2	4.7461
0.71	2	4	1	0.60	0.2	**6**	0.2	0.1	0.2	4.3076

0.71	2	4	1	0.60	0.2	**8**	0.2	0.1	0.2	3.9838
0.71	2	4	1	0.60	0.2	2	**0.4**	0.1	0.2	3.9645
0.71	2	4	1	0.60	0.2	2	**0.6**	0.1	0.2	3.9782
0.71	2	4	1	0.60	0.2	2	**0.8**	0.1	0.2	4.0374
0.71	2	4	1	0.60	0.2	2	0.2	**0.3**	0.2	4.2575
0.71	2	4	1	0.60	0.2	2	0.2	**0.5**	0.2	4.1361
0.71	2	4	1	0.60	0.2	2	0.2	**0.7**	0.2	4.0803
0.71	2	4	1	0.60	0.2	2	0.2	0.1	**0.2**	3.9003
0.71	2	4	1	0.60	0.2	2	0.2	0.1	**0.4**	4.0260
0.71	2	4	1	0.60	0.2	2	0.2	0.1	**0.6**	4.1394

Table 2.2: Influence *Pr, & R in* Nusselt number

Pr	*R*	N_w
0.71	1	-1.8481
1.0	1	-2.5801
3.0	1	-12.6354
0.71	**2**	-1.3343
0.71	**4**	-1.0765
0.71	**6**	-0.9903

Table 2.3: Influence Sc & *Kr in* Sherwood number

Sc	*Kr*	S_h
0.60	0.2	-0.3507
0.78	0.2	-0.4778
0.95	0.2	-0.6012
0.60	**0.4**	-0.4195
0.60	**0.6**	-0.5314
0.60	**0.8**	-0.6236

Conclusions:

- It is noticed that Velocity declined as the magnetic parameter raises
- Temperature strengthened as the radiation parameter growing up.

IOP Conf. Series: Journal of Physics: Conf. Series **2765** (2024) 012017 doi:10.1088/1742-6596/2765/1/012017

- It is noticed that Velocity and Concentration differences reduced due to the growing chemical reaction parameter
- Velocity and temperature diminished with the progressive values of Prandtl number

Appendix:

$$m_1 = \frac{Sc + \sqrt{(Sc)^2 + 4ScKr}}{2}, m_2 = \frac{Sc + \sqrt{(Sc)^2 + 4\gamma}}{2}, m_3 = \frac{\lambda + \sqrt{\lambda^2 + 4\xi}}{2}$$

$$m_4 = \frac{1 + \sqrt{1 + 4N}}{2}, m_5 = \frac{1 + \sqrt{1 + 4Q}}{2}$$

Nomenclature	
A	Real positive constant
B_0	Magnetic induction (A.m^{-1})
C	Non dimensional Concentration
C_p	Specific heat for constant pressure(J. kg $^{-1}$.K)
D	Chemical molecular diffusivity (m^2.S^{-1})
ω^*	Frequency of oscillation
T_w	Fluid temperature at walls *(K)*
u	components of velocity vector in x direction
K	Permeability of porous medium
Kr	Chemical reaction parameter *(m.S^{-1})*
M	Magnetic parameter
T	Temperature of the fluid *(K)*
k_e^*	mean absorption coefficient
σ_s^*	Stefan-Boltzmann constant($Wm^{-2}K^{-4}$)
L^*	Velocity slip factor
N	Dimensionless material parameter
N_u	Nusselt number
Pr	Prandtl number
S_h	Sherwood number
T_*	Dimensional Temperature *(K)*
t^*	dimensional time *(S)*
e	electron charge*(C)*
v	Velocity component in y direction (m.S^{-1})
T_∞	Dimensional free stream temperature (K)
q_r^*	Radiation heat flux density (W.m^{-2})
Gc	Solutal Grashof number

Sc	Schmidt number
τ_w	Skin-friction coefficient
R^*	Radiation parameter
V_0	Scale of suction velocity $(m.S^{-1})$
U_0	Scale of free stream velocity
Greek Symbols	
β	Spin gradient viscosity (K^{-1})
β^*	Coefficient of volumetric expansion $(m^3.kg^{-1})$
μ	Fluid dynamic viscosity
θ	Non dimensional temperature K
ε	Scalar constant(<<1)
σ	Electrical conductivity $(\Omega^{-1}m^{-1})$
Subscripts	
P	Plate
∞	Free stream condition
W	Wall condition
*	Dimensionless properties

References:

[1]. Shah, N. A., Ebaid, A., Oreyeni, T., & Yook, S. J. (2023). MHD and porous effects on free convection flow of viscous fluid between vertical parallel plates: advance thermal analysis. *Waves in Random and Complex Media*, 1-13. DOI: https://doi.org/10.1080/17455030.2023.2186717

[2]. Kabir, T. M., Umar, A. M., & Umar, L. (2023). Magnetohydrodynamics Free Convection Couette Flow and Heat Transfer Through a Vertical Porous Plate with Heat Generation and Absorption Effect. *Dutse Journal of Pure and Applied Sciences (DUJOPAS)*, **9,** 133-145. DOI: https://dx.doi.org/10.4314/dujopas.v9i2a.13

[3]. Mahmuda Maya, M. U., Alam, M. N., & Refaie Ali, A. (2023). Influence of magnetic field on MHD mixed convection in lid-driven cavity with heated wavy bottom surface. *Scientific Reports*, ***13*(1),** 18959. DOI: https://doi.org/10.1038/s41598-023-45707-x

[4]. Alzabut, J., Nadeem, S., Noor, S., & Eldin, S. M. (2023). Numerical analysis of Magnetohydrodynamic convection heat flow in an enclosure. *Results in Physics*, 106618. DOI: https://doi.org/10.1016/j.rinp.2023.106618

[5]. Appidi, L., Malga, B. S., & Pramod Kumar, P. (2022). Effect of thermal radiation on an unsteady MHD flow over an impulse vertical infinite plate with variant temperature in existence of Hall current. *Heat Transfer*, *51*(3), 2367-2382. DOI: https://doi.org/10.1002/htj.22402

[6]. Raghunath, K., Gulle, N., Vaddemani, R. R., & Mopuri, O. (2022). Unsteady MHD fluid flow past an inclined vertical porous plate in the presence of chemical reaction with aligned magnetic field, radiation, and Soret effects. *Heat Transfer*, *51*(3), 2742-2760.
DOI : https://doi.org/10.1002/htj.22423

[7]. Prasad, T. R. K. D., Raju, T. L., & Rajakumar, K. V. B. (2022, November). Influence of viscous dissipation on MHD free convective flow past a semi-infinite moving inclined porous plate with variable suction and radiation absorption. In *AIP Conference Proceedings*, **2516(1)**.
DOI: https://doi.org/10.1063/5.0108864

[8]. Malapati, V., & Polarapu, P. (2015). Unsteady MHD free convective heat and mass transfer in a boundary layer flow past a vertical permeable plate with thermal radiation and chemical reaction. *Procedia Engineering*, ***127*,** 791-799.
DOI: https://doi.org/10.1016/j.proeng.2015.11.414

[9]. Ali, B., Liu, S., Jubair, S., Khalifa, H. A. E. W., & Abd El-Rahman, M. (2023). Exploring the impact of Hall and ion slip effects on mixed convective flow of Casson fluid Model: A stochastic investigation through non-Fourier double diffusion theories using ANNs techniques. *Thermal Science and Engineering Progress*, ***46***, 102237.
DOI: https://doi.org/10.1016/j.tsep.2023.102237

[10]. Narasimha Rao, N. S. L. V., & Gangadhar, K. (2022). Hall and ion-slip effects on MHD natural convective flow past an unbounded vertical porous channel with thermodiffusion. *Heat Transfer*, ***51*(2),** 1501-1523.
DOI: https://doi.org/10.1002/htj.22361

[11]. Jha, B. K., & Malgwi, P. B. (2022). Computational analysis and heat transfer of MHD transient free convection flow in a vertical microchannel in presence of Hall and ion slip effects. *International Journal of Applied and Computational Mathematics*, ***8*(4)**, 156.
DOI: https://doi.org/10.1007/s40819-022-01352-y

[12]. Kanimozhi, N., Vijayaragavan, R., & Shanmugam, K. (2022). Impacts of thermal radiation, viscous dissipation, ohmic heating, and diffusion-thermo effects on unsteady MHD free convective rotating flow of second-grade fluid with Hall and ion-slip currents. *Heat Transfer*, ***51*(8)**, 7435-7461.
DOI: https://doi.org/10.1002/htj.22651

[13]. Yasin, M., Hina, S., Naz, R., Abdeljawad, T., & Sohail, M. (2023). Numerical Examination on impact of hall current on peristaltic flow of eyring-powell fluid under ohmic-thermal effect with slip conditions. *Current Nanoscience*, ***19*(1)**, 49-62.
DOI: https://doi.org/10.2174/1573413718666220104124038

[14]. Hari Babu, B., Srinivasa Rao, P., & Varma, S. V. K. (2021). Hall and ion-slip effects on MHD free convection flow of rotating Jeffrey fluid over an infinite vertical porous surface. *Heat Transfer*, ***50*(2)**, 1776-1798.
DOI: https://doi.org/10.1002/htj.21954

[15]. Saikia D. J., & Ahmed, N. (2024). Chemical reaction and radiation impact on unsteady free convective flow through a porous medium past an infinite oscillatory vertical

plate. *ZAMM-Journal of Applied Mathematics and Mechanics/Zeitschrift für Angewandte Mathematik und Mechanik*, ***104*(1),** e202200579.
DOI: https://doi.org/10.1002/zamm.202200579

[16]. Hossain, M. M., Laskar, A. R., & Bhuiyan, M. A. (2023). Effect of chemical reaction on unsteady MHD convective transport passing a vertical porous sheet. *Journal of Engineering Science*, ***14*(1),** 85-93.
DOI: https://doi.org/10.3329/jes.v14i1.67638

[17]. Chandra Sekhar, B., & Vijaya Kumar, P. (2023). Chemical reaction impacts the unsteady MHD convective flow of an incompressible viscous fluid past an infinite vertical porous plate. *Heat Transfer*, *52*(2), 1474-1496.
DOI: https://doi.org/10.1002/htj.22751

[18]. Goud, B. S., Srilatha, P., Mahendar, D., Srinivasulu, T., & Reddy, Y. D. (2023). Thermal radiation effect on thermostatically stratified MHD fluid flow through an accelerated vertical porous plate with viscous dissipation impact. *Partial Differential Equations in Applied Mathematics*, *7*, 100488.
DOI: https://doi.org/10.1016/j.padiff.2023.100488

[19]. Nagaraju, L., & Naikoti, K. (2023). The effects of thermal radiation on unsteady MHD free convective flow through the porous medium between two vertical plates with Hall effects. *Heat Transfer*, ***52*(4)**, 3074-3093.
DOI ; https://doi.org/10.1002/htj.22817

[20]. Rajakumar, K. V. B., Murthy, P. P., Govinda Rao, T., & Ranganath, N. (2022). Significance of Casson dissipative fluid and spanwise cosinusoidally fluctuating temperature on unsteady MHD free convective flow past a hot vertical permeable plate with radiation absorption and chemical reaction. *Heat Transfer*, *51*(8), 7725-7752.
DOI: https://doi.org/10.1002/htj.22663

[21]. Prasad TRKD Vara, T. Linga Raju, and K. V. B. Rajakumar. "Viscous dissipation and radiation absorption effects on unsteady MHD free convective fluid flow past an inclined porous plate with chemical reaction." *Heat Transfer* 52, no. 2 (2023): 1254-1274.
DOI: https://doi.org/10.1002/htj.22739

[22]. Raghunath K., Gulle, N., Vaddemani, R. R., & Mopuri, O. (2022). Unsteady MHD fluid flow past an inclined vertical porous plate in the presence of chemical reaction with aligned magnetic field, radiation, and Soret effects. *Heat Transfer*, *51*(3), 2742-2760.
DOI : https://doi.org/10.1002/htj.22663

Acoustical-Debye Temperature-Surface Tension & Redlich Kister Coefficient Studies In Multifunctional Material Combination Of Glycerol

G.Sridevi[1] and Sk. Fakruddin Babavali[1,2]

[1] Department of Physics, V.R.Siddhartha Engineering College, Vijayawada,A.P., India.
[2] Corresponding Author email id: sfdnspnl@gmail.com

Abstract. Debye temperature and surface tension analyses were conducted on mixtures of glycerol and paraxylene materials, consisting of the whole range of mole fractions of glycerol. T= (318.15, 313.15, 308.15, and 303.15,) K was the temperature range in studies. Through a meticulous examination of the similarities in Debye temperatures and the disparities in viscosity and surface tension, they demonstrate interactions between molecules between the individual molecules that made up these liquid mixtures and the molecular contributions to the fluctuations in thermal energy in binary liquid mixtures. These interactions are explained by the derived results.

1. Introduction

Extensive research has been conducted on paper ultrasound tests for velocity estimation of mixtures, these are used to investigate the thermal, acoustic, and other interactions between molecules in liquid mixtures. Molecular interactions in mixtures are best described by common transport properties such as mass transfer, surface tension, viscosity, heat transfer, and fluid flow [1-11]. A range of experimental techniques, such as Raman spectroscopy, permittivity measurements, and nuclear magnetic resonance, used as the purpose of investigating the variations in molecular interactions caused by structural changes in these interactions. However, no single technique can precisely define how intra- and intermolecular interactions function. Ultrasound technology offers the advantage of being less expensive while still being effective. Obtaining understanding of liquid mixtures molecular made up and intermolecular forces can be achieved both qualitatively and quantitatively by examining variations in mixturesthermodynamic characteristics and the extent of separation from the ideal. The excessive thermodynamic properties of liquid mixtures have been the topic of theoretical and experimental investigations as a result. NMR, Raman, infrared, nuclear magnetic resonance, and dielectric property measurements are just a few of the experimental methods that have been utilised to investigate that changes in structure affect molecular interactions. However, a single approach might not be sufficient to obtain a complete understanding of the characteristics of interactions between and within molecules. In addition, compared to other methods, ultrasonic methods are equally efficient and cost-effective. Consequently, numerous researchers have conducted tests using ultrasound methods. The chemical industry relies on reliable thermodynamic data from binary systems, particularly in the context of mixed solvents. These solvents possess a wide range of thermodynamic parameters that contribute to a comprehensive understanding of their physicochemical properties. Ultrasound technology finds numerous applications in various fields such as physics, chemistry, biology, and medicine. In the chemical and food processing industries, ultrasonic measurement proves to be highly valuable, along with its utility in material testing, underwater measurement, cleaning, and more. Mechanical materials processing, colloid and emulsion production, seed pregermination, biological tissue imaging, and non-destructive testing (NDT) also commonly utilize ultrasonic vibrations. The implementation of Ultrasonic AFM (atomic force microscopy) at the nanometer scale can enhance manufacturing techniques. By eliminating friction in the nanometer range, ultrasound plays a crucial role in this regard. Sound velocity measurements have recently emerged as a valuable tool in polymer technology, aiding in the understanding of polymer solvents, polymer interactions, and polymer structure. These measurements are particularly useful for studying chemical and biological processes in liquid mixtures. Among the various physical interactions, an important factor affecting the thermodynamic characteristics of a solution is the creation of hydrogen bonds between molecules in a liquid mixture. Hydrogen molecules connected to electronegative atoms have a tendency to form bonds with each other and to interact with other molecules that have accessible electronegative atoms. Strong hydrogen bond dilution and a progressive decline in aggregate concentration are two aspects of the mixture's composition that affect the degree of intermolecular bonding and self-association between various molecules. Testing for food quality can benefit from the effectiveness of an identification technique that has been approved by the American Oil Chemists Society for the detection of unsaturated aldehydes. Paraxylene, an aromatic hydrocarbon and one of the three isomers of dimethylbenzene, is used as a precursor in the production of various acids and esters, which are used in the production of polyesters, plastics, and rubber products. Simple triol compound glycerol is a colourless, odourless, viscous liquid that tastes sweet and is safe to use. It is widely used in wound and burn treatment due to its antibacterial and antiviral properties and is also used as a culture medium for bacteria. It is also frequently used as a humectant in pharmaceutical formulations and as a sweetener in the food industry. The viscosity of a liquid mixture of glycerol and paraxylene was experimentally tested, and Glycerol's surface tension was computed theoretically at 4 different temperatures and for the full mole fraction range. The findings were tabulated and discussed.

2. Materials and Methods

The protocol involves the use of Glycerol and paraxylene as AR grade and SDFCL chemicals, respectively, which are purified prior to use. Liquid mixtures of varying concentrations and mole fractions are created using Job's continuous variation method. The ultrasound velocity is measured using an ultrasound interferometer with pulse-echo (Mittal Enterprises, u3 MHz, New Delhi, India) with a variable path single crystal interferometer. The measuring cell and the high frequency generator are the two components of the experiment. The measuring cell's experimental liquid generates ultrasonic waves due to the High Frequency Generator. This double-walled measuring cell, which has an internal diameter of 0.9 cm and an internal capacity of 12 cm3, is made of stainless steel and is intended to keep the temperature constant throughout the experiment. The reflector plate can be adjusted a quartz crystal could be placed at the bottom of the cell and fastened at the top with a micrometre screw. The two buttons on the High Frequency Generator: an "ADJ" button that changes the position of the ammeter needle and a "GAIN" button that increases the deflection and sensitivity of the device. The High Frequency Generator has frequencies of 1, 3, and 5 MHz, while the measuring cell has a frequency of 1 MHz. Ultrasound waves propagate through this crystal until they are reflected by a moving reflector. The melted rod acts as a reflector with a reflective surface diameter of 1.4cm. This reflective surface is a Teflon-coupled micrometre screw system that measures the steel rod to a precision of 0.001 mm is attached. This spring-loaded steel rod is able to handle any back case when the micrometre screw is inserted. Using a threaded cap and Teflon ring, the micrometre assembly and reflector are attached to the liquid cell. Remove the knurled cell top by unscrewing it and lifting it away from the cell's twin walls. Put the liquid solution or experimental liquid in the centre, then screw on the knurled lid. If there is too much liquid to fill the cell, it gets rinsed out. Once in the square base socket, the cell is next placed and screwed in place. Through the coaxial line of the instrument, the cell and high frequency generator are connected. Until it reaches a maximum or minimum reading, the high frequency generator continuously rotates its anode current ammeter in either a clockwise or anticlockwise direction. Micrometer readings were recorded that matched the maximum and minimum micrometer values. To average all deviations, 50 consecutive maximum or minimum measurements (5/2) must be consider. The generator frequency, or 3 MHz, is represented by the equation U = 5/f, can be used to calculate the speed of sound. Sound velocity measurements have an accuracy of 0.5 ms-1. It is possible to calculate the density of both pure and mixed liquids using density bottles. Standardisation of the specific gravity bottle volumes was conducted at 318.15, 313.15, 308.15, and 303.15 K. The relative density of liquids and liquid mixtures to double-distilled water is calculated using the provided ratios.

Relative density (ρ) = $(Wt_L / Wt_W)*\rho w$

The corresponding weights of liquid and water are represented by WtL and WtW, respectively, while the density of water at a specific temperature is represented by ρw. The density of water at different temperatures was determined based on literature. Fill the experimental solution and specific gravity bottle all the way to the thermostat's neck in order to keep the temperature steady with a thermal stability of 0.01K. When specific gravity bottles are used, density calculations generate results that are more accurate than 0.5%. One electronic balance (Shimadzu AUY220, Japan) the weight of the liquid and liquid mixture is measured in this experiment using a resolution of 0.1 mg. Two different liquid types were measured for viscosity using an Ostwald glass capillary viscometer: pure and mixed. There are two legs on the viscometer, A and B, with leg A having a large bowl at the bottom and leg B consisting of two marks above and below the capillary and bulb. After cleaning and drying the viscometer, add approximately 10 ml of a solution of known concentration to part A and use the rubber ball at the top of link B to draw the solution into flask B. As soon as the solution starts to fall because of gravity, set a stopwatch to record the time it takes to reach the lower limit mark (1/100 second). After that, acetone is used to clean the viscometer. To keep the temperature constant, the thermostat in which the viscometer is placed has a 0.01K thermal stability. The relationship between the liquid and water flow times (t and t0) and their respective densities (ρ and ρ0) it's possible to determine a liquid's absolute viscosity in solutions.

$$\eta = [(\rho \times t) / (\rho w \times tw)] \times \eta w$$

At the experimental temperature, the viscosity of pure distilled water was calculated and shown as ηw. Utilising an electronic stopwatch, the flow time was determined, and water's viscosity at 318.15, 313.15, 308.15, and 303.15 K was calculated using published data. The viscosity measurement has an accuracy of 0.001 mPaS. A temperature-controlled water bath was used to measure the liquid mixture's temperature.

3. Theory

Umamaheswara Rao and Sreehari Sastry[12] determined the excess thermoacoustic parameters, by ultrasonic examination, such as excess free volume, excess enthalpy, and excess adiabatic compressibility, and excess intermolecular free length of propylene glycol in hexanol with various mole fractions at 303 K. The results were analyzed based on intermolecular interactions between components in liquid mixtures. Devi et al.[13] measured the viscosity, density, and ultrasound velocity of a ternary mixture of glycerin and ethylene glycol in octanol at 303.15 K. Molecular association was established by calculating thermoacoustic parameters using experimental values, including free volume, intermolecular free length, absorption coefficient, adiabatic compressibility, acoustic impedance, internal pressure, bulk modulus, Gibbs free energy, relaxation time, and molar sound velocity. Sowjanya and Sreehari Sastry measured the density, sound velocity, and viscosity values of a binary mixture containing pure liquid and propylene glycol and 1-pentanol at 303.15 K throughout the whole mole fraction range [14]. The calculated thermoacoustic parameters include excess free length, excess molar volume, excess enthalpy, and excess isentropic compressibility using experimental data. The examination of the changes in these parameters resulting from molecular interactions allowed for the identification of hydrogen bond formation between mixtures. Sherif and Vinayagam[15] measured the density, viscosity, and ultrasound velocity of solutions containing ethylene glycol and transition metal acetates at 303 K. From experimental values, acoustic parameters including cohesive energy, they calculated the following: free length, acoustic impedance, absorption coefficient, free volume, internal pressure, effective volume, and adiabatic compressibility. Strong interactions between the solute and solvent were found by using trends in the acoustic parameters.Srinivasan et al.[16] measured the ultrasonic velocity across the composition range of a ternary liquid mixture consisting of glycerol, ethylene glycol, and ethyl alcohol at temperatures of 303.15, 308.15, and 313.15 K. The resulting ultrasound velocities were analyzed for various relationships, and molecular interaction parameters were calculated using experimental and theoretical ultrasound velocity values. Birajdar et al[17] determined the structure of ZnS nanoparticles using X-ray diffraction technique, showing a cubic zinc Berende structure with a lattice constant of 5.380Å0 and an average particle size in the nanometer range. After the ZnS nanopowder was prepared, it was dissolved in the base liquid ethylene glycol to create ZnS nanofluid. This was then examined utilising the ultrasonic interferometer technique, ultrasonically at a frequency of 2 MHz. Raut and Smrutiprava[18] At 303.15 K in both glycol + water and glycol + water, the density and ultrasound velocity of NaCl, NaBr, and NaI were measured. Sound velocity, acoustic impedance, adiabatic compressibility, apparent molar volume, intermolecular free length, apparent molar compressibility, and limit of apparent molar compressibility are some of their thermodynamic properties, and their relationship to constants, were also examined. By measuring ultrasound velocity and density in a binary mixture of isopropanol, ethylene glycol/propylene glycol, and 308.15 K throughout the whole range of composition, Kondaia et al. [19] established a quantitative relationship. The acquired experimental data were used to calculate a number of excess parameters, such as excess acoustic impedance, excess molar volume, and excess intermolecular free length. It was concluded that there is a strong interaction, considering the geometrical placement of small molecules in the spaces left by larger molecules in a liquid mixture Additionally, ultrasound velocity measurements were compared to various empirical relationships. Sheba and Priakumari[20] carried out ultrasonic studies of the molecular interactions between polyethylene glycol and ethanol in binary liquid mixtures. At room temperature, they measured the ultrasound velocity, density, and viscosity at various mole fractions of ethanol in polyethylene glycol ranging from 0 to 0.055. Using the experimental data, acoustic parameters such as intermolecular free length, adiabatic compressibility, relaxation time, free volume, and acoustic impedance were calculated. Gupta et al.[21] determined, at 308.15 K, the viscosity, density, and ethylene glycol and glycerol aqueous solution ultrasonic velocity as a function of concentration. Numerous parameters, including acoustic impedance, relaxation time, adiabatic compressibility, and intermolecular free length, additionally to the Gibbs free energy, free volume, absorption coefficient, Rao and Wada constants, and relative association, and the volume available to assess molecular association in mixtures, were computed using the experimental values.

The Auerbach's empirical relation[22-24] that has been used to calculate Surface tension (S) is,

$$\mathbf{S = 6.3x10\text{-}4\,.U3/2.\rho \qquad N/m \qquad \text{---------}\ (1)}$$

The effective Debye temperature can be calculated using this relationship [7].

$$\theta_D = \frac{h}{k_B}\left[\frac{9N/4\pi V}{\left(\frac{1}{U_l^3}\right)+\left(\frac{2}{U_t^3}\right)}\right]^{1/3} \qquad \text{---------}\ (2)$$

U_l and Ut, respectively, stand for the longitudinal and transverse mode propagation velocities. The molar volume is represented by V, and Avogadro's number, boltzmann's constant, and Planck's constant are represented by the letters h, kB, and N, respectively.

4. Results and Discussion

Over a range of mole fractions, T = (318.15, 313.15, 308.15, and 303.15) K, experimental measurements of glycerol viscosity (η), ultrasonic velocity (U) and density (ρ) have been carried out. Although the effective Debye temperature is calculated using equation (2) based on the measured ultrasonic density (ρ), viscosity (η), and velocity (U), equation (1) is used for the theoretical estimation of surface anxiety values. At T = (303.15, 308.15, 313.15, or 318.15 K), Table 1 displays all of the measured values for surface tension (S), viscosity (η), density (ρ), ultrasonic velocity (U), and derived Debye temperature (θD). Figures 1 to 5 demonstrate the equivalent versions with the mole fraction of glycerol.

Table 1. The readings below were recorded during mole-fraction in glycerol and propylene liquid mixtures at T = (318.15, 313.15, 308.15, and 303.15K): density (ρ), ultrasonic velocity (U), Debye temperature (θ_D), viscosity (η), and surface tension (S).

Mole.fraction(X)	303.15K	308.15K	313.15K	318.15K
		Ultrasonic.velocity (U) m/s		
0.0000	1334.67	1314.81	1293.25	1276.89
0.1564	1358.83	1338.97	1317.41	1301.05
0.2581	1384.04	1364.18	1342.62	1326.26
0.3593	1407.18	1387.32	1365.76	1349.4
0.4600	1432.3	1412.44	1390.88	1374.52
0.5602	1456.59	1436.73	1415.17	1398.81
0.6600	1479.71	1459.85	1438.29	1421.93
0.7592	1506.15	1486.29	1464.73	1448.37
0.8580	1531.34	1511.48	1489.92	1473.56
0.9563	1558.21	1538.35	1516.79	1500.43
1.0000	1567.32	1548.87	1527.34	1512.23
		Density. (ρ) Kg/m³		
0.0000	868.93	864.12	860.18	856.64
0.1564	905.27	899.35	897.52	893.98
0.2581	940.58	936.27	932.38	930.92
0.3593	978.7	975.7	969.72	968.26
0.4600	1018.03	1014.69	1007.06	1005.6
0.5602	1054.68	1051.03	1044.4	1044.49
0.6600	1091.22	1088.37	1081.74	1081.83
0.7592	1130.17	1125.71	1119.08	1120.74
0.8580	1166.83	1163.05	1157.11	1158.08
0.9563	1203.76	1200.39	1194.45	1197.99
1.0000	1251.43	1246.62	1241.23	1238.39
		Viscosity.(η) x 10-3 Pa S		
0.0000	0.2418	0.2393	0.2217	0.2176
0.1564	0.3718	0.3693	0.3517	0.3476
0.2581	0.5018	0.4993	0.4817	0.4776
0.3593	0.6318	0.6293	0.6117	0.6076

IOP Conf. Series: Journal of Physics: Conf. Series **2765** (2024) 012018 doi:10.1088/1742-6596/2765/1/012018

0.4600	0.7618	0.7593	0.7417	0.7376
0.5602	0.8918	0.8893	0.8717	0.8676
0.6600	1.0218	1.0193	1.0017	0.9976
0.7592	1.1518	1.1493	1.1317	1.1276
0.8580	1.2818	1.2793	1.2617	1.2576
0.9563	1.4118	1.4093	1.3917	1.3876
1.0000	1.4734	1.46871	1.4523	1.4489
Surface.Tension (S) x10^3 N/m				
0.0000	26.69	25.95	25.20	24.62
0.1564	28.57	27.76	27.04	26.43
0.2581	30.51	29.72	28.90	28.33
0.3593	32.55	31.76	30.84	30.24
0.4600	34.77	33.93	32.91	32.28
0.5602	36.94	36.06	35.03	34.43
0.6600	39.13	38.25	37.17	36.54
0.7592	41.62	40.64	39.52	38.92
0.8580	44.05	43.06	41.92	41.27
0.9563	46.65	45.63	44.45	43.86
1.0000	48.92	47.87	46.68	45.88
Debye.Temperature (θ_D) K				
0.0000	50.18	48.93	46.54	44.22
0.1564	53.16	51.91	49.52	47.2
0.2581	56.07	54.82	52.49	50.11
0.3593	59.00	57.75	55.42	53.04
0.4600	61.94	60.66	58.38	55.98
0.5602	64.91	63.63	61.35	58.93
0.6600	67.89	66.61	64.33	61.91
0.7592	70.75	69.47	67.19	64.9
0.8580	73.62	72.34	70.08	67.79
0.9563	76.58	75.30	73.04	70.75
1.0000	79.57	78.26	75.91	73.71

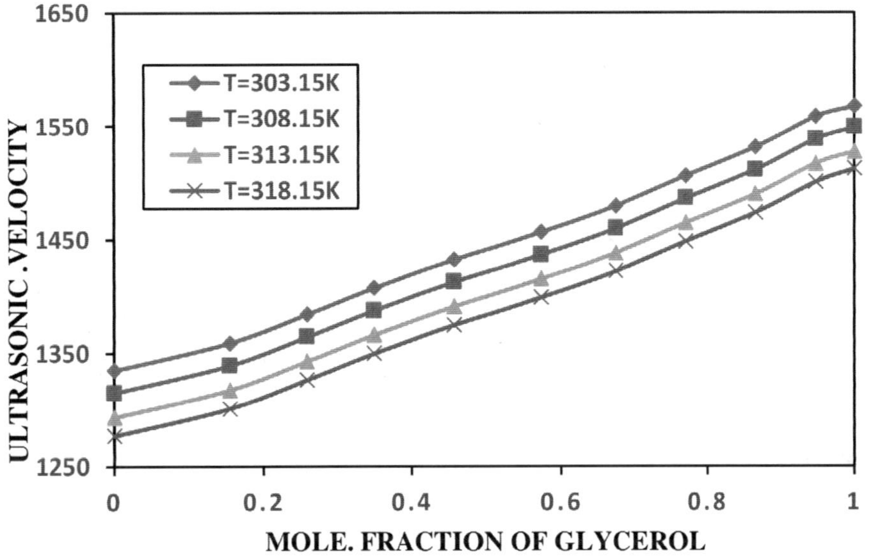

Figure 1. Variation of Ultrasonic Velocity with Mole. fraction of Glycerol

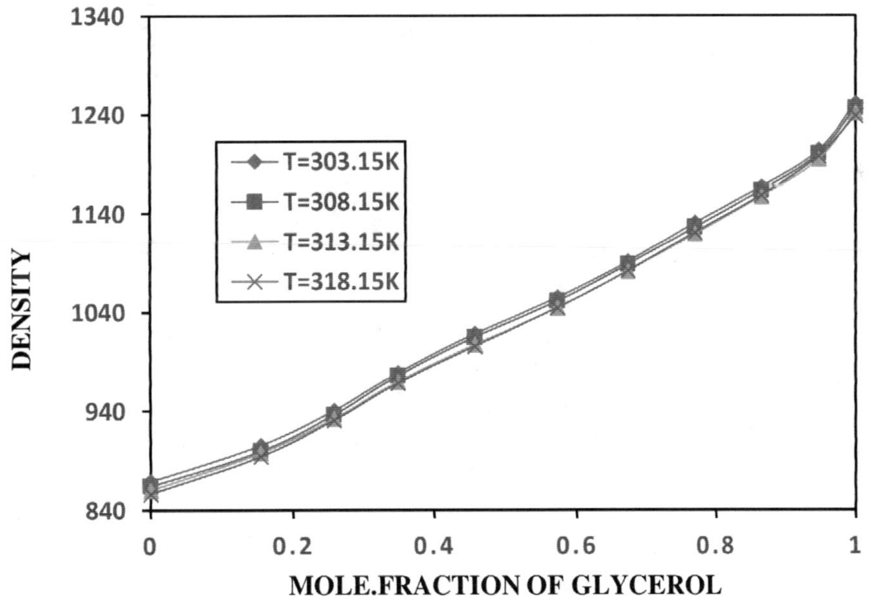

Figure 2. Variation.of. Density with Mole. fraction of Glycerol

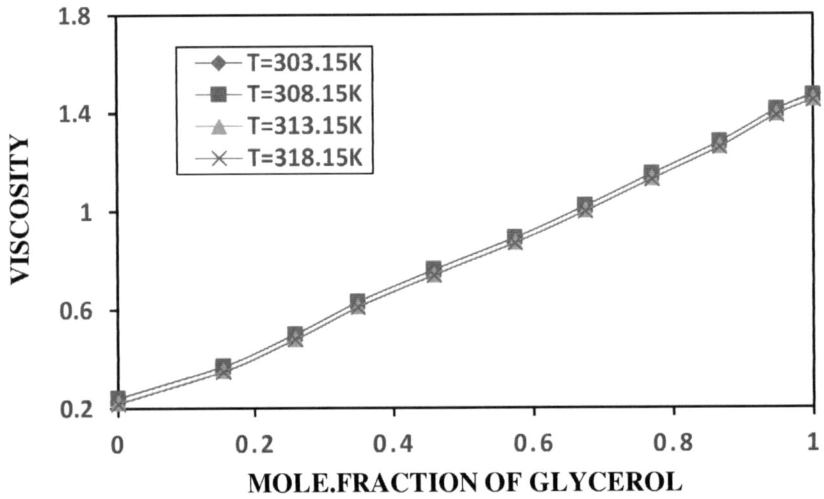

Figure 3. Variation of Viscosity with Mole. fraction of Glycerol

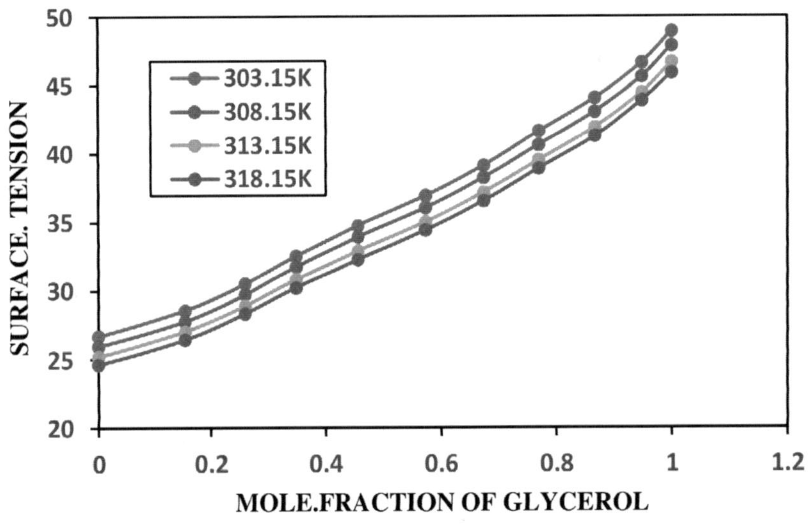

Figure 4. Variation of Surface Tension with Mole. fraction of Glycerol

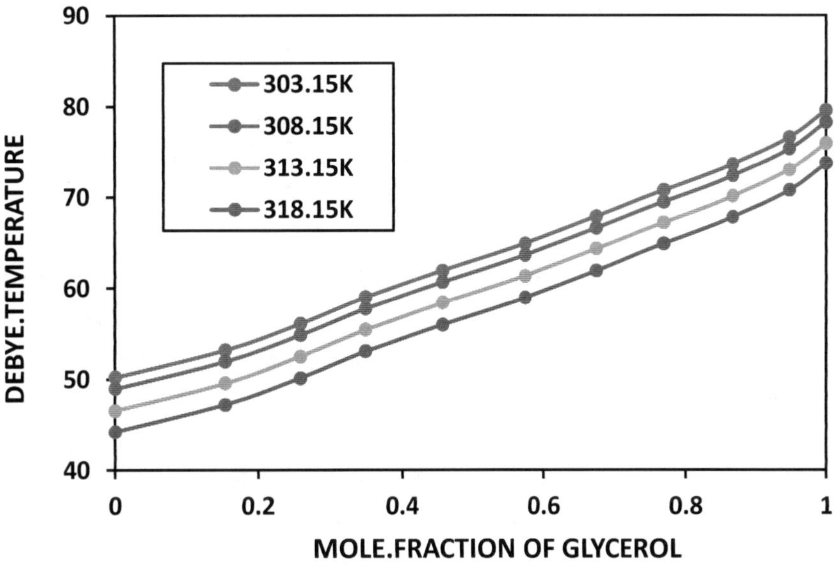

Figure 5. Variation of Debye Temperature with Mole. fraction of Glycerol

"When an ultrasonic pulse returns to the same transducer after passing through another (pitch catch) or vice versa (pulse echo), it can be detected using an ultrasonic flaw detector. The non-destructive assessment of material properties often involves observing changes in the propagation velocity of ultrasound waves and the energy loss resulting from their interaction with the microstructure of the material. Ultrasonic attenuation and sound velocity measurements of polycrystalline metals can be characterised by their elastic properties. Surface tension refers to the tendency of a stationary liquid surface to condense over a small area. Certain animals, such as toucans and water striders, can float on the water surface due to their density being greater than that of water, which is possible because of surface tension. When a liquid and air come into contact, surface tension occurs due to the cohesive force between liquid molecules being stronger than the force of attraction (adhesion force) between air molecules. Two main mechanisms contribute to surface tension: the inward pressure applied to the surface molecules, causing the liquid to contract, and the force applied tangentially and parallel to the liquid surface. This tangential force is commonly referred to as surface tension. One feature of the liquid-vapor or liquid-air interface is surface tension, in contrast, the degree of deformation determines the tension in elastic membranes; this is important to keep remember even though the behaviour of a liquid with surface tension is comparable to the stretch of an elastic membrane over its surfaces. Developed by Peter Debye in 1912 in the fields of thermodynamics and solid state physics, the Debye model is a amount that phonons contribute to a solid's specific heat. As opposed to the optoelectronic model proposed by Einstein, that considers a solid as a combination of harmonic quantum oscillators that are noninteracting, the Debye model views atomic lattice vibrations as grouped phonons. A solid's low temperature dependence on its heat capacity is described by the Debye Cube Law, is precisely predicted by the Debye model. At high temperatures, the Dulong-Pitit law, similar to Einstein's optoelectronic model, is restored. However, due to the simplifications made in the assumptions, the accuracy of the Debye model decreases at intermediate temperatures. Figure-1 illustrates the change in ultrasonic velocity of Glycerol with molecular disruption. The results demonstrate that the ultrasonic velocity increases linearly with molecular disruption, demonstrating the presence of molecular interactions among the constituents of the liquid mixture. Strong support for these findings is also given by Thivagarajan and Palaniappan [25, 26]. The changes in the combination's viscosity and density of glycerol and para-xylene are shown in Figures 2 and 3, respectively. According to these results, density and viscosity increase linearly with molecular fraction, indicating the presence of substantial interactions in liquid mixtures [25, 26]. Surface tension, a phenomenon of molecular association, is greatly affected by the intermolecular interactions within the system. The change in surface tension with molecular fraction is shown in Figure 4. According to the observations, surface tension decreases with temperature and rises with molecular fraction. A study conducted by Karla Granados further confirms that strong interactions develop in liquid mixtures as the 'S' value of the mixture increases, supporting the formation of interfaces with high surface tension transfer, this is a basic characteristic of surface tension. The thermal characteristics of liquid mixtures are expressed in terms of specific heats using the Debye theory. The strong interactions between molecules strengthen the bonds between atoms, that indicates that increasing a substance's temperature needs more energy. It follows that the mixture's Debye temperature inceases as well. Whether the mole fraction of glycerol affects the variation in Debye temperature in Figure 5 demonstrates that the predicting as the mole fraction of glycerol in the liquid mixture under investigation increases, so does the Debye temperature.. However, as the temperature rises, the contribution of specific heat diminishes, resulting in a lower Debye temperature [27,28]. The Debye temperature and surface tension results show that there is strong molecular interaction in the mixtures.

5. Conclusions

For all mole fractions of glycerol, density, the ultrasonic velocity, and viscosity of binary combinations of glycerol and four different temperatures are used to measure paraxylene. (T = 318.15, 313.15, 308.15, and 303.15 K). These measurements provide the surface tension and Debye temperature values are computed. A correlation has been observed between the mole fraction and the Debye temperature. Moreover, there are parallels between the mole fraction increase and surface tension increase. Furthermore, correlations with experimental values have been established by these results. These results indicate that there is strong molecular interaction in the liquid mixture.

6. Acknowledgments

The authors of this paper express their sincere gratitude to the administrative authorities of Velagapudi Ramakrishna Siddhartha Engineering Institute in Andhra Pradesh, India, for generously providing the necessary research facilities.

7. References

[1] Sarma N T , Fakruddin Babavali SK and Srinivasu Ch 2021 *Mater.Today: Proc.* **43 Part5** 3125
[2] Fakruddin Babavali SK, Sarma N T and Srinivasu Ch 2020 *Mater.Today: Proc.*, **46 Part1**, 369
[3] Reed T M and Taylor T E 1959 *J.Phys.* **63** 58
[4] Redlich O and Kister A T 1948 *Indian Eng. Chem.* **40** 345
[5] Palanippan L and Thiyagarajan R 2008 *Ind. J. Chem.* **47B** 1906
[6] Pandey J D 1979 *Acoustic Letter.* **3** 90
[7] Ibrahim S M, Fakruddin Babavali SK and Sarma N T 2020 *Mater.Today:Proc.* 46 **Part1** 376
[8] Rama Rao G V, Vishwanatha Sarma A, Ramachandran D and Rambabu C 2007 *Indian J. Chem.* **46A** 1972
[9] Alculea Palofax M 1993 *Ind. J. Pure Appl. Phys.* **31** 90
[10] Sumathi T and Govindarajan S 2012 *Int. J. Bio Pharm Allied Sci.* **1(8)** 1153
[11] Eswari Bai M, Subha M C S, Narayana Swami G and Chowdoji Rao K 2004 *J. Pure App. Ultrason.* **26** 79
[12] Umamaheswara Rao D and Sreehari Sastry S 2017 *Int. J. Innovative research in Sci., Engg. and Tech.* **6(3)** 4444
[13] Devi D, Arasu A M M and Girija A M 2017 *Int. Research J. Engg. & Tech.* **4(9)** 36
[14] Sowjanya Ch and Sreehari Sastry S 2017 *Int. J. Engg. Research & Tech.* **6(4)** 17
[15] Sherif A and Chidambara Vinayagam S 2016 *J.Chem. Pharm. Research.* **8(8)** 948
[16] Srinivasan G Palani R and Balakrishnan S 2015 *Int. J. Innovative research in Tech. Sci. & Engg.* **1(7)** 1
[17] Birajdar S D, Khirade P P, Humbe A V, Kshirsagar P M , Aghav P S and Jadhav K M 2015 *Int.J. Science and Research.* **4(2)** 497
[18] Raut N C and Smrutiprava D *Int. J. Pharma Research and Health Science.* 2015 **3(2)** 6
[19] Kondaiah M, Sreekanth K, Nayeem Md and Krishna Rao D 2014 *J.Chem.Pharma. Research.* **6(4)** 1243
[20] Sheba S G S and Priakumari O R 2014 *Int. J. Chemical and Molecular Engg.* **8(2)** 156
[21] Gupta V, Sharma A K and Sharma M 2014 *J.Chem.Pharma Research.*, **6(1)** 714
[22] Auerbach N 1948 *Experientia.* **4** 473
[23] Ramadoss P and Buvaneswari N 2011 *E-Journal of Chem.* **8(3)** 1246
[24] Oswal S and Tathnam M V 1984 *Can J. Chem.* **62** 2851
[25] Prajapati A N 2017 *Indian J. Pure Appl. Phys.* **55** 297
[26] Thivagarajan and Palaniappan L 2008 Indian *J. Pure Appl. Phys.* **46** 852
[27] Rajan Dass, Prakash K, Muhuri and Dilip Hazra K 1994 *Acoustics Letters.* **18** 69
[28] Palaniappan L 2002 *Indian J. Pure Appl. Phys.* **40** 828

Influence of Photonics and Plasmonics Effect on Ultrathin Amorphous Silicon Thin Film Solar Cell

S. Saravanan[1*], R.S. Dubey[2], S. Srikanth[3], D. Jagadeeswara rao[4], M.V. Someswararao[5]
[1]Department of Physics (S&H), Swarnandhra College of Engineering & Technology (A), Seetharampuram, Narsapur-534275, Andhra Pradesh
[2]University Institute of Engineering and Technology, Guru Nanak University, Ibrahimpatnam, RR District, Telangana, India
[3,5]Department of Physics, SRKR Engineering College, Bhimavaram, (WG), Andhra Pradesh, INDIA
[4]Department of Physics, S V Government Degree College, Parvathipuram-535501, A.P. INDIA
Email ID: shasa86@gmail.com

Abstract

Our recent surge in silicon derived materials has major demand in thin film photovoltaic (PV) modules and enhancing the significant numbers. The rigorous coupled wave analysis method is a simple and fast method also known as developed to determine the light absorption and cell efficiency. The optics of thin film solar cells (amorphous silicon) garnering crucial role in the photovoltaic market. In this work, an ultrathin thin film amorphous silicon solar cell PV performance investigated by periodically textured surfaces at the nanoscale level. This analysis of periodic textured substrate was deriving optimal surface textures. The nanogratings lead to light scattering mechanism with the higher order diffraction angle and enhanced the light absorption of the incidence spectrum. The influence of grating period and height, the collection of the charge carriers increased at various wavelengths from ultraviolet, visible and infrared spectral regions are discussed with the assistance of photonic and plasmonic modes. Finally, nanoscale engineering mechanism the optimized thin film amorphous silicon solar cell yielded the highest current densities (22.6 and 23.8 mA/cm^2) in both polarization modes.

Keywords: RCWA, absorption, amorphous silicon, Photonics, Plasmonics

1.Introduction

The economical, flexible and more efficient thin film solar cells have been a great demand in photovoltaic (PV) market. Because of rising energy demand as well as atmospheric condition and global issue associated with the fossil fuels. For that, research communities still believe that solar energy is the major, substantial and dependable renewable energy sources as compare to others. For that the amorphous silicon thin film solar cells are the better choice and cheaper to manufacture than the other semiconductor materials. Because of its availability (naturally abundant), non-toxic and relatively low cost. However, the Photonics and plasmonic effects are improving the absorption by dispersing the light (or photon) with the support of semiconductor and metallic nanostructures. Because, the incoming light has certain frequency is stimulating the electron oscillations on the surface of the nanostructures. Also, the nanostructured interfaces increasing the optical path length which leads to the enhanced the optical performance as compared to the smooth surface. The size and shapes are playing crucial role in the fabrication of solar cells. Also, the major impacts various nanostructures are nanoparticles, nanopillar, distributed Bragg reflector, triangular grating, rectangular, and square shaped nanostructures. These nanostructures are distinctly improving the current density, and cell efficiency of the prepared solar cells as reported by various research communities. Recently, ***Elkhamisy et al.,*** presented the comprehensive survey on silicon based thin film solar cells. Also, they have discussed the current challenges in nanomaterials or structures and novel trends in the photovoltaic market. These review works including the development (advanced) of PV technology, cell configuration, circuit model, cell performance along with that the global market [**1**]. ***Abolfazl Jangjoy et al., (2019)*** reported an ultrathin film Si solar cell by embedded paired nanoparticles. With the influence of plasmonics, they reached the maximum current densities of 21.68 and 21.84 mA/cm^2 by two different radius nanoparticles integrated solar cells [**2**]. **Preeti and Sushil Kumar (2023)** experimentally demonstrated the recycling (waste thin film PV module) of thin film PV module is based on a very easy and simple approach under various chemical, thermal and mechanical treatments. According to the TCO properties, extracted FTO has more potential to reuse for the fabrication of new thin film PV module and shown better performance [**3**]. Similarly, **Saravanan and Dubey (2022)** proposed the enhancement of amorphous silicon solar cell performance by integrating one-dimensional photonic crystal back side reflector using rigorous coupled wave analysis (RCWA) method. This thin film solar cell performance enhanced due to the various incident wavelengths (400, 600 and 800 nm). Also, the nanostructures are incorporated (nanogratings) and noticed increased the light harvesting [**4**]. **Ghahremani and Fathy (2016)** presented the high efficiency thin film based amorphous silicon solar cells. The metallic nanoparticles were embedded as textured back side reflectors. These simulations were compared to the relevant to the other results and showed 30% improvement in the conversion cell efficiency [**5**]. Considerable research is still needed to optimize the advanced technological studies to enrich its full potential with commercialization of photon and plasmon enhanced solar cells

In this communication, we focused the design and simulation of ultrathin amorphous silicon solar cells by integrating silver back reflector and ITO antireflection coating with gratings. In section 2, the various parameters of nanostructures and methods are explained. The optical absorption in both (TE/TM) polarizations and field distributions at different incident wavelengths (UV-Vis-IR) are discussed in section 3. Finally, the summarized the work in conclusion in section 4.

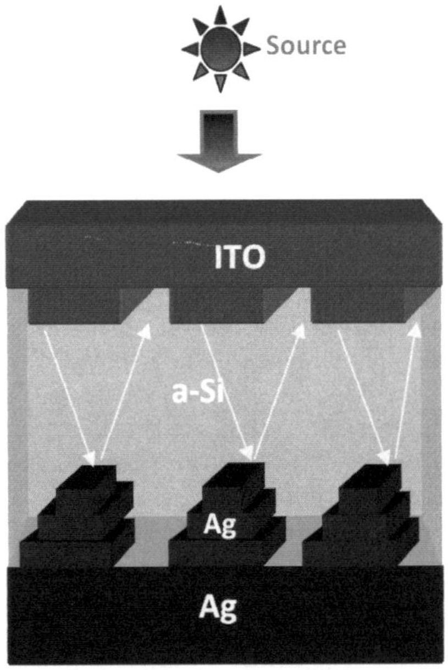

Figure 1. The schematic diagram of ultrathin amorphous silicon solar cells

2. Designing Approach

The optical structure has been investigated by various techniques like finite integration technique (FIT), rigorous coupled wave analysis (RCWA), finite difference time domain (FDTD), and finite difference frequency domain (FDFD) methods. These numerical techniques are strongly dependent on Maxwell's equations. The proposed thin film amorphous silicon solar cell is designed and investigated by using RCWA method also known as a simple and fast method. This RCWA approach numerically solving the various Maxwell's equations with the complex device design structures (or layers). So, the complex design simulated and can be optimized. In photovoltaic market, this amorphous silicon thin film cells are majorly used. In this work, the structure type describes by different layers integrated or incorporated as a thin film solar cell. First, the background in this case consists of semi-infinite region and referred as substrate and indium tin-oxide (ITO) antireflection coating (ARC) material layer and top grating thicknesses are 70 nm and 30 nm. Next, the amorphous silicon absorber layer thickness varies at 40 nm and bottom gratings (10+10+10 = 30nm). The thickness (width and height) of the gratings are maintained by the cross section in X-axis (horizontal) and Y-axis (vertical) directions. Also the silver (Ag) substrate used as a back reflector and the thickness is 90 nm. The back reflector is the combination of silver triangular nanogratings (3 different gratings) with the substrate. For the above materials, the real and imaginary part of the refractive indices are defined. The real part of the refractive index of the particular materials defined in between the refractive index of the core (n_x) and background materials (n_0).

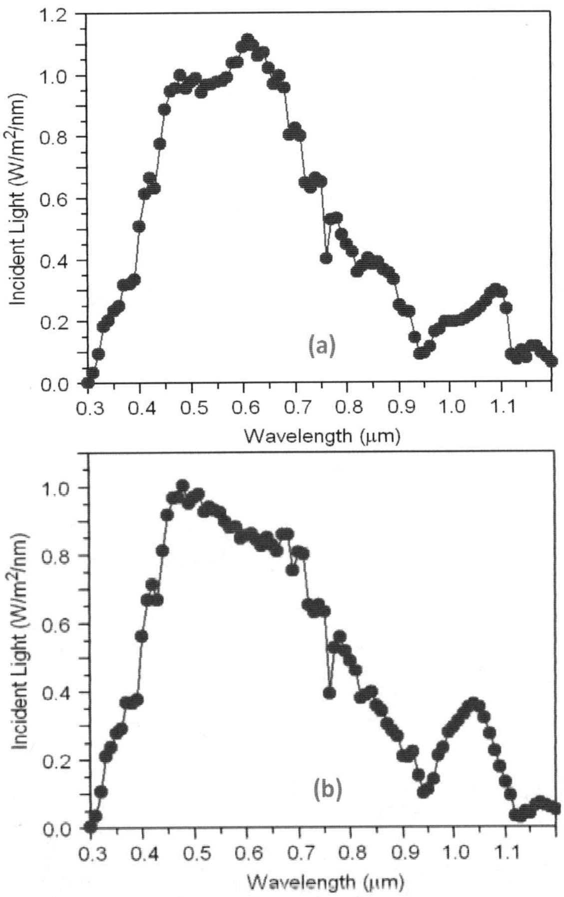

Figure 2. The absorption spectra of ultrathin amorphous silicon solar cells (TE & TM)

According to the materials and their property can be changed. Also, the nanogratings is one of the strongest light trapping strategies that can be used to enhance not only the photo-generated current density.

3. Results and Discussions

The absorption of the light is the key factor which plays a significant role by converting light into electricity. **Figure 2(a) -(b)** shows the light absorption spectrum of transverse electric and magnetic field polarization. The incident light absorption is enhanced in the ultraviolet and visible (400-700 nm) spectral region. With the assistance of photonic and plasmonic modes, the light harvesting mechanism enhanced. **Figure 2(a)** shows the light absorption of amorphous silicon solar cells under transverse electric field (TE) mode polarizations. The high absorption noticed at 649 nm and shows the reduced the collection of the light in the longer spectral spectrum [**6**]. Because of bottom integrated triangular grating the major absorption noticed in the visible spectral region. Here, the Fabri-Perot resonance and Surface guided modes plays a crucial role to trap the photons inside the absorber region.

IOP Conf. Series: Journal of Physics: Conf. Series **2765** (2024) 012019 doi:10.1088/1742-6596/2765/1/012019

For that, we need optimized and novel nanostructures and materials. In longer wavelength, the collection of the photons decreasing due to the heat as reported by the various researchers [**7-8**]. Also, the various incidences of light, the photon interaction with the nanostructures which integrated with the thin film solar cells are shown in **Figure 3(a) -(l)**. **Figure 2(b)** shows the light absorption spectrum under transverse magnetic (TM) field polarization modes. With the incident solar spectrum (0.3 to 1.2 μm), the light harvesting improved in the near infrared region as compared to the TE absorption. Here, the light trapping mechanism improved due to the plasmonic effect, for example, localized surface plasmon resonance (LSPR), surface excitation, and surface guided mode.

Table 1: thin film amorphous silicon solar cell optical performance

S.N.	Structure	TE		TM	
		%	mA/cm^2	%	mA/cm^2
1	Dual GRA+ ARC	15.67	23.82	14.96	22.57

3.1 Photonic Effect

The optical model is used to know analyze field distributions with the assistance of periodic triangular and rectangular groves. **Figure 3 (a) -(i)** shows the transverse electric (TE) field distribution at visible and infrared spectrum. A comparison of optical field distributions is shown in **figure 3**, the incident center wavelength for the field distributions are 479 nm (a), 610 nm (b), 670 nm (c), 700 nm (d), 739 nm (e), 780 nm (f), 840 nm (g), 990 nm (h), 1080 nm (i), 1160 nm (j), 1170 nm (k), 1200 nm (l). Among these field distributions, yellow color shows the strong field in between the nanogratings.

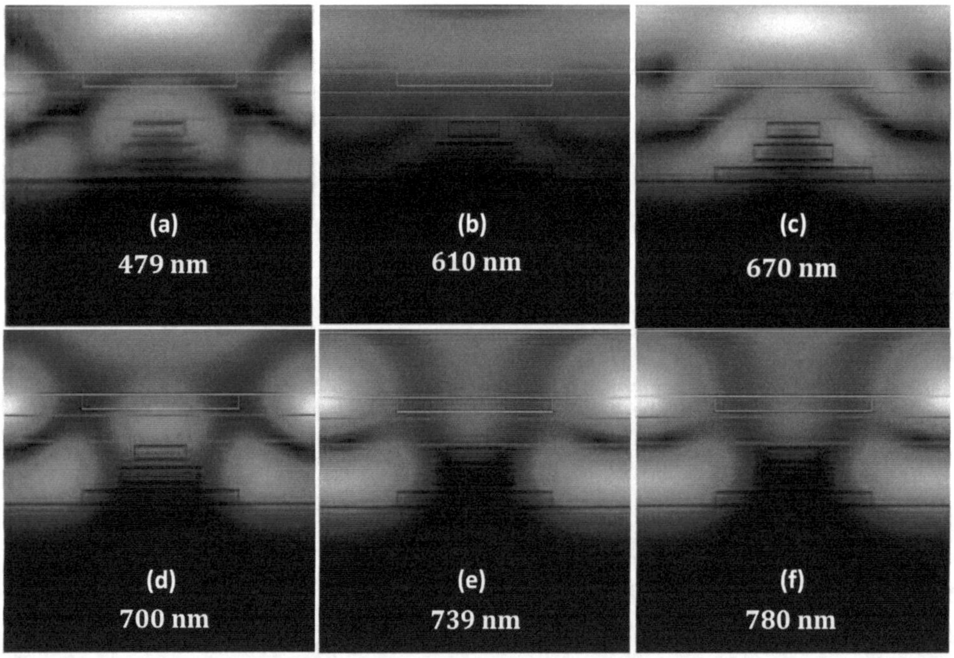

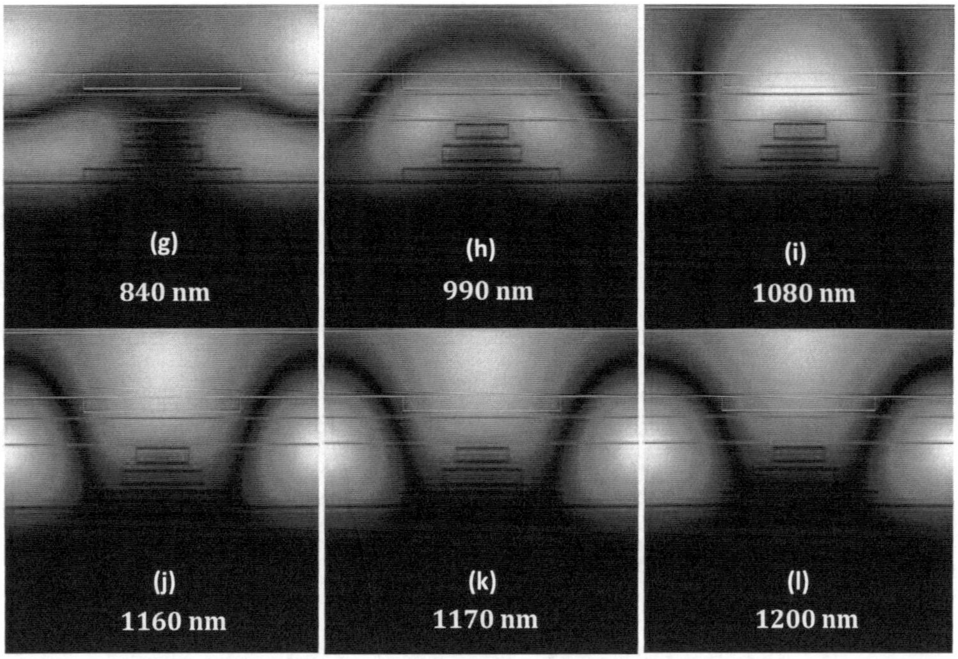

Figure 3. The transverse electric (TE) field distribution at visible and infrared spectral region.

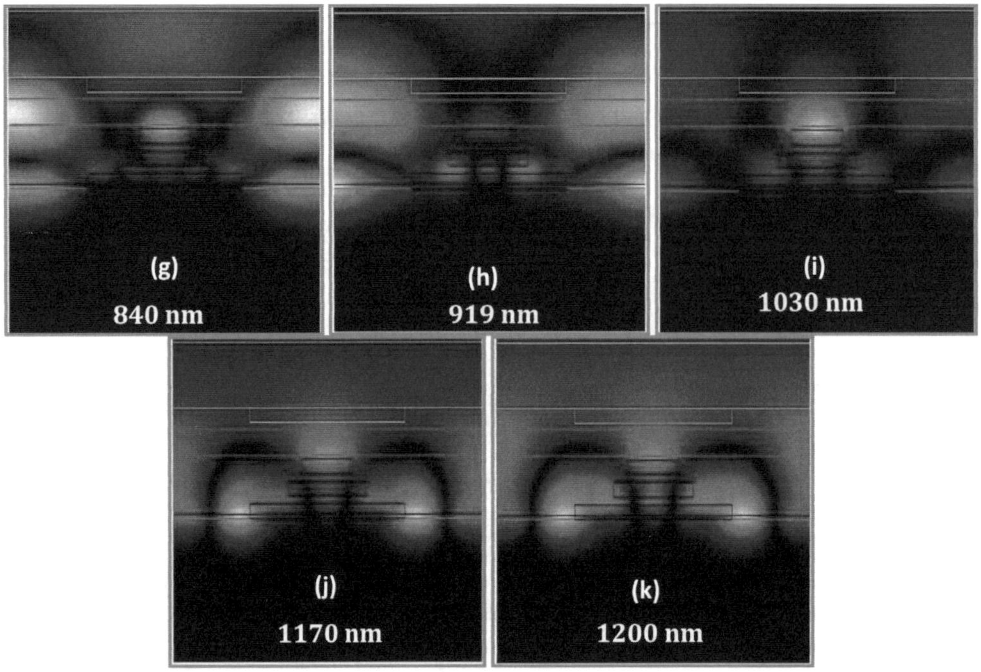

Figure 4. The transverse magnetic (TM) field distribution at visible and infrared spectral region.

The top gratings are increasing the visibility field distributions (479, 610, and 670 nm) towards the Ag-back reflector structure includes substrate and triangular nanogratings. These field distributions are demonstrating light trapping mechanism in between the active layer region also known as Fabry-Perot resonance effect. The near infrared spectral region (700, 739, 780, and 840 nm) shows the decreasing the light trapping mechanism using surface guiding modes. Also, the strong surface guided mode noticed in the active region and noticed from far infrared spectral region such as 990, 1080, 1160, 1170 and 1200 nm.

3.2 Plasmonic effects

Figure 4 (a) -(k) shows the transverse magnetic (TM) field distribution of various incidents (or centered), wavelength (λc). The centered wavelengths are selected from the light absorption spectrum of **Figure 2(a)**. The light trapping mechanism and field distributions are discussed in **Figure 4 (a) -(k).** The irradiation of light excites the charge carriers by scattering resonance frequency and increasing the light absorption. When the nanostructures are embedded inside the absorber region (or active), improved the absorption due to a maximum density of states of the photons around the nanostructures. The light striking on the silver triangular nanogratings and scattering in omni-directional ways. This nanogratings are allowing the light (or photon) scattering and diffraction which helps to enhance the effective path length of the light particles within the absorber region. It could be travelling along the solar cell and bounce back in between the absorber region and absorbing more light. Also, small gratings heights supporting the larger absorption by diffraction methods as compare to larger gratings as reported by **Rahul Dewan and Knipp (2009) [9]**. Under TM field distributions, the surface guided mode resonance was noticed in absorber region such as 419, 479, 679 nm (visible spectral region). These light trapping mechanism is a two way process by reflecting back to the absorber region due to the backside reflector. Next, the surface excitation modes appeared in between dielectric and metallic nanostructures as evidenced in 700, 739, 1170 and 1200 nm (figure (d), (e), (j), (k)). The strong light localization filed noticed in the centre wavelength of 780, 840, and 1030 nm as shown in **figure (f), (g), (i)**. In thin film solar cells, with increasing the incidence wavelength, the collection of the photons decreasing (refer in figure 2). Consequently, the light photons or plasmon interaction reduced with the dielectric or metallic nanostructure based solar cells and reported by various researchers [**9-10**]. Experimentally, the amorphous silicon solar cells can be fabricated by low temperatures (100-250°C) using plasma enhanced chemical vapor deposition (PECVD), and radio frequency PECVD techniques [**11-13**]. Further, these thin film and flexible solar cells are finding to use in various energy sectors such as automobiles, defense or energy storage devices [**14**]. **Morawiec et al.,** experimentally quantification on plasmon enhanced silicon thin film based solar cells using PECVD method. With the assistance of photonic and plasmonic modes the absorption enhancement achieved up to 91% as compare reference cell [**15**]. The hybrid thin film nanostructure could be enhanced the optical performance up to 80% as reported by Van Tan Tran [**16**].

4. Conclusions

In this summary, the rigorous coupled wave analysis (RCWA) method used for the investigation of ultrathin amorphous silicon solar cells. The improved light absorption was noticed in the visible and infrared spectral region in both polarization modes (electric and magnetic). Also, the transverse electric and magnetic field distributions were demonstrated and discussed. An influence of photonic and plasmonic effects is observed by Fabri-Perot resonance, surface plasmon polarization, localized surface plasmon and of an ultrathin thin film amorphous silicon solar cell photovoltaic performance investigated by periodically textured surfaces at nanoscale. This analysis of periodic textured substrate was deriving optimal surface textures. The light trapping mechanisms at various wavelengths from ultraviolet, visible and infrared spectral regions are discussed with the assistance of photonic (transverse electric, TE) and plasmonic (transverse magnetic, TM) modes. At nanoscale engineering, the optimal thin film solar cell yielded highest current densities (Jsc) of 22.6 and 23.8 mA/cm^2.

References

1. ElKhamisy, K., Abdelhamid, H., El-Rabaie, ES.M. et al. A Comprehensive Survey of Silicon Thin-film Solar Cell: Challenges and Novel Trends. Plasmonics (2023). https://doi.org/10.1007/s11468-023-01905-x
2. Abolfazl Jangjoy, H. Bahador, H. Heidarzadeh "Design of an ultra-thin silicon solar cell using localized surface plasmonic effects of embedded paired nanoparticles", Optics Communications, 450, 216-221 (2019). https://doi.org/10.1016/j.optcom.2019.06.007
3. Preeti and Sushil Kumar "Extraction and analysis of TCO coated glass from waste amorphous silicon thin film solar module" Solar energy Materials and Solar cells, 253, 112227 (2023). https://doi.org/10.1016/j.solmat.2023.112227

4. S. Saravanan and R.S. Dubey "Performance enhancement of amorphous silicon solar cell using 1D photonic crystal as back reflector", Materials Today: Proceedings, 49(7), 2822-2825 (2022)
5. A. Ghahremani, A.E. Fathy "High efficiency thin-film amorphous silicon solar cells", Energy Science & Engineering, 4(5), 334-343 (2016).
6. R. Dewan, D. Knipp "Light trapping in thin film silicon solar cells with integrated diffraction grating", J. Appl. Phys., 106, 074901 (2009).
7. R.W. Wood "Anamalous diffraction gratings", Phys. Rev., 48 (12), 928-936 (1935)
8. Sigamani Saravanan and Raghvendra Sarvjeet Dubey "Study of ultrathin film amorphous silicon solar cell performance using photonic and plasmonic nanostructures", International Journal of Energy Research, 2022.
9. Rahul Dewan and D. Knipp "Light trapping in thin-film silicon solar cells with integrated diffraction grating", Journal of applied Physics, 106, 074901 (2009).
10. R. S. Dubey, S. Saravanan "Performance of ultrathin amorphous silicon solar cells: An influence of plasmonic effect", Progress in electromagnetic Research M, 112, 29-39 (2022).
11. Dong-Won Kang, Jun Ryu, Makoto Konagai "High-performance amorphous silicon thin film solar cells prepared toward flexible building-integrated photovoltaics", Electronic Materials Letters, 15(5), 623-629 (2019).
12. D. Hamdani, S. Prayogi, Y. Cahyono, G. Yudoyono, D. Darminto "The influences of the front work function and intrinsic bilayer (i1, i2) on p-i-n based amorphous silicon solar cell's performance: A numerical study", Cogent Engineering, 9(1), 2022.
13. C. Lee, H. Jiyeon, J. Nam, S.H. Jeong, H. Song et al. "Amorphous silicon thin film deposition for poly-Si/SiO2 contact cells to minimize parasitic absorption in the near-infrared region", Enegies, 14 (4), 8199 (2021).
14. S. Santhosh, R. Rathanasamy, G. Velu Kaliyannan, H. Panchal, A. J. Alrubaie, M. M. Jaber, Z. Said, and S. Memon "A Comprehensive Review on current performance, challenges and progress in thin film solar cells", Energies, 15(22), 8688 (2022).
15. S. Morawiec, J. Holovsky, M.J. Mendes, M. Muller, K. Ganzerova, A. Vetushka et al. "Experimental quantification of useful and parasitic absorption of light in plasmon-enhanced thin silicon films for solar cells application", Sci. Rep., 6, 22481 (2016). DOI:10.1038/srep22481.
16. Van Tan Tran, Huu-Quang Nguyen, Young-Mi Kim, G. Ok, and J. Lee, "Photonic-Plasmonic nanostructures for solar energy utilization and emerging biosensors", Nanomaterials, 10(11), 2248 (2020).

Exploring Entropy Minimization in Hybrid Nanofluid Flow Past an Exponentially Stretching Sheet with Significant Quadratic Radiation and Lorentz Forces: A Non-Fourier Heat Flux Analysis

Poojari Prakash Gowd[1], Gandrakota Kathyayani[2]*

[1,2]Department of Applied Mathematics, Yogi Vemana University, Kadapa-516005, India

*Corresponding author email: kathyagk@yogivemanauniversity.ac.in

Abstract. The real-world applications of fluid flow across an exponentially extended sheet are manifold, encompassing crystal formation, paper manufacture, and the cooling of metallic sheets. The primary aim of this study is to conduct a comprehensive theoretical analysis on the behaviour of a hybrid nanofluid flow through an exponentially extended sheet under the influence of quadratic thermal radiation, non-Fourier heat flux and magnetic field. The initially presented equations have been simplified to a set of ODEs, and the bvp4c solver has expertly found solutions to these. Validated the results (of engineering parameters including friction coefficient) obtained using the bar graphs by using Multiple linear regression. It has been noted that a greater magnitude of magnetic field is associated with a temperature enrichment. It is found that higher values of Brinkman number lead to a greater rate of entropy generation. It has been shown that the thermal relaxation parameter (Γ) and magnetic field parameters (M) exert distinct influences on the rate of heat transmission. It is detected that the Nusselt number enhances by 0.700996 (when $0 \leq \Gamma \leq 0.6$) and the same declines by 0.14086 (when $0 \leq M \leq 3.5$). Within the range of $0 \leq M \leq 3.5$, it is seen that the friction factor exhibits a decline with a notable rate of 1.41843.

Keywords: Entropy Generation, bvp4c, Multiple linear regression, Nusselt number, Stretching sheet.

1. Background Information

The investigation of nanofluid flow over an exponentially extending sheet carries substantial significance in numerous disciplines owing to its multifaceted applications and the invaluable elucidation it offers about heat and mass transport phenomena. As an illustration, the use of nanofluids possessing augmented thermal conductivity serves the purpose of cooling molten polymers in the course of casting procedures, hence resulting in enhanced product quality and elevated production rates. Venkata Subba Rao et al. [1] conducted a thorough investigation on the movement of magnetohydrodynamic nanofluid through an exponentially stretched sheet, employing the SRM (spectral relaxation method). An observation has been made that when the Brownian motion parameter upsurges, the temperature also rises. According to Hussain et al. [2], in the slip regime, a water-based nanofluid's dissipative flow was studied across a porous exponentially extending sheet. It has been discovered that the velocity for Cu-water nanofluid lowers as the volume percent of nanoparticles rises. Anuar et al. [3] conducted a comprehensive stability analysis of a hybrid nanofluid flow across an exponentially extending/dwindling sheet, taking into account the slip parameter. Ghosh and Mukhopadhyay [4] employed the Tiwari and Das model to investigate the behavior of water-based nanofluid flows through an exponentially extended sheet. Their findings revealed a positive correlation between the percentage of nanoparticle volume and the consequent rise in fluid temperature. Waini et al. [5] examined hybrid ($\boldsymbol{H_2O + Al_2O_3 + Cu}$) nanofluid flow with thermal radiation and observed that there is a reduction in the heat transmission rate with the rise in that parameter for both solutions. Waini et al. [5] carried out an investigation on the flow of a hybrid nanofluid ($\boldsymbol{H_2O + Al_2O_3 + Cu}$) with thermal radiation. Their findings revealed a decline in the heat transmission rate as this parameter amplified, for both solutions. In their study, Rashid et al. [6] performed an analysis on the impact of suction on a radiative nanofluid (composed of $\boldsymbol{H_2O + Cu}$) across an exponentially extending surface. Their findings revealed a positive correlation between the solid volume percentage and the velocity profile, indicating an upsurge in the latter with the former. Khan et al. [7] scrutinized the characteristics of a blood-based HNF (hybrid nanofluid) flow, taking into account suction and slip factors, employing the HAM (homotopy analysis method). Waini et al. [8] meticulously inspected the flow of an HNF with buoyancy effects at the stagnation point. It was their contention that the utilization of hybrid nanoparticles resulted in an enhancement in the heat transmission rate. Afterwards, a number of scholars [9-18] delved into the realm of exponentially stretching sheets, meticulously examining diverse HNF flows under varying parameters, encompassing the intricate phenomenon of heat generation.

The impact of thermal radiation on fluid dynamics has extensive utility across many domains, spanning from engineering and technology to astronomy and environmental research. For instance, during polymer extrusion and moulding, heat generated by friction must be effectively dissipated to prevent product defects. Understanding the role of radiation in this process helps optimize cooling systems and improve product quality. Venkateswarlu et al. [19] conducted a thorough analysis on the impact of thermal radiation on a stable fluid flow in an upward channel, taking into account heat source and suction characteristics. An intriguing observation is that the fluid velocity experiences a noticeable augmentation when the thermal radiation parameter rises. The MHD movement of a Casson nanofluid across an irregularly expanding sheet with thermal radiation and nonconstant (variable) viscosity was meticulously examined by Mabood et al. [20]. It has been shown that there is a positive correlation between the radiation parameter and an upsurge in transfer of heat. In their study on the 2D flow of a viscous fluid over an exponentially stretched sheet with thermal radiation and suction parameters, Rajput et al. [21] observed a decline in temperature as the suction parameter elevated. The radiative flow of a micropolar fluid across an exponential elastic sheet with velocity slip was studied by Mandal and Mukhopadhyay [22] along with its flow and heat transmission aspects. Asjad et al. [23] conducted a computational investigation on the bioconvective flow of a nanofluid, taking into account thermal radiation and non-Fourier heat flux using the same geometry. Subsequently, a number of scholars [24-29] engaged in a comprehensive examination of diverse radiative fluid flows, taking into account a range of geometrical configurations.

Given the existing research, it is noteworthy that there has been a lack of effort in elucidating the significance of non-Fourier heat flux, quadratic thermal radiation, and Lorentz force effects in the context of hybrid nanofluid flow (namely, Ethylene Glycol + $ZnO + TiO_2$) over a permeable sheet that extends exponentially. The relevant physical characteristics are elucidated through the utilization of bar graphs and subsequently verified through the application of multiple linear regression. Incorporating the Cattaneo-Christov (C-C) Heat Flux (H - F) into the energy equation allows for a more comprehensive examination of heat transport parameters.

2. Research Methodology

A steady and laminar flow of a hybrid nanofluid via an exponentially elastic sheet with quadratic thermal radiation and magnetic field parameters is theoretically investigated in this study. The present investigation is based on the subsequent suppositions:

(i) The utilization of the non-Fourier heat flux idea is employed for the purpose of scrutinizing the intricate occurrence of thermal transfer.

(ii) See Table 1 for precise numerical approximations of the thermodynamic attributes of the base fluid and nanoparticles employed in this investigation.

(iii) The flow is influenced by an external magnetic field with a magnitude $B = B_0 e^{x/2L}$, which is applied at a right angle to the flow.

(iv) Induced magnetic field impact is disregarded in this work.

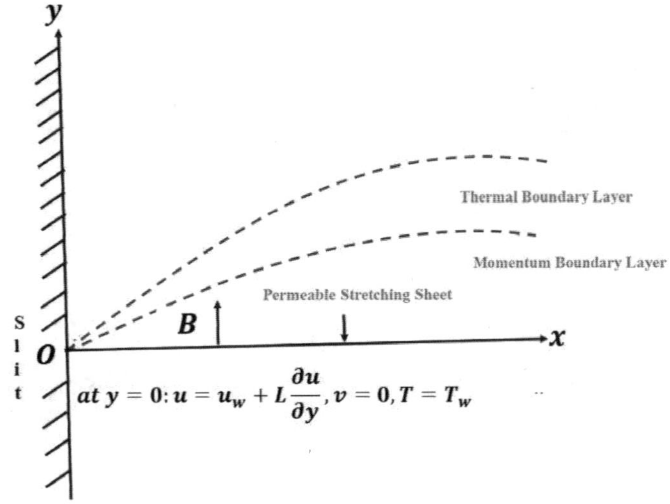

Figure 1. Schematic diagram

Given these assumptions, the requisite conditions and fundamental equations required for this investigation can be described as below: (Rao et al. [30])

$$\frac{\partial}{\partial x}(u)+\frac{\partial}{\partial y}(v)=0 \tag{1}$$

$$u\frac{\partial}{\partial x}(u)+v\frac{\partial}{\partial y}(u)=\frac{\mu_{hnf}}{\rho_{hnf}}\frac{\partial}{\partial y}\left(\frac{\partial}{\partial y}(u)\right)-\frac{\upsilon_{hnf}}{\kappa}u+g\beta_{1T}\left(T-T_\infty\right)^2-\frac{1}{\rho_{hnf}}u\sigma_{hnf}B^2 +g\beta_T\left(T-T_\infty\right) \tag{2}$$

$$u\frac{\partial}{\partial x}(T)+v\frac{\partial}{\partial y}(T)=\frac{1}{\left(\rho C_p\right)_{hnf}}\left(k_{hnf}\frac{\partial}{\partial y}\left(\frac{\partial}{\partial y}(T)\right)-\frac{32\sigma^* T_\infty^3}{3k^*}\frac{\partial}{\partial y}\left(\frac{\partial}{\partial y}(T)\right)+\frac{24\upsilon^* T_\infty^2}{3k^*}\frac{\partial}{\partial y}\left(\frac{\partial}{\partial y}(T^2)\right)\right)$$
$$-\Gamma^*\begin{pmatrix} u^2\frac{\partial}{\partial x}\left(\frac{\partial}{\partial x}(T)\right)+v^2\frac{\partial}{\partial y}\left(\frac{\partial}{\partial y}(T)\right)+2uv\frac{\partial}{\partial x}\left(\frac{\partial}{\partial y}(T)\right) \\ +\frac{\partial}{\partial x}(T)\left(u\frac{\partial}{\partial x}(u)+v\frac{\partial}{\partial y}(u)\right)+\frac{\partial}{\partial y}(T)\left(u\frac{\partial}{\partial x}(v)+v\frac{\partial}{\partial y}(v)\right)\end{pmatrix}$$

(3)

$$\left.\begin{array}{l}\text{at } y=0: v=0, u=u_w=ae^{x/L}, T=T_w,\\ \text{at } y=\infty: u=0, T=T_\infty.\end{array}\right\} \quad (4)$$

2.1. Hybrid nanofluid thermodynamic attributes

$$\left.\begin{array}{l}k_{hnf}=k_{nf}\times\dfrac{k_{s_2}+2k_{nf}-2k_{nf}\phi_2+2k_{s_2}\phi_2}{k_{s_2}+2k_{nf}+k_{nf}\phi_2-k_{s_2}\phi_2},\\ (\rho C_p)_{hnf}=\left[(1-\phi_1)(\rho C_p)_f+\phi_1(\rho C_p)_{s_1}\right](1-\phi_2)+(\rho C_p)_{s_2}\phi_2,\\ \sigma_{nf}=\sigma_f\times\dfrac{\sigma_{s_1}+2\sigma_f-2\sigma_f\phi_1+2\sigma_{s_1}\phi_1}{\sigma_{s_1}+2\sigma_f+\sigma_f\phi_1-\sigma_{s_1}\phi_1}, \rho_{hnf}=\left[(1-\phi_1)\rho_f+\phi_1\rho_{s_1}\right](1-\phi_2)+\rho_{s_2}\phi_2,\\ k_{nf}=k_f\times\dfrac{k_{s_1}+2k_f-2k_f\phi_1+2k_{s_1}\phi_1}{k_{s_1}+2k_f+k_f\phi_1-k_{s_1}\phi_1}, \mu_{hnf}=\dfrac{\mu_f}{(1-\phi_1)^{2.5}(1-\phi_2)^{2.5}},\\ \sigma_{hnf}=\sigma_{nf}\times\dfrac{\sigma_{s_2}+2\sigma_{nf}-2\sigma_{nf}\phi_2+2\sigma_{s_2}\phi_2}{\sigma_{s_2}+2\sigma_{nf}+\sigma_{nf}\phi_2-\sigma_{s_2}\phi_2}.\end{array}\right\}$$

Table 1 Assessments regarding different attributes of EG, ZnO and TiO_2 (Jamshed and Aziz [31])

Properties	(EG) (f)	ZnO (s_1)	TiO_2 (s_2)
$\sigma(Sm^{-1})$	0.0000055	0.01	2380000
$k(Wm^{-1}K^{-1})$	0.252	13	8.953
$\rho(kgm^{-3})$	1114	5660	4250
$C_p(Jkg^{-1}K^{-1})$	2415	495.2	686.2

The subsequent similarity transmutations were suggested for transforming governing equations by Jamaludin et al. [32]:

$$\left.\begin{array}{l}\eta=\sqrt{\dfrac{a}{2\upsilon L}}e^{x/2L}y, v=-\sqrt{\dfrac{a\upsilon}{2L}}e^{x/2L}\left[f(\eta)+\eta f'(\eta)\right], u=ae^{x/L}f'(\eta),\\ \theta(\eta)=\dfrac{T-T_\infty}{T_w-T_\infty}, T_w-T_\infty=T_0e^{2x/L}.\end{array}\right\} \quad (5)$$

The continuity equation (1) is easily fulfilled when considering (5), and equations (2-4) can be transformed accordingly using equation (5) in the following manner:

$$\frac{1}{S_1S_2}f'''+ff''-2f'^2+2\tau\theta(1+\tau^*\theta)-\frac{2}{S_1S_2}\lambda f'-\frac{S_3}{S_1}Mf'=0 \quad (6)$$

$$\frac{S_4}{S_5}\frac{1}{\Pr}\theta''+\frac{4}{S_5}\frac{Rd}{\Pr}\theta''+\frac{12}{S_5}\frac{Rd(t-1)\theta}{\Pr}\theta''+\frac{12}{S_5}\frac{Rd(t-1)}{\Pr}\theta'^2 \\ -\Gamma\left(f^2\theta''-ff'\theta'+2\eta f'^2\theta'\right)-4f'\theta+f\theta'=0 \tag{7}$$

$$\left.\begin{array}{l} f(\eta)=0, f'(\eta)=1, \theta(\eta)=1 \text{ at } \eta=0, \\ f'(\eta)=0, \theta(\eta)=0 \text{ at } \eta=\infty. \end{array}\right\} \tag{8}$$

where

$$\left.\begin{array}{l} M=\dfrac{2L\sigma_f B_0^2}{\rho a}, Gr=\dfrac{g\beta_T(T_w-T_\infty)L^3}{\upsilon^2}, \tau=\dfrac{Gr}{\mathrm{Re}_x^2}, \tau^*=\dfrac{\beta_{1T}(T_w-T_\infty)}{\beta_T}, \\ \mathrm{Re}_x=\dfrac{aLe^{x/L}}{\upsilon}, Rd=\dfrac{4}{3}\dfrac{\sigma^* T_\infty^3}{kk^*}, \Gamma=\dfrac{a}{2L}\Gamma_0, \Gamma^*=\dfrac{\Gamma_0}{e^{x/L}}, \lambda=\dfrac{\upsilon L}{\kappa_0 a}, \\ t=\dfrac{T_w}{T_\infty}, \kappa=\kappa_0 x, \Pr=\dfrac{\mu C_p}{k}. \end{array}\right\}$$

and

$$\left.\begin{array}{l} S_3=\dfrac{\sigma_2+2S_{31}\sigma_f-2\phi_2(S_{31}\sigma_f-\sigma_2)}{\sigma_2+2S_{31}\sigma_f+\phi_2(S_{31}\sigma_f-\sigma_2)}S_{31}, S_{41}=\dfrac{k_1+2k_f-2\phi_1(k_f-k_1)}{k_1+2k_f+\phi_1(k_f-k_1)}, \\ S_2=(1-\phi_1)^{2.5}(1-\phi_2)^{2.5}, S_5=(1-\phi_2)\left[(1-\phi_1)+\phi_1\dfrac{(\rho C_p)_1}{(\rho C_p)_f}\right]+\phi_2\dfrac{(\rho C_p)_2}{(\rho C_p)_f}, \\ S_{31}=\dfrac{\sigma_1+2\sigma_f-2\phi_1(\sigma_f-\sigma_1)}{\sigma_1+2\sigma_f+\phi_1(\sigma_f-\sigma_1)}, S_1=(1-\phi_2)\left[(1-\phi_1)+\phi_1\dfrac{\rho_1}{\rho_f}\right]+\phi_2\dfrac{\rho_2}{\rho_f}, \\ S_4=\dfrac{k_2+2S_{41}k_f-2\phi_2(S_{41}k_f-k_2)}{k_2+2S_{41}k_f+\phi_2(S_{41}k_f-k_2)}S_{41}. \end{array}\right\}.$$

Friction factor is defined as:

$$Cf=\frac{\mu_{hnf}}{\rho u_w^2}\left(\frac{\partial u}{\partial y}\right)\bigg|_{y=0}, \tag{9}$$

The expression in equation (9) can be reformulated using (5) as $\sqrt{\mathrm{Re}_x}Cf=\frac{1}{\sqrt{2}}\frac{1}{S_2}f''(0)$.

The following can be used to find the rate of heat transmission:

$$Nu=\frac{-Lq_w}{k_f(T_w-T_\infty)}\bigg|_{y=0}, \quad q_w=\left(k_{hnf}+\frac{4}{3}\frac{4\sigma^* T_\infty^3}{k^*}\right)\frac{\partial T}{\partial y}. \tag{10}$$

The expression in (10) can be reformulated using equation (5) as

$(\mathrm{Re}_x)^{-1/2}Nu=-\frac{1}{\sqrt{2}}(S_4+4Rd)\theta'(0)$.

The formula for determining the generation of entropy in the present flow problem is expressed in dimensional form as follows:

$$S_g = \frac{k_{hnf}}{T_\infty^2}\left(\frac{\partial T}{\partial y}\right)^2 + \frac{\mu_{hnf}}{T_\infty}\left(\frac{\partial u}{\partial y}\right)^2 + u^2\frac{\sigma_{hnf}B^2}{T_\infty}. \quad (11)$$

By employing (5), equation (10) can be reformulated in the following manner:

$$E_G = S_4\alpha\theta'^2 + \frac{Br}{S_2}f''^2 + S_3 Brf'^2 M, \quad (12)$$

where

$$E_G = \frac{2\upsilon L T_\infty S_g}{ae^{x/L}(T_w - T_\infty)k_f}, \alpha = \frac{T_w - T_\infty}{T_\infty}, Br = \frac{\mu a^2}{kT_0}.$$

The following equation can be used to compute the Bejan number:

$$Bn = \frac{\text{The irreversibility arising from the transmission of heat}}{\text{Complete irreversibility}}.$$

Bn can be rephrased by utilising equation (12) in the following manner:

$$Bn = \frac{S_4\alpha\theta'^2}{S_4\alpha\theta'^2 + \frac{Br}{S_2}f''^2 + Brf'^2 M}.$$

3. Discussion of the outcomes

Results Equations (6-7) are solved using the MATLAB inherent function bvp4c under conditions (8). Fig. 2 shows the streamlines of the current flow analysis.

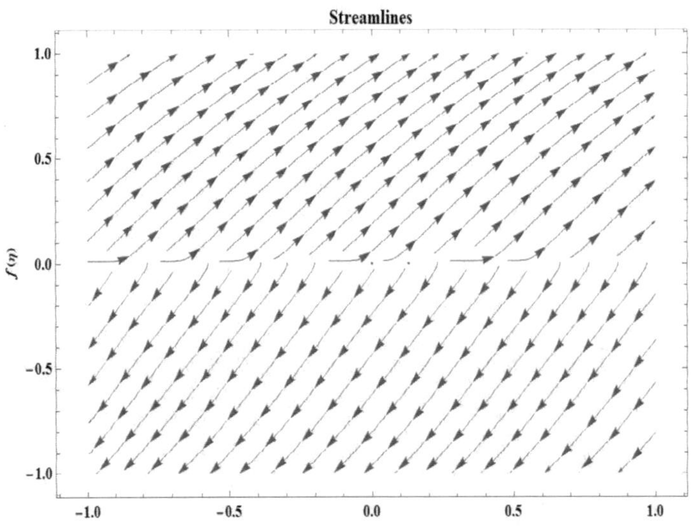

Figure 2. Streamlines of the current flow analysis

3.1. *Regular profiles including velocity profile*

With the rise in M, the linear momentum of the electromagnetic field is transferred to the fluid particle at a faster pace. This signifies the fluid particle's capacity to take up energy from its surrounding electromagnetic field has increased. Hence, higher values of magnetic field are linked to the enrichment in the temperature (see Fig. 3). By augmenting the nanoparticle volume fraction, one can observe a corresponding elevation in the collision rate of fluid particles. Consequently, the fluid temperature experiences a noticeable rise, as shown in Fig. 4. Upon careful examination of Fig. 5, it becomes evident that there is a discernible augmentation in the fluid temperature as the value of Rd rises. An upsurge in Γ leads to a fall in the temperature of the fluid, as seen in Fig. 6. It has been found, as shown in Fig. 7, that when the magnetic field parameter upsurges, the velocity declines. When a moving, electrically conducting fluid encounters a magnetic field, the charged particles within the fluid experience a force perpendicular to both their motion and the magnetic field direction. This force, called the Lorentz force, acts against the fluid's flow, essentially resisting its movement. As shown in Fig. 8, fluid velocity declines as inverse porosity parameter upsurges.

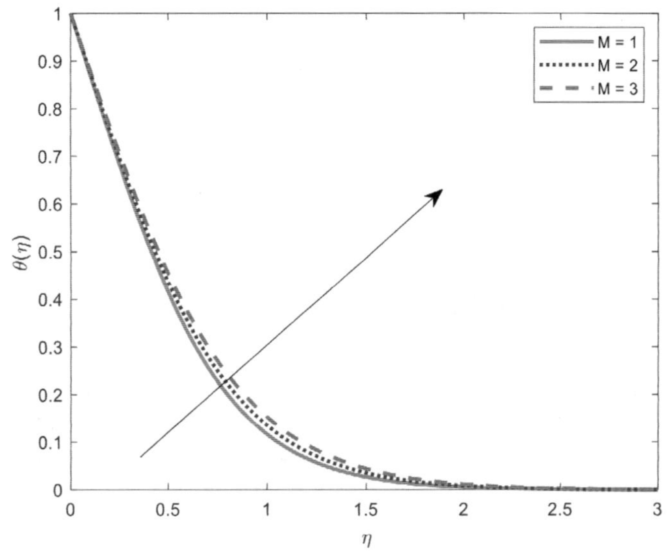

Figure 3. Scenario in which $\theta(\eta)$ is affected by M

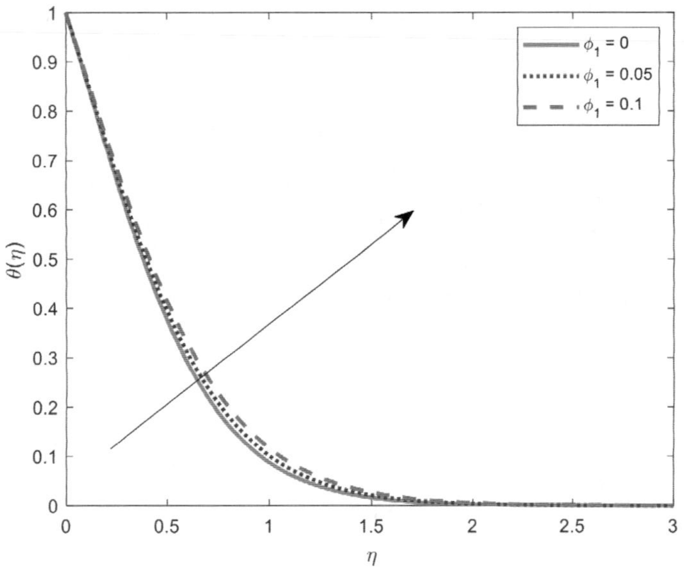

Figure 4. Scenario in which $\theta(\eta)$ is affected by ϕ_1

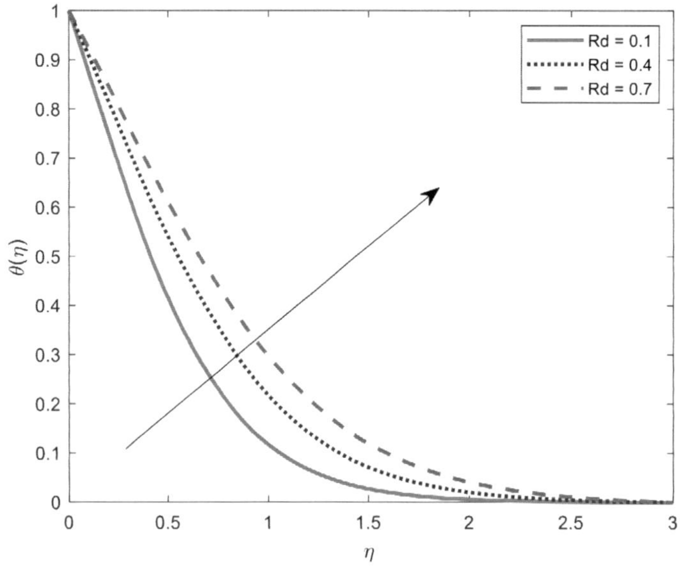

Figure 5. Scenario in which $\theta(\eta)$ is affected by Rd

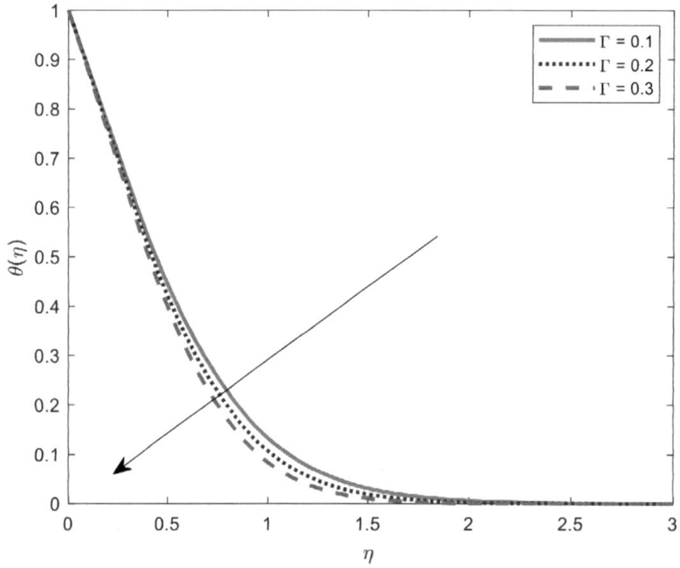

Figure 6. Scenario in which $\theta(\eta)$ is affected by Γ

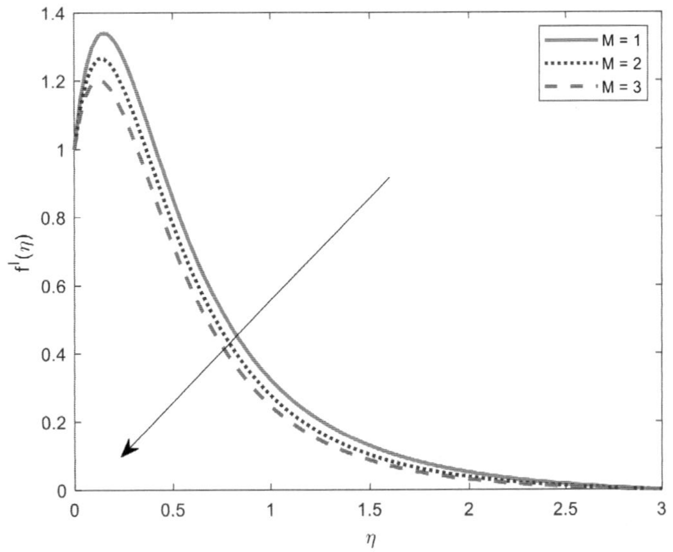

Figure 7. Scenario in which $f'(\eta)$ is affected by M

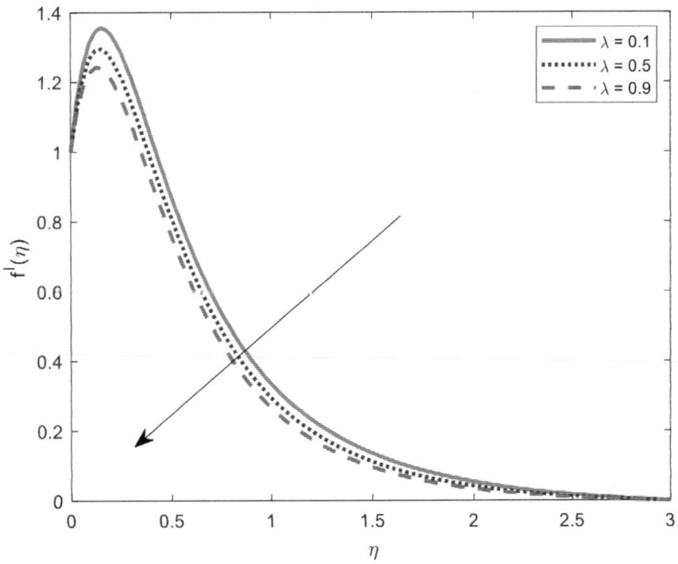

Figure 8. Scenario in which $f'(\eta)$ is affected by λ

3.2. *Irreversibility Analysis*

It is evident from the observation of Fig. 9 that an increase in Br values result in a corresponding augmentation in the rate of entropy formation. It is evident from the observation of Fig. 10 that an augmentation in the parameter ϕ leads to a reduction in the formation of entropy. Figs. 11–12 show that as Br increases, the Bejan number declines, and as the volume fraction of nanoparticle parameter upsurges, the Bejan number rises.

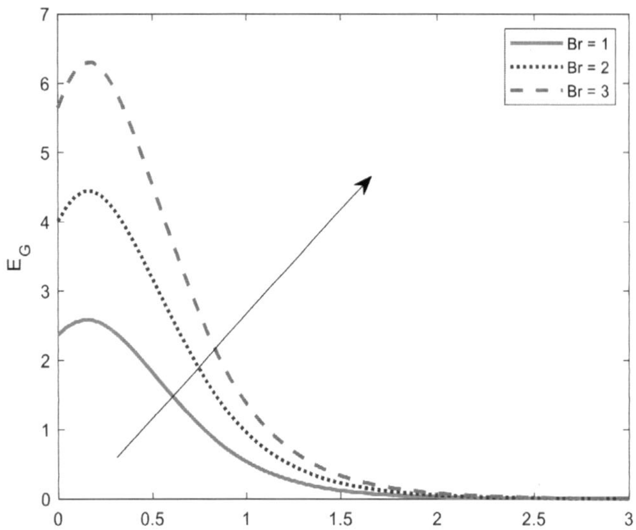

Figure 9. Scenario in which E_G is affected by Br

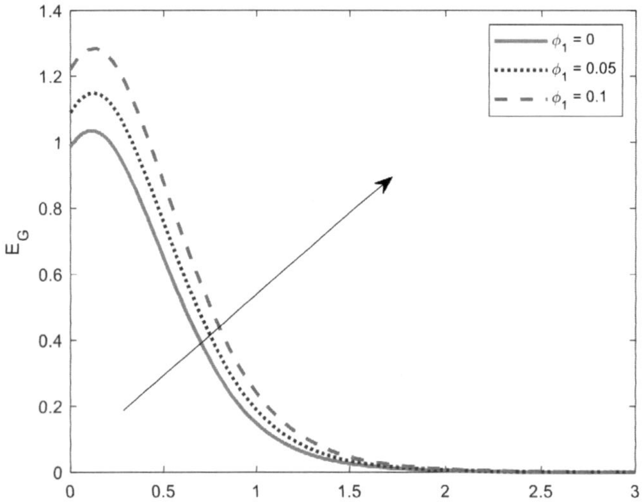

Figure 10. Scenario in which E_G is affected by ϕ_1

IOP Conf. Series: Journal of Physics: Conf. Series **2765** (2018) 012020 doi:10.1088/1742-6596/2765/1/012020

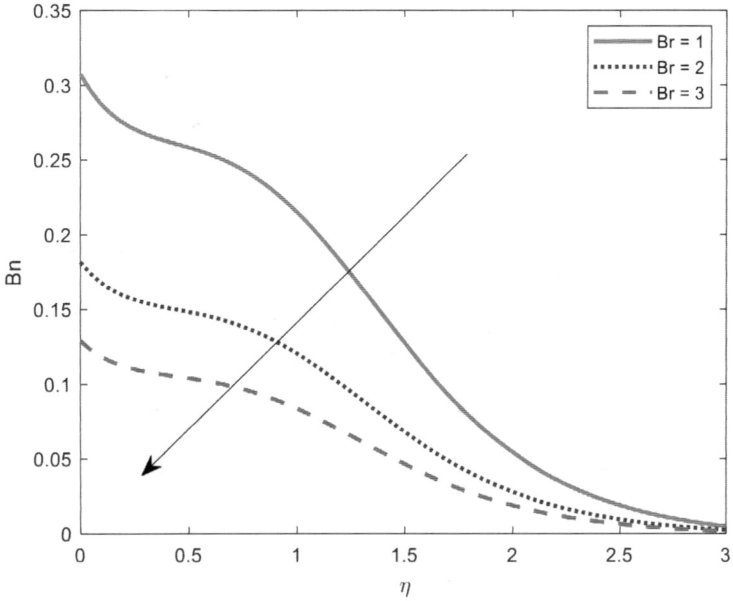

Figure 11. Scenario in which Bn is affected by Br

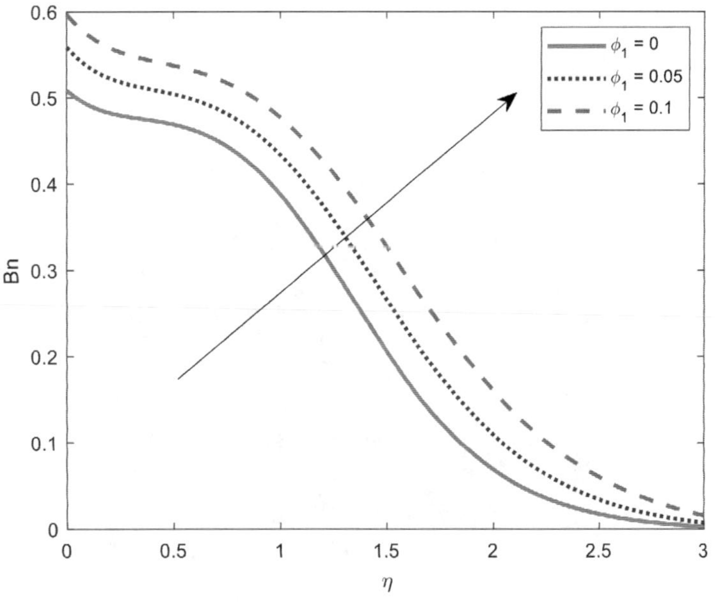

Figure 12. Scenario in which Bn is affected by ϕ_1

3.3. *Nusselt number and Skin Friction Coefficient Profiles*

It is clear from a closer inspection of Figs. 13-14, that as friction factor enhances with the escalation in ϕ_1, and the same declines as M raises. Within the range of $0 \le M \le 3.5$, it is seen that the friction factor exhibits a decline with a notable rate of 1.41843. It has been shown that Γ and M exert distinct influences on the rate of heat transmission (Figs. 15-16). It is detected that the Nusselt number enhances by 0.700996 (when $0 \le \Gamma \le 0.6$) and the same declines by 0.14086 (when $0 \le M \le 3.5$).

IOP Conf. Series: Journal of Physics: Conf. Series **2765** (2018) 012020 doi:10.1088/1742-6596/2765/1/012020

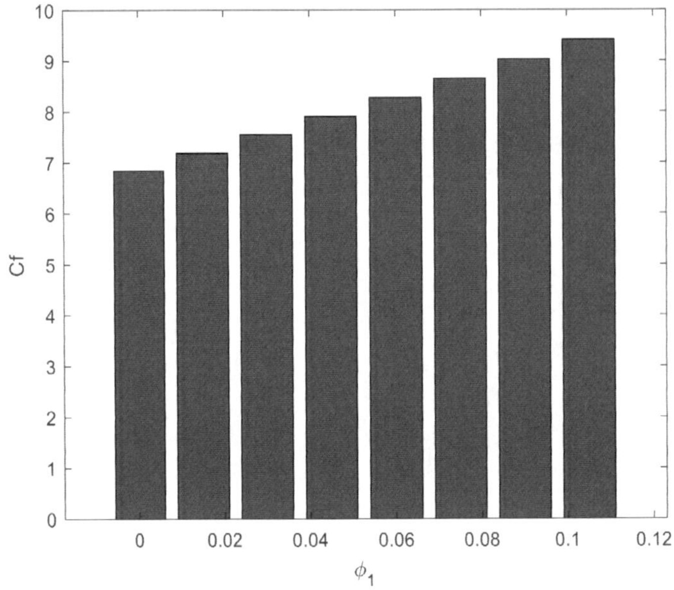

Figure 13. Scenario in which Cf is affected by ϕ_1

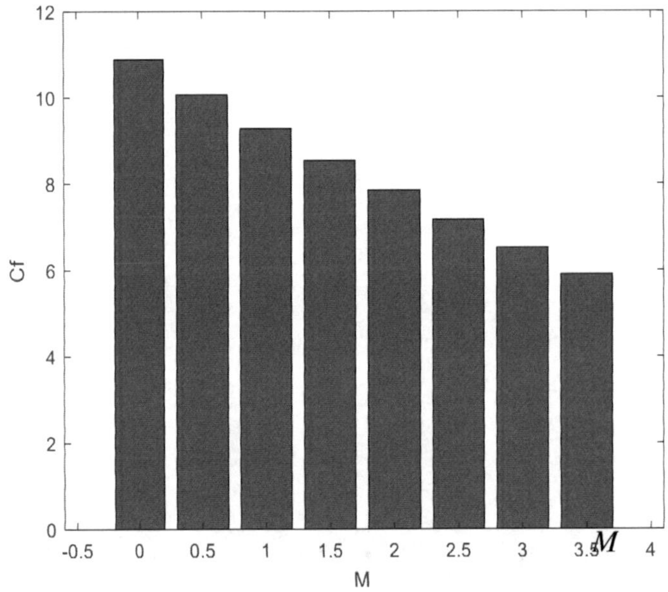

Figure 14. Scenario in which Cf is affected by

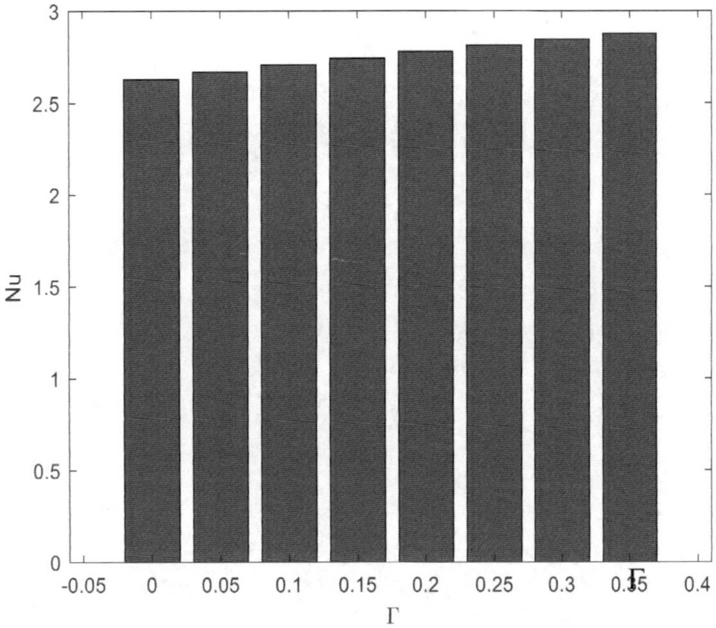

Figure 15. Scenario in which Nu is affected by

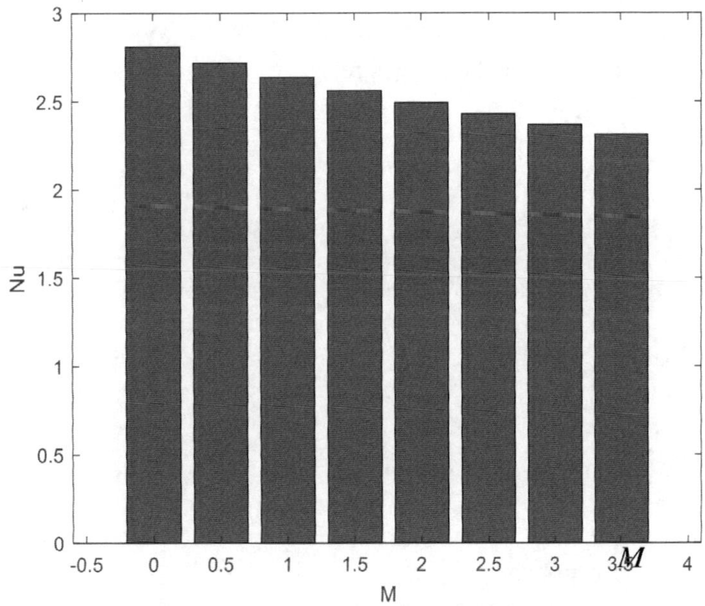

Figure 16. Scenario in which Nu is affected by

3.4. Multiple linear regression

Multiple linear regression, a statistical technique, delves into the intricate interplay between a dependent variable, which represents the desired prediction, and a multitude of independent variables, which encompass the potential factors that could exert impact on the outcome. The objective is to determine a linear equation that optimally corresponds to the data, enabling the ability to forecast the dependent variable by considering the values of the independent variables. To calculate multiple linear regression, use this formula:

$\hat{y} = \beta_0 + \beta_1 x_1 + \beta_2 x_2 + ... + \beta_p x_p$,

where,

$\hat{y}$ represents the predicted value of the dependent variable,

β_0 represents the intercept, or the value of the dependent variable when all independent variables are 0,

$\beta_1, \beta_2, ..., \beta_p$ represent the regression coefficients, which indicate the change in the dependent variable associated with a one-unit change in each independent variable,

$x_1, x_2, ..., x_p$ represent the independent variables.

This study employed the following models to identify correlations between important engineering properties, such as Nusselt number, and certain factors, such as the thermal relaxation parameter.

$$Cf = p_0 + p_1\phi_1 + p_2 M \tag{13}$$

$$Nu = q_0 + q_2\Gamma + q_3 M \tag{14}$$

In order to facilitate the process of drawing conclusions, we employed a total of 25 distinct sets of alternative values for each computation.

$$Cf = 8.2156 + 25.1241\phi_1 - 1.4093M \tag{15}$$

$$Nu = 2.7732 + 0.691\Gamma - 0.1397M \tag{16}$$

According to equation (15), rise in M lessens the friction factor (see Fig. 14). And also, it is discovered that there is an enhancement in the same with the rise in ϕ_1 (see Fig. 13). It is perceived that, equation (16) displayed the context that the thermal relaxation parameter had a positive impact on heat transmission rate (see Fig.15). Furthermore, it is observed that, when there is an increment in M, there is a reduction in the same (see Fig. 16).

4. Conclusions

The primary aim of this study is to conduct a comprehensive theoretical analysis on the behaviour of a hybrid nanofluid under the influence of quadratic thermal radiation and non-Fourier heat flux. The flow occurs through an exponentially extended sheet and is subjected to magnetic field and entropy generation. The findings of the research are briefly articulated as follows:

- It is detected that higher Brinkman number values lead to a greater rate of entropy generation and as the volume fraction of nanoparticle parameter upsurges, the Bejan number rises.
- Friction factor enhances with the escalation in ϕ_1, and the same declines as M raises.
- At $0 \le M \le 3.5$, the skin friction coefficient is seen to drop at a rate of 1.41843.
- It is evident that Γ and M both have different effects on the Nusselt number.
- It is discovered that the Nusselt number enhances by 0.700996 (when $0 \le \Gamma \le 0.6$) and the same declines by 0.14086 (when $0 \le M \le 3.5$).
- An upsurge in the thermal radiation parameter causes the fluid's temperature to rise.
- A drop in fluid velocity is observed as the inverse porosity parameter rises.
- there is a discernible augmentation in the fluid temperature as the value of Rd rises.

Nomenclature

M	Magnetic field Parameter	σ^*	Stefan-Boltzmann constant
Pr	Prandtl number	$\Gamma^* (= \Gamma_0 x)$	Thermal relaxation time of heat

υ	Kinematic viscosity		Flux
ϕ_1	Volume fraction of ZnO	k^*	Mean absorption coefficient
k	Thermal conductivity	Rd	Thermal radiation parameter
ϕ_2	Volume fraction of TiO_2	t	Temperature ratio parameter
Γ	Thermal relaxation parameter	Br	Brinkman number
K, d	Constants and $d > 0$ for liquids	τ, τ^*	Convection parameters (mixed and nonlinear)
ρ	Density	Re_x	Reynolds number
β_T, β_{1T}	Thermal expansion coefficients	σ	Electrical conductivity
α	Temperature difference parameter		
λ	Inverse porosity parameter		

References

[1] M. Venkata Subba Rao, K. Gangadhar, and P. L. N. Varma, "A spectral relaxation method for three-dimensional MHD flow of nanofluid flow over an exponentially stretching sheet due to convective heating: an application to solar energy," *Ind. J. Phys.,* vol. 92, No. 12, pp. 1577-1588, Dec 2018. DOI: https://doi.org/10.1007/s12648-018-1226-0.

[2] M.S. Hussain, R. Sharma, G.S. Seth, and M.R. Mishra, "Thermal radiation impact on boundary layer dissipative flow of magneto-nanofluid over an exponentially stretching sheet, " *Int. J. Heat Technol.,* vol. 36, no. 4, pp. 1163-1173, Dec 2018. DOI: https://doi.org/10.18280/ijht.360402.

[3] N.S. Anuar, N. Bachok, N. M. Arifin, H. Rosali, and I. Pop, "Stagnation-point flow and heat transfer over an exponentially stretching/shrinking sheet in hybrid nanofluid with slip velocity effect: stability analysis," *J. Phys. Conf. Ser.,* vol. 1366, no. 1, pp. 012002. Nov 2019. DOI: 10.1088/1742-6596/1366/1/012002.

[4] S. Ghosh, and S. Mukhopadhyay, "Effects of slip on Cu–water or Fe 3 O 4–water nanofluid flow over an exponentially stretched sheet," *Pramana,* vol. 92, no. 6, pp. 93, Jun 2019. DOI: https://doi.org/10.1007/s12043-019-1754-y.

[5] I. Waini, A. Ishak, and I. Pop. "Hybrid nanofluid flow induced by an exponentially shrinking sheet," *Chin. J. Phys.,* vol. 68, pp. 468-482, Dec 2020. DOI: https://doi.org/10.1016/j.cjph.2019.12.015.

[6] I. Rashid, M. Sagheer, and S. Hussain, "Exact solution of stagnation point flow of MHD Cu–H2O nanofluid induced by an exponential stretching sheet with thermal conductivity," *Phys. Scr.,* vol. 95, no. 2, pp. 025207, Jan 2020. DOI: https://doi.org/10.1088/1402-4896/ab444d.

[7] A.S. Khan, He-Yong Xu, and W. Khan, "Magnetohydrodynamic hybrid nanofluid flow past an exponentially stretching sheet with slip conditions," *Mathematics,* vol. 9, no. 24, pp. 3291, Dec 2021. DOI: https://doi.org/10.3390/math9243291.

[8] I. Waini, A. Ishak, and I. Pop, "Hybrid nanofluid flow towards a stagnation point on an exponentially stretching/shrinking vertical sheet with buoyancy effects," *INT J NUMER METHOD H.,* vol. 31, no. 1, pp. 216-235, Jan 2021. DOI: https://doi.org/10.1108/HFF-02-2020-0086.

[9] K.S.S. Babu, A. Parandhama, and R. Bhuvana Vijaya. "Non-linear MHD convective flow of Carreau nanofluid over an exponentially stretching surface with activation energy and viscous dissipation," *S.N Appl. Sci..* vol. 3, no. 3, pp. 382, Mar 2021. DOI: https://doi.org/10.1007/s42452-021-04339-4.

[10] K.S.S. Babu, A. Parandhama, and R. Bhuvana Vijaya, "Significance of heat source/sink on the radiative flow of Cross nanofluid across an exponentially stretching surface towards a stagnation point with chemical reaction," *Heat Transf.* vol. 51, no. 4, pp. 2885-2904, Jun 2022. DOI: https://doi.org/10.1002/htj.22428.

[11] A.S. Oke, B. C. Prasannakumara, W. N. Mutuku, RJ Punith Gowda, B. A. Juma, R. Naveen Kumar, and O. I. Bada, "Exploration of the effects of Coriolis force and thermal radiation on water-

based hybrid nanofluid flow over an exponentially stretching plate," *Sci. Rep.,* vol. 12, no. 1, pp. 21733, Dec 2022. DOI: https://doi.org/10.1038/s41598-022-21799-9.

[12] T.S. Neethu, A. S. Sabu, A. Mathew, A. Wakif, and S. Areekara, "Multiple linear regression on bioconvective MHD hybrid nanofluid flow past an exponential stretching sheet with radiation and dissipation effects," *Int. Commun. Heat Mass Transf.,* vol. 135, pp.106115, Jun 2022. DOI: https://doi.org/10.1016/j.icheatmasstransfer.2022.106115.

[13] S.B. Ramakrishna, S.R. Avula, M.U. Sadananda, J.B. Macherla, and S.K. Raju Chakravarthula, "Significance of non-Fourier flux on a chemically reactive ternary hybrid nanofluid flow (water+ Al2O3+ ZnO+ Fe3O4) by a quadratically radiated tended elongating surface," *Z Angew Math Mech.,* vol. 102, no. 12, pp. e202200103, Dec 2022. DOI: https://doi.org/10.1002/zamm.202200103.

[14] G. Revathi, K.R. Gujjula, and M. Jayachandra Babu, "Influence of shape factor on a chemically reactive hybrid nanofluid flow via a moving plate when Soret and Dufour effects are significant: An irreversibility analysis with Cattaneo-Christov heat flux model," *NUMER HEAT TR A-APPL.,* pp. 1-19, Oct 2023. DOI: https://doi.org/10.1080/10407782.2023.2266570.

[15] A.M. Alqahtani, M. Bilal, M. Usman, T.R. Alsenani, Aatif Ali, and Samy Refahy Mahmuod, "Heat and mass transfer through MHD Darcy Forchheimer Casson hybrid nanofluid flow across an exponential stretching sheet," *Z Angew Math Mech.,* pp. e202200213, Jan 2023. DOI: https://doi.org/10.1002/zamm.202200213.

[16] K. Ramesh, K.K. Asogwa, T. Oreyeni, M. Gnaneswara Reddy, and A. Verma, "EMHD radiative titanium oxide-iron oxide/ethylene glycol hybrid nanofluid flow over an exponentially stretching sheet," *Biomass Convers. Biorefin.,* 1-10, Mar 2023. DOI: https://doi.org/10.1007/s13399-023-04033-y.

[17] P. Chandrakala, "Effect of Heat and Mass Transfer over Mixed Convective Hybrid Nanofluids past an Exponentially Stretching Sheet," *CFD Letters,* vol. 16, no. 3, pp. 125-140, Mar 2024. DOI: https://doi.org/10.37934/cfdl.16.3.125140.

[18] T. Gupta, A. Kumar Pandey, and M. Kumar, "Effect of Thompson and Troian slip on CNT-Fe3O4/kerosene oil hybrid nanofluid flow over an exponential stretching sheet with Reynolds viscosity model," *Mod. Phys. Lett. B.,* vol. 38, no. 02, 2350209, Jan 2024. DOI: https://doi.org/10.1142/S0217984923502093.

[19] M. Venkateswarlu, O. D. Makinde, and D. V. Lakshmi, "Influence of thermal radiation and heat generation on steady hydromagnetic flow in a vertical micro-porous-channel in presence of suction/injection," *J. Nanofluids.,* vol. 8, no. 5, pp. 1010-1019, May 2019. DOI: https://doi.org/10.1166/jon.2019.1647.

[20] F. Mabood, S. M. Ibrahim, P. V. Kumar, and G. Lorenzini, "Effects of slip and radiation on convective MHD Casson nanofluid flow over a stretching sheet influenced by variable viscosity," *J. Eng. Thermophys.,* vol. 29, pp. 303-315, Apr 2020. DOI: https://doi.org/10.1134/S1810232820020125.

[21] G.R. Rajput, B.P. Jadhav, and S. N. Salunkhe, "Magnetohydrodynamics boundary layer flow and heat transfer in porous medium past an exponentially stretching sheet under the influence of radiation," *Heat Transf.,* vol. 49, no. 5, pp. 2906-2920, Jul 2020. DOI: https://doi.org/10.1002/htj.21752.

[22] I.C. Mandal, Swati Mukhopadhyay, and Kuppalapalle Vajravelu, "Melting heat transfer of MHD micropolar fluid flow past an exponentially stretching sheet with slip and thermal radiation," *Int. J. Appl. Comput. Math.,* vol. 7, no. 31, pp. 1-18, Apr 2021. DOI: https://doi.org/10.1007/s40819-021-00955-1.

[23] M.I. Asjad, Noman Sarwar, Bagh Ali, Sajjad Hussain, Thanin Sitthiwirattham, and Jiraporn Reunsumrit, "Impact of bioconvection and chemical reaction on MHD nanofluid flow due to exponential stretching sheet," *Symmetry,* 13, no. 12, pp. 2334, Dec 2021. DOI: https://doi.org/10.3390/sym13122334.

[24] V.S. Sajja, R. Gadamsetty, P. Muthu, M. Jayachandra Babu, and I. L. Animasaun, "Significance of Lorentz force and viscous dissipation on the dynamics of propylene glycol: Water subject to joule heating conveying paraffin wax and sand nanoparticles over an object with a variable thickness," *Arab J Sci Eng.,* 47, no. 12 (2022): 15505-15518, Dec 2022. DOI: https://doi.org/10.1007/s13369-022-06658-z.

[25] H. Waqas, U. Farooq, D. Liu, M. Abid, M. Imran, and T. Muhammad, "Heat transfer analysis of hybrid nanofluid flow with thermal radiation through a stretching sheet: A comparative study," *Int. Commun. Heat Mass Transf.,* vol. 138, pp. 106303, Nov 2022. DOI: https://doi.org/10.1016/j.icheatmasstransfer.2022.106303.

[26] B. Shankar Goud, Y. Dharmendar Reddy, and S. Mishra, "Joule heating and thermal radiation impact on MHD boundary layer Nanofluid flow along an exponentially stretching surface with thermal stratified medium," *Proc. Inst. Mech. Eng.,* vol. 237, no. 3-4, pp. 107-119, Sep 2023. DOI: https://doi.org/10.1177/23977914221100961.

[27] G. Revathi, G. Veeram, M. Jayachandra Babu, KS Srinivasa Babu, and A. Suneel Kumar, "Darcy–Forchheimer flow of power-law (Ostwald-de Waele type) nanofluid over an inclined plate with thermal radiation and activation energy: an irreversibility analysis," *Int. J. Ambient Energy,* pp. 1-10, Apr 2023. DOI: https://doi.org/10.1080/01430750.2023.2200434.

[28] K. Karthik, J.K. Madhukesh S. Kiran, K.V. Nagaraja, B. C. Prasannakumara, and Gamaoun Fehmi, "Impacts of thermophoretic deposition and thermal radiation on heat and mass transfer analysis of ternary nanofluid flow across a wedge," *Int. J. Model. Simul.,* pp. 1-13, Jan 2024. DOI: https://doi.org/10.1080/02286203.2023.2298234.

[29] G. Revathi, M. Jayachandra Babu, K.S. Srinivasa Babu, and C. Bapanayya, "Dynamics of different heat sources and activation energy on the hybrid nanofluid (EG+ MgO+ MWCNT) flow in a microchannel with thermal radiation: An irreversibility analysis," *Numer. Heat Transf.; A: Appl.,* pp. 1-23, Feb 2024. DOI: https://doi.org/10.1080/10407782.2024.2316222.

[30] D. Purna Chandar Rao, S. Thiagarajan, and V. Srinivasa Kumar, "Significance of quadratic thermal radiation on the bioconvective flow of Ree-Eyring fluid through an inclined plate with viscous dissipation and chemical reaction: Non-Fourier heat flux model," *Int. J. Ambient Energy.,* Vol. 43, no. 1, pp. 6436-6448, Dec 2022. DOI: https://doi.org/10.1080/01430750.2021.2023043.

[31] W. Jamshed and A. Aziz, "Cattaneo–Christov based study of TiO 2–CuO/EG Casson hybrid nanofluid flow over a stretching surface with entropy generation," *Appl. Nanosci.,* vol. 8, no. 4, pp. 685-698, Apr 2018. DOI: https://doi.org/10.1007/s13204-018-0820-y.

[32] A. Jamaludin, R. Nazar, K. Naganthran, and I. Pop, "Mixed convection hybrid nanofluid flow over an exponentially accelerating surface in a porous media," *Neural. Comput. Appl.,* vol. 33, no. 22, pp.15719-15729, Nov 2021. DOI: https://doi.org/10.1007/s00521-021-06191-4.

Facile and an efficient one-pot synthesis of 2-arylbenzoxazoles promoted by transition metal oxide nano catalyst

Koppolu Prathap[1], ShaikLakshman[2], M.V. Lakshmi[3], B.V. Durgarao[4], D. V. L. Sirisha[5], N. Krishnarao[5]*
[1] Department of Chemistry, Government Degree College, Dornala, A.P, India
[2] Department of Chemistry, GITAM (Deemed to be University), Visakhapatnam, A.P, India.
[3] Department of Engineering Chemistry, SRKR Engineering College, Bhimavaram, A.P, India.
[4] Department of DFS (Chemistry), ONGC, RJY, AP, India.
[5] Department of Chemistry, PRISM P.G & D.G College, Visakhapatnam, A.P, India.
*Corresponding Author Email: naallakrishnarao@gmail.com,

Abstract: The advantages and an efficient path way of process for 2-arylbenzoxazole can be obtained by transition metal oxide Nano catalyst. The transition metal nanoTiO_2 catalysed and a tandem reaction between substituted 2-aminophenols and substituted aromatic aldehydes to produces scaffold series of 15 smoothly forming benzo heterocycles with good to excellent yields (80–94%).The synthesized benzoxazole compounds were confirmed by FT-IR, ^{1}H, ^{13}CNMR, and mass spectroscopy. This article describes a one-stage coupling technique that provides a unique way to access variously functionalized benzoxazole scaffolds. This methodology consists of a one-pot combination of condensation followed by oxidative addition reactions. It was shown that nano TiO_2 in ethanol encouraged the formation of 2-aryl benzoxazole scaffolds and this catalyst quickly boosted the cascade reaction that produced a variety of collections of named derivatives. These protocols are atom-economic, efficient, reusable and modular means of accessing restricted named derivatives that have a substantial substrate tolerance and over ten points of diversity.
Keywords: 2-aminophenol, substituted aromatic aldehydes,2-arylbenzoxazolesNanoTiO_2.

1.Introduction:

Over the past ten years, research and development in the fields of nanoscience and nanotechnology have been highly regarded. Nowadays, industries and business depend heavily on nanotechnology. A new chapter has been opened in application sectors such as optoelectronics, sensing, medicine, and catalysis due to effect of the chemical and physical features of nanoscale materials compared to macro and micro systems. In both homogeneous and heterogeneous catalysis, chemists have accomplished noteworthy feats. For an extended period, heterogeneous catalysis, which has the advantage of simple catalyst material removal and the ability to operate at elevated temperatures impact from deficiency of selectivity and comprehension of the mechanistic elements that are crucial for measurement enhancement.

The primary building blocks of synthetic organic chemistry are N-heterocyclic compounds, which are frequently regarded as preferred target molecules in the pharmaceutical industry for the development of lead molecules with certain activities. Therefore, it has always been interesting to watch the development of increasingly effective and adaptable techniques for synthesizing N-heterocycles in organic synthesis. In particular, benzoxazole derivatives have gathered a plenty of application in various field covered by medicinally chemistry agrochemicals, natural products and material science

The benzoxazole are fused benzene and five membered heterocyclic having two hetero atoms like" Nitrogen and oxygen" atoms. These moieties occur in nature as well as synthetic compounds and preparing single step by two components [1-4]. The biological features exhibited by molecules containing benzoxazole moieties make them desirable candidates for synthesis. The synthesis of benzoxazoles has been the subject of many attempts due to their widespread presence in pharmaceuticals [5,6], natural products, and functional materials. The pharmacological activity of these heterocycles include Antimicrobial agents [7-8], anti-inflammatory, analgesic [6,9], Alzheimer, anticancer [10-11], Antiprotozoal Agents[12], A distinct and extensive class of antirhino / enteroviral agents, benzomidazole derivatives have the ability to selectively block the neuropeptide Y Y1 receptor and have anti-ulcerative and anti-allergic properties[13-15]. They have also been shown to be effective against the human cytomegalovirus.

The majority of benzoxazole synthesis methods that have been reported have several drawbacks, including the use of hazardous solvents, elevated reaction temperatures, prolong reaction time with low yield of the product, extension of work-up procedures, and the application of chemicals that are rather expensive. Because of this, there has been a constant interest in catalyst research. The improvement of catalyst is being sought to create benzoxazole along with straightforward work-up process, minimized toxicity, more recovery and reusability, safety for around environments and atom economy. For the intended synthesis, a variety of organic development reaction has also been used metal.

The synthesis of series of 2-substituted benzoxazole can be accomplished primarily in two ways. First one is the reaction coupling of acyl chloride and derivatives of carboxylic acid with o-substituted amino aromatics, which is either microwave- or strong acid-catalyzed. The another process is the oxidative cyclization of Phenolic imine group which are produced when aldehydes and o-substituted amino aromatics condense. The conventional method of thermal condensation of 2-aminophenols with substituted aromatic aldehydes, is one of the techniques that have been employed to gain access to benzoxazoles promoted by nano catalyst. Recently synthesis of benzoxazole promoted by Nano materials $SiO_2/MgSnO_3$[16], Nickel Supported Silica [17], nano-solid acid [18], Nano-$NiFe_2O_4$[19], ZnS nano [20], nano ZnO [21], Ag@TiO_2 Nanocomposite [22], SBA-15asananocatalyst [23], nanoSnO_2 [24], nanoPd/Fe_3O_4/ZnO [25], Magnetic Nanoparticles [26].

The majority of these documented techniques have numerous shortcomings when it comes to the synthesis of benzoxazole derivatives. Some of these shortcomings include the use of hazardous solvents during reactions; prolong time of reaction with lower product yields, laborious work-up procedures, and the need for costly reagents, which require more attention to overcome. Transition-metal nanoparticles (NPs) are materials with tens to thousands of metal atoms stabilized as surface-protecting agents by ligands, polymers, or dendrimers. TiO_2 was the most promising of the several types of metal oxide nanoparticles. TiO_2 exhibits exceptional electrical and optical characteristics, including low resistivity, unusual optical transparency, and high theoretical specific capacity of a stable, n-type broad band gap semiconductor.

For the intended synthesis, a variety of organic moieties transformations have also been used to transition metal oxide as a catalyst function. In particularly, TiO_2Nano catalysts have been applied to a variety of chemical processes, including epoxidation, dehydrogenation, isomerization, and dehydrogenation. In addition, a variety of fresh and creative approaches have been devised, which have simplified the formerly difficult process of benzoxazole core synthesis. Hence, the purpose of TiO_2 metal oxide as a mesoporous heterogeneous catalyst, we report here a straightforward and effective process for the improvement of substituted2-arylbenzoxazole from 2-aminophenol and substituted aromatic aldehydes (**Scheme-1**). According to our model research, the quantity of metal oxide as catalyst is required for this reaction to happen. a quick response time very few minutes.

2.Experimental:

All chemicals (AR grade) were used without additional purification after being acquired from commercial sources. An Agilent GC-6820 chromatograph fitted with a flame ionization detector and an HP-Innowax (30 m × 0.32 mm × 0.5 μ m) capillary column was used to perform gas chromatography (GC) analysis. GC-MS spectra were captured using a TRB-5MS (30 m × 0.25 mm × 0.25 μ m) column on a Thermo Trace DSQ GC-MS spectrometer. The melting points are uncorrected and were calculated with a Yamato melting point instrument Model MP-21. A Bruker DRX 500 (500-MHz) spectrometer was used to produce 1 H NMR spectra in $CDCl_3$, using TMS as the internal standard. TLC was used to track the reactions' progress using silica-gel polygrams (SIL G/UV 254 plates). Silicycle (40–60 mm) silica gel was used for column chromatography. Items were recognized.

2.1. Materials and Methods

Titanium isopropoxide [Ti{$OCH(CH_3)_2$}$_4$], Ethanol (C_2H_5OH), were purchased from Sigma Aldrich Chemicals (India) Ltd. All the reagents were analytically grade and used

2.2. General procedure for the synthesis of 2-aryl substituted benzoxazole:

The starting materials 2-amino phenol (1.2 mmol), aromatic aldehydes (1.5mmol) are dissolved in ethanol (15ml) and introduced in 50mL of RBF. The catalyst TiO_2(10m mol %) in added into above solution gradually and continued stirring at 70^{0}C for appropriate time (2h). The development of the reaction was identified thin layer chromatography (TLC) Hexane: EtOAc (5:5)and observed until reaction was completed. The reaction mixture was cooled in crushed ice and neutralized with a saturated solution of $NaHCO_3$. The crude was extracted with EtOAc (2x100) and that organic layer was washed with water (2x100) and followed by brine (1x100) and dried over Na_2SO_4 and evaporated under vacuum distillation. Obtained crude residue was purified by column chromatography to give derivatives of 2-substituted benzoxazole.

2.3. Synthesis of TiO_2 nanoparticles:

9 mL of TTIP (Titanium tetra isopropoxide) is added drop wise to the homogeneous mixture of 150 mL of ethanol and 3.72 mL of distilled water. This mixture is stirred for half an hour and maintained at 85˚C for 4 h. This solution was kept under continuous magnetic stirring and later it was sent to centrifugation and finally dried at 60˚C in a hot air oven for 30 min. Finally, it was calcined at 400˚C for 3 h to yield Anatase TiO2 nanoparticles.

1. **2-phenylbenzo[d]oxazole (3a):**

Colourless solid; ^{1}H NMR (400MHz, $CDCl_3$) δ ppm: 8.145–7.841 (m, 2H,Ar-H), 7.754–7.280 (m, 7H, Ar-H); ^{13}C NMR (100 MHz, $CDCl_3$) δ ppm: 159.23, 148.28, 140.55, 130.06, 128.88, 128.24, 127.67, 124.75, 122.46, 117.34, 112.43; LCMS (m/z) = 196.49 [M+H]. Molecular formulae: $C_{13}H_9NO$: C- 79.98; H- 4.65; N- 7.17. Found : C- 79.91; H- 4.64; N- 7.25.

2. **4-(benzo[d]oxazol-2-yl) phenol (3b);**

Colourless solid; ^{1}H NMR (400MHz, $CDCl_3$) δ ppm: 9.127(s, 1H,-OH), 7.885–7.718 (m,2H, Ar-H), 7.841–7.296 (m, 6H, Ar-H);^{13}C NMR (100 MHz, $CDCl_3$) δ ppm: 162.07, 156.28, 147.79, 140.09, 128.02, 125.41, 123.22, 121.01, 118.27, 111.38.LCMS (m/z) = 211.47 [M+]. Molecular formulae: $C_{13}H_9NO_2$: C- 73.92; H- 4.30; N- 6.63. Found : C- 73.85,H- 4.28;N- 6.69.

3. **2-(3, 4, 5-trimethoxyphenyl) benzo[d]oxazole (3c):**

Colourless solid; ^{1}HNMR (400MHz, $CDCl_3$) δ ppm: 7.710–7.7325 (m, 6H, Ar-H), 3.743 ((OCH_3)2, 6H, s), 3.607 (OCH_3, 3H, s); ^{13}C NMR (100 MHz, $CDCl_3$) δppm:160.09, 157.11, 148.37, 140.05, 138.24, 122.07,120.18, 107.45, 57.45, 53.72;.LCMS (m/z) = 286.17 [M+H]. Molecular formulae: $C_{16}H_{15}NO_4$: C- 67.36; H- 5.30; N- 4.91. Found : C- 67.30; H- 5.28;N- 4.98.

4. **2-(3,5-dichlorophenyl)benzo[d]oxazole (3d):**

Colourless solid; ^{1}HNMR (400MHz, $CDCl_3$) δ ppm: 7.971–7.7755 (m, 2H, Ar-H), 7.254 (Ar-H, 1H, s), 7.621-7.315(Ar-H, 4H, s); 13C NMR (100 MHz, $CDCl_3$) δ ppm: 163.21, 149.05, 140.24, 132.87, 128.79, 128.15, 127.76, 124.09, 122.88, 120.02, 111.61;.LCMS (m/z) = 264.32 [M+2]. Molecular formulae: $C_{13}H_7Cl_2NO$: C- 59.12; H- 2.67; N- 5.30. Found: C- 59.05; H-2.65, N- 5.36

5. **2-(4-nitrophenyl) benzo[d]oxazole (3e):**

Colorless solid; ^{1}HNMR (400MHz, $CDCl_3$) δppm: 8.258(d, J= 8.8Hz, 2H,Ar-H), 8.046(d,J=10.2Hz,Ar-H), 7.547–7.732 (m, 4H, Ar-H), ^{13}C NMR (100 MHz, $CDCl_3$) δppm:.164.28, 149.74, 146.31, 140.08, 135.11, 129.46, 128.68, 124.15, 122.86, 120.08, 112.65; LCMS (m/z) = 240.73 [M+]. Molecular formulae: $C_{13}H_8N_2O_3$: C- 65.00; H- 3.36; N- 11.66. Found: C- 64.93, H-3.35; N-11.71

6. **6-methoxy-2-phenylbenzo[d]oxazole (3f):**

Colourlesssolid;^{1}HNMR (400MHz, $CDCl_3$) δ ppm: 8.145(d, J= 8.0Hz, 2H, Ar-H), 7.604-7.128(Ar-H,5H,m),6.871(s,Ar-H,1H),3.773(OCH3,3H,s);13CNMR(100MHz,$CDCl_3$)δppm:158.65,156.06, 148. 39,130.38,129.46,129.09,128.77,128.13,120.05,115.52,99.19,54.07;;LCMS(m/z)=226.66 [M+H]. Molecular formulae: $C_{14}H_{11}NO_2$: C- 76.65; H- 4.92; N-6.22. Found : C- 76.59, H-4.90,N-6.27.

7. **4-(6-methoxybenzo[d]oxazol-2-yl) phenol (3g):**

Colourless solid; ^{1}HNMR (400MHz, $CDCl_3$) δ ppm: 9.247(-OH,s,1H), 7.784(d, J= 9.2Hz, 2H,Ar-H), 7.517(d, J=8.0Hz,1H,Ar-H), 7.124 (Ar-H,1H, d=6.4Hz), 6.943 (Ar-H,s,1H), 6.805 (d, J= 8.0Hz, 2H,Ar-H), 3.675 (OCH3, 3H,s); ^{13}C NMR (100 MHz, $CDCl_3$) δppm:.159.11, 154.26, 150.09, 146.72, 131.54, 124.46, 121.65, 118.77, 116.03, 113.41,98.35,56.29;LCMS (m/z) = 241.17 [M+]. Molecular formulae: $C_{14}H_{11}NO_3$: C- 69.70; H- 4.60; N-5.81.Found : C- 69.64, H-4.58;N-5.88.

8. 6-methoxy-2-(3,4,5-trimethoxyphenyl)benzo[d]oxazole (3h);

Colourless compound;^{1}HNMR (400MHz, $CDCl_3$) ppm: 7.567 (Ar-H, 1H, d, J= 8.8 Hz), 7.369 (d, J= 10.6, 2H, Ar-H), 7.152 (Ar-H, 1H, d, J= 6.8Hz), 6.974 (s, 1H, Ar-H),3.810(OCH_3, 3H, s),3.756((OCH_3)$_2$,6H,s),3.681 (OCH_3,3H,s),^{13}CNMR (100Hz,$CDCl_3$) ppm: 160.74, 156.43,150.65, 148.17, 138.04, 132.35, 126.67, 124.75, 120.03, 113.39, 101.05, 58.77, 54.65, 52.43. LCMS (m/z): 316.14 (M+H); Molecular formulae:$C_{17}H_{17}NO_5$; Analysis of elements: Calculated: C- 64.75, H- 5.43, N- 4.44, Found: C- 64.67, H- 5.42, N- 4.53.

9. 6-methoxy-2-(4-nitrophenyl) benzo[d]oxazole (3i):

Yellow compound; ^{1}HNMR (400MHz, $CDCl_3$) ppm: 8.254 (Ar-H, 2H, d, J= 7.2 Hz), 7.942 (d, J= 10.6, 2H, Ar-H), 7.457 (Ar-H, 1H, d, J= 6.4Hz), 7.241 (Ar-H, 1H, d,J=8.0Hz),6.875(s,1H,Ar-H),3.647(OCH3,3H,s);^{13}CNMR(100Hz,$CDCl_3$)ppm:163.71,158.04,150.69,146.43,137.29,135.36,132.06,128.96,128.47,121.18,115.83,102.55,54.35;LCMS(m/z):370.56(M^+);Molecular formulae: $C_{14}H_{10}N_2O_4$; Analysis of elements: Calculated: C- 62.22, H- 3.73, N- 10.37, Found: C- 62.15, H- 3.71, N- 10.43.

10. 2-(3,5-dichlorophenyl)-6-methoxybenzo[d]oxazole (3j):

Pale-yellow compound; ^{1}HNMR (400MHz, $CDCl_3$) ppm: 7.872 (d, J= 8.0, 2H,Ar-H),7.681(Ar-H,1H,s),7.573(Ar-H,1H,d,J=7.6Hz),7.073(Ar-H,1H,d,J=10.6Hz),6.892(s,1H,Ar-H),3.708(OCH_3, 3H, s);^{13}CNMR(100Hz,$CDCl_3$)ppm:160.08,155.77,150.09,134.48,131.05,128.84,128.06,127.69,122.69,115.39,103.34,55.02;LCMS(m/z):295.29(M+2);Moleculaformulae:$C_{14}H_9Cl_2NO_2$; Analysis of elements: Calculated: C- 57.17, H- 3.08, N- 4.76, Found: C- 57.11, H- 3.06, N- 4.19.

11. 2-phenyl-6-(trifluoromethyl) benzo[d]oxazole (3k)

Colourlescompound;^{1}HNMR(400MHz,$CDCl_3$)ppm:8.058(Ar-H,s,1H),8.058(Ar-H,2H,d,J= 12.2Hz), 7.714(Ar-H,s,1H),7.612(Ar-H,2H,d,J=8.0Hz),7.547(Ar-H,1H,t=8.8Hz),7.367(Ar-H, 1H,d,J=8.0Hz), 7.145(Ar-H,1H,d=6.48Hz),^{13}CNMR(100Hz,$CDCl_3$)ppm:161.39,148.24,141. 65,131.33, 129.05, 128.76, 127.44, 125.75, 123.56, 121.66, 118.73, 110.09.LCMS (m/z): 262.14(M-2); Molecular formulae: $C_{14}H_8F_3NO$; Analysis of elements: Calculated: C- 63.88, H- 3.06, N- 5.32, Found: C- 63.81, H- 3.05, N- 5.39.

12. 4-(6-(trifluoromethyl) benzo[d]oxazol-2-yl) phenol (3l):

Colourlescompound;^{1}HNMR(400MHz,$CDCl_3$)ppm:9.547(-OH,1H,s),7.884(Ar-H,2H,d,J=7.6Hz), 7.693(Ar-H,s,1H),7.455(Ar-H,1H,d,J=8.0Hz),7.184(Ar-H,1H,d=8.8Hz),6.946(d, J=12.4Hz, 1H,Ar-H);^{13}CNMR(100Hz,$CDCl_3$)ppm:160.78, 155.65, 148.38, 128.46, 127.07, 124.02, 121.41, 118.06, 115.19, 108.04;.LCMS (m/z): 280.59(M+H); Molecular formulae: $C_{14}H_8F_3NO_2$; Analysis of elements: Calculated: C- 60.22, H- 2.89, N- 5.02, Found: C- 60.15, H- 2.87, N- 5.10.

13. 6-(trifluoromethyl)-2-(3,4,5-trimethoxyphenyl)benzo[d]oxazole (3m):

Colourlescompound;^{1}HNMR(400MHz,$CDCl_3$)ppm:7.648(Ar-H,s,1H),7.448(Ar-H,1H,d,J=7.2Hz), 7.315(Ar-H,1H,d,J=8.0Hz),7.147(Ar-H,1H,d,J=8.8Hz),3.747(OCH_3,3H,s),3.658(OCH_3,3H,s); ^{13}CNMR(100Hz,$CDCl_3$)ppm:163.76,150.07,147.64,142.27,138.15,128.22,127.53,125.03,123.48, 121.48, 119.67, 108.88, 58.91, 54.06;LCMS (m/z): 355.45(M+2); Molecular formulae: $C_{17}H_{14}F_3NO_4$; Analysis of elements: Calculated: C- 57.79, H- 3.99, N- 3.96, Found: C- 57.72, H- 3.97, N- 4.07.

14. 2-(3,5-dichlorophenyl)-6-(trifluoromethyl)benzo[d]oxazole (3n) :

Colourlescompound;1HNMR(400MHz,CDCl3)ppm:7.854(Ar-H,2H,d,J=6.0Hz),7.707(Ar-H, s,1H), 7.658(Ar-H,s,1H),7.341(Ar-H,1H,d,J=7.6Hz),7.057(Ar-H,1H,d=8.8Hz);^{13}CNMR(100Hz,$CDCl_3$) ppm:160.72,147.64,143.38,132.24,129.02,128.49,127.95,126.66,123.85,121.64,118.96,108.47; LCMS (m/z): 332.13(M+H); Molecular formulae: $C_{14}H_6F_3NO$; Analysis of elements: Calculated: C- 50.63, H- 1.82, N- 4.22, Found: C- 50.56, H- 2.26, N- 9.22.

15. 2-(4-nitrophenyl)-6-(trifluoromethyl) benzo[d]oxazole (3o):
Colourlescompound;^{1}HNMR(400MHz,$CDCl_3$)ppm:8.247(Ar-H,2H,d,J=9.6Hz),8.075 (Ar-H,2H,d, J= 8.8Hz),7.753(Ar-H,s,1H),7.412(Ar-H,1H,d,J=6.4Hz),7.178(Ar-H,1H,d=7.6Hz);^{13}CNMR (100Hz, $CDCl_3$) ppm:163.04, 149.26, 145.51, 142.68, 135.08, 129.44,128.96, 127.74, 123.33, 121.71, 119.46, 109.35;LCMS (m/z): 309.15(M+H); Molecular formulae: $C_{14}H_7F_3N_2O_3$; Analysis of elements: Calculated: C- 54.56, H- 2.29, N- 9.09, Found: C- 54.50, H- 2.27, N- 9.14.

Results and discussion:

XRD analysis:

The preliminary characterization of TiO_2 NPs was confirmed by XRD (Figure 1). XRD peaks for obtained adsorbent with diffraction angles at 2θ = 25.28°, 37.5°, 47.5°, 55°, 62.4°, 70°, and 76.5° respectively. Sharp peaks at diffraction angles of 25.28°, 37.5°, 47.5°, and 55° in XRD pattern confirmed crystalline structure of Anatase a mineral form of TiO_2 NPs [27]. Further, the size of the nanoparticles was calculated by using Debye–Scherrer $D = K\lambda / \beta \cos\theta$, where D represents crystalline size, K represents the Scherrer constant (0.98), λ denotes the wavelength (1.54Å), β denotes the full width at half maximum (FWHM). From this equation, the size of the TiO_2 NPs was found to be 24.1 nm.

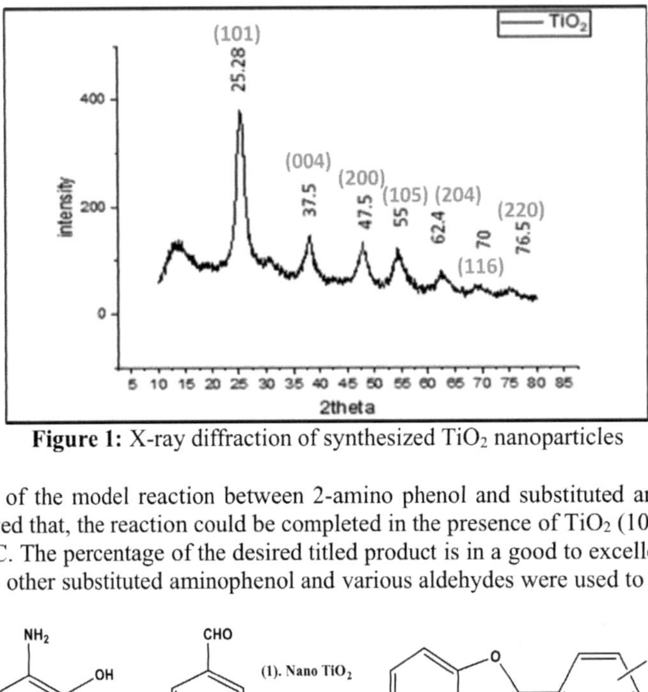

Figure 1: X-ray diffraction of synthesized TiO_2 nanoparticles

Our initial study of the model reaction between 2-amino phenol and substituted aromatic aldehydes (Scheme 1) showed that, the reaction could be completed in the presence of TiO_2 (10 mmol%), ethanol as solvent at 76^{0}C. The percentage of the desired titled product is in a good to excellent yields. In light of these findings, other substituted aminophenol and various aldehydes were used to explore the extent of the reaction.

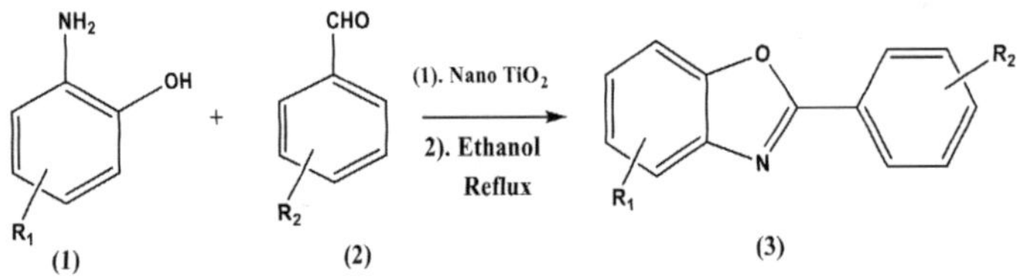

R_1 = H, OCH_3, CF_3

R_2 = H, 4-OH, 3,4,5$(OCH_3)_3$, 3,5$(Cl)_2$, 4-NO_2

(Scheme-1)

Initially, the reaction was monitored by various other nano catalysts in order to understand the influence of synthesized nano catalyst. The results were presented in table 1. From the table, it was observed that among the various catalysts were applied during the reaction and maintained constant temperature, the function of the nano TiO_2 catalyst was played an important role than the other nano catalyst viz; ZnO, MgO and SnO_2. The evidence of the data present in Table-1 reveals that the yield of titled derivatives was obtained in good to excellent by using TiO_2 nanoparticles.

Table 1: Comparison of the influence of TiO_2 with that of other catalysts in the synthesis of benzoxazole

Entry	**Catalyst**	**Time (min)**	**Yield (%)**
1	ZnO	120	48
2	MgO	120	56
3	TiO_2	120	94
4	SnO_2	120	74

Later, it was studied the effect of quantity of the catalyst on the yields of the products. The results of various catalyst loads were presented in table 2. From the table, it was observed that there was no further development of the product, even an excess of catalyst was applied. The results showed that when a 2.5 mmol catalyst yields 45% of the product, 5 mmol of catalyst yields 68%, and when 10 mmol of catalyst yields 94% of desired product. Also, the reaction was monitored without using the catalyst and the product was not obtained. From these observations it was evident that the reaction using an excessive amount of catalyst did not result an improved yield. It was evident that 10 mmol of catalyst is an optimized quantity for the completion of the reaction.

Table 2: Investigation of amount of catalyst on the synthesis of 2- phenyl benzoxazole

S.No	**Amount of catalytic**	**Yield (%)**	**Time(min)**
1	NO catalyst	no	120
2	2.5mmol	45	120
3	5mmol	68	120
4	10mmol	94	120
5	15mmol	94	120

Table 3 represented the impact of solvent, and time on the reaction. Different types of solvents (protic and aprotic) were tested to understand the suitability of the solvent for the reaction. From the results it was noted that, ethanol was the most effective solvent for the reaction and it was further facilitating the reaction to get the product in 94% yield within 150 min.

Table 3: The solvent comparison for the synthesis of 2-ary benzoxazole

Entry	**solvent**	**Time(min)***	**Yield (%)#**
1	Water	210	10
2	Toluene	240	71
3	DCM	300	52
4	Ethanol	150	94
5	Methanol	180	54
6	THF	120	65

*-appropriate time, #- isolated yield

Our initial study of the model reaction between 2-amino phenol and substituted aromatic aldehydes showed that, in the presence of TiO_2 (10 mmol%) in the ethanol solvent at 76^0C, the reaction could be completed under very basic conditions, the percentage of the desired titled product in a good to excellent yield. In light of these findings, other aminophenol and modified aldehydes were used to explore the extent of the reaction. The structure of the series titled derivatives was constructed by the evidence of spectral data such as ^{1}HNMR, ^{13}CNMR, LCMS. In this study, proton NMR of titled derivatives was exhibited by various values of respective groups viz; hydroxyl proton appears at 9.547-9.127ppm, methoxy protons appears at 3.810-3.681ppm, as well as aromatic protons appeared at range between 8.258-6.871 ppm. The various range of values. ^{13}CNMR of these derivatives appeared at different values appeared at 163.21ppm.

After completion of the reaction, the reusability of the nanocatalyst was tested and the results were presented in table 4. From the table, it was noted that the nano TiO_2 particles showed its efficacy until the three cycles and these findings were matched with scanning electron microscope (SEM) results

(Figure 2). From the figure, it was observed that the nano TiO_2 before participating in the reaction (figure 2a) has defined crystalline structure with sharp edges. Whereas, the nano TiO_2 after three cycles (figure 2b) represented the conversion of crystalline to semi-crystalline nature with agglomeration. This observation attested the efficiency of the nanocatalsyt has been decreased with the reusability.

Table 4: Reusability of the nanocatalyst

S.NO	Cycle	Yield %
1	Cycle 1	94
2	Cycle 2	82
3	Cycle 3	67

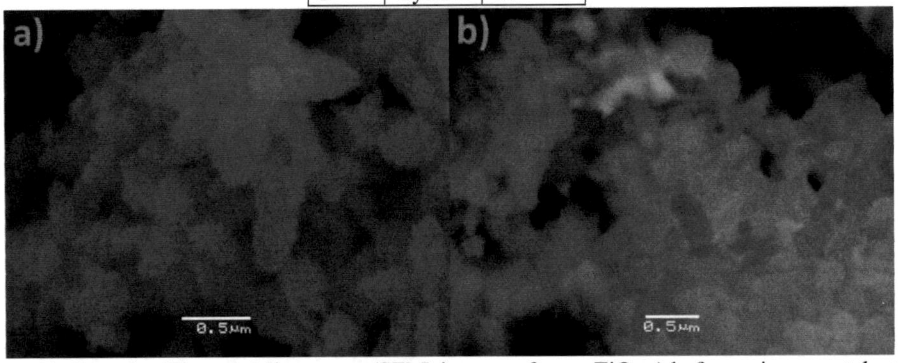

Figure 2: Scanning Electron Microscope (SEM) images of nano TiO_2 a) before using as catalyst b) after 3cycles as catalyst

Table-5 provides an overview of the reaction with different substituted aminophenol and benazaldehyde derivatives. Preliminary observations show that the particular position of the substituent on the aromatic ring of the aldehydes affects the yield of benzoxazole, since the substituted aromatic aldehydes easily cyclized with 2-aminophenol during the reaction. The yield was reduced when using solely para-substituted aryl aldehydes as opposed to both Meta- and para-substituted aryl aldehydes (Entry-3, 4, 8, 9, 13 and 14). Additionally, the results in (Entry-3, 8, 13) would suggest that the substituent's electrical nature affects the yield as well.

In this reaction, both of the reactants (1) and (2) combined to corresponding desired product by using nano catalyst (TiO_2) and showed Table-5. The yield of titled adduct mainly impact of the nature of substituent on the aryaldehyde and 2-amino phenol and both of the reactants are bearing electron releasing group and electron attracting group. If both of the reactants having electron releasing group as well as electron attracting group generate low yield of titled product (Entry–5, 10 ,14 and 15) than the reactants are containing either electron releasing and electron attracting group (Entry-5 and 9).

Table 5: Synthesis of series of titled derivatives in the presence of TiO_2.

Entry	2-substituted aminophenol	Aromatic aldehyde (R_2)	Product	Time (min)	Yield (%)	Melting point(^{0}C)
1	H	C_6H_5CHO	$C_{13}H_9NO$	170	86	152-154
2	H	4-OH-C_6H_4CHO	$C_{13}H_9NO_2$	150	85	187-189
3	H	3,4,5- $(OCH_3)_3$ C_6H_2CHO	$C_{16}H_{15}NO_4$	120	86	194-196
4	H	3,5 $(Cl)_2$ C_6H_3CHO	$C_{13}H_7Cl_2NO$	180	89	163-165
5	H	4-NO_2-C_6H_4CHO	$C_{13}H_8N_2O_3$	210	90	204-206

6	-OCH_3	C_6H_5CHO	$C_{14}H_{11}NO_2$	170	85	169-171
7	-OCH_3	4-OH-C6H4CHO	$C_{14}H_{11}NO_3$	150	84	178-180
8	-OCH_3	3,4,5- $(OCH_3)_3$ C_6H_2CHO	$C_{17}H1_7NO_5$	120	82	218-220
9	-OCH_3	$3,5(Cl)_2$ C_6H_3CHO	$C_{14}H_{10}N_2O_4$	180	90	247-249
10	-OCH_3	4-NO_2-C_6H_4CHO	$C_{14}H_9Cl_2NO_2$	210	94	239-241
11	-CF_3	C_6H_5CHO	$C_{14}H_8F_3NO$	170	85	216-218
12	-CF_3	4-OH-C_6H_4CHO	$C_{14}H_8F_3NO_2$	150	82	231-233
13	-CF_3	3,4,5- $(OCH_3)_3$ C_6H_2CHO	$C_{17}H_{14}F_3NO_4$	120	87	240-242
14	-CF_3	$3,5(Cl)_2$ C_6H_3CHO	$C_{14}H_6F_3NO$	180	86	235-237
15	-CF_3	4-NO_2-C_6H_4CHO	$C_{14}H_7F_3N_2O_3$	210	84	222.224

Reaction conditions: **1** mmol, 2-aminophenol (**1** mmol), aromatic aldehyde, catalytic TiO_2, 4 m L solvent at 76^0C.

Conclusion:

In this study important of heterocyclic analogous such as the structure 2-arylbenzoxazoles from 2-amino phenol combined with substituted aromatic aldehyde promoted by Nano transition metal oxide such as TiO_2as intramolecular cyclizing in ethanol as a solvent agent at 76^0C. The structures of the new compounds were evaluated spectral methods (^1H-NMR, ^{13}C-NMR, LCMS). These derivatives were obtained by applying different nature of aryl aldehyde means electron donating groups as well as electron donating groups.

References:

[1]. Zhengyu Li, Jingnan Dong, Zihang Yuan, Ding-Yah Yang, Zhiqiang Weng. 2018 *Organic Letters*, **20** (20), 6407-6410
[2]. Sharghi H, Aboonajmi J, Aberi M 2020**,** *J Org Chem.* **85**(10)**,** 6567-6577.
[3]. Thirupathi Rapolu, Polkam Naveen, K.V.P. Pavan Kumar, Krishnakanth Reddy Leleti, Raghubabu korapolu, 2022, *Results in Chemistry*, **4**, 100351.
[4]. Liu YK, Mao DJ, Lou SJ, Qian JQ, Xu ZY. 2009, *J Zhejiang Univ Sci B*. **10**(6):472-8.
[5]. C Praveena, A Nandakumara, P Dheenkumar, D Muralidharan, P T Perumal, 2012, *J. Chem. Sci.* **124** (3), 609–624.
[6]. Zahra Moniri, Farahnaz K. Behbahani 2019 *Sch Int J Chem Mater Sci*, **2**(7): 115-126.
[7]. Gajanan S. Shanbhag, Amit Bhargava, Giridhar Pal Singh, Shrinivas D. Joshi, and Narendra Singh Chundawat Turk 2023 *J Chem*,**47**(1): 263–279.
[8]. Nuaman F Alheety 2019, *Baghdad Science Journal*, **16**(3), 616.
[9]. Sham M. Sondhi, Nirupma Singh, Ashok Kumar, Olivier Lozach, Laurent Meijer, 2006 *Bioorganic & Medicinal Chemistry*, **14**(11), 3758-3765.
[10]. Abdel.Ghany A. El.Helby , Helmy Sakr , Ibrahim H. Eissa , Hamada Abulkhair Ahmed A. Al Karmalawy Khaled El-Adl1, 2019 *Arch Pharm Chem Life Sci*, e1900113.
[11]. ZaferAsım Kaplancıklı, 2021 *European Journal of Medicinal Chemistry*, **210**, 112979.

[12]. Mohamed A Abdelgawad, Mohammad M Al-Sanea, Mohamed A Zaki, Enas IA Mohamed, Shabana I Khan, Babu L Tekwani, Amar G Chittiboyina, Ikhlas A Khan, Tarfah Al-Warhi, Nada Aljaeed, Ohoud J Alotaibi, Abdulsalam AM Alkhaldi, Heba AH Elshemy, 2021, *Journal of Chemistry*, **2021**, 1-11
[13]. Asit K. Chakraborti, 2014 *ACS Med Chem Lett,* **5**(5): 512–516.
[14]. RafaqatHussain a, Fazal Rahim a, WajidRehman a, Shoaib Khan, 2023*, Arabian Journal of chemistry* **,11**, 105244.
[15]. Zhi-Ying Gong,Cheng-Li Yang,DanWang,Lang Huang Andzhi-BingDong, 2023, *The Journal Of Oraganic Chemistry*, **13**(4), 658.
[16]. S. S. Sagar1, and R. P. Chavan, 2021, Rasayan. J.Chem, **1,** 186-193.
[17]. Suresh Maddila, Sreekanth. B Jonnalagadda 2012, *Journal of the Chilean chemical society*, **57,** 1099-1100.
[18]. Abbas Teimouri, AlirezaNajafi Chermahini, Hossien Salavati, Leila Ghorbania,2013, *Journal of Molecular Catalysis A: Chemical*, **373**, 38-45.
[19]. Geetika Patel, Ashok Raj Patel, Trimurti. Lambat, Subhash Banerjee,2021, *Current Research in Green and sustainable chemistry* ,100149.
[20]. Fatemeh Hakimi, Elham Golrasan, 2023, *Asian Journal of green chemistry*, **7**, 47-53.
[21]. S. Khameneh Asl, S.K. Sadrnezhaad, M. Kianpour rad, 2010, *Mater. Lett.* **64** 1935.
[22]. Mehdi ghayeri, Seyed Mohammad Vahdat, Abbas Mohammadzadeh and Somaieh Akhoondi,2015, *RSC Advance*, **5**, 46545-46551.
[23]. Ghasem Rezanejade Bardajee, Marzieh Mohammadi, Hasan Yari, Aseyeh Ghaedi, 2016 *Chinese Chemical Letters*, **27**, 265-270.
[24]. Seyed Mohammad Vahdata, Shima Ghafouri Raza and Saeed Baghery, 2014, *J. Chem. Sci*, **3**, 579–585.
[25]. Mona Hosseini-Sarvari, 2015, *Journal of Materials Science* **50(8),** 3065–3074.
[26]. Firouz Matloubi Moghaddam, VahidSaberi, NazilaVeisi , 2018, *Appl Organometal Chem.* **32**:e4240.
[27]. Toro, Roberta Grazia, Mohamed Diab, Tilde de Caro, Mona Al-Shemy, Abeer Adel, and Daniela Caschera. 2020. *Materials* **13**,1326.

Impact of Shape Factor on a Couple Stress Flow of a Ternary Hybrid Nanofluid over an Inclined Flat Plate when Non-Uniform Heat Source and Thermal Radiation are Significant: A Darcy-Forchheimer Model

Gandrakota Kathyayani[1]*, Satuluri Satya Nagendra Rao[2]

[1,2]Department of Applied Mathematics, Yogi Vemana University, Kadapa-516005, India

*Corresponding author email: kathyagk@yogivemanauniversity.ac.in

Abstract. The study of fluid flow over an inclined flat plate finds applications in a diverse range of engineering fields including aerodynamics, energy production and automotive design. This study theoretically investigates the steady and radiative flow of a ternary hybrid nanofluid (Water + TiO_2 + $CoFe_2O_4$ + MgO) with couple stress, using the Darcy-Forchheimer model. The flow occurs through a tilted flat plate and is subjected to irregular heat source parameter and entropy generation. The problem's equations have been transformed into a collection of ordinary differential equations (ODEs), which has been skillfully resolved using the bvp4c solver. Graphs are utilized to elucidate outcomes for two instances of shape components, namely platelet and spherical. An escalation in the couple stress parameter (S) is demonstrated to be inversely related to the fluid velocity, resulting in a drop. Specifically, when $0.5 \le S \le 3$, the friction factor exhibits a decline, with rates of 0.306201851 (for Platelet shape) and 0.304466755 (for Spherical shape). An intriguing observation reveals an augmentation in the generation of entropy as the volumetric fraction of TiO_2 rises. Upon investigation, it has been determined that when the Eckert number (Ecn) increases within the range of $0 \le Ecn \le 0.3$, there is a significant reduction in the Nusselt number. Specifically, the decline is measured to be 0.328685192 for the platelet shape and 0.308939422 for the spherical shape. The utility of the Forchheimer number in regulating the fluid's motion has been unveiled.

Keywords Shape factor, bvp4c, Couple Stress, Irreversibility, Heat source, Viscous dissipation.

1. Introduction

The phenomenon of fluid flow through a flat plate is a quintessential quandary in the realm of fluid mechanics, boasting a plethora of practical implications across several branches of engineering. The study of this phenomenon provides insights into the behavior of fluids past solid surfaces and has contributed to the evolution of important engineering concepts such as drag reduction and heat transfer enhancement. Some of the applications of fluid flow over a flat plate are aerospace engineering, surface coatings, and thermal engineering. Daniel [1] performed a mathematical investigation of the laminar movement of a fluid through a flat plate subjected to convective heating. Safaei et al. [2] inspected the forced convection of a nanofluid consisting of water and nanotubes flowing across a flat plate. They noticed that increasing the weight % (percentage) of nanotubes resulted in an increase in heat transfer. Utilising the Keller-box approach, Ziaei-Rad et al. [3] investigated the mixed convective nanofluid flow over a porous slanted surface with wall suction parameter. They suggested that by introducing nanoparticles into the base fluid, it is feasible to increase both the heat transmission and skin friction coefficient. Hakeem et al. [4] discussed two types of nanofluid flows over a slippery flat plate with velocity slip. It has been observed that the friction factor upsurges as the magnetic field parameter rises. Saranya et al. [5] expanded on the previous research by incorporating nonlinear thermal radiation. Nandeppanavar et al. [6] examined the magnetohydrodynamic (MHD) Blasius flow of a nanofluid on a flat plate by employing a combination of the 4th Runge-Kutta method and shooting procedures. Single-walled carbon nanotubes have been observed to exhibit superior heat transfer capabilities in comparison to multi-walled carbon nanotubes. Rashed et al. [7] investigated nanofluid radiative convective flow across a magnetic field parameter-controlled permeable inclined flat plate. By considering the same geometry, Reddy and Reddy [8] studied the dynamics of mass and heat transport in chemically reactive nanofluid flow with thermal radiation. Jha and Samaila [9] inspected the affect of thermal radiation on various nanofluid flows on a flat plate that is heated by convection. Khademi et al. [10] simulated the laminar movement of a fluid containing nanoparticles over a flat surface that is inclined, while also considering the effect of a magnetic field that is perpendicular to the flow direction. They used the Darcy-Forchheimer model to describe the flow behaviour. It has been noted that when the Richardson number grows, the impact of nanoparticle dispersion on reducing skin friction becomes more pronounced. Following this, several research papers [11-20] delved into the flat plate as a geometric form, investigating a range of fluid dynamics scenarios with different parameters, encompassing thermal radiation.

Heat sources correspond to areas where thermal energy is introduced into the fluid, resulting in a rise in temperature. Such occurrences may arise as a consequence of diverse sources, including but not limited to combustion, chemical reactions, or electrical heating. The existence of heat sources has the potential to give rise to thermal plumes and boundary layers, so exerting an influence on the flow patterns and rates of heat transmission. The use of fluid flow with heat sources spans across a diverse array of disciplines, encompassing engineering, science, and medicine. For instance, heat sources are often used to control chemical reactions by providing the necessary activation energy. They can also be used to maintain desired temperature profiles in reactors, ensuring optimal product yield and quality. In their study, Macha et al. [21] explored how the heat source affects the flow of a power-law nanofluid over an irregular elastic sheet, while also taking into account the influence of thermal radiation. Dhanai et al. [22] addressed the mixed convective flow of nanofluid (using the Buongiorno model) with the inclusion of viscous dissipation and heat source in the slip regime. It has been observed that the rate of heat transmission reduces as the thermal slip parameter rises. Thumma et al. [23] investigated the flow of a nanofluid across an irregular tilted extending/contracting sheet with heat source parameter using the Keller-Box method. They analysed both the dissipative and radiative aspects of the flow. An observation has been made that the surface drag factor exhibits an improvement when the power-law parameter rises. Narayana et al. [24] conducted a study investigating the influence of an uneven heat source on the radiative flow of a nanofluid over an extended sheet. They employed the shooting process to analyse the unsteady nature of the flow. The temperature profile lowers due to the influence of space and temperature-dependent heat sink factors. Gireesha et al. [25] delved into the effectiveness of different heat source characteristics on the flow of Casson nanofluid across an elongated surface with activation energy. The heat transmission rate is shown to be significantly higher in the case of dynamic radiative heat activity comparing to regular radiative heat activity. Saleem et al. [26] examined the dynamics of magnetohydrodynamic (MHD) and chemically reactive flow of a Jeffrey fluid on a rotating upright cone with the presence of a heat source using OHAM. With an unequal heat absorption/generation and Ohmic heating, Ramadevi et al. [27] numerically solved the three-dimensional magnetohydrodynamic and dissipative flow of Carreau fluid through an elastic sheet. It has been observed that the temperature of the fluid raises as the Eckert number grows. Kumar et al. [28] conducted an investigation on the properties of non-uniform thermal radiation and slip factor on the movement of micropolar fluid across an elongated surface. The microrotation profile is shown to be a declining function of the micropolar parameter. Mahanthesh et al. [29] utilised the probable error to evaluate the dependability of the correlation factor in their examination of Williamson fluid flow close to a stagnation point featuring a variable heat source (ESHS) parameter. A positive correlation is evident among the Nusselt number and the Biot number. Subsequently, a number of studies [30-36] examined diverse fluid flows with varying parameters, including heat source.

Ternary hybrid nanofluids can be engineered to possess multiple functionalities simultaneously, such as enhanced heat transfer, improved thermal stability, and rheological control. This makes them versatile for a range of applications. The considerable importance of the flow of ternary hybrid nanofluid over a flat plate has given rise to numerous prospective applications across various industries. For instance, it can be applied in cooling systems for electronics, thermal management in aerospace, and energy systems. In the realm of the couple stress and Darcy-Forchheimer flow of ternary hybrid nanofluid (Water + TiO_2+ $CoFe_2O_4$ + MgO), the intricate interplay of non-uniform heat source/sink, thermal radiation, and viscous dissipation effects has remained unexplored till now. This work, however, boldly presents the uniqueness of shedding light on these significant factors. Outcomes are explained for two instances of shape factors i.e., platelet and spherical. Influences of pertinent parameters on Nusselt number and friction coefficient are explained using graphs and slope of linear regression.

2. Research Methodology

A radiative and couple stress flow of a THNF (ternary hybrid nanofluid) (Water + TiO_2+ $CoFe_2O_4$ + MgO) via an inclined flat plate with irregular heat sink/source is theoretically investigated in this study. The present investigation is based on the subsequent suppositions:

(i) The quantitative estimates of the material properties pertaining to the base fluid and nanoparticles utilized in this investigation can be found in Table 1.

(ii) The fluid across the plate's surface is in a state of constant velocity U_∞ (see Fig. 1).

(iii) The flow experiences an influence from an external magnetic field applied vertically with an intensity denoted as $B = \dfrac{B_0}{\sqrt{x}}$.

(iv) Induced magnetic field impact is disregarded in this work.

Given the aforementioned presumptions, the appropriate conditions and fundamental equations required for this investigation are outlined below (Sneha et al. [37] and Pal and Mondal [38]):

$$\frac{\partial u}{\partial x}+\frac{\partial v}{\partial y}=0 \tag{1}$$

$$u\frac{\partial u}{\partial x}+v\frac{\partial u}{\partial y}=\frac{\mu_{thnf}}{\rho_{thnf}}\frac{\partial}{\partial y}\left(\frac{\partial u}{\partial y}\right)-\frac{\sigma_{thnf}}{\rho_{thnf}}Bu-\frac{\gamma_0}{\rho_{thnf}}\frac{\partial^4 u}{\partial y^4}-\frac{\upsilon_{thnf}}{\kappa}u-Fu^2+g\beta_T\left(T-T_\infty\right)\mathrm{Cos}\,\alpha \tag{2}$$

$$u\frac{\partial T}{\partial x}+v\frac{\partial T}{\partial y}=\frac{k_{thnf}}{\left(\rho C_p\right)_{thnf}}\frac{\partial^2 T}{\partial y^2}+\frac{1}{\left(\rho C_p\right)_{thnf}}q'''+\frac{1}{\left(\rho C_p\right)_{thnf}}\frac{16\sigma^* T_\infty^3}{3k^*}\frac{\partial^2 T}{\partial y^2}+\frac{\mu_{thnf}}{\left(\rho C_p\right)_{thnf}}\left(\frac{\partial u}{\partial y}\right)^2 \tag{3}$$

$$\left.\begin{aligned} &\text{at } y=0: v=0, u=0, \frac{\partial^2 u}{\partial y^2}=0, -k_{thnf}\left.\frac{\partial T}{\partial y}\right|_{(x,0)} = h_f\left(T_w - T\right), \\ &\text{at } y=\infty: u=U_\infty, \frac{\partial u}{\partial y}=0, T=T_\infty . \end{aligned}\right\} \tag{4}$$

Note that $q''' = \frac{kU_\infty}{2\upsilon x}\left[Q\left(T-T_\infty\right)+Q^*\left(T_w-T_\infty\right)\exp\left(-y\sqrt{\frac{U_\infty}{2\upsilon x}}\right)\right]$ (5)

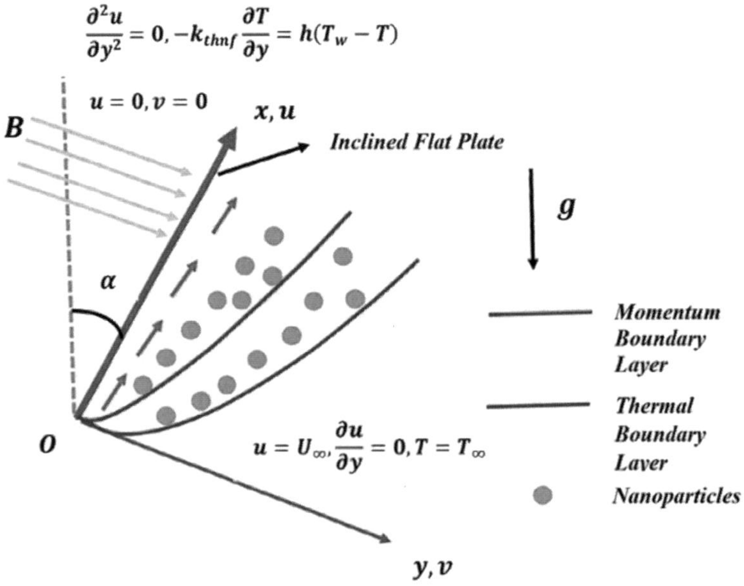

Figure 1. Diagrammatic depiction of the current investigation

where $Q<0$, $Q^*<0$ resembles heat sink and $Q>0$, $Q^*>0$ resembles heat source parameters.

Through the utilisation of (5), the expression (3) may be reformulated as

$$\begin{aligned} u\frac{\partial T}{\partial x}+v\frac{\partial T}{\partial y} &= \frac{k_{thnf}}{\left(\rho C_p\right)_{thnf}}\frac{\partial^2 T}{\partial y^2}+\frac{1}{\left(\rho C_p\right)_{thnf}}\frac{kU_\infty}{2\upsilon x}\left[Q\left(T-T_\infty\right)+Q^*\left(T_w-T_\infty\right)\exp\left(-y\sqrt{\frac{U_\infty}{2\upsilon x}}\right)\right] \\ &+\frac{1}{\left(\rho C_p\right)_{thnf}}\frac{16\sigma^* T_\infty^3}{3k^*}\frac{\partial^2 T}{\partial y^2}+\frac{\mu_{thnf}}{\left(\rho C_p\right)_{thnf}}\left(\frac{\partial u}{\partial y}\right)^2 \end{aligned} \tag{6}$$

Table 1 The fluid and nanomaterial properties utilised in this study along with their respective numerical values (Alharbi et al. [39])

Properties	(H_2O) (f)	TiO_2 (ϕ_T)	$CoFe_2O_4$ (ϕ_C)	MgO (ϕ_M)

$\rho\left(kgm^{-3}\right)$	997.1	4250	4907	3560
$k\left(Wm^{-1}K^{-1}\right)$	0.613	8.9538	3.7	45
$C_p\left(Jkg^{-1}K^{-1}\right)$	4179	686.2	700	955
$\sigma\left(Sm^{-1}\right)$	0.0000055	2.38×10^{6}	5.51×10^{9}	1.42×10^{-3}

2.1. *Thermal and physical properties of THNF (Manjunatha et al. [40])*

$$\left.\begin{aligned}
&\left(\rho C_p\right)_{thnf}=\left[\left(1-\phi_C\right)\left(\left(1-\phi_M\right)\left(\rho C_p\right)_f+\phi_M\left(\rho C_p\right)_M\right)+\phi_C\left(\rho C_p\right)_C\right]\left(1-\phi_T\right)+\left(\rho C_p\right)_T\phi_T,\\
&\sigma_{hnf}=\sigma_{nf}\times\frac{\sigma_C+2\sigma_{nf}-2\sigma_{nf}\phi_C+2\sigma_C\phi_C}{\sigma_C+2\sigma_{nf}+\sigma_{nf}\phi_C-\sigma_C\phi_C},\ \mu_{thnf}=\frac{\mu_f}{\left(1-\phi_T\right)^{2.5}\left(1-\phi_C\right)^{2.5}\left(1-\phi_M\right)^{2.5}},\\
&\rho_{thnf}=\left[\left(1-\phi_C\right)\left(\left(1-\phi_M\right)\rho_f+\phi_M\rho_M\right)+\phi_C\rho_C\right]\left(1-\phi_T\right)+\rho_T\phi_T,\\
&\sigma_{thnf}=\sigma_{hnf}\times\frac{\sigma_T+2\sigma_{hnf}-2\sigma_{hnf}\phi_T+2\sigma_T\phi_T}{\sigma_T+2\sigma_{hnf}+\sigma_{hnf}\phi_T-\sigma_T\phi_T},\ k_{nf}=k_f\times\frac{k_M+(n+1)k_f-(n+1)k_f\phi_M+(n+1)k_M\phi_M}{k_M+(n+1)k_f+k_f\phi_M-k_M\phi_M},\\
&\sigma_{nf}=\sigma_f\times\frac{\sigma_M+2\sigma_f-2\sigma_f\phi_M+2\sigma_M\phi_M}{\sigma_M+2\sigma_f+\sigma_f\phi_M-\sigma_M\phi_M},\ k_{hnf}=k_{nf}\times\frac{k_C+(n+1)k_{nf}-(n+1)k_{nf}\phi_C+(n+1)k_C\phi_C}{k_C+(n+1)k_{nf}+k_{nf}\phi_C-k_C\phi_C},\\
&k_{thnf}=k_{hnf}\times\frac{k_T+(n+1)k_{hnf}-(n+1)k_{hnf}\phi_T+(n+1)k_T\phi_T}{k_T+(n+1)k_{hnf}+k_{hnf}\phi_T-k_T\phi_T}.
\end{aligned}\right\}$$

n indicates the shape parameter. $n=3$ relates to a spherical shape, while $n=5.7$ relates to a platelet shape.

Pal and Mondal [38] provided the succeeding similarity transmutations for converting governing equations:

$$u=U_\infty\frac{df}{d\eta}=U_\infty f'(\eta),v=\sqrt{\frac{\upsilon U_\infty}{2x}}\left[-f(\eta)+\eta f'(\eta)\right],T=T_\infty+\theta(\eta)\left(T_w-T_\infty\right),\eta=y\sqrt{\frac{U_\infty}{2\upsilon x}}.\quad(7)$$

By leveraging equation (7), the continuity equation (1) seamlessly aligns, allowing for a straightforward fulfilment. Equations (2, 4 & 6) can then be adeptly adjusted by incorporating equation (7) in the subsequent process:

$$\frac{2}{A_1A_2}f'''+ff''+2\frac{A_3}{A_2}Mnf'-\frac{2}{A_2}Sf^{v}-\frac{2}{A_1A_2}\Gamma f'-2Fnf'^2+2\lambda\theta\,\mathrm{Cos}\,\alpha=0\quad(8)$$

$$\frac{A_4}{A_5}\frac{1}{\Pr}\theta''+\frac{1}{A_5}\frac{Rd}{\Pr}\theta''+f\theta'+\frac{1}{A_5}\left(Q\theta+Q^*\exp(-\eta)\right)+\frac{1}{A_2A_5}Ecnf''^2=0\quad(9)$$

$$\left.\begin{aligned}
&\text{at }\eta=0: f(\eta)=0, f'(\eta)=0, f'''(\eta)=0, \theta'(\eta)=-\frac{Bi}{A_4}\left(1-\theta(\eta)\right),\\
&\text{at }\eta=\infty: f'(\eta)=1, f''(\eta)=0, \theta(\eta)=0.
\end{aligned}\right\}\quad(10)$$

where

$$\left.\begin{array}{l} S=\dfrac{\kappa_1 U_\infty}{\rho \upsilon^2}, \kappa_1=\dfrac{\gamma}{x}, Rd=\dfrac{16}{3}\dfrac{\sigma^* T_\infty^3}{kk^*}, \Pr=\dfrac{\mu C_p}{k}, \Gamma=\dfrac{\upsilon}{\kappa_0 U_\infty}, \kappa=\kappa_0 x, Mn=\dfrac{\sigma_f B_0^2}{\rho_f U_\infty}, Fn=\dfrac{C_b}{\sqrt{\kappa_0}}, \\ Bi=\dfrac{d}{k_f\sqrt{\dfrac{U_\infty}{\upsilon}}}, d=h_f\sqrt{x}, F=\dfrac{C_b}{\sqrt{x\kappa}}, \lambda=\dfrac{Gr}{\mathrm{Re}_x^2}, Gr=\dfrac{g\beta_T\left(T_w-T_\infty\right)x^3}{\upsilon^2}, \mathrm{Re}_x=\dfrac{xU_\infty}{\upsilon}, Ecn=\dfrac{U_\infty^2}{C_p\left(T_w-T_\infty\right)}. \end{array}\right\}$$

and

$$A_1=\left(1-\phi_M\right)\left(\left(1-\phi_C\right)\left[\left(1-\phi_T\right)+\phi_T\frac{\rho_T}{\rho_f}\right]+\phi_C\frac{\rho_C}{\rho_f}\right)+\phi_M\frac{\rho_M}{\rho_f}, A_2=\left(1-\phi_T\right)^{2.5}\left(1-\phi_C\right)^{2.5}\left(1-\phi_M\right)^{2.5},$$

$$A_{311}=\frac{k_T+(n+1)k_f-(n+1)\phi_T\left(k_f-k_T\right)}{k_T+(n+1)k_f+\phi_T\left(k_f-k_T\right)}, A_{31}=\frac{k_C+(n+1)A_{311}k_f-(n+1)\phi_C\left(A_{311}k_f-k_C\right)}{k_C+(n+1)A_{311}k_f+\phi_C\left(A_{311}k_f-k_C\right)}\times A_{311},$$

$$A_3=\frac{k_M+(n+1)A_{31}k_f-(n+1)\phi_M\left(A_{31}k_f-k_M\right)}{k_M+(n+1)A_{31}k_f+\phi_M\left(A_{31}k_f-k_M\right)}\times A_{31}, A_{411}=\frac{\sigma_T+2k_f-2\phi_T\left(\sigma_f-\sigma_T\right)}{\sigma_T+2\sigma_f+\phi_T\left(\sigma_f-\sigma_T\right)},$$

$$A_5=\left(1-\phi_M\right)\left(\left(1-\phi_C\right)\left[\left(1-\phi_T\right)+\phi_T\frac{\left(\rho C_p\right)_T}{\left(\rho C_p\right)_f}\right]+\phi_C\frac{\left(\rho C_p\right)_C}{\left(\rho C_p\right)_f}\right)+\phi_M\frac{\left(\rho C_p\right)_M}{\left(\rho C_p\right)_f},$$

$$A_{41}=\frac{\sigma_C+2A_{411}\sigma_f-2\phi_C\left(A_{411}\sigma_f-\sigma_C\right)}{\sigma_C+2A_{411}\sigma_f+\phi_C\left(A_{411}\sigma_f-\sigma_C\right)}\times A_{411}, A_4=\frac{\sigma_M+2A_{41}\sigma_f-2\phi_M\left(A_{41}\sigma_f-\sigma_M\right)}{\sigma_M+2A_{41}\sigma_f+\phi_M\left(A_{41}\sigma_f-\sigma_M\right)}\times A_{41}.$$

Friction factor is defined as:

$$Cf=\frac{1}{\rho}\left.\frac{\mu_{thnf}\dfrac{\partial u}{\partial y}}{U_\infty^{\;2}}\right|_{y=0}. \tag{11}$$

By utilising equation (7), the formula used in (11) can be elegantly restated as:

$$\sqrt{\mathrm{Re}_x}Cf=\frac{1}{A_2}f''(0).$$

The following can be used to find the rate of heat transmission:

$$Nu=-\left.\frac{x\left(\dfrac{16}{3}\dfrac{T_\infty^3\sigma^*}{k^*}+k_{thnf}\right)\dfrac{\partial T}{\partial y}}{k_f\left(T_w-T_\infty\right)}\right|_{y=0}. \tag{12}$$

By utilising equation (7), the formula used in (12) can be elegantly restated as:

$$\sqrt{\frac{1}{\mathrm{Re}_x}}Nu=-\left(A_4+Rd\right)\theta'(0).$$

The dimensional version of the entropy formation formula in the context of current flow is articulated as follows:

$$S_G=\frac{1}{T_\infty^2}k_{thnf}\left(1+\frac{16}{3}\frac{\sigma^*T_\infty^3}{kk^*}\right)\left(\frac{\partial T}{\partial y}\right)^2+\frac{1}{T_\infty}\mu_{thnf}\left(\frac{\partial u}{\partial y}\right)^2+\frac{1}{T_\infty}\sigma_{thnf}u^2B^2 \tag{13}$$

By employing (7), equation (13) can be reformulated in the following manner:

$$EG = \left(A_4 + Rd\right)\alpha_1\theta'^2 + \frac{1}{A_2}Brf''^2 + A_3MnBrf'^2 \quad (14)$$

where

$$EG = \frac{x\upsilon T_\infty S_G}{U_\infty\left(T_w - T_\infty\right)k_f}, Br = \frac{\mu U_\infty^{\ 2}}{\left(T_w - T_\infty\right)k_f}, \alpha_1 = \frac{T_w - T_\infty}{T_\infty}.\Bigg\}$$

The Bejan number can be determined by use this mathematical expression:

$$Bn = \frac{The\ generation\ of\ entropy\ as\ a\ consequence\ of\ heat\ transmission}{aggregate\ generation\ of\ entropy}.$$

Bn can be restated by employing (14) in the subsequent fashion:

$$Bn = \frac{\left(A_4 + Rd\right)\alpha_1\theta'^2}{\left(A_4 + Rd\right)\alpha_1\theta'^2 + \frac{1}{A_2}Brf''^2 + A_3MnBrf'^2}.$$

3. Results and Discussion

Equations (8-9) are solved using the MATLAB inherent function bvp4c under conditions (10). Explanations are provided for the outcomes in two scenarios, namely platelet and spherical shapes.

3.1. Regular profiles including velocity profile

It is evident from Fig.2 that augmenting the volume fraction of nanoparticles in a fluid leads to a decrease in the velocity of the fluid. The reason for this is because nanoparticles enhance the viscosity of a fluid by augmenting the resistance to flow. The analysis of Fig.3 clearly demonstrates that enhancing the magnetic field parameter leads to a decline in velocity. This is due to a force called the Lorentz force, which acts on any electrically conductive fluid that is moving through a magnetic field. The Lorentz force, which is orthogonal to both the magnetic field and the fluid's velocity, functions in a manner that counters the fluid's movement. Upon examining Fig.4, it becomes apparent that a rise in the Forchheimer number results in a reduction in the fluid velocity. It is crucial to observe that as the Forchheimer number escalates, the impact of the inertial term is accentuated, thereby hindering the flow. It can be inferred that an augmentation in the Forchheimer number will result in an elevation of the flow impedance and a concomitant decrease in the fluid velocity. It is evident from Fig. 5 that an increase in the couple stress parameter results in a reduction in fluid velocity. The inclusion of the couple stress component in the equation is justified by its ability to accurately capture the heightened opposition to fluid movement resulting from the rotating dynamics of individual fluid particles. As the couple stress parameter escalates, the rotating motion of fluid particles becomes increasingly constrained, resulting in a reduction in fluid velocity. It is evident from this outcome that the velocity profile exhibits a greater magnitude in the case of the platelet shape factor as compared to the other shape factors. Based on the observation from Fig.6, an elevation in thermal radiation generally results in an increase in the temperature of the fluid. This phenomenon occurs due to thermal radiation, which is a method of energy transmission characterised by the propagation of electromagnetic waves across space. Upon impact with the fluid molecules, these waves effectively impart their energy, leading to heightened molecular vibrations and thereby elevating the temperature of the fluid. Fig.7 exhibits a noticeable relationship among the heat source parameter (Q) and the augmentation in fluid temperature. The phenomenon arises as a consequence of the transference of thermal energy from the heat source to the fluid, hence causing a discernible increase in the temperature of said fluid. As Ecn increases, a greater proportion of the kinetic energy is converted into thermal energy due to the effects of viscous forces. As depicted in Fig. 8, the temperature of the fluid consequently rises. Upon careful examination of Fig. 9, it becomes apparent that an augmentation in Bi results in an elevation of the temperature of the fluid. This is because the Biot number is a measure of the relative importance of conduction heat transfer inside a solid to convection heat transfer at the surface of the solid. A greater Biot number signifies the prevalence of convective heat transfer, resulting in a more substantial flow of heat from the solid to the fluid.

3.2. Irreversibility Analysis

Fig. 10 clearly demonstrates a noticeable increase in entropy formation as Br levels increase. With an increase in Br, the significance of viscous forces inside the system becomes more pronounced. Such circumstances may lead to elevated energy losses caused by fluid friction, hence resulting in heightened viscous dissipation. These supplementary energy losses serve to augment the overall entropy buildup inside the system. Upon careful examination of Fig. 11, it becomes evident that the augmentation of ϕ_T leads to a corresponding escalation in the formation of entropy. This is because the nanoparticles upsurge the viscosity of the fluid, which leads to more friction and heat generation. Additionally, the nanoparticles can also cause the fluid to become more turbulent, which further increases the entropy generation. The increase in entropy generation can be a problem in some applications, such as heat transfer systems. In these applications, minimising entropy formation is crucial for maximising system efficiency. In order to do this, it is important to use a low volume fraction of nanoparticles in the fluid. Fig. 12 clearly demonstrates the correlation between the rise in S and an upsurge in the generation of entropy. When the couple stress parameter is increased, the fluid becomes more viscous and resistant to flow. Consequently, there is a notable escalation in the production of entropy as a result of the occurrence of viscous heating. Figs.13–15 show that as the value of the ϕ_T increases, the Bejan number rises, while the same decrease with the rise in Br and S. With an escalation in the Brinkman number, there is a corresponding escalation in the heat generated by viscous dissipation, resulting in a concomitant reduction in the Bejan number.

3.3. Other profiles including Nusselt number

Upon closer examination of Figs. 16-18, it becomes evident that the friction factor exhibits a decline as the value of Mn increases, whereas the same increases with the rise in ϕ_T and S. The application of a magnetic field can suppress the fluid motion by generating a Lorentz force that acts against the flow direction. This flow suppression reduces the momentum transfer near the solid boundary, resulting in a diminution in the friction factor. The Lorentz force acts as additional resistance, counteracting the fluid flow and reducing the frictional drag. Within the range of $0.5 \le S \le 3$, it is observed that the friction factor exhibits a decrease with a rate of 0.306201851 (for Platelet shape) and 0.304466755 (for Spherical shape). Figs. 19-21 show that the Nusselt number upsurges with the escalation in S but the same decline with the elevation in Ecn and Q. Note that, when the couple stress parameter is increased, the fluid becomes more resistant to deformation, which leads to an increase in the shear stress at the wall. The augmentation of shear stress serves to intensify the fluid's intermingling, hence amplifying both the heat transmission coefficient and the Nusselt number. Upon investigation, it has been ascertained that for values of $0 \le Ecn \le 0.3$, the Nusselt number experiences a reduction of 0.328685192 (in the case of Platelet shape) and 0.308939422 (in the case of Spherical shape). Upon comparing the Platelet to other shape, it becomes apparent that its heat transmission rate exhibits a more rapid decline.

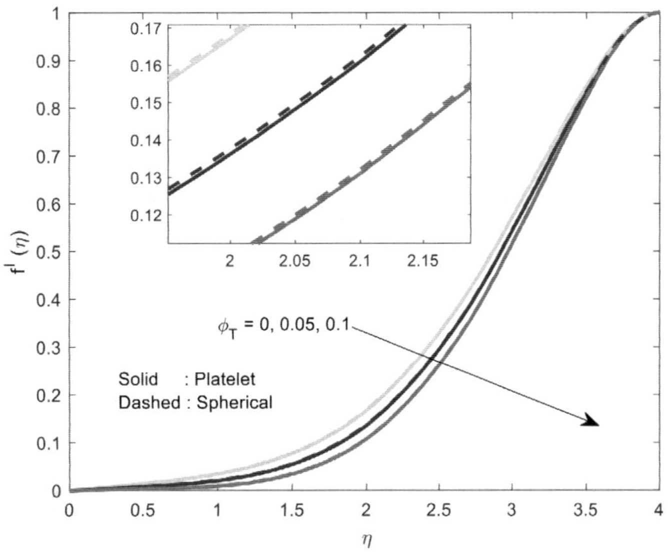

Figure 2. Scenario in which $f'(\eta)$ is affected by ϕ_T

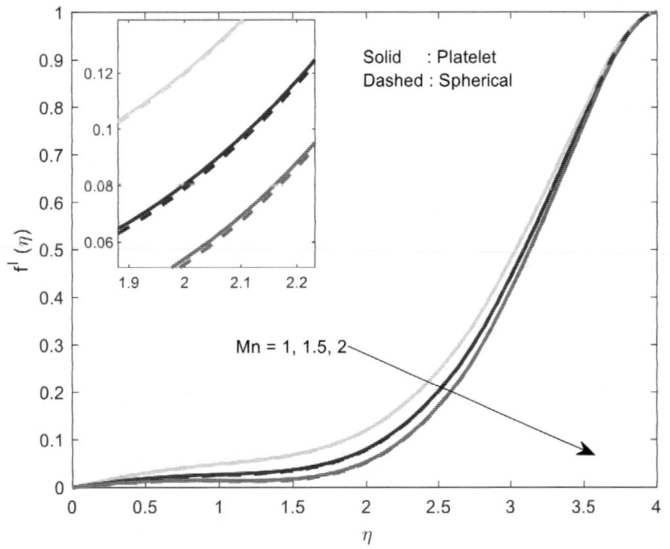

Figure 3. Scenario in which $f'(\eta)$ is affected by Mn

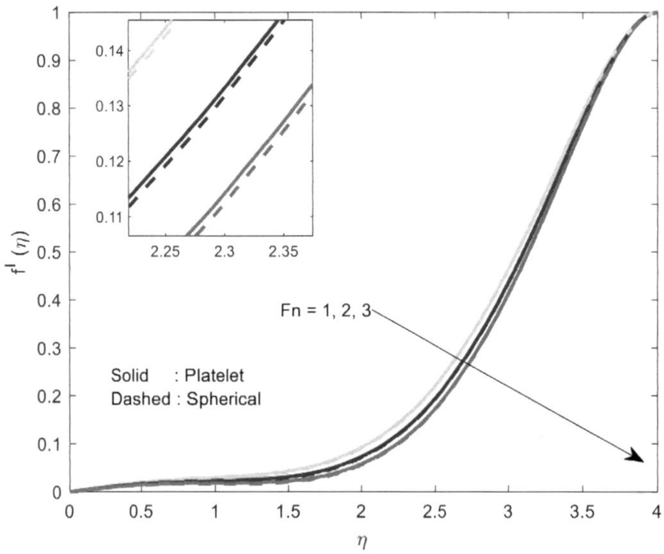

Figure 4. Scenario in which $f'(\eta)$ is affected by Fn

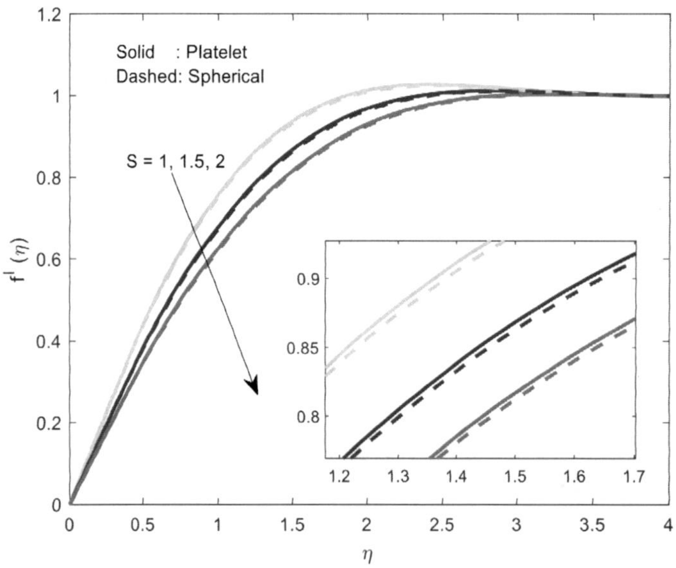

Figure 5. Scenario in which $f'(\eta)$ is affected by S

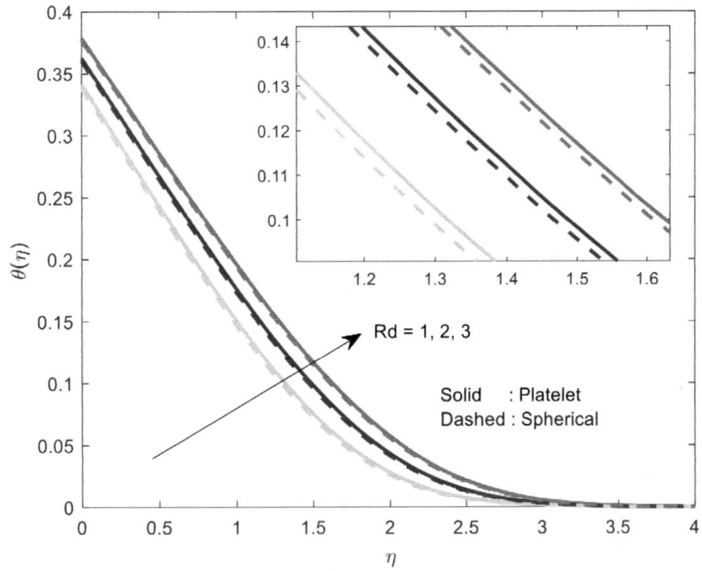

Figure 6. Scenario in which $\theta(\eta)$ is affected by Rd

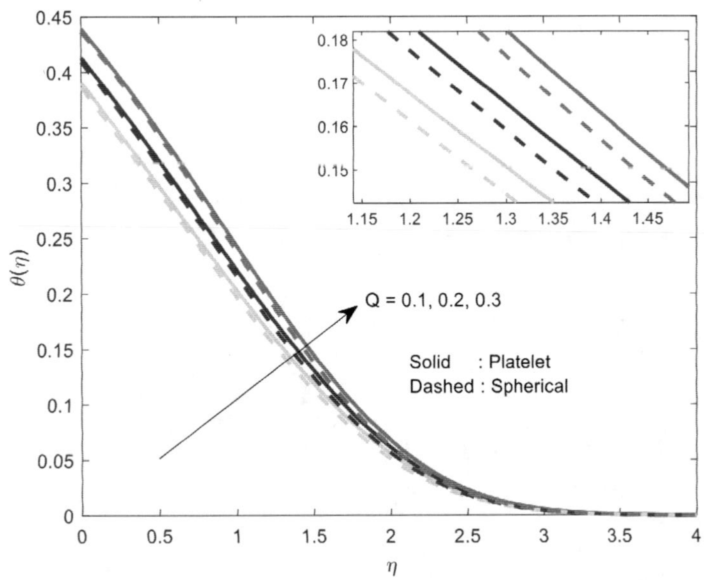

Figure 7. Scenario in which $\theta(\eta)$ is affected by Q

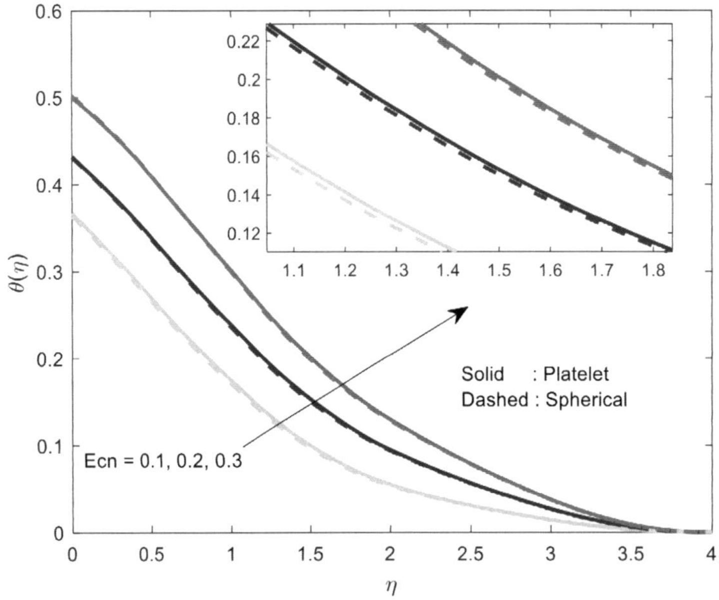

Figure 8. Scenario in which $\theta(\eta)$ is affected by Ecn

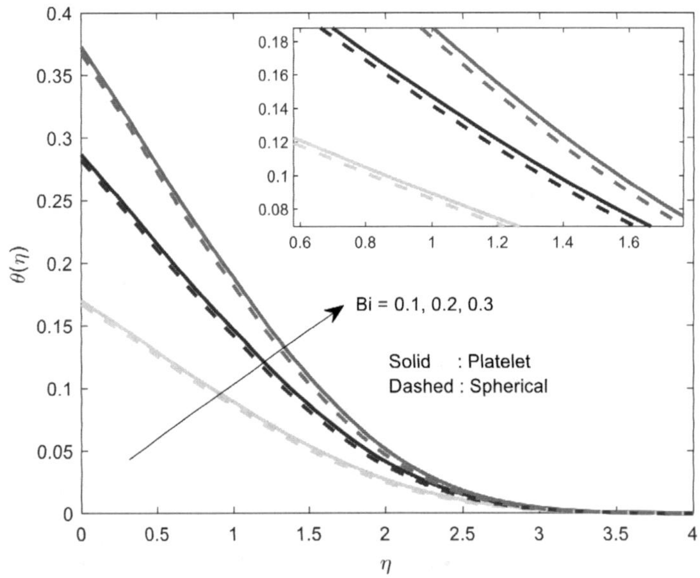

Figure 9. Scenario in which $\theta(\eta)$ is affected by Bi

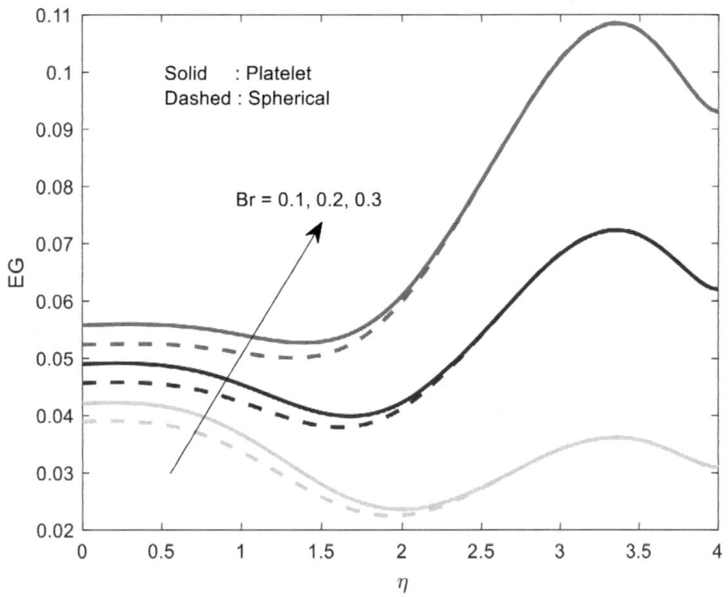

Figure 10. Scenario in which EG is affected by Br

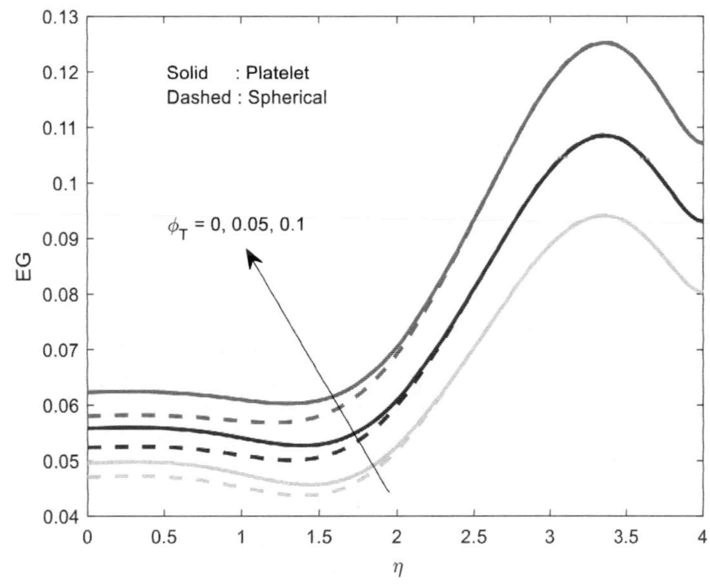

Figure 11. Scenario in which EG is affected by ϕ_T

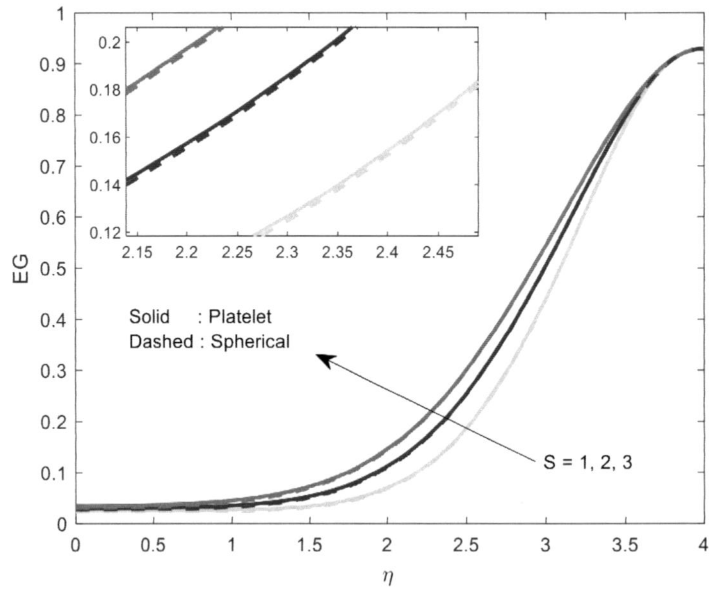

Figure 12. Scenario in which EG is affected by S

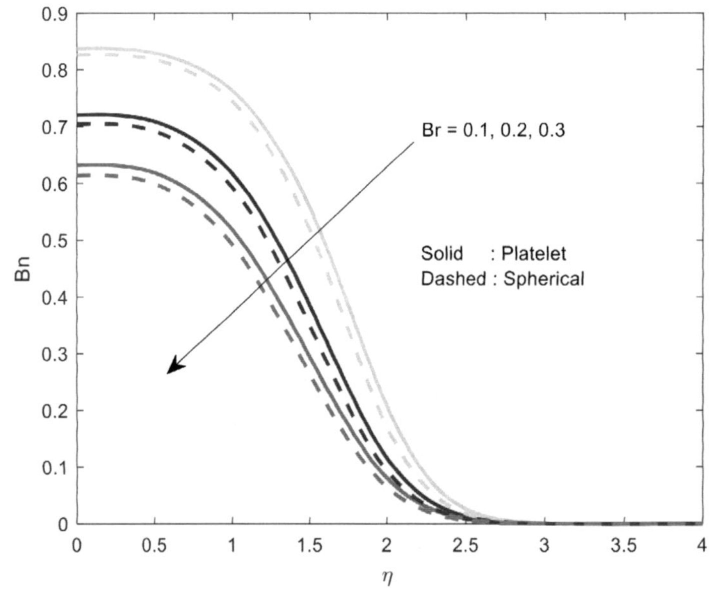

Figure 13. Scenario in which Bn is affected by Br

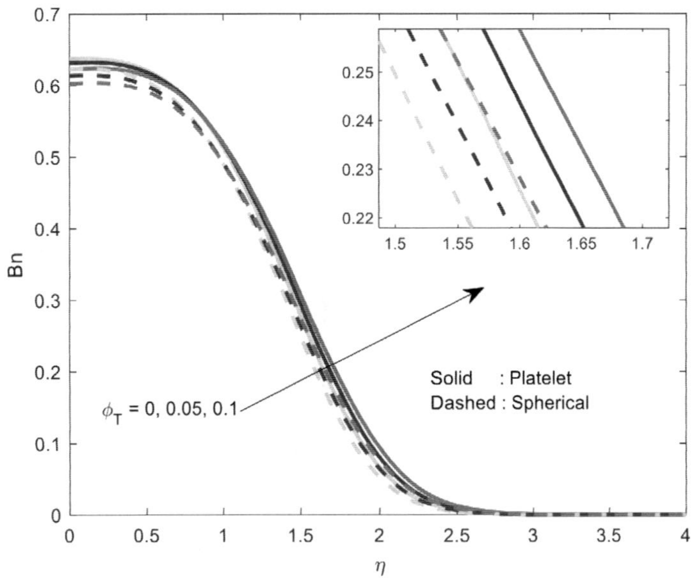

Figure 14. Scenario in which Bn is affected by ϕ_T

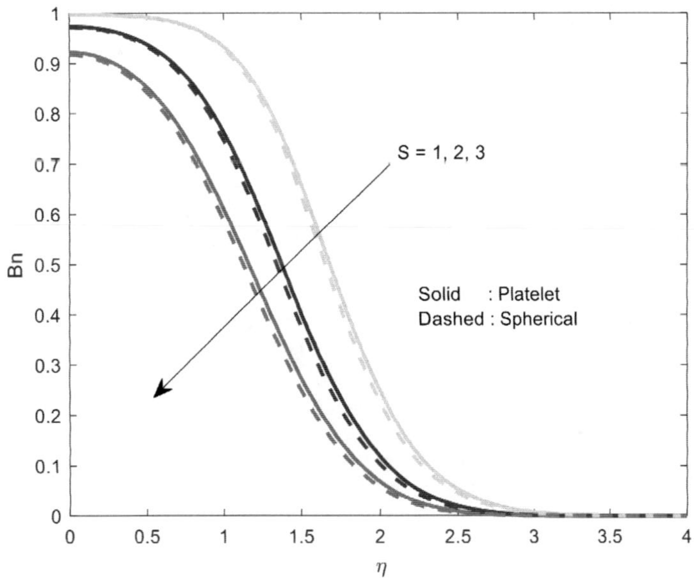

Figure 15. Scenario in which Bn is affected by S

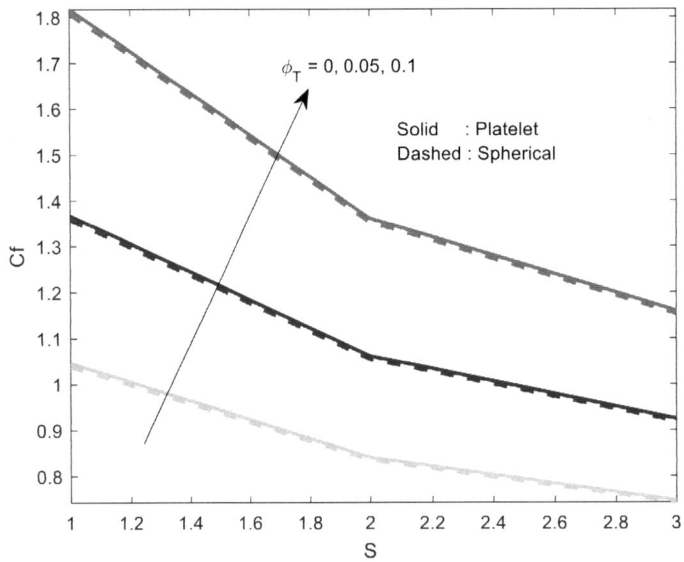

Figure 16. Scenario in which Cf is affected by ϕ_T

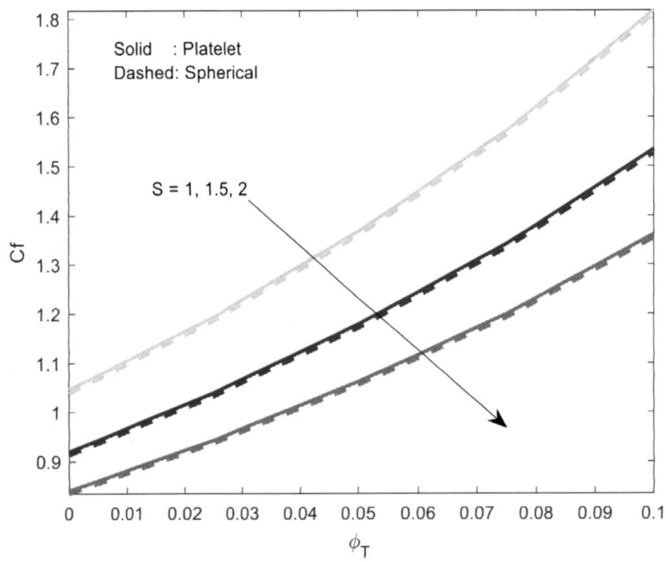

Figure 17. Scenario in which Cf is affected by S

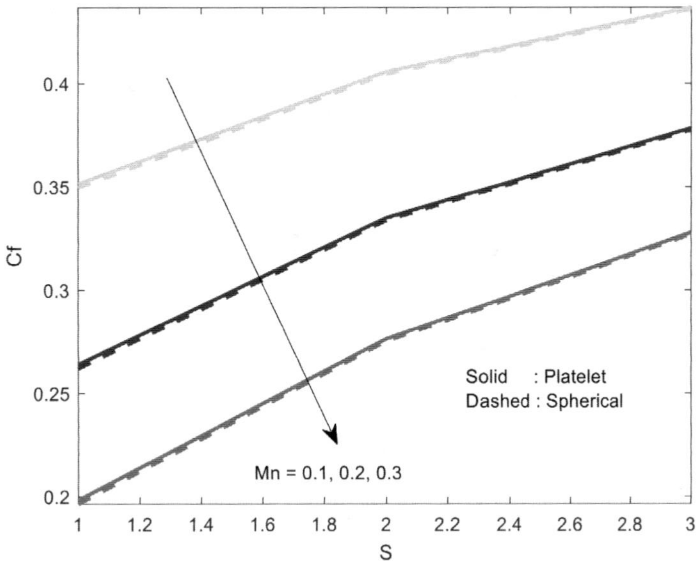

Figure 18. Scenario in which Cf is affected by Mn

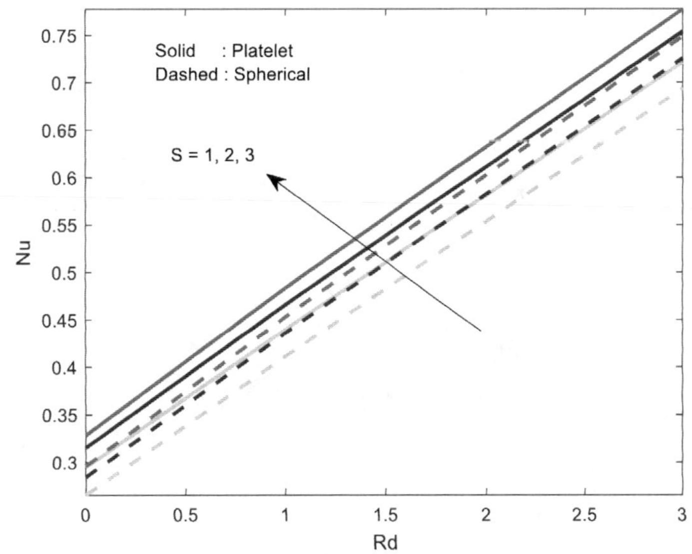

Figure 19. Scenario in which Nu is affected by S

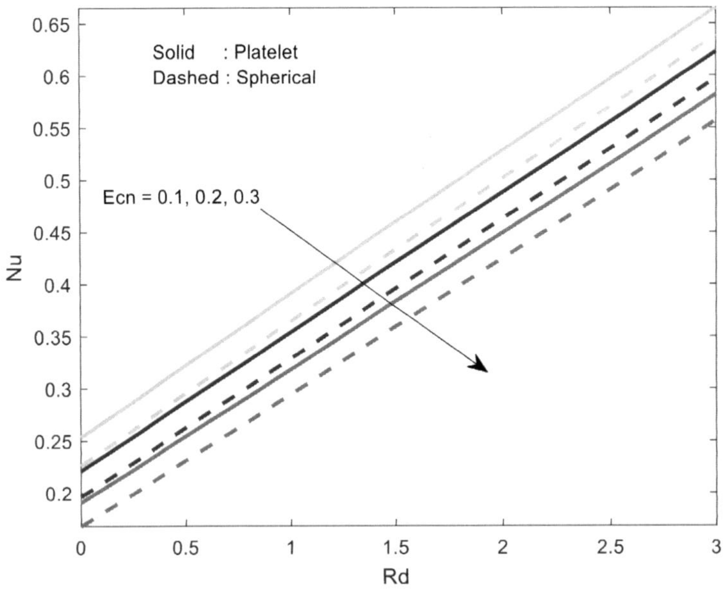

Figure 20. Scenario in which Nu is affected by Ecn

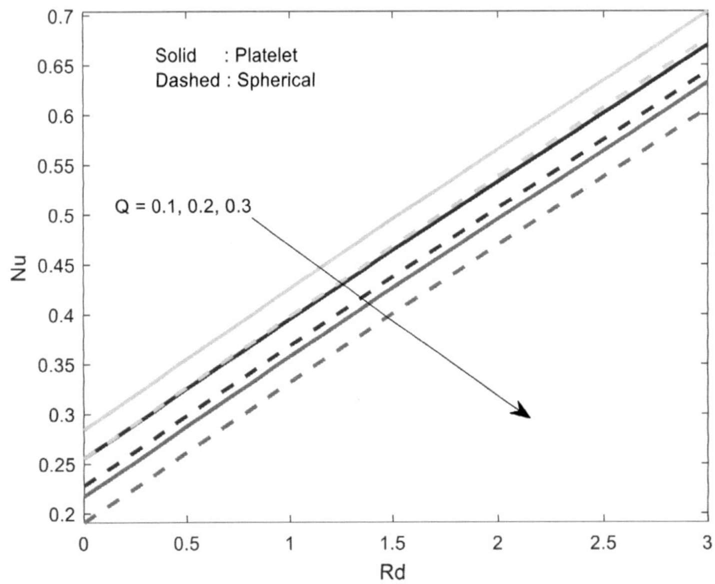

Figure 21. Scenario in which Nu is affected by Q

4. Conclusions

This study theoretically investigates the steady and radiative flow of a ternary hybrid nanofluid (Water + TiO_2 + $CoFe_2O_4$ + MgO) with couple stress, using the Darcy-Forchheimer model. The flow occurs through a tilted flat plate and is subjected to irregular heat source/sink parameter and entropy generation. The findings of the research are briefly articulated as follows:

- It can be noted that the rise in ϕ_T leads to a drop in the velocity of the fluid.
- There is an upsurge in entropy generation associated with the growth in Br.
- When Ecn grows higher, more of the kinetic energy is transformed into thermal energy by viscous forces. The fluid's temperature therefore increases
- Within the range of $0.5 \le S \le 3$, it is observed that the friction factor exhibits a decrease with a rate of 0.306201851 (for Platelet shape) and 0.304466755 (for Spherical shape).
- It has been determined that within the range of $0 \le Ecn \le 0.3$, the Nusselt number lowers by 0.328685192 for Platelet form and by 0.308939422 for Spherical shape.
- Upon comparing the Platelet to other shape, it becomes apparent that its heat transmission rate exhibits a more rapid decline.
- As the value of the ϕ_T increases, the Bejan number rises, while the same decrease with the rise in Br and S.

Nomenclature

σ	Electrical conductivity $\left[Sm^{-1}\right]$	H	Diffusion parameter
$\Pr$	Prandtl number	Rd	Thermal radiation parameter
Br	Brinkman number	Re_x	Reynolds number
υ	Kinematic viscosity $\left[m^2s^{-1}\right]$	χ	Permeability parameter $\left[m^2\right]$
k	Thermal conductivity $\left[Wm^{-1}K^{-1}\right]$	Γ	Inverse porosity parameter (dimensionless)
ϕ_T	Volume fraction of TiO_2	u, v	Velocity components in x, y directions $\left[ms^{-1}\right]$
ϕ_C	Volume fraction of $CoFe_2O_4$	ϕ_M	Volume fraction of MgO
ρ	Density $\left[Kgm^{-3}\right]$	C_p	Specific heat capacitance $\left[JKgK^{-1}\right]$
k^*	Mean absorption coefficient $\left[m^2Kg^{-1}\right]$	Bi	Biot number
β_T	Thermal expansion Coefficient $\left[K^{-1}\right]$	Fn	Forchheimer number
α_1	Temperature difference parameter	h_f	Convective heat transfer Coefficient
σ^*	Stefan-Boltzmann constant $\left[Wm^{-2}K^{-4}\right]$	S	Couple stress parameter
Mn	Magnetic field Parameter	γ	Couple stress coefficient
		Ecn	Eckert number
		Q, Q^*	Non-uniform (space and temperature dependent) heat source/sink parameters

References

[1] Y.S. Daniel, "Laminar convective boundary layer slip flow over a flat plate using homotopy analysis method," J. Inst. Eng. (India): E., vol. 97, pp. 115-121, Oct 2016. DOI: https://doi.org/10.1007/s40034-016-0084-6.

[2] M.R. Safaei, G. Ahmadi, M.S. Goodarzi, A. Kamyar, and S. N. Kazi, "Boundary layer flow and heat transfer of FMWCNT/water nanofluids over a flat plate," Fluids, vol. 1, no. 4, pp. 31, Sep 2016. DOI: https://doi.org/10.3390/fluids1040031.

[3] M. Ziaei-Rad, A. Kasaeipoor, M.M. Rashidi, and G. Lorenzini, "A similarity solution for mixed-convection boundary layer nanofluid flow on an inclined permeable surface," J Therm Sci Eng Appl., vol. 9, no. 2, pp. 021015, Jun 2017. DOI: https://doi.org/10.1115/1.4035733.

[4] A.A. Hakeem, S. Saranya, and B. Ganga, "Comparative study on Newtonian/non-Newtonian base fluids with magnetic/non-magnetic nanoparticles over a flat plate with uniform heat flux," J. Mol. Liq., vol. 230, pp. 445-452, Mar 2017. DOI: https://doi.org/10.1016/j.molliq.2016.12.087.

[5] S. Saranya, P. Ragupathi, B. Ganga, R.P. Sharma, and A.A. Hakeem, "Non-linear radiation effects on magnetic/non-magnetic nanoparticles with different base fluids over a flat plate," Adv Powder Technol., vol. 29, no. 9, pp. 1977-1990, Sep 2018. DOI: https://doi.org/10.1016/j.apt.2018.05.002.

[6] M.M. Nandeppanavar, G.R.S. Reddy Gorla, and S. Shakunthala, "Magneto-hydrodynamic Blasius flow and heat transfer from a flat plate in the presence of suspended carbon nanofluids," Proc. Inst. Mech. Eng., vol. 232, no. 1, pp. 31-40, Mar 2018. DOI: https://doi.org/10.1177/2397791417744702.

[7] Z.Z. Rashed, S.E. Ahmed, and M.A. Sheremet, "MHD buoyancy flow of nanofluids over an inclined plate immersed in uniform porous medium in the presence of solar radiation," J. Mech., vol. 35, no. 4, pp. 563-576, Aug 2019. DOI: https://doi.org/10.1017/jmech.2018.40.

[8] Y. Reddy, and M.S. Reddy, "Heat and mass transfer analysis of single walled carbon nanotubes and multi walled carbon nanotubes-water nanofluid flow over porous inclined plate with heat generation/absorption," J. Nanofluids., vol. 8, no. 5, pp. 1147-1157, Mar 2019. DOI: https://doi.org/10.1166/jon.2019.1667.

[9] B.K. Jha, and G. Samaila, "Thermal radiation effect on boundary layer over a flat plate having convective surface boundary condition," SN Appl. Sci., vol. 2, no. 3, pp. 381, Mar 2020. DOI: https://doi.org/10.1007/s42452-020-2167-8.

[10] R. Khademi, A. Razminia, V.I. Shiryaev, "Conjugate-mixed convection of nanofluid flow over an inclined flat plate in porous media," Appl. Math. Comput., vol. 366, pp. 124761, Feb 2020. DOI: https://doi.org/10.1016/j.amc.2019.124761.

[11] N.H. Abu-Hamdeh, A.A. Aljinaidi, M.A. Eltaher, K.H. Almitani, K.A. Alnefaie, A.M. Abusorrah, and M.R. Safaei, "Implicit finite difference simulation of Prandtl-Eyring nanofluid over a flat plate with variable thermal conductivity: a Tiwari and das model," Mathematics, vol. 9, no. 24, pp. 3153, Dec 2021. DOI: https://doi.org/10.3390/math9243153.

[12] K.S.S. Babu, A. Parandhama, and R. Bhuvana Vijaya, "Non-linear MHD convective flow of Carreau nanofluid over an exponentially stretching surface with activation energy and viscous dissipation," *SN Appl. Sci.,* vol. 3, no. 3, pp. 382, Mar 2021. DOI: https://doi.org/10.1007/s42452-021-04339-4.

[13] A.K. Alzahrani, M.Z. Ullah, A.S. Alshomrani, and T. Gul, "Hybrid nanofluid flow in a Darcy-Forchheimer permeable medium over a flat plate due to solar radiation," Case Stud. Therm. Eng., vol. 26, pp. 100955, Aug 2021. DOI: https://doi.org/10.1016/j.csite.2021.100955.

[14] A.S. Sabu, J. Mackolil, B. Mahanthesh, and A. Mathew, "Nanoparticle aggregation kinematics on the quadratic convective magnetohydrodynamic flow of nanomaterial past an inclined flat plate with sensitivity analysis," Proc. Inst. Mech. Eng. E., vol. 236, no. 3, pp. 1056-1066, Jun 2022. DOI: https://doi.org/10.1177/09544089211056235.

[15] K.S.S. Babu, A. Parandhama, and R. Bhuvana Vijaya, "Significance of heat source/sink on the radiative flow of Cross nanofluid across an exponentially stretching surface towards a stagnation

point with chemical reaction," *Heat Transf.,* vol. 51, no. 4, pp. 2885-2904, Jun 2022. DOI: https://doi.org/10.1002/htj.22428.

[16] H.X. Bao, M.B. Arain, S. Shaheen, H.I. Khan, M. Inc, S.W. Yao, "Boundary-layer flow of heat and mass for tiwari-das nanofluid model over a flat plate with variable wall temperature," Thermal Science, vol. 26, Spec. issue 1, pp. 39-47, Dec 2022. DOI: https://doi.org/10.2298/TSCI22S1039B.

[17] N. Fatima, W. Belhadj, K.S. Nisar, M.K. Alaoui, M.B. Arain, and N. Ijaz, "Heat and mass transmission in a boundary layer flow due to swimming of motile gyrotactic microorganisms with variable wall temperature over a flat plate," Case Stud. Therm. Eng., vol. 45, pp. 102953, May 2023. DOI: https://doi.org/10.1016/j.csite.2023.102953.

[18] M. Yaseen, S.K. Rawat, and M. Kumar, "Linear and quadratic thermal radiation influence on Marangoni convective flow of hybrid nanofluid over a flat surface in a Darcy-Forchheimer porous medium," J. Porous Media., vol. 26, no. 5, pp. 15-36, May 2023. DOI: 10.1615/JPorMedia.2022042246.

[19] G. Revathi, G. Veeram, M. Jayachandra Babu, K.S. Srinivasa Babu, and A. Suneel Kumar, "Darcy–Forchheimer flow of power-law (Ostwald-de Waele type) nanofluid over an inclined plate with thermal radiation and activation energy: an irreversibility analysis," *Int. J. Ambient Energy*, pp. 1-10, 2023. DOI: https://doi.org/10.1080/01430750.2023.2200434.

[20] G. Revathi, M. Jayachandra Babu, K.S. Srinivasa Babu, and C. Bapanayya, "Dynamics of different heat sources and activation energy on the hybrid nanofluid (EG+ MgO+ MWCNT) flow in a microchannel with thermal radiation: An irreversibility analysis," *Numer. Heat Transf.; A: Appl.,* pp. 1-23, Feb 2024. DOI: https://doi.org/10.1080/10407782.2024.2316222.

[21] M. Macha, K. Naikoti, and A.J. Chamkha, "MHD flow of a non-Newtonian nanofluid over a non-linearly stretching sheet in the presence of thermal radiation with heat source/sink," Eng. Comput., vol. 33, no. 5, pp. 1610-1626, Jul 2016. DOI: https://doi.org/10.1108/EC-06-2015-0174.

[22] R. Dhanai, P. Rana, and L. Kumar, "MHD mixed convection nanofluid flow and heat transfer over an inclined cylinder due to velocity and thermal slip effects: Buongiorno's model," Powder Technol., vol. 288, pp. 140-150, Jan 2016. DOI: https://doi.org/10.1016/j.powtec.2015.11.004.

[23] T. Thumma, O.A. Bég, and A. Kadir, "Numerical study of heat source/sink effects on dissipative magnetic nanofluid flow from a non-linear inclined stretching/shrinking sheet," J. Mol. Liq., vol. 232, pp. 159-173, Apr 2017. DOI: https://doi.org/10.1016/j.molliq.2017.02.032.

[24] P.V. Narayana, S.M. Akshit, J.P. Ghori, and B. Venkateswarlu, "Thermal radiation effects on an unsteady MHD nanofluid flow over a stretching sheet with non-uniform heat source/sink," J. Nanofluids., vol. 6, no. 5, pp. 899-907, Oct 2017. DOI: https://doi.org/10.1166/jon.2017.1374.

[25] B.J. Gireesha, M. Archana, B. Mahanthesh, and B.C. Prasannakumara, "Exploration of activation energy and binary chemical reaction effects on nano Casson fluid flow with thermal and exponential space-based heat source," Multidiscip. Model. Mater. Struct., vol. 15, no. 1, pp. 227-245, Nov 2018. DOI: https://doi.org/10.1108/MMMS-03-2018-0051.

[26] S. Saleem, M.M. Al-Qarni, S. Nadeem, and N. Sandeep, "Convective heat and mass transfer in magneto Jeffrey fluid flow on a rotating cone with heat source and chemical reaction," Commun. Theor. Phys., vol. 70, no. 5, pp. 534, Nov 2018. DOI: 10.1088/0253-6102/70/5/534.

[27] B. Ramadevi, K.A. Kumar, V. Sugunamma, and N. Sandeep, "Influence of non-uniform heat source/sink on the three-dimensional magnetohydrodynamic Carreau fluid flow past a stretching surface with modified Fourier's law," Pramana, vol. 93, pp. 86, Dec 2019. DOI: https://doi.org/10.1007/s12043-019-1847-7.

[28] K.A. Kumar, V. Sugunamma, N. Sandeep, and M. Mustafa, "Simultaneous solutions for first order and second order slips on micropolar fluid flow across a convective surface in the presence of Lorentz force and variable heat source/sink," Sci. Rep., vol. 9, no. 1, pp. 14706, Oct 2019. DOI: https://doi.org/10.1038/s41598-019-51242-5.

[29] B. Mahanthesh, J. Mackolil, and S.A. Shehzad, "Statistical analysis of stagnation-point heat flow in Williamson fluid with viscous dissipation and exponential heat source effects," Heat Transf., vol. 49, no. 8, pp. 4580-4591, Dec 2020. DOI: https://doi.org/10.1002/htj.21842.

[30] S.S.K. Raju, M.J. Babu, and C.S.K. Raju, "Irreversibility analysis in hybrid nanofluid flow between two rotating disks with activation energy and cross-diffusion effects," Chin. J. Phys., vol. 72, pp. 499-529, Aug 2021. DOI: https://doi.org/10.1016/j.cjph.2021.03.016.

[31] A. Dadheech, A. Parmar, and A. Olkha, "Inclined MHD and radiative Maxwell slip fluid flow and heat transfer due to permeable melting surface with a Non-linear heat source," Int. J. Appl. Comput. Math., vol. 7, no. 3, pp. 89, Jun 2021. DOI: https://doi.org/10.1007/s40819-021-01021-6.

[32] A. Venkateswarlu, N. Murshid, H. Mulki, M. Abu-samha, S. Suneetha, M.J. Babu, C.S.K. Raju, R.Z. Homod, and W. Al-Kouz, "A Significant Role of Activation Energy and Fourier Flux on the Quadratically Radiated Sphere in Low and High Conductivity of Hybrid Nanoparticles," Symmetry, vol. 14, no. 11, pp. 2335, Nov 2022. DOI: https://doi.org/10.3390/sym14112335.

[33] M. Gayatri, K. Jayaramireddy, and M. Jayachandra Babu, "Nonlinear convective flow of Maxwell fluid over a slendering stretching sheet with heat source/sink," J. Appl. Comput. Mech., vol. 8, no. 1, pp. 60-70, Jan 2022. DOI: 10.22055/JACM.2019.31394.1868.

[34] B.K. Sharma, A. Kumar, R. Gandhi, and M.M. Bhatti, "Exponential space and thermal-dependent heat source effects on electro-magneto-hydrodynamic Jeffrey fluid flow over a vertical stretching surface," Int. J. Mod. Phys. B., vol. 36, no. 30, pp. 2250220, Dec 2022. DOI: https://doi.org/10.1142/S0217979222502204.

[35] A.C. Venkata Ramudu, K. Anantha Kumar, and V. Sugunamma, "Application of heat transfer on MHD shear thickening fluid flow past a stretched surface with variable heat source/sink," Heat Transf., vol. 52, no. 2, pp. 1161-1177, Mar 2023. DOI: https://doi.org/10.1002/htj.22734.

[36] T. Sajid, W. Jamshed, R.W. Ibrahim, M.R. Eid, A. Abd-Elmonem, and M. Arshad, "Quadratic regression analysis for nonlinear heat source/sink and mathematical Fourier heat law influences on Reiner-Philippoff hybrid nanofluid flow applying Galerkin finite element method," J. Magn. Magn. Mater., vol. 568, pp. 170383, Feb 2023. DOI: https://doi.org/10.1016/j.jmmm.2023.170383.

[37] K.N. Sneha, G.P. Vanitha, U.S. Mahabaleshwar, and D. Laroze, "Effect of couple stress and mass transpiration on ternary hybrid nanoliquid over a stretching/shrinking sheet with heat transfer," Micromachines, vol. 13, no. 10, 1694, Oct 2022. DOI: https://doi.org/10.3390/mi13101694.

[38] D. Pal, and H. Mondal, "Influence of Soret-Dufour and thermophoresis on hydromagnetic mixed convection heat and mass transfer over an inclined flat plate with non-uniform heat source/sink and chemical reaction," Int. J. Comput. Methods Eng. Sci. Mech., vol. 19, no. 2, pp. 49-60, Mar 2018. DOI: https://doi.org/10.1080/15502287.2018.1430073.

[39] K.A.M. Alharbi, A.E.S. Ahmed, M. Ould Sidi, N.A. Ahammad, A. Mohamed, M.A. El-Shorbagy, M. Bilal, and R. Marzouki, "Computational valuation of Darcy ternary-hybrid nanofluid flow across an extending cylinder with induction effects," Micromachines, vol. 13, no. 4, pp. 588, Apr 2022. DOI: https://doi.org/10.3390/mi13040588.

[40] S. Manjunatha, V. Puneeth, B.J. Gireesha, and A. Chamkha, "Theoretical study of convective heat transfer in ternary nanofluid flowing past a stretching sheet," J. Appl. Comput. Mech., vol. 8, no. 4, pp. 1279-1286, Oct 2022. DOI: 10.22055/JACM.2021.37698.3067.

A Review on Distribution of Reinforcement and tensile Strength of Aluminum Lithium Alloy based Nano Metal Matrix Composites Fabricated by STIR Casting

Sk M Z M S Ali [a,c*], K Krishna Bhaskar [b], K Sitarama Raju [c]
[a] Research Scholar, Dept. of Mechanical Engineering,UCEK,JNTUK,Kakinada- 533003,India
[b] Dept. of Mechanical Engineering,UCEK,JNTUK,Kakinada-533003,India
[c] Dept. of Mechanical Engineering ,S.R.K.R. Engineering. College, Bhimavaram – 534204, India

*Corresponding Author email: saqheebali@gmail.com

Abstract. Aluminum Alloy (AA) parts are employed in lightweight engineering fields and have poor mechanical, wear, and corrosion resistance. Aluminum Matrix Composites (AMCs) are materials created by incorporating hard reinforcing particles into an aluminum matrix alloy. The new material has improved characteristics, including higher specific stiffness and strength, as well as higher corrosion resistance, elastic modulus, wear resistance, and low weight. Metal Matrix Composites (MMCs) are composite materials in which metal serves as the matrix and hard particles or ceramics serve as reinforcement. MMCs are used to improve the functionality of a wide range of technical materials. It could be used in many different fields, including aerospace, automotive, space travel, and medicine. In this paper, the reviewer investigates the distribution of reinforcement and Tensile Strength (TS) of Aluminum Lithium (Al-Li) Alloy/Nano Al_20_3 MMC fabricated by stir casting. Finally, the authors provide a comparative analysis of prior reviewed research papers. After comparative analysis, TiB_2/2195 composite has higher Tensile Strength (115.4%) than the other composite such as AA7010- TiB_2 (260%), Al_2O_3 + SiC (72%), A356/SiC (21.87%), and Al-SiC (37%).

Keywords: Stir Casting, Tensile Strength, Metal Matrix Composites, Al-Li Alloy, Composite Materials.

1. Introduction

The aircraft, automotive, military, and structural industries could all benefit from AA nanocomposite. Hybrid nanocomposites each have their unique characteristics. Lightweight AA has their mechanical characteristics improved during production by including ceramic powders such as Boron Carbide (B_4C), zirconia, Aluminum Oxide (Al_2O_3), and Silicon Carbide (SiC). In recent years, hybrid composites with improved mechanical characteristics have been developed. Controlling factors like stirring duration, pace, and temperature ensure a consistent mix and thorough wetting. A low-cost and appealing method of making Aluminum Metal Matrix Composites (AMMCs) is regular stir casting, which is only one of the various production processes. SiC and Al_2O_3 are two reinforcements that are chemically appropriate for lightweight aluminum and enable optimal bonding with the matrix [1-2].

Aerospace and automotive materials must meet demanding requirements for weight, specific strength, fatigue load, corrosion/crack resistance, and wear rate [3]. Alloy steels were formerly employed mostly in aircraft and automobile parts. There are several advantages to using ferrous metals, such as their low cost and widespread availability, but they also have certain disadvantages, such as their high density, high wear rate, and low corrosion resistance. According to their low densities relative to ferrous metals, nonferrous metals such as aluminum, Magnesium (Mg), Titanium (Ti), etc., have attracted widespread attention for their potential uses. Mg is a lightweight nonferrous metal; however, it can't be used to make airplanes since it catches fire easily [4]. Ti weighs the same as Mg and costs 20–30 times more than Al.

Pure Al has a TS of about 90 MPa, which could be enhanced by around twice that by rolling or other cold working operations. The TS could be enhanced to the strength range of structural steel through alloying with various metals or through heat treatment techniques. In many applications, AA is preferred over Mg alloys due to its greater strength. AA is made by several processes by adding components like copper, manganese, silicon, Mg, zinc, and lithium to the basic metal.

1.1 Classification of Composite Materials

Composites are a kind of engineered materials that are formed from a variety of different materials that are individually unique and give several benefits that would be impossible to get with just one material separately [5-6]. The matrix phase is continuous in composite materials, whereas reinforcing phases are often discontinuous and could take on a variety of shapes depending on their aspect ratio. The composite material could be adapted to meet specific needs with the help of a product designer by selecting the most suitable materials [7]. Matrix and reinforcing materials are used to categorize composites [8-9]. Figure 1 provides a general overview of the many types of composites. Composites could be broken down into the following categories based on the kind of matrix material that employ. A composite material is the result of the careful combination of two or more constituent materials. The properties of these aggregates are distinct from those of the constituent materials. There are two phases in composite materials [10].

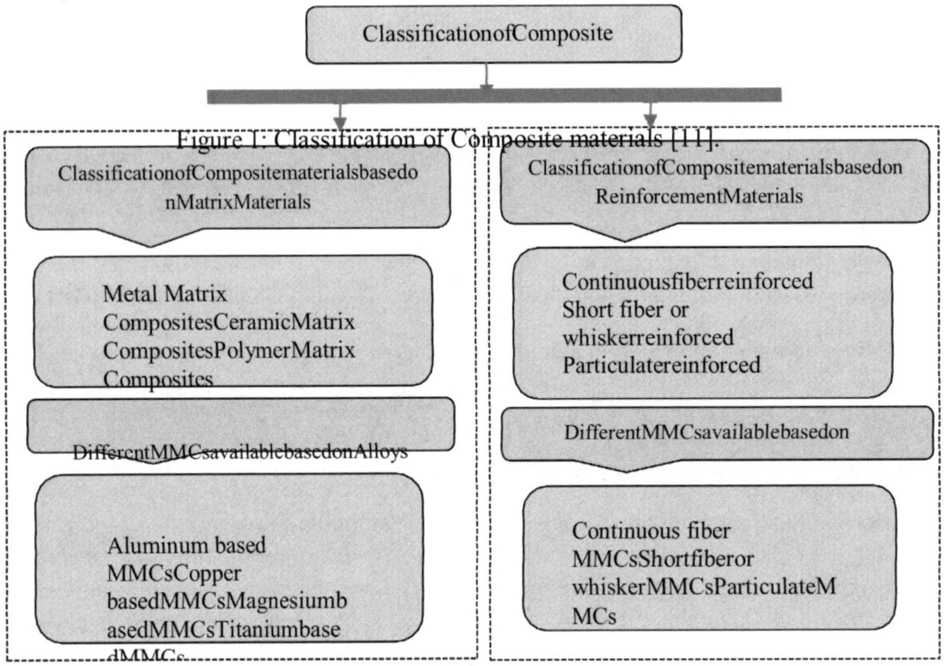

Figure 1: Classification of Composite materials [11].

A. Matrix phase

The matrix is the primary material from which the reinforced particle (ceramics/fiber/whisker) is derived and used to enhance the material's properties. Matrix materials could be selected depending on the composite material application. The metal matrix is important because any grain structure change would modify the composite material. Composite materials are divided into MMCs, Ceramic-Matrix Composites (CMCs), and Polymer-Matrix Composites (PMCs) depending on the kind of matrix phase. Figure 2 depicts examples of matrix materials used in composites.

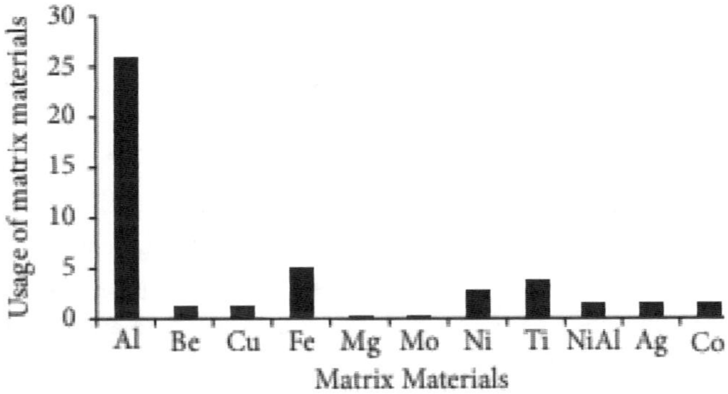

Figure 2: Matrix materials [12].

- *Metal Matrix Composites (MMCs)*

Matrix materials of MMCs must be metallic (often Al, Mg, Copper (Cu), and Ti), whereas reinforcements may be either scattered ceramics such as oxides and carbides or metallic (i.e., tungsten, molybdenum, lead). The number of reinforced composites ranges from a few percent to almost half of the entire volume. Al-based MMCs are commonly used in the automotive and aerospace industries because reinforcement compounds such as SiC and Al_2O_3 could be simply and efficiently mixed in molten Al to accomplish preferred characteristics such as superior strength, improved stiffness, reduced density, controlled thermal expansion, and improved wear resistance [13-15]. MMCs have significant tool wear rates because of their stiffness and abrasive structures, hence they are frequently machined using unusual methods that prevent tool-material contact [16].

- *Ceramic Matrix Composites (CMCs)*

CMCs are a kind of ceramics characterized by fibers of carbon, SiC, Al_2O_3, or Silicon Nitride (SiN) inserted in a CMC [17]. They are intended to mitigate the brittleness of monolithic ceramics. CMCs are classified as inverse composites since the matrix failure strain is less than the fiber failure strain. This effect occurs in a backward direction in most polymers or MMCs. Therefore, it is the matrix that breaks under loading circumstances before the brittle fibers do [18-19]. CMCs are manufactured using two distinct types of processing methods: gas and liquid phase. The inner phase and matrix are generated from gaseous or liquid precursors surrounding the fibers in this process [20-21].

- *Polymer Matrix Composites (PMCs)*

PMCs are made up of a matrix containing thermosetting or thermoplastic material reinforced using carbon, glass, Kevlar, or metal fibers [22]. Thermosets are highly common than thermoplastics owing to their superior strength and stability to high temperatures [23]. Resin and hardener are combined to make thermosets. The most common kind of composite is a laminar one, which is fabricated by stacking and gluing thin layers of fiber and polymer to the appropriate thickness. PMCs are inexpensive composites because of their cheap material cost and ease of handling and production [24].

B. Reinforcement Phase

The reinforcement phase of composite material is the second phase. The high stiffness and high strength of the composite material could be attributed to the reinforcement. Load (stress) is transferred from the matrix material to the reinforcing material, which shields and maintains the matrix material's form. The major composite family member that transfers stress to the matrix material is reinforced material. Strength, stiffness, hardness, etc. are only a few of the mechanical characteristics that could be improved in a composite material by using reinforcement material. Fiber, particle, whisker, and flake composites are the four main types based on reinforcement morphology. Three kinds of composites are explained below in the reinforcing phase. Figure 3 illustrates the use of reinforcing materials in composites.

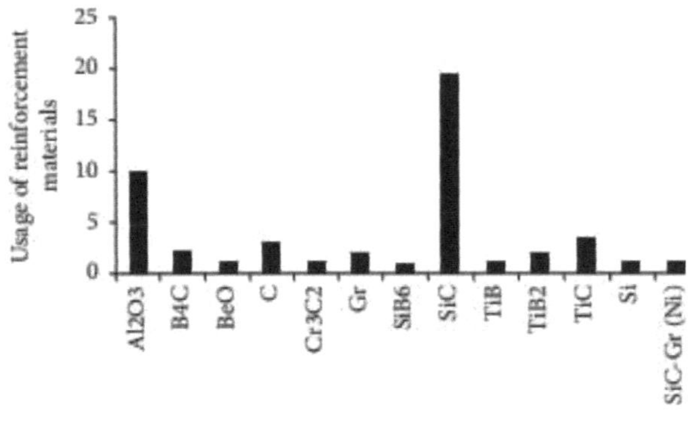

Figure 3: Reinforcement materials [12].

- *Fiber-Reinforcement Composite (FRC)*

Fiber length is a useful metric for categorizing composites, which are made up of fibers in a matrix structure. Continuous FRC is those in which the fiber reinforcements are long, whereas discontinuous FRC is those in which the fiber reinforcements are short. A hybrid FRC uses reinforcement from more than one fiber type in a single matrix [25]. Fibers could be arranged unidirectionally or bidirectionally in continuous fiber composite matrix structures to transfer loads from the matrix to the fiber easily and effectively. In the case of brittle matrices, load transmission is maximized, and material failure is prevented by ensuring that the length of the discontinuous fibers is long enough to prevent the propagation of fractures. The structure and characteristics of composite materials are determined by the direction and organization of their fibers [26-27]. The utilization of chemically treated natural fibers could enhance characteristics like as impact toughness and fatigue strength.

- *Particle-Reinforced Composite (PRC)*

PRC is superior to FRC in terms of strength and fracture resistance. A material's deformation could be limited, and its rigidity improved by the incorporation of ceramic, metallic, or inorganic particles. PRCs have recently gained popularity due to their isotropic features and low cost. Additionally, these composites are produced using the same methods as monolithic materials. PRCs are used in a variety of civil applications, including roads and concrete buildings, because of their exceptional wear resistance in these environments. Cement operates as a binder in concrete, while coarse rock or gravel aggregates serve as a filler to provide strength and rigidity to the material [28].

- *Sheet-Model Composites (SMC)*

SMCs are made by a compression molding technique that bonds homogenous layers of materials to create nonhomogeneous composite laminates. Laminated materials are constructed from many layers, and the buckling stability of Fiber Reinforced Polymers (FRP) composites manufactured from fiber sheets increases with the number of layers in the laminate. SMC is used in big structural components such as automobile body sections that have a high strength-to-weight ratio . Natural fibers' tensile characteristics are determined by their chemical compositions. Fibers with a higher cellulose concentration have a higher TS, whereas fibers with a higher lignin content have a lower TS [30].

1.2 Fabrication methods of MMCs

There are three primary categories of fabrication processes for MMCs are explained in given below:

A. Liquid State Fabrication

Liquid phase processing of MMCs involves incorporating the reinforcing phase into the molten metal itself. Increased mechanical characteristics could be achieved by obtaining fine interfacial bonding between the reinforcements and the molten metal. Wetting improvement could be obtained by coating the reinforcing particles (fibers).

- *Stir Casting*

Mechanical stirring is used to incorporate the reinforcing phase into the metal matrix in this procedure. The stir-casting method is widely used by scientists who study matrix composites for a variety of purposes. Initially, the base metal or alloy is stored in a crucible and warmed in a furnace until it reaches melting. After the matrix has melted, the liquid metal is evenly heated and distributed with the help of a stirrer. The reinforcing particles or fibers are now added, and the mixture is stirred again to ensure that they are distributed evenly. The required MMC is produced when the melting process has cooled and solidified. The produced MMC is more uniform and defect-free because of the constant stirring [31]. The improved mechanical properties of MMCs produced using this method would find use in a wide range of technical fields [32]. The most important part of this technique is the stirring phase, which combines the liquid metal with the dispersion phase [33-34]. The stir-casting process for producing MMC is shown in Figure 4.

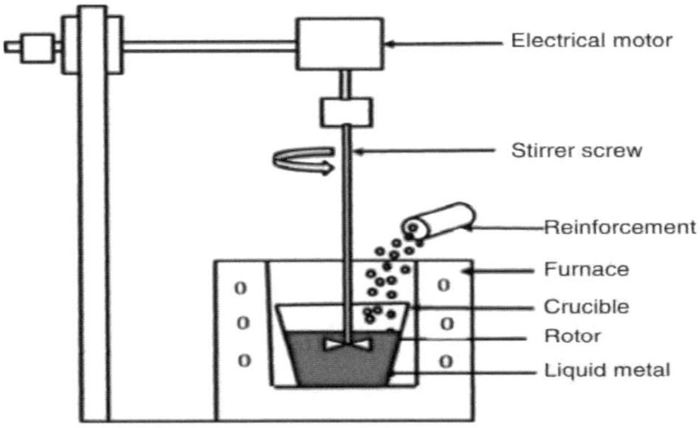

Figure 4: Stir Casting method [35].

- *Infiltration*

Composites are produced using an infiltration technique in which the reinforcing phase (particulates, fibers) is submerged in a liquid metal matrix. Liquid metal covers the voids among the reinforcing phase defects in the constructed structure. A molten matrix phase could be infiltrated by using the capillary force of the reinforcement phase or by applying external pressure. The authors [36] observed limitations to this approach, including void creation following particle penetration into reinforced metals. High volume percentages of particles and temperature expansion variations between the matrix and reinforcement both contribute to the formation of voids [37].

B. Solid State Fabrication

MMCs are produced using procedures where the metal matrix and reinforcement phase bond. High-pressure, high-temperature bonding among the metal matrix with reinforcement phase occurs by mutual diffusion. It is the two types of fabrication such as:

- *Diffusion Bonding*

Diffusion bonding is a process in which atoms diffuse in a solid state because of pressure and heat. Traditional welding is challenging to connect various AA because of the lack of adequate filler metals due to varied chemical makeup. Diffusion bonding is used to connect alloys and eliminates fracture formation, deformation of grain boundaries, and other undesirable effects by heating the materials to temperatures between half and eight times their melting points. The holding duration must be kept to a minimum to ensure a chemical connection is formed across the whole contact without allowing an excessively thick inter-diffusion layer to form. It's a time-consuming, expensive, and laborious process [38].

- *Powder Metallurgy (PM)*

PM is a metal-working process in which finished products are fabricated from powdered metal using a sequence of operations, including the mixing of powders (matrix and reinforcement) in a predefined ratio, the compression of the powder into the desired shape, as well as the heating of the powder to a temperature just under its melting point. PM improves mechanical characteristics, particle dispersion, and matrix-reinforcement ratio, allowing reinforcement material to approach 50% by weight. PM is a complicated procedure that requires a controlled environment such as a vacuum or protective atmosphere. It is expensive to produce because of the complexity of the required technology and the small size and shape of the available pieces [39-40].

C. In-Situ Fabrication

The use of in-situ techniques to create MMC is a more recent development. The matrix reinforcing phase is created by directed metallurgical processes. The molten alloy matrix is a reactant in this process. Powder or gaseous forms of another chemical could be added to the melt [41]. Particle size, dispersion, and thermodynamic stability are only a few of the benefits of this approach . The mechanical characteristics of composites are enhanced by the fineness and uniform distribution of reinforcing particles. Additionally, since the particles are not exposed to air or impurities, a clean matrix-reinforcement contact is formed, which increases the strength. But alignment and geometry during the reinforcing step are typically difficult to manage. A simplified diagram of in-situ composites manufacture is shown in Figure 5 [42].

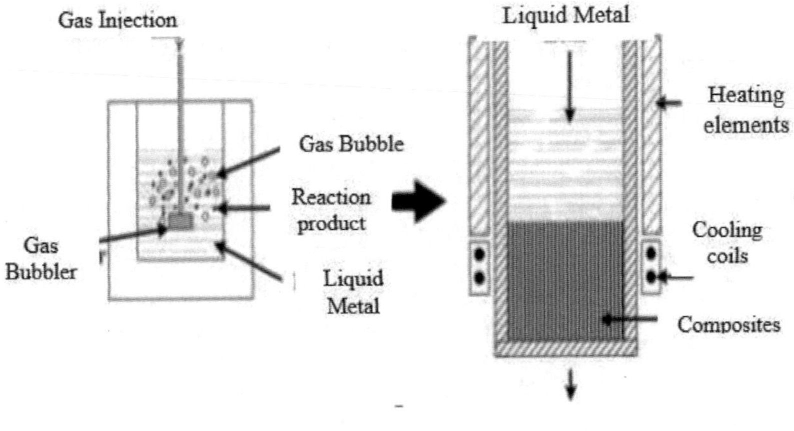

Figure 5: In-situ process [42].

2. Literature of Review

The following study expands reinforcement and TS of Al Lithium Alloy/Nano Al203 MMC fabricated by stir Casting. Several researchers explained their findings as seen below.

Xie Z. et al., (2022) [43] evaluate the phase composition, microstructure, and mechanical characteristics of the suggested 2wt.% TiB_2/2195 composites, specimens were made using ultrasonic assisted in-situ casting method. The findings indicate that, in comparison to the remelted matrix, the stomatal flaws in the composites vanished, and the granules were polished. However, the second phase structures and TiB_2 particles accumulated considerably while no Ultrasonic Treatment (UT) was performed. TS, Yield Strength (YS), and elongation were all improved by 115.4%, 49.8%, and 342.9%, respectively, compared to the remelted matrix, and by 30.9%, 21.8%, and 67.2%, respectively, comparing to the composite without UT.

Narasimmalu R. et al., (2022) [44] focus of this work is to enhance the material's resistance to wear. Stir casting techniques were used to create composites of AA8090, B_4C, and Al_2O_3. The findings revealed that the composite materials had improved characteristics compared with AA8090 because of the reinforcing impact of B_4C and Al_2O_3 particles. Additionally, the microhardness and TS of the AA/3BC composite improved by 12.9% and 10.8%, accordingly.

Huda A. et al., (2020) [45] purpose of this study is to better understand how the Al6061/Al_2O_3 nanocomposite formed during the stir casting synthesis method. Results from experiments showed that the mechanical characteristics had been significantly improved, with Yield Tensile Strength (YTS) increasing from 52 to 69 MPa with Ultimate Tensile Strength (UTS) increasing from 122 to 158 MPa. The use of 10 wt% aluminas increased the hardness test results from 28 to 38 Brinell Hardness (BHN). There were observable increase ratios of 32.69, 29.51, and 35.71 percent.

Mallireddy N. et al., (2020) [46] evaluated that the exothermic reaction of molten AA7010 with halide salts K2TiF6 and KBF4 injected in 120% excesses to the stoichiometric ratio resulted in the production of AA7010-TiB2 composites with reinforcing levels of 5, 7.5, and 10 wt%. Optical micrographs show that the in-situ stir casting technique is effective for producing ceramic-reinforced MMC with uniform particle dispersion and little permeability. The findings suggest that the composites' hardness, TS, and YS were increased by about 35%, 260%, and 240%, respectively.

Suresh S. et al., (2019) [47] investigated the effects of Al2O3 (20-30 nm), SiC (50 nm), and Mg (micro) on the mechanical and wear characteristics of Al7075. Al metal matrix nanocomposites are fabricated using a stir casting process with compositions ranging from 1 to 4 wt. The microhardness, Compression Strength (CS), and TS of Al7075 alloy improve by integrating Al_2O_3 and SiC reinforcing.

Dhaneswara D. et al., (2018) [48] utilized the stir casting process, and nanoscale SiC reinforcements were effectively produced in A356/SiC. Testing for TS, hardness, and wear was used to evaluate the material's mechanical characteristics, while an optical microscope was used to study the material's microstructure. Mechanical characteristics were maximized at 0.25 wt.% Nano SiC, with a value of 175.57 MPa (21.87%) for strength, 0.0287 Joules/mm3 (14.8%) for toughness, 56.54 HBN (50% for hardness), and 1.75 x 10-5 mm3/mm for wear resistance (21.13%).

Reddy M. et al., (2017) [49] explore the Al matrix composite was produced by microwave-assisted rapid sintering and hot extrusion, and then enhanced with 5, 10, and 15 vol. Microhardness, YS, UTS, and Young's modulus all increased with increasing hard ceramic particle concentration, whereas flexibility and thermal expansion coefficient decreased. Al-15vol.%Al_2O_3 composite has a higher UTS and YS than pure Al (33% and 32.4% increases, respectively) due to the presence of the reinforcing particles.

Matli P. et al., (2017) [50] analyze the produced Al-Sic nanocomposites' structural, microstructural, and mechanical characteristics as a function of SiC addition. When compared to pure Al, the reinforcing of SiC nanoparticles (1.5%SiC) increases hardness, CS, and TS by 121, 11, and 37%, respectively, while reducing ductility from 13.5 to 7.3. The inclusion of hard ceramic phases into the soft metal matrix is credited for the observed improvements in hardness and strength.

Malleswararao et.al [51] investigated the effect of varying process parameters on the metallurgical behaviour and particle distribution and their influence on the mechanical properties, viz., ultimate tensile strength and tensile elongation. The results obtained gained a clear perspective of friction stir welding process parameters on AA6061-T6 and AA2017- T6 joints with optimal mechanical properties.

There is a wide range of authors who used the technique and presented their discoveries, as given in table 1.

Table 1. Summarize the table of the reviewed literature

Authors	Technique Used	Composites	Outcomes
Xie Z. et al., (2022) [43]	Ultrasonic-Assisted In-Situ Casting	TiB_2/2195	TS, YS, and elongation were all improved by a factor of 115.4%, 49.8%, and 342.9% in comparison to the remelted matrix, and by a factor of 30.9%, 21.8%, and 67.2% in comparison to the composite without UT.
Narasimmalu R. et al., (2022) [44]	Stir Casting	AA8090	Microhardness and TS were both improved by 12.9% and 10.8% in the AA/3BC composite.
Huda A. et al., (2020) [45]	Stir Casting	Al6061/Al_2O_3	The testing findings showed that YTS improved from 52 to 69 MPa and UTS increased from 122 to 158 Mpa.

Malliredd y N. et al., (2020) [46]	In-situ process	AA7010-TiB_2	The findings indicate that the composites' hardness, TS, and YS were increased by roughly 35%, 260%, and 240%.
Suresh S. et al., (2019) [47]	Stir Casting	Al_2O_3 + SiC	Al2O3 and SiC reinforcing boost the microhardness, CS, and TS of Al7075 alloy.
Dhanesw ara D. et al., (2018) [48]	Stir Casting	A356/SiC	The mechanical characteristics were optimal at 0.25wt.% Nano SiC, with strength of 21.87%, toughness of 14.8%, hardness of 50%, and wear resistance of 21.13%.
Reddy M. et al., (2017) [49]	Microwave -assisted rapid sintering.	Al-Al_2O_3	The Al-15vol.%Al_2O_3 composite has 154±6 MPa UTS and 139±8 MPa YS, 33% and 32.4% higher than pure Al.
Matli P. et al., (2017) [50]	Microwave Sintering	Al-SiC	The hardness, CS, and TS have all been increased by 121, 11%, and 37% due to the use of SiC nanoparticles (1.5%SiC) for reinforcing.

3. Comparison Analysis

This part of the review provides a comparative analysis of the literature that was examined above. By using a variety of datasets, it determined the TS that had been found in all the prior reviewed research papers. After comparative analysis, TiB_2/2195 composite has higher tensile strength (115.4%) than the other composite such as AA7010-TiB_2, Al_2O_3 + SiC, A356/SiC, and Al-SiC. Table 2 indicates the comparative analysis of LOR and Figure 6 indicates the graph of comparative analysis.

Table 2. Evaluation of existing studies

Authors	Technique Used	Composites	Tensile Strength (%)
Xie Z. et al., (2022) [43]	In-situ Casting	TiB_2/2195	115.4%

Mallireddy N. et al., (2020) [46]	In-situ process	AA7010-TiB_2	260%
Suresh S. et al., (2019) [47]	Stir Casting	Al_2O_3 + SiC	72%
Dhaneswara D. et al., (2018) [48]	Stir Casting	A356/SiC	21.87%
Matli P. et al., (2017) [50]	Microwave Sintering	Al-SiC	37%

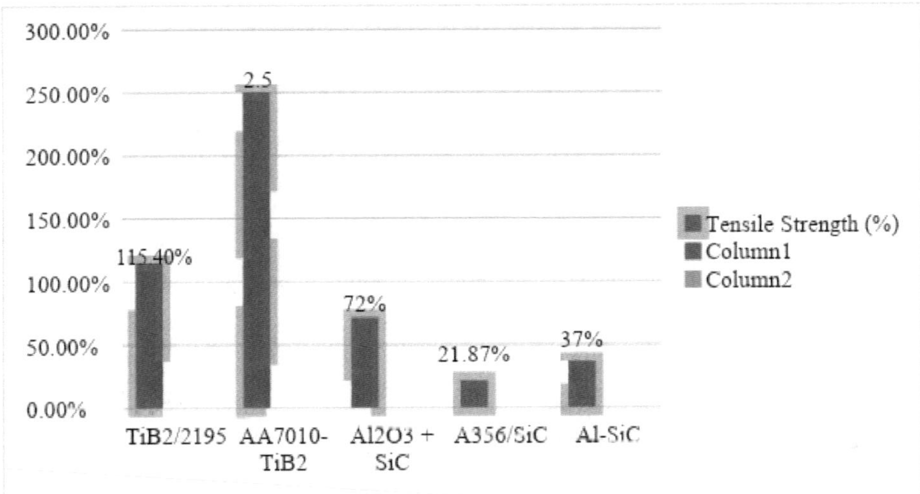

Figure 6: Graph of tensile strength

4. Conclusion and Future Scope

The increase in demand for the new materials is occurring on a global scale. The lightweight and cost-effective nature of metal matrix nanocomposites makes them ideal for a wide range of applications, including those in the aerospace, automotive, armed forces, and commercial sectors. MMCs are best fabricated using the stir cast technique in liquid phase manufacturing. The incorporation of reinforcing particle boosted the enhancement in the composite material's mechanical characteristics. This review suggested an investigation on the distribution of reinforcement and TS of Al-Li Alloy/Nano Al_2O_3 MMC fabricated by stir casting. By using a variety of datasets, it determined the TS that had been found in all the prior reviewed research papers. After comparative analysis, TiB_2/2195 composite has higher tensile strength (115.4%) than the other composite such as AA7010-TiB_2 (260%), Al_2O_3 + SiC (72%), A356/SiC (21.87%), and Al-SiC (37%).

In the future, Additional study and development are needed to simplify the process of combining MMCs, and specific tool improvements are required for the manufacturing of MMCs.

References

[1] A. Baradeswaran, S.C. Vettivel, A. Elaya Perumal, N. Selvakumar, R. Issac, Experimental investigation on mechanical behaviour, modelling, and optimization of wear parameters of B4C and graphite reinforced aluminium hybrid composites. Mater. Des. **63**, 620–632 (2014).

[2] Suresh, S., G. Harinath Gowd, and M. L. S. Deva Kumar. "Mechanical properties of AA 7075/Al2O3/SiC nano-metal matrix composites by stir-casting method." *Journal of The Institution of Engineers (India): Series D* **100**, no. 1 (2019): 43-53.

[3] F. C. Campbell Jr, Manufacturing Technology for Aerospace Structural Materials, Elsevier, Amsterdam, Netherlands, 2011.

[4] R. Jojith, N. Radhika, and M. Govindaraju, —Reciprocating wear behavioural analysis of heat-treated aluminium ZrO2/ Al7Si0. 3Mg functionally graded composite through taguchi's optimization method,‖ Silicon, pp. 1–18, 2022.

[5] Vencl, A.; Rac, A.; Bobi´c, I. Tribological Behaviour of Al-Based MMCs and their Application in Automotive Industry. Tribol. Ind. 2004, **26**, 31–38.

[6] Basak, A.K.; Pramanik, A.; Prakash, C. Deformation and strengthening of SiCreinforced Al-MMCs during in-situ micro-pillar compression. **Mater. Sci. Eng.** A 2019, **763**, 138141.

[7] Shanmugavel, R.; Chinthakndi, N.; Selvam, M.; Madasamy, N.; Shanmugakani, S.K.; Nair, A.; Prakash, C.; Buddhi, D.; Dixit, S. Al-Mg-MoS2 Reinforced Metal Matrix Composites: Machinability Characteristics. **Materials** 2022, **15**, 4548.

[8] Subramaniam, S.; Raju, N.; Shanmugam, E.; Rajavel, N.; Pramanik, A.; Basak, A.K.; Prakash, C.; Gur, A.K. Tribo-mechanical behaviour of aluminium alloy (AlSi10Mg) reinforced with palmyra shell ash and silicon carbide particles. **Metall. Res. Technol**. 2022, **119**, 1–9.

[9] Dawoud, M.M.; Saleh, H.M. Introductory Chapter: Background on Composite Materials. In Characterizations of Some Composite Materials; **IntechOpen: London**, UK, 2019.

[10] P. K. J. &. M. M. M. Belete Sirahbizu Yigezu, "The Key Attributes of Synthesizing Ceramic Particulate Reinforced Al-Based Matrix Composites through Stir Casting Process: A Review," **Materials and Manufacturing Processes (Taylor & Francis Group)**, pp. 969-979, 2013.

[11] Lakshmikanthan, Avinash, Santosh Angadi, Vinayak Malik, Kuldeep K. Saxena, Chandar Prakash, Saurav Dixit, and Kahtan A. Mohammed. "Mechanical and Tribological Properties of Aluminum-Based Metal-Matrix Composites." ***Materials*** 15, no. **17** (2022): 6111.

[12] Sambathkumar, M., R. Gukendran, T. Mohanraj, D. K. Karupannasamy, N. Natarajan, and David Santosh Christopher. "A Systematic Review on the Mechanical, Tribological, and Corrosion Properties of Al 7075 Metal Matrix Composites Fabricated through Stir Casting Process." **Advances in Materials Science and Engineering** 2023 (2023).

[13] Mallick PK. Materials, design, and manufacturing for lightweight vehicles. **Wood head Publishing**; 2010,

[14] Mallick PK. Advanced materials for automotive applications: an overview. **Advanced Materials in Automotive Engineering** 2012:5–27.

[15] Haghshenas M. Reference module in materials science and materials engineering. Metal–Matrix Composites 2015.

[16] Li J, Laghari RA. A review on machining and optimization of particle-reinforced metal matrix composites. **Int J Adv Manuf Technol** 2018.

[17] Zhang C. Understanding the wear and tribological properties of ceramic matrix composites. **Advances in Ceramic Matrix Composites** 2014:312–39.

[18] Ying S, Naveen GM, US 2019 / 0017704 A1 Anaheim. **Ceramic matrix composite manufacturing**; 2019.

[19] David WR, William EL. Modern ceramic engineering: properties, processing, and use in design. 4th ed. **Taylor & Francis Group**; 2018.

[20] Kopeliovich D. Advances in the manufacture of ceramic matrix composites using infiltration techniques. **Advances in Ceramic Matrix Composites** 2014:79–108.

[21] Naslain RR, Pomeroy M. Ceramic matrix composites: matrices and processing. **Reference Module in Materials Science and Materials Engineering** 2016.

[22] Ray BC. Temperature effect during humid ageing on interfaces of glass and carbon fibers reinforced epoxy composites. **J Colloid Interface Sci** 2006;**298(1)**: 111–7.

[23] Dang ZM, Yuan JK, Zha JW, Zhou T, Li ST, Hu GH. Fundamentals, processes, and applications of high-permittivity polymer–matrix composites. **Prog Mater Sci** 2012;**57(4)**:660–723.

[24] Thoppul SD, Finegan J, Gibson RF. Mechanics of mechanically fastened joints in polymer–matrix composite structures—a review. **Compos Sci Technol** 2009;**69(3–4)**:301–29.

[25] Panthapulakkal, S.; Raghunanan, L.; Sain, M.; Birat, K.C.; Tjong, J. Natural fiber, and hybrid fiber thermoplastic composites. **Green Compos**. 2017.

[26] Nair, A.B.; Joseph, R. Eco-friendly bio-composites using natural rubber (NR) matrices and natural fiber reinforcements. In Chemistry, **Manufacture and Applications of Natural Rubber; Woodhead Publishing: Sawston, UK; Cambridge, UK**, 2014.

[27] Agarwal, B.D.; Broutman, L.J.; Chandrashekhara, K. Analysis and Performance of Fiber Composites; John Wiley & Sons: Hoboken, NJ, USA, 2017.

[28] Tanzi, M.C.; Farè, S. Foundations of Biomaterials Engineering; Matthew, D., Ed.; **Elsevier**: Amsterdam, The Netherlands; Academic Press: London, UK, 2019.

[29] Fang, K. Encapsulation Technologies for Electronic Applications, 2nd ed.; **Elsevier**: Amsterdam, The Netherlands, 2019; pp. 123–181.

[30] Rajak, Dipen Kumar, Durgesh D. Pagar, Pradeep L. Menezes, and Emanoil Linul. "Fiber-reinforced polymer composites: Manufacturing, properties, and applications." ***Polymers*** 11, no. **10** (2019): 1667.

[31] T. Ozben, E. Kilickap, O. Çakır, Investigation of mechanical and machinability properties of SiC particle reinforced Al- MMC, J. Mater. Process. **Technol. 198 (1-3)** (2008) 220–225.

[32] T. Rajmohan, S. Ranganathan, T.S.A. Suryakumari, Experimental Study on fabrication of hybrid (Al+TiO2+Gr) metal matrix composites, Int. J. Adv. Eng. Appl. **7 (2)** (2014) 11–14.

[33] S. Dixit, A. Mahata, D.R. Mahapatra, S.V. Kailas, K. Chattopadhyay, Multi-layer graphene reinforced aluminium– manufacturing of high strength composite by friction stir alloying, Compos. B Eng. **136** (2018) 63–71.

[34] S. Suresh, N.S.V. Moorthi, Process development in stir casting and investigation on microstructures and wear behavior of TiB2 on Al6061 MMC, Procedia Eng. **64** (2013) 1183–1190.

[35] Thandalam, Satish Kumar, Subramanian Ramanathan, and Shalini Sundarrajan. "Synthesis, microstructural and mechanical properties of ex situ zircon particles (ZrSiO4) reinforced Metal Matrix Composites (MMCs): a review." ***Journal of Materials Research and Technology*** 4, no. **3** (2015): 333-347.

[36] M. Schöbel, G. Requena, G. Fiedler, D. Tolnai, S. Vaucher, H.P. Degischer, Void formation in metal matrix composites by solidification and shrinkage of an AlSi7 matrix between densely packed particles, **Compos. A Appl. Sci. Manuf. 66** (2014) 103–108.

[37] Kumar, Ashwini, Omkar Vichare, Kishore Debnath, and Manikant Paswan. "Fabrication methods of metal matrix composites (MMCs)." ***Materials Today: Proceedings*** **46** (2021): 6840-6846.

[38] F.B.A. Sagai, R.E. Raj, M.C. Mabel, Prediction of shear and tensile strength of diffusion bonded AA5083 and AA7075 aluminum alloy using ANN, **Mater. Sci. Eng.**: **A 692** (2017) 1–8.

[39] C. Pinca-Bretotean, A.L. Craciun, A. Josan, M. Ardelean, Friction and wear characteristics of organic brake pads material, **IOP Conf. Ser: Mater. Sci. Eng. 477** (2019).

[39] N. Panwar, A. Chauhan, Fabrication Methods of particulate reinforced Aluminium metal matrix composite, **Mater. Today: Proc. 5** (2018) 5933–5939.

[40] Reddy BS, Das K, Das S. A review on the synthesis of in situ aluminum-based composites by thermal, mechanical, and mechanical–thermal activation of chemical reactions. **Journal of Materials Science, 1;42(22):**9366-78, 2007 Nov.

[41] Wang H, Li G, Zhao Y, Chen G. In situ fabrication and microstructure of Al2O3 particles reinforced aluminum matrix composites. **Materials Science and Engineering: A, 15;527(12)**:2881-5, 2010 May.

[42] https://www.researchgate.net/publication/333787226_The_production_and_sustainable_role_of_particle_composites_in_engineering_structures.

[43] Xie, Ziming, Ripeng Jiang, Xiaoqian Li, Lihua Zhang, Anqing Li, and Zhuoli He. "Microstructural evolution and mechanical properties of TiB2/2195 composites fabricated by ultrasonic-assisted in-situ casting." ***Ultrasonics Sonochemistry* 90** (2022): 106203.

[44] Narasimmalu, Ramasubbu, and Ramabalan Sundaresan. "Wear behavior and mechanical properties of AA8090 aluminum alloy reinforced with B4C-Al2O3 hybrid nanoparticles." ***Industrial Lubrication and Tribology* 74**, no. 2 (2022): 237-242.

[45] Al-Salihi, Huda A., and Hadia Kadhim Judran. "Effect of Al2O3 reinforcement nanoparticles on the tribological behaviour and mechanical properties of Al6061 alloy." ***AIMS Mater. Sci*** 7 (2020): 486-498.

[46] Mallireddy, Naresh, and K. Siva. "Investigation of microstructural, mechanical and corrosion properties of AA7010-TiB2 in-situ metal matrix composite." ***Science and Engineering of Composite Materials* 27, no. 1** (2020): 97-107.

[47] Suresh, S., G. Harinath Gowd, and MLS Deva Kumar. "Mechanical and wear behavior of Al 7075/Al 2 O 3/SiC/mg metal matrix nanocomposite by liquid state process." ***Advanced Composites and Hybrid Materials*** 2 (2019): 530-539.

[48] Dhaneswara, Donanta, Luthfi Noviardi Andani, Nofrijon Sofyan, and Anne Zulfia. "Mechanical Properties of Nano SiC-Reinforced Aluminum A356 Fabricated by Stir Casting Method." In ***Materials Science Forum*, vol. 929**, pp. 86-92. Trans Tech Publications Ltd, 2018.

[49] Reddy, M. Penchal, F. Ubaid, R. A. Shakoor, Gururaj Parande, Vyasaraj Manakari, A. M. A. Mohamed, and Manoj Gupta. "Effect of reinforcement concentration on the properties of hot extruded Al-Al2O3 composites synthesized through microwave sintering process." ***Materials Science and Engineering: A* 696** (2017): 60-69.

[50] Matli, P.R.; Fareeha, U.; Shakoor, R.A.; Yusuf, M.; Mohamed, A.M.A.; Gupta, M. Fabrication, and mechanical properties of extruded Al-SiC nanocomposites. NH Nano Hybrids. 2017, **16**, 9–12.

[51] N Malleswararao Battina, Varaha Siva Prasad Vanthala and Hari Krishna Ch, Effect of friction stir welding process parameters on mechanical and metallurgical behavior of AA6061-T6 and AA2017-T6 tailored blanks, **Engineering Research Express, 2021, 3 (2)**

Synthesis, characterization and magnetic properties of iron oxide nanoparticles from reverse chemical co-precipitation method

K Satyam Naidu[1], Nowduri Annapurna[2*]
[1]*Department of Chemistry, Raghu Engineering College (A), Visakhapatnam, Andhra Pradesh, (India)*
[2*]*Department of Engineering Chemistry, AU College of Engineering (A), Andhra University, Visakhapatnam, Andhra Pradesh, (India)*

Abstract: Studying magnetic characteristics of iron oxide nanoparticles was highlighted as a key step in creating potential powder nanomaterials. The occurrence of electron spins parallel to the crystallographic orientations and single domain nanopartilces are the recognized causes of superparamagnetism in bulk magnetite. Nanoparticles whose size could be tuned exhibit this superparamagnetism effect. The reverse co-precipitation process is used to create the current magnetite nanoparticles. The iron oxide nanoparticles were synthesized by reverse chemical coprecipitation method and were confirmed based on XRD data. The room temperature magnetic property was studied and the tiny levels of coercivity and remanence indicate that the samples may exhibit superparamagnetic nature.

1. Introduction:

Nanotechnology has been regarded as one of the most significant contemporary advances in science and technology. Nanoparticles are a key building component in the production and development of nanomaterials [1]. As a result, the nanoparticle has sparked widespread attention among academics due to unique properties such as form, size, and dispersion, which might be used in a variety of applications. Nanoparticles of iron oxide have an important role in many chemical, physical, and material sciences [2].

Due to their abundance, quick reaction, super-paramagnetic nature, high competence, nontoxicity, improved stability, and effectiveness in the chemical and physical adsorption of organic and inorganic pollutants, including heavy metals from contaminated waters, iron magnetic nanoparticles have drawn the most attention among nanoparticles [3]. Fe_3O_4-NPs' special qualities enable their widespread usage in a variety of applications, including targeted drug delivery [8], environmental treatment [6], magnetic resonance imaging (MRI) [7], catalysis [4], magnetic storage media [5], and environmental treatment [6]. The size distributions and particle sizes of nanoparticles can be used to fit their magnetic characteristics. The synthesis path sequentially affects the magnetic nanoparticle sizes and size distributions.

For these purposes, numerous synthesis techniques have been created to generate magnetic iron oxide nanoparticles with the desired properties [9], as reported in other papers, including gas-phase deposition and mechanical techniques (physical methods) [10], green synthesis (biological method) [11], co-precipitation method [12], microwave assisted synthesis [13], and chemical methods.
Fe_3O_4 nanoparticles are most frequently prepared via the chemical technique. Comparatively speaking, chemical preparation techniques used less energy than physical ones. One may choose pick the reaction medium, as well as the physical parameters of the reaction (precursors, reactant concentration, base (NaOH and ammonium hydroxide), temperature, and pH) to regulate the size and shape of the nanoparticles.
The biological approach is an excellent production methodology because to its high yield, cheap prices, and minimal energy consumption, however the process of fermenting takes a long period. [14]. In other study describes the synthesis of various forms of Fe_3O_4 nanoparticles as spherical, plate, and nano flowers by chemical approach by solving thermal method aided by microwave radiation, utilizing $FeSO_4{\cdot}(NH_4)_2SO_4{\cdot}6H_2O$ as an iron precursor, ethanol and NaOH [15]. In this study, the production and characterisation of Fe_3O_4 nanoparticles is done with inexpensive raw materials such as ferrous sulphate hexahydrate and ammonium hydroxide utilizing the reverse co-precipitation technique.

2. Experimental Work:

2.1 Materials:

Every component utilized in the manufacture of Fe_3O_4 was analytical grade and did not require additional purification. Sigma supplied Ferrous Sulphate hexahydrate (98%), and ammonium hydroxide [NH_4OH] (98%), which were utilized without purification.

2.2 The Procedure of Fe_3O_4 Synthesis[16]:

In this process initially a 0.05M ferrous sulphate solution was made by mixing 0.695g $FeSO_4.7H_2O$ with deionized water under ultrasonication for 15 mins. The process is continued until the salt dissolved completely. Meanwhile in another 250 mL beaker, mix 50 mL of ammonia and 50 mL of water until the solution turned to pH 13. To this base solution add the iron precursor solution at once and stirring was continued under 600 rpm for one hour. After completion of the reaction, the solution turned to dark color and it can be washed three times with deionized water to remove the traces of the base.

2.3 Characterization:

The specification and characteristics of nano Fe_3O_4 were measured using a variety of characterisation methods. A Shimadzu 6000 (Japan) was used for XRD analysis at ambient temperature, with a CuKα radiation Nickel filter (λ= 1.5418A°) at a rate of 0.5° between 2 and 80° 2θ.

3. Results and Discussion:

3.1 X-Ray Diffraction (XRD):

XRD study was used to verify crystalline structure of synthesized iron oxide nanoparticles and they have followed the required pattern. The results of XRD were presented in Figure 1. From the figure, it was noticed that the produced nano Fe3O4 X-ray diffraction pattern is about equivalent to the typical lattice spacing and angle. From the XRD data the size of the nanoparticles are calculated by using debye scherrers equation $D = \frac{K\lambda}{\beta cos\theta}$ where D is the nanoparticles crystalline size, K represents the Scherrer constant (0.98), λ denotes the wavelength (1.54), β denotes the full width at half maximum (FWHM). The size of the nanoparticles was determined to be 29.6 nm using this equation.

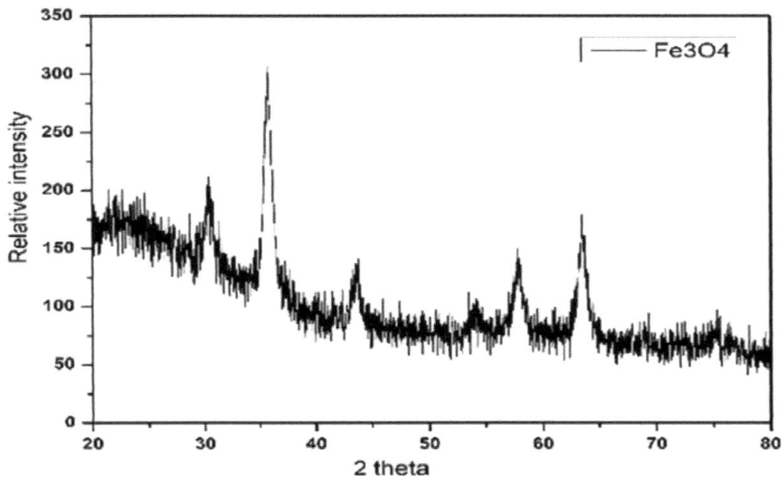

Fig.1: XRD pattern of prepared nano Fe_3O_4

3.2 Scanning electron microscopy study:

Figure 2 shows the SEM images acquired for the colloid. The SEM pictures in Figure 2 clearly show that the particles are crystalline. The SEM picture clearly demonstrates the presence of the Scherrer's rings, a distinctive feature of fcc Fe_3O_4 nanoparticles, indicating that the structure is nano-crystalline in nature. Fe3O4 nanoparticles are shown to be dispersed throughout the surface, and a SEM analysis reveals the absence of any aggregates. It's possible that the varied timeframes at which the nanoparticles are forming account for the size disparity.

Fig 2: SEM images of surface of Fe_3O_4 nanoparticles

3.3 TEM studies

Fe3O4 nanoparticles exhibited a spherical shape with uneven or non-uniform distribution. Similarly, the nano-materials were found to be crystalline in form, with sharp edges in a spherical shape (the crystal planes were also seen in the materials). The Fe3O4 nanoparticles are formed as core-shell nanoparticles with spherical shapes. The mean diameter was used to calculate the grain size and which is the same value as the particle size calculated from the XRD data.

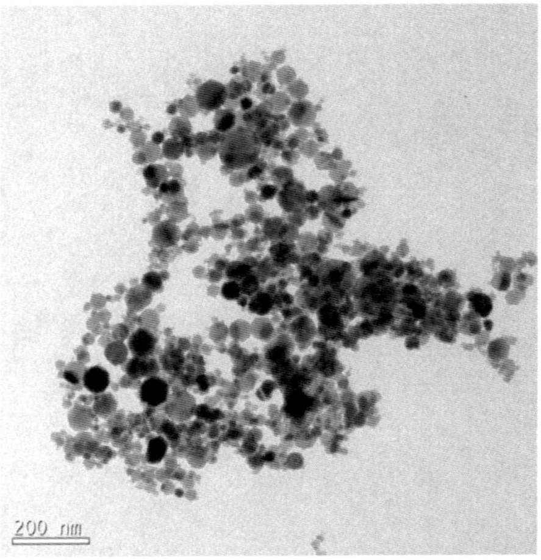

Figure 3: Transmission electron micrographs of the Fe_3O_4 nanoparticles. The bar marker represents 200 nm

3.4 M-H Studies:

In order to create potential nanocomposite powders, the magnetic properties of iron oxide nanoparticles were the focus of the investigation. Figure 4 displays the magnetic characteristics of the produced nanoparticles, while Table 1 displays the magnetic data. It can be shown from figure 5 that the samples behave super-paramagnetically at normal temperature. It is commonly known that single domain nanopartilces and the electron spins that are parallel to the crystallographic orientations cause bulk magnetite's super-paramagnetic behavior. It's possible that super-paramagnetism relaxation prevents the temperature from falling. This effect is often seen in particles smaller than 10 nm. XRD data indicates that the iron oxide (Fe_3O_4) nanoparticles are 29.6 nm in size. The magnetic qualities are supported by this value. The tiny remanence and coercivity values in Figure 5 indicate that the samples are super-paramagnetic. There are possible uses for these kinds of low coercive materials in magnetic hyperthermia

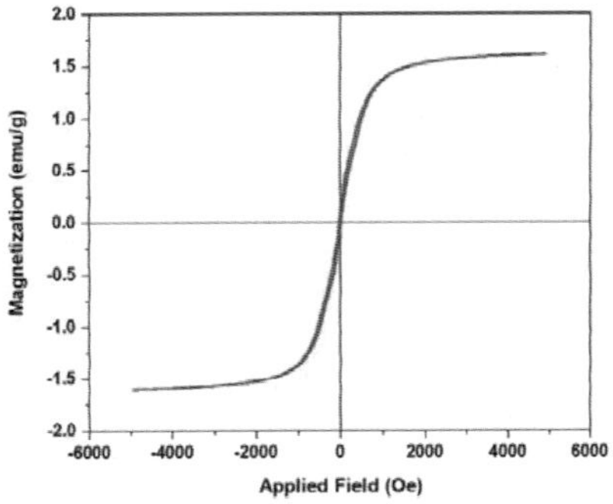

Fig. 4: Magnetic hysteresis of a iron oxide nanoparticles

Table 1: Magnetic characteristics of the produced nanocomposites

S. No	Sample	Saturation Magnetization (Ms) emu/g	Remanence Magnetization (MR) emu/g	Coercivity (Oe)
1	Iron oxide Np	1.67	0.137	5342

Fig. 5: Magnetic response of water dispersed iron oxide nanoparticles

4. Conclusions:

The development of magnetite nanomaterials employing Reverse co-precipitation method in a open vessel process. The structural characterization of the iron oxide nanoparticles was in coincidence with literature reports. The average particle size was 29.6 nm and the result was in accordance with TEM reports. The particles were in spherical in shape and were not agglomerated. Further, the magnetic studies of the iron oxide nanoparticles were studied at room temperature. The results showed that super-paramagnetic nature in the prepared sample due to small remanence and coercivity values. This type of low coercive materials is found to have potential applications in magnetic hyper thermia.

5. References:

[1] Begum, K. M. S., & Anantharaman, N., 2009. "Removal of chromium (VI) ions from aqueous solutions and industrial effluents using magnetic Fe_3O_4 nano-particles". Adsorption Science & Technology, 27, pp. 701-722.

[2] Yew, Y. P., Shameli, K., Miyake, M., Kuwano, N., Khairudin, N. B. B. A., Mohamad, S. E. B., & Lee, K. X. , 2016." Green synthesis of magnetite (Fe_3O_4) nanoparticles using seaweed (Kappaphycus alvarezii) extract". Nanoscale research letters, 11, pp. 1-7.

[3] Malakootian, M., Askarpuor, A., Amirmahani, N., Nasiri, Z., & Nasiri, A., 2015, "Removal of Hexavalent Chromium from Aqueous Solutions Using Magnetic Nanoparticles Coated with Alumina and Modified by Cetyl Trimethyl Ammonium Bromide", Journals of Community Health Research, 4, PP. 177-193.

[4] Campos, E. A., Pinto, D. V. B. S., Oliveira, J. I. S. D., Mattos, E. D. C., & Dutra, R. D. C. L. ,2015. "Synthesis, characterization and applications of iron oxide nanoparticles-A short review". Journal of Aerospace Technology and Management, 7, PP. 267-276.

[5] Ríos-Hurtado, J. C., Múzquiz-Ramos, E. M., Zugasti Cruz, A., & Cortés-Hernández, D. A., 2016. "Mechanosynthesis as a Simple Method to Obtain a Magnetic Composite (Activated Carbon/ Fe_3O_4) for Hyperthermia Treatment". Journal of Biomaterials and Nanobiotechnology, 7, PP. 19-28

[6] Xu, P., Zeng, G. M., Huang, D. L., Feng, C. L., Hu, S., Zhao, M. H., & Liu, Z. F. ,2012." Use of iron oxide nanomaterials in wastewater treatment: a review". Science of the Total Environment, 424, PP. 1-10.

[7] Zhang, Y., Liu, J. Y., Ma, S., Zhang, Y. J., Zhao, X., Zhang, X. D., & Zhang, Z. D., 2010. "Synthesis of PVP-coated ultra-small Fe3O4 nanoparticles as a MRI contrast agent". Journal of Materials Science: Materials in Medicine, 21(4), pp. 1205-1210.

[8] Chomoucka, J., Drbohlavova, J., Huska, D., Adam, V., Kizek, R., & Hubalek, J., 2010. "Magnetic nanoparticles and targeted drug delivering". Pharmacological Research, 62(2), pp. 144-149.

[9] Chin, S., Pang, S. & Tan, C., 2016. "Green Synthesis of Magnetite Nanoparticles (via Thermal Decomposition Method) with Controllable Size and Shape". Environ. Sci. 2, PP. 299-302.

[10] Reddy, L. H., Arias, J. L., Nicolas, J., & Couvreur, P., 2012."Magnetic nanoparticles: design and characterization, toxicity and biocompatibility, pharmaceutical and biomedical application". Chemical reviews, 112, PP. 5818-5878.

[11] Pranita, L. & Preeti, J., 2015. "Preparation and Characterization of Magnetite Nanoparticle using Green Synthesis". Int. J. Res. Chem. Environ., 5, PP. 38-43.

[12] Khalighyan, N., Hooshmand, N., Razzaghi-Asl, N., Zare, K., & Miri, R. ,2014. "Response surface strategy in the synthesis of Fe3O4 nanoparticles". International Journal of Nano Dimension , 5,PP. 421-430.

[13] Herlekar ,M. & Barvel,S., 2015. "Optimization of microwave assisted green synthesis protocol for iron oxide nanoparticles and its application for simultaneous removal of multiple pollutants from domestic sewage". International Journal of Advanced Research, 3, PP. 331-345.

[14] Xu, J. K., Zhang, F. F., Sun, J. J., Sheng, J., Wang, F., & Sun, M. 2014. "Bio and nanomaterials based on Fe3O4". Molecules, 19, PP. 21506-21528.

[15] Abdulghani, A. J., & Al-Ogedy, W. M. "Synthesis and Characterization of Multishapes of Fe_3O_4 Nanoparticle by Solve-Hydrothermal Method Using Microwave Radiation". Baghdad Science Journal, 13, pp. 331-339.

[16] Chandra Sekhar Devarapu, S. Diwakar Bhagavathula, Bhavani Penmatsa, Madhavi Nannapaneni, Development of multifunctional metal oxide coated magnetite nanocomposites for advanced applications Materials Today: Proceedings 18, 2019, 2089-2094

Optimization of Process Parameters in Under Water Friction Stir Welded AA2014 and AA6082 joints.

***Singuru Madhavarao[1], Gopalakrıshnam Raju P[2], Rama Bhadri Raju Chekuri[1],Manikanth Vatala[1], B. Soundaryasanthoshi[1]**

[1,]Assistant Professor, Sagi Rama krishnam Raju Engineering College(A),Bhimavaram,W.G-534204,A.P,India

[2]Master study program, Riga Technical University,Riga ,latvia

[1]madhava690@gmail.com ; [2]gopalakrishnamraju259@gmail.com; [1]bhadri.mech@gmail.com , [1]manikanth.srkrec@gmail.com, [1]soundaryasanthosi@gmail.com

Corresponding Author Name: Singuru Madhavarao
*Corresponding Author email: madhava690@gmail.com

Abstract. This work focuses on optimizing process parameters in underwater friction stir welding to study their impact on the mechanical characteristics of aluminum 2014 T6 and 6082 T6 joints. The aluminum plates are precisely machined to provide the necessary polish for the underwater friction stir welding process (UFSW). Experiments are designed using Taguchi L-9 orthogonal array, considering three process parameters: speed, transverse speed, and tool angles at three levels each. The aluminum plates of series 2014-T6 and 6082-T6 are securely installed in the machine. The plates will be welded using the UFSW process using a taper pin tool. The welded plates are next subjected to tests to assess their mechanical qualities. Tests for tensile strength, hardness, and impact strength are conducted on the welded area of the plate to determine its mechanical qualities. An ANOVA was conducted to determine the significant influence of each variable. The optimal values achieved are Tensile Strength of 185.592 N/mm^2, Hardness of 75.1 (BHN), and Impact Strength of 21.236 Joules at a speed of 1120 rpm, transverse speed of 80 mm/min, and tilt angle of 2 degrees.
Keywords: AA2014-T6 plate, AA6082-T6 plate, Under water FSW, Mechanical properties, ANOVA and Taguchi

1. Introduction

Friction Stir Welding (FSW) is a process or method which is used to join two different materials by plasticization the welded zone of solid bodies to be connected and by its solidification after cooling. This process is increasingly use in world widely in ship building industries, Rail industries, Electronics and Electrical appliances, etc. Underwater Friction Stir Welding has a rotating and a penetrating tool which penetrates the place where it melts and mixes the semi molten metal with friction heat, in the place of welded place an onion ring structure is formed. Then the welded plates are exposed to mechanical tastings like Tensile, Hardness and Impact test, and the results are recorded which will be practical side of experiment.

FSW aims to examine and summarize several elements of UFSW. Emphasis is placed on fundamental concepts such as material flow, temperature production, process parameters, microstructure, and mechanical qualities. [1-3]. The rotating welding device is slowly inserted into the workpiece until the shoulder of the welding tool makes firm contact with the top surface of the material. [4]. Underwater friction stir welding is a preferred method to address the challenges posed by low heat input and short dwell time in processing dissimilar joints of AA6082/AA8011. The pin profile has a significant impact on the microstructure and mechanical properties of the welded joint. Hard and brittle precipitates occur at the weld bead due to the difference in peak temperature of UFSW compared to air friction stir welding (AFSW), making it more problematic. [5-6]. A curved interface forms due to increased materials mixing at the advancing side, resulting in a linear hardness gradient across the dissimilar weldment. The computational time for the firefly optimization-based texture in the heat affected zone was found to be nearly equivalent to that of the base material and simple shear [7]. The mechanical properties of the aluminum alloy 6082-T6 joint are influenced by the process parameters of underwater friction stir welding. Various evolutionary optimization approaches are used to describe this process [8]. Electron backscatter diffraction technique was utilized to analyze microstructural changes and texture evolution in different 7003/6060 aluminum alloys during underwater friction stir welding [9] Microstructural analysis showed that most fractures are located in the thermomechanically damaged zone [10]. An experimental study examined the ultimate tensile strength (UTS) of weld joints created by Friction Stir Welding (FSW) and Ultrasonic Welding with Friction Stir Welding (UWFSW) utilizing a newly developed fixture to tackle post-process problems [11]. Examining the weldability of aluminum and silicon carbide metal matrix composite components containing 10% silicon carbide by utilizing a TiAlN coated tool in the friction stir welding process. A CNC milling machine is used for Friction Stir Welding (FSW). Aluminum plates from series 6061 and 7075 are firmly fastened to the machine [12-13]. The study examines the recrystallization mechanism, grain boundary characteristics, phase transformation, texture evolution, characteristic microstructures, and their effects on the hardness, tensile strength, fatigue properties, and superplastic behavior of aluminum alloys processed using friction stir welding/friction stir processing [14]. An underwater friction stir welding technique is used on aluminum alloy 6082-T6 to develop a mathematical model for optimizing process parameters and obtaining the highest tensile strength [15]. Optimizing friction stir welding (FSW) parameters to achieve both high tensile strength and impact toughness using Taguchi-based Greyish Relational Analysis (GRA) and Artificial Neural Network (ANN) [16-18]. An investigation is conducted on how input operating process factors, such as tool rotating speed (TRS), feed rate, and pitch values, affect output response parameters like ultimate tensile strength (UTS) and hardness of welded connections. [19,20] Regression models in mathematics were employed to predict the ultimate tensile strength and tensile elongation of custom welded joints [20-21]. Friction stir welded dissimilar alloys may be shaped by modifying three crucial process parameters: tool rotating speed, weld speed, and tool tilt angle. Various combinations of sheets were used for experimental and finite element modelling testing to assess the maximum dome height [22-24].

2.Experimentation

Friction stir welding (FSW) of aluminum alloy was performed using a modified milling machine. The gadget would have operated at an optimal speed of 1800 revolutions per minute with a power output of 5.5 kilowatts per revolution. The materials used for this project are Al 6082 T6 alloy and Al 2014 T6 rolled plates with specified compositions as shown in the table. The counter plates have been ground at the optimal location with welding completed prior to further processing. Work out parts has been ready with respective duration, thickness, width. The metal plates are securely installed into the apparatus shown in Figure 1. A hardened High-Speed Steel (HSS) tool was used to machine a shoulder with a diameter of 16mm, a pin with a taper diameter of 6mm, and a length of 2mm. This device is equipped in the equipment holder as well as career portion is rigidly clamped to printer kitchen table making use of fixtures. The water is flow through the welded surface the heat transfer through it. It is act as a coolant for under water friction stir welding.

2.1 Selection of materials

Aluminium alloy 6082 is a moderately robust alloy with excellent resistance to oxidation. It has the highest strength among the 6000 series alloys. Alloy 6082 is acknowledged as a structural alloy.

Aluminum alloy 6082 has good machinability and forms tight coils of metal chips when chip breakers are used in the T651 and T6 temper conditions. aluminum 2014 may be tempered during machining and is likely one of the strongest lightweight aluminum alloys, boasting high hardness. The chemical composition of aluminum alloys is displayed in Table 1. It is the second most recognized lightweight aluminum alloy in the 2000 series.

Table 1. The chemical composition of 6082 T6 &2014 T6

Elements of 6082 T6	Silicon	iron	Copper	Manganese	Magnesium	Zinc	Titanium	Chromium	Aluminum
Wt.% of elements	0.6 – 1.4	0.0 – 0.6	0.0 – 0.1	0.4 – 1.0	0.6– 1.2	0.0 – 0.2	0.0 – 0.1	0.0 – 0.25	95.2 – 98.3
Elemens of 2014 T6	Silicon	iron	Copper	Manganese	Magnesium	Zinc	Titanium	Chromium	Aluminum
Wt.% of elements	0.5 – 1.3	0.7 max	3.8 – 4.9	0.4 – 1.3	0.2– 0.8	0.25 max	0.15 max	0.1 max	93.5

2.2 Selection of Machining Paramteres

The procedure parameters are chosen based on the literature review. The input variable is "selected Speed" in revolutions per minute (rpm). Transverse Speed (TS) in mm/min, Tool angle in degrees Various levels of each process parameter is mentioned. Welded area undergoes mechanical property testing. Tensile strength, hardness, and impact strength Table 2 displays the UFSW parameters utilized in the experiment. The process parameter experimental design is conducted using MINITAB software. The Taguchi approach involves designing trials using an L-9 orthogonal array..

Table 2. UFSW parameters used in experimentation.

S.NO	**Process parameters**	**Level 1**	**Level 2**	**Level 3**
1	Speed in rpm	560	900	1120
2	Transverse Speed in mm/min	40	60	80
3	Tool angle in degrees	1	2	3

Friction Stir Welding is solid state welding process where friction acts to weld, in this aluminium 2014 T6 and aluminium 6082 T6 is taken because aluminium is the one of the lightest materials in present living world.

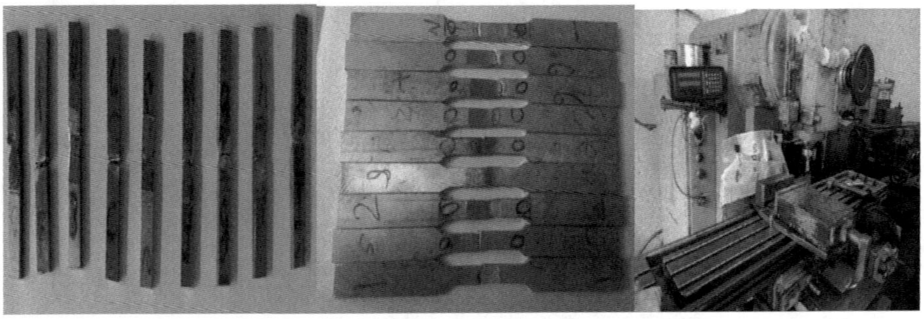

Figure 1. Tested specimens and Underwater FSW

First the plates are welded by process called FSW and they cut in ASTM standards and sent to mechanical tests like Tensile, Hardness and Impact. The underwater friction stir welding machine and welded joint After the mechanical testing is performed the outcome of those are saved and values are noted. Then particle and predicted values are compared, for predicted values we are using L9 orthogonal array where there should be 9 specimens for each testing. We use a software called MINITAB where TAGUCHI and ANOVA module is used and made some numerical calculations are performed. Hence showing that practice and predicted are approx. same

3.Mechanical Properties Evaluation

The mechanical properties are evaluated on FSW welded joints. The performed tests are Hardness test, Tensile test, and Impact test.

3.1 Hardness Test

This basic sequence of examination pressure program proved to become a significant advance within the realm of hardness tests. The user was enabled by it to do a precise hardness test on an assortment of sized areas inside only several seconds.

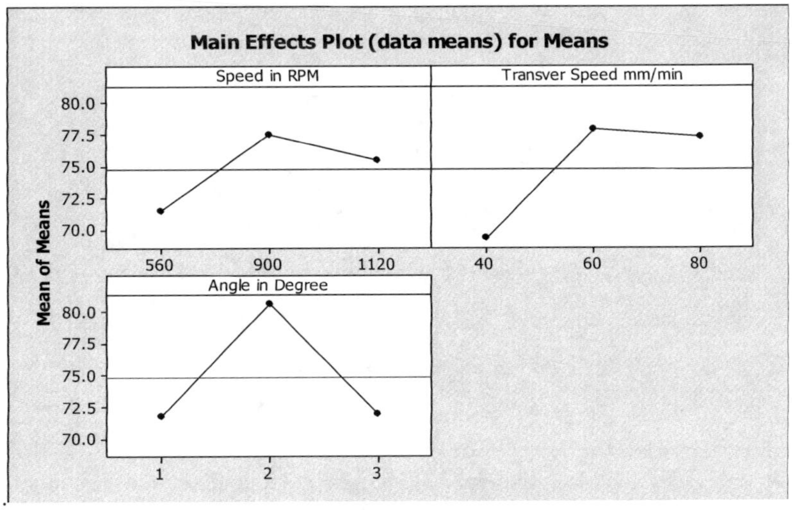

Figure 2. Main effects plot for Hardness

The analysis of Hardness with help of Main effects plot from increase of speed, transverse speed and tool angle the hardness increase up to 900rpm,60 mm/min and 2^0later decreases as shown in Figure 2

Table 3. Percentage contribution of parameters in hardness test

Source	DF	Seq SS	Adj MS	F	P	%Cont.
Speed in rpm	2	997.4	498.7	0.06	0.946	74.23
Transvers speed mm/min	2	136.9	68.5	0.14	0.879	10.18
Angle in Degree	2	152.7	76.4	0.15	0.867	11.36
Error	2	56.5	28.2			4.20
Total	8	1343.6				100

From Table3.we can acknowledge that percentage of contribution by the three parameters like speed in rpm is 74.23%, transverse speed is 10.18%, angle in degree is 11.36% and we can calculate percentage of error recorded.

By observing the graph of SN ratio and mean of mean value graph in Figure3.we choose optimum values at larger point in hardness test we consider larger value is better, so we have tool speed in rpm at 1120, transverse speed of feed at 80 mm/min and tool angle is at 2°

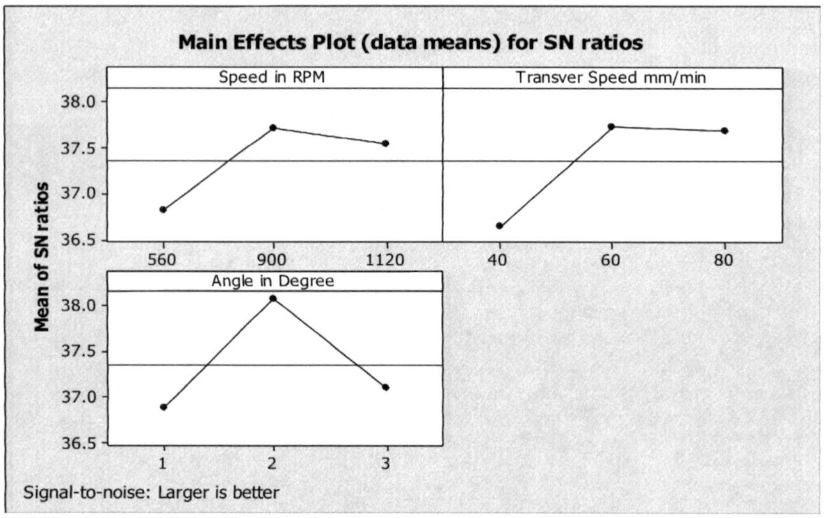

Figure 3. SN ratio plot for hardness

3.2Tensile Test

A tensile check, likewise, recognized like a stress check, is among the majority of essential as well as typical kinds of physical tests. A tensile check is applicable tensile (pulling) pressure to a substance and measures the specimen 's reaction on the anxiety.

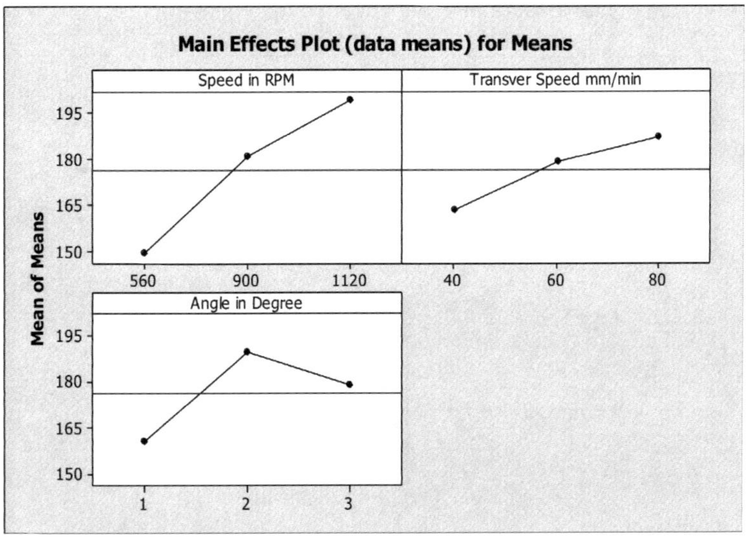

Figure 4. Tested specimens and Main effects plot for tensile

The analysis of Tensile strength with help of Main effects plot from increases of speed and transverse speed the tensile strength increases and increase of tool angle the tensile strength is increase up to angle of 2^0 later decreases as shown in Figure 4.

Table 4. Percentage contribution of parameters in Tensile test

Source	DF	Seq SS	Adj MS	F	P	%Cont.
Speed in rpm	2	131.56	1.78	0.03	0.974	39.27
Transverse speed mm/min	2	96.89	48.44	0.74	0.576	29.22
Angle in Degree	2	99.56	49.78	0.76	0.569	30.02
Error	2	3.56	65.78			1.07
Total	8	331.56				100

From Table 4. We can acknowledge that percentage of contribution by the three parameters like tool speed in rpm is 39.27%, transverse speed of feed is 29.22%, angle in degree is 30.02% and we can calculate percentage of error recorded.

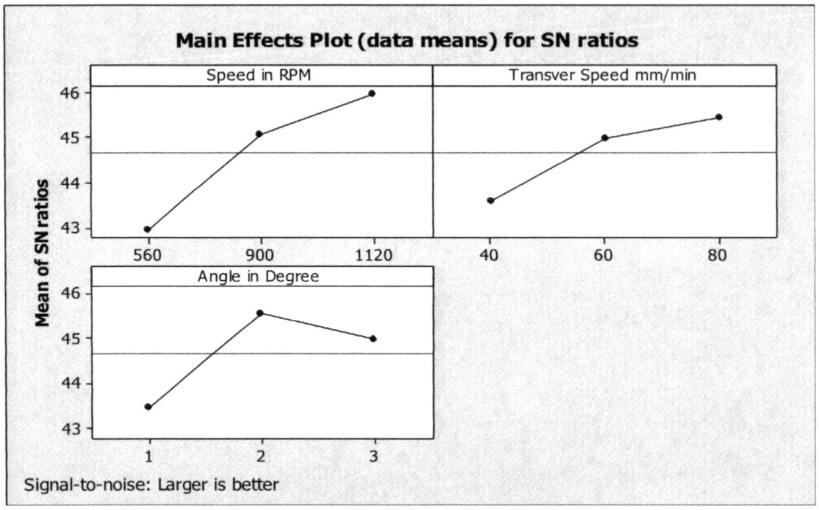

Figure 5. S/N ratio plot for tensile strength

By observing the SN ratio and mean of mean value graph in Figure 5. we choose optimum values at larger point as in tensile test we consider larger value is better, so we have tool speed in rpm at 1120, transverse speed at 80 mm/min and tool angle is at 2°

3.3Impact Test

A greater strength material is going to absorb a lot more power upon influence. The analysis of impact strength with help of Main effects plot from increase of speed the impact strength is linear up to 900rpm later decreases, the increase of transverse speed the impact strength increases up to 60 mm/min later decreases and increase of tool angle the impact strength is increase up to 20 later decreases as shown in Figure 6.

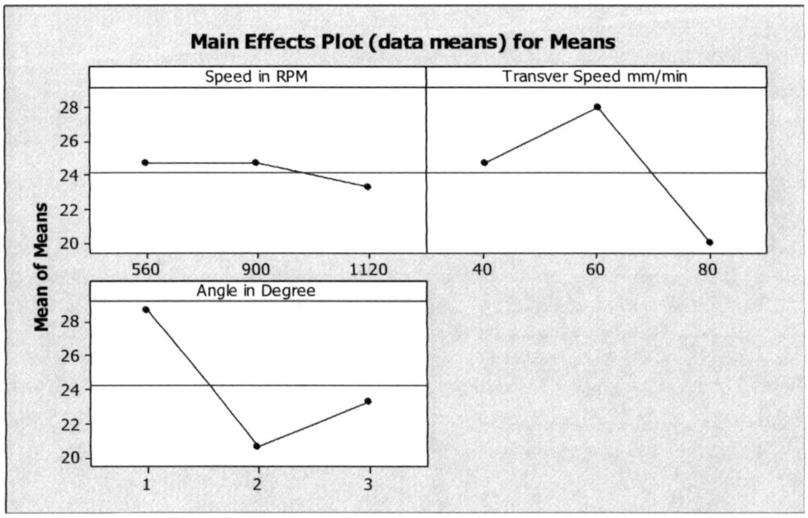

Figure 6. Tested specimens and Main effects plot for tensile

Table 5. Percentage contribution of parameters in Impact test

Source	DF	Seq SS	Adj MS	F	P	%Cont.
Speed in rpm	2	1480.9	564.6	0.62	0.618	59.83
Transverse speed mm/min	2	462.9	296.4	0.26	0.791	18.70
Angle in Degree	2	350.9	245.4	0.21	0.825	14.17
Error	2	179.3	131			7.24
Total	8	2474.9				100

From this Table 5 we can acknowledge that percentage of contribution by the three parameters like speed in rpm is 59.83%, transverse speed is 18.70%, angle in degree is 14.17% and we can calculate percentage of error recorded.

By observing the graph of SN ratio and mean of mean value graph in Figure 7 we choose optimum values at smaller point in impact test we consider smaller value is better, so we have tool speed in rpm at 1120, transverse speed of feed at 80mm/min and tool angle is at 2°

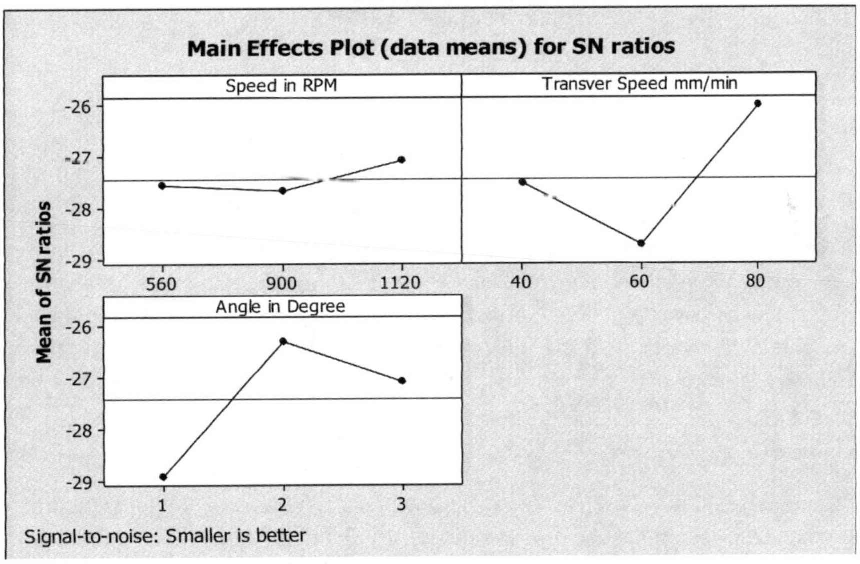

Figure 7. S/N ratio plot for impact strength

3.4 Validation

To validate the experiment, we need to compare the practical values and predicted optimum values from which we can justify optimum parameters of friction stir welding of aluminum 2014 T6 and aluminum 6082 T6 are Table 6. By the observation from the above tables of practical and theoretical values we have obtained optimized value at 1120 speed of rpm 80mm/sec transverse speed of feed and 2° degree angle of tool i.e., sample No 9

The regression equations are.
Hardness = 55.9 + 0.0080 Speed in rpm + 0.198 Transverse Speed mm/min + 0.06 Angle in degree (1)
UTS N/mm^2 = 159 + 0.0329 Speed in rpm - 0.156 Transverse Speed mm/min - 5.77Angle in degree (2)
Impact = 38.40 - 0.0022 Speed in rpm - 0.117 Transverse Speed mm/min - 2.67 Angle in degree (3)

Validation studies are required since the optimal circumstances for different responses are found to be beyond the scope of the L-9 orthogonal array. The validation run is conducted under the ideal circumstances specified in Table 6. To compare with the projected optimal response levels. The optimal anticipated values for hardness, ultimate tensile strength (UTS), and impact determined from regression analysis for experimental data are derived from equations 1-3. The average discrepancy between the projected and experimental results is 10.42%, indicating that the accuracy of the development models is 89.57%.

Table 6. Validation of Process parameters

Mechanical Tests	Optimum conditions	Predicted optimum values	Practical optimum values
Tensile	Speed in rpm 1120 Transverse speed 80 Angle in degree 2	185.592 N/mm^2	186.889 N/mm^2
Hardness	Speed in rpm 1120 Transverse speed 80 Angle in degree 2	75.1(BHN)	72.16(BHN)
Impact	Speed in rpm 1120 transverse speed 80 Angle in degree 2	21.236 Joules	20 Joules

We tabulate the predicted optimum values corresponding to practical optimum value for mechanical testing according to their parameters. From these the optimized best parameter value for welding of aluminum plates is at speed of tool in rpm 1120, transverse speed of feed 80mm/sec and angle in degree 2°. At which predicted values and practical optimum values are relevant equal, so I conclude these parameters of tensile, hardness and impact test are optimal values for underwater friction stir welding.

4.Conclusions

The experiment was conducted to establish the optimal conditions for underwater friction stir welding. The methodology presented in the thesis can be implemented in manufacturing units or production lines to determine the most effective values of welding parameters (speed, transverse speed, tool angle) in order to enhance weld strength, improve surface quality of the welded area, ensure weld quality, and prevent potential deterioration of weld mechanical properties.

1. Optimum solutions for tool speed in rpm and transverse speed of feed are obtained by Taguchi method. And it is shown that the angle of tool should not be higher than 2°.
2. It is shown that with the increasing of tool speed and temperature in welding zone, the weld becomes stronger (for the case of joining aluminum 2014 T6 with aluminum 6082 T6).

3. It can be concluded that optimal welding parameters predicted by ANOVA (speed in rpm, transverse speed, tool angle) and practical values of welding parameters from experiments are approximately the same. Realization of these optimal parameters will make it possible to receive the best welding surface and increased strength characteristics.
4. Optimal welding parameters are determined with the Taguchi method. These optimal parameters are as follows: speed of tool is 1120 rpm; transverse speed of feed is 80mm/sec and angle of tool is 2° (for the case of Underwater friction stir welding of details from aluminum 2014 T6and aluminum 6082 T6).
5. Finally, from validation part we have predicted value of tensile test is 185.592 N/mm^2 and practical value is about 186.889 N/mm^2 are approximately equal. The value of hardness test is 75.1 (BHN) and practical value is about 72.16 (BHN) are relatively equal and value of Impact test is 21.236 Joules and practical value is about 20 Joules are approximately equal.

References

1. Wahid, M. A., & Siddiquee, A. N. (2018). Review on underwater friction stir welding: A variant of friction stir welding with great potential of improving joint properties. Transactions of Nonferrous Metals Society of China, 28(2), 193-219.https://doi.org/10.1016/S1003-6326(18)64653-9
2. Heirani, F., Abbasi, A., & Ardestani, M. (2017). Effects of processing parameters on microstructure and mechanical behaviors of underwater friction stir welding of Al5083 alloy. Journal of Manufacturing Processes, 25, 77-84. https://doi.org/10.1016/j.jmapro.2016.11.002
3. Sabry, I., & Zaafarani, N. (2021, February). Dry and underwater friction stir welding of aa6061 pipes-a comparative study. In IOP conference series: materials science and engineering (Vol. 1091, No. 1, p. 012032). IOP Publishing. DOI 10.1088/1757-899X/1091/1/012032
4. Singh, R. P., Dubey, S., Singh, A., & Kumar, S. (2021). A review paper on friction stir welding process. Materials Today: Proceedings, 38, 6-11.https://doi.org/10.1016/j.matpr.2020.05.208
5. Msomi, V., Mabuwa, S., Muribwathoho, O., & Motshwanedi, S. S. (2021). Effect of tool geometry on microstructure and mechanical properties of submerged friction stir processed AA6082/AA8011 joints. Materials Today: Proceedings, 46, 638-644. https://doi.org/10.1016/j.matpr.2020.11.580
6. Lader, S. K., Baruah, M., & Ballav, R. (2021). Underwater Friction Stir Welding of Aluminum Alloy. AIJR Abstracts, 13.
7. Eyvazian, A., Hamouda, A., Tarlochan, F., Derazkola, H. A., & Khodabakhshi, F. (2020). Simulation and experimental study of underwater dissimilar friction-stir welding between aluminium and steel. Journal of Materials Research and Technology, 9(3), 3767-3781. https://doi.org/10.1016/j.jmrt.2020.02.003
8. Wahid, M. A., Masood, S., Khan, Z. A., Siddiquee, A. N., Badruddin, I. A., & Algahtani, A. (2020). A simulation-based study on the effect of underwater friction stir welding process parameters using different evolutionary optimization algorithms. Proceedings of the Institution of Mechanical Engineers, Part C: Journal of Mechanical Engineering Science, 234(2), 643-657. https://doi.org/10.1177/095440621988390
9. Dong, J., Zhang, D., Luo, X., Zhang, W., Zhang, W., & Qiu, C. (2020). EBSD study of underwater friction stir welded AA7003-T4 and AA6060-T4 dissimilar joint. Journal of Materials Research and Technology, 9(3), 4309-4318. https://doi.org/10.1016/j.jmrt.2020.02.056
10. Pedapati, S. R., Paramaguru, D., Awang, M., Mohebbi, H., & Korada, S. V. (2020). Effect of process parameters on mechanical properties of AA5052 joints using underwater friction stir welding. Journal of Mechanical Engineering and Sciences, 14(1), 6259-6271. https://doi.org/10.15282/jmes.14.1.2020.05.0490
11. Sabry, I., & Zaafarani, N. (2021, February). Dry and underwater friction stir welding of aa6061 pipes-a comparative study. In IOP conference series: materials science and engineering (Vol. 1091, No. 1, p. 012032). IOP Publishing. doi:10.1088/1757-899X/1091/1/012032

12. Madhavarao, S., & Raju, C. R. B. (2018). Investigation of friction stir welding of metal matrix composites using a coated tool. Materials Today: Proceedings, 5(2), 7735-7742. https://doi.org/10.1016/j.matpr.2017.11.450
13. Madhavarao, S., Penmetsa, R. V., Kumar, G. S., Raju, C. R., & Kumar, K. T. (2019). Investigation On Mechanical Properties Of Friction Stir Welded Aa7075 & Aa6061 Joints. Materials Today: Proceedings, 18, 2288-2297.https://doi.org/10.1016/j.matpr.2019.07.011
14. Ma, Z. Y., Feng, A. H., Chen, D. L., & Shen, J. (2018). Recent advances in friction stir welding/processing of aluminum alloys: microstructural evolution and mechanical properties. Critical Reviews in Solid State and Materials Sciences, 43(4), 269-333. https://doi.org/10.1080/10408436.2017.1358145
15. Wahid, M. A., Khan, Z. A., Siddiquee, A. N., Shandley, R., & Sharma, N. (2019). Analysis of process parameters effects on underwater friction stir welding of aluminum alloy 6082-T6. Proceedings of the Institution of Mechanical Engineers, Part B: Journal of Engineering Manufacture, 233(6), 1700-1710. https://doi.org/10.1177/0954405418789982
16. Wakchaure, K. N., Thakur, A. G., Gadakh, V., & Kumar, A. (2018). Multi-objective optimization of friction stir welding of aluminium alloy 6082-T6 Using hybrid Taguchi-Grey relation analysis-ANN method. Materials Today: Proceedings, 5(2), 7150-7159. https://doi.org/10.1016/j.matpr.2017.11.380
17. Aravindkumar, D., & Balamurugan, A. (2016). A review on friction stir welding of dissimilar materials between aluminium alloys to copper. Int. J. Latest Res. Eng. Technol, 2(2), 9-15.
18. Kumar, G. S. V. S., Kumar, A., Rajesh, S., Chekuri, R. B. R., & Sundaramurthy, V. P. (2021). Experimental and thermal investigation with optimization on friction stir welding of nylon 6A using Taguchi and microstructural analysis. Advances in Mechanical Engineering, 13(10), https://doi.org/10.1177/168781402110507
19. Parikh, V. K., Badgujar, A. D., & Ghetiya, N. D. (2019). Joining of metal matrix composites using friction stir welding: a review. Materials and Manufacturing Processes, 34(2), 123-146. https://doi.org/10.1080/10426914.2018.1532094
20. Battina, N. M., Vanthala, V. S. P., & Ch, H. K. (2021). Effect of friction stir welding process parameters on mechanical and metallurgical behavior of AA6061-T6 and AA2017-T6 tailored blanks. Engineering Research Express, 3(2), 025043. https://doi.org/10.1088/2631-8695/ac09d9
21. Battina, N. M., Vanthala, V. S. P., & Chirala, H. K. (2021). Influence of tool pin profile on mechanical and metallurgical behavior of friction stir welded AA6061-T6 and AA2017-T6 tailored blanks. Engineering Research Express, 3(3), 035026. https://doi.org/10.1088/2631-8695/ac1a5a
22. Battina, N. M., Hari Krishna, C., & Vanthala, V. S. P. (2022). Influence of pin profile on formability of friction stir-welded aluminum tailor-welded blanks: an experimental and finite element simulation analysis. Transactions of the Canadian Society for Mechanical Engineering, 46(3), 602-613. https://doi.org/10.1139/tcsme-2022-0031
23. Battina, N. M., Chirala, H. K., Vanthala, V. S. P., Javvadi, E. M., Nukathoti, R. S., & Mohammed, A. (2023). Experimental and finite element investigations on formability of friction stir welded tailor welded blanks of AA6061 and AA2017. Materials Today: Proceedings. https://doi.org/10.1016/j.matpr.2023.04.357
24. Battina, N. M., Chirala, H. K., Inala, R., Kummitha, O. R., Veeravalli, L. N., & Pericherla, S. R. (2022). Influence of coefficient of friction between punch-blank interface on formability of friction stir welded aluminum tailor welded blanks–An experimental and finite element simulation investigations. Proceedings of the Institution of Mechanical Engineers, Part C: Journal of Mechanical Engineering Science, 236(17), 9677-9686. https://doi.org/10.1177/09544062221096271

TiO_2 nanoparticles as reusable catalysts for the efficient synthesis of 2, 4, 5-trisubstituted-1H-imidazoles

Shaik Lakshman[1], N. Srinivasa Rao[2], Sridevi. Kadgamala[3], B. V. Durgarao[4], N. Krishnarao[5], Bhagavathula S Diwakar[6]

[1]Department of chemistry, GITAM (Deemed to be University), Visakhapatnam, A.P, India
[2]Department of chemistry, Government Degree College, Cumbum, A.P, India.
[3]TDR-HUB-Andhra university, Visakhapatnam, AP, India.
[4]Department of DFS (chemistry), ONGC, RJY, AP, India.
[5]Department of chemistry, PRISM P.G & D.G College, Visakhapatnam, A.P, India.
[6]SRKR Engineering College, Bimavarm, A.P, India
Email : naallakrishnarao@gmail.com ,

Abstract: A simple, highly efficient and eco-friendly protocol for the synthesis of 2,4,5-trisubstituted-1H-imidazoles via one-pot three component condensation of benzil, aldehydes and ammonium acetate under solvent-free conditions has been achieved utilizing TiO_2.The applicability of this procedure was investigated further in relation to the synthesis of fused imidazoles that exhibit structural versatility. The remarkable features of this methodology are excellent yields in shorter reaction times, cleaner reaction profile, and environmentally benign nature, use of non-toxic, readily synthesizable, inexpensive and recyclable catalyst. The titled compounds can be confirmed advanced spectroscopic evaluation by ^{1}HNMR, ^{13}CNMR and LCMS.The synthesis of green nanoparticles utilizing green protocols is a very good, safe, effective, and environmentally friendly technique in Nano science research.These are employed in many scientific fields. These are used in the catalysis

Keywords: Benzil, aryl aldehydes, NH_4OAc,2, 4, 5-trisubstituted-1H-imidazoles, TiO_2

1. Introduction:

Numerous pharmacological characteristics and a significant role in biological processes are possessed by heterocyclic compounds that include an imidazole moiety. Multicomponent reactions, or MCRs for short, are One-Pot techniques that use three or more components and have outstanding atom economy and great selectivity. They are highly regarded in modern organic synthesis and medicinal chemistry. Multifunctional carbon rings (MCRs) play a major role in the convergent synthesis of important and complex chemical compounds from simple and easily accessible starting materials, in addition to being helpful tools for drug development.

Numerous pharmacological characteristics and a significant role in biological processes are possessed by heterocyclic compounds that include an imidazole moiety. Imidazole molecules are essential for several biological processes [1-3] and are mostly used as organic resources . Since imidazole has several uses, it is important to have access to efficient synthetic routes that result in highly substituted, well-planned imidazole derivatives.

One significant basic organic compound is imidazole. Numerous naturally occurring substances, including biotin, histidine, histamine, pilocarpine alkaloids, and vitamin B_{12}, include it. Additionally, it has been discovered that imidazole functions as an active ingredient in antifungal substances such as clotrimazole, ketoconazole, miconazole, isoconazole, and Econozole, therapeutic potential (4), Analgesic and Anti-Inflammatory Activity (5), Antibacterial And Anti-Inflammatory (6), antimicrobial agents (7), Antioxidant, and Antiproliferative (8), Cytotoxic evaluation (9), anticonvulsant activity (10,11), α-amylase inhibition activity (13), COVID-19 main protease (14). The wide range of applications for substituted imidazoles has led to the development of numerous synthetic processes for their synthesis. Most of the time, a three-component cyclocondensation process using various catalysts combines 1,2-diketone, α-hydroxyketone, or α-ketomonoxime with an aldehydes and ammonium acetate to produce 2,4,5-trisubstituted imidazoles. These catalysts and their methods were all plagued by drawbacks, such as the requirement for specialized equipment, time-consuming procedures for recovering and reusing the catalysts, expensive, poisonous, and moisture-sensitive catalysts, as well as volatile organic solvents and labor-intensive setup.

Several methods have been developed for the synthesis of substituted imidazole sand its derivatives have been devised by three-component cyclocondensation method promoting with Nano catalystviz RGO Supported Au Nano-catalyst (15), Nano carbon rods (16), Naano-TiCl4 • SiO2 (17), $CdFe_{12}O_{19}$@SiO_2nanocomposite (18), $CoFe_2O_4$ (19), SiO_2-Pr-SO_3H(20), Magnetic Fe_3O_4@FU (21), zirconia-coated magnetic nanoparticles (22), Nano catalyst (FHS/SiO_2)(23), MoO_3/SiO_2 (24), Nano-$SbCl_5$/SiO_2(25).Numerous of these artificial techniques have one or more significant disadvantages, such as high temperatures, large volumes of non-recyclable waste, extremely acidic environments, low yields, side reactions, low substrate tolerance, and labor-intensive and time-consuming work-up and purification procedures.

The current focus of research is on improving already-existing multicomponent reactions and finding new MCRs. While each of the previously listed techniques for this reaction has benefits of its own, some of them have disadvantages including low yield, long reaction periods, difficult work-up, and effluent contamination. Here, we describe a simple, cost-efficient, and high-yield multicomponent one-pot synthesis of 2, 4, 5-trisubstituted imidazoles utilizing recyclable solid catalyst Nano TiO_2. Here, we report the first instance of an aqueous phase synthesis of substituted imidazoles supported on TiO2 using Au nanoparticles as a reusable catalyst with a broad range of substrate compatibility, as part of our continuous efforts to develop an aqueous phase process. Here, we report the first instance of an ethanol as a solvent for synthesis of substituted imidazoles supported on TiO_2 nanoparticles as a reusable catalyst with a broad range of substrate compatibility, as part of our continuous efforts to develop an aqueous phase process.

2. Experimental:

All the reagents were purchased from Fine chemicals and before used without further purification. The melting points of the newly prepared compounds were determined in open capillaries using Aggarwal thermometer apparatus and are uncorrected. The progress of the reactions as well as purity of compounds was identified by thin layer chromatography with F254 silica-gel precoated sheets using (hexane: ethyl acetate 8:2 as eluent); UV light was used for detection. Products were characterized by comparing with authentic samples and by spectroscopy data (IR, 1 H NMR and 13C NMR). IR spectra were recorded on Perkin– Elmer 100S spectrophotometer using KBr pellet, values are expressed in cm1. NMR spectra were recorded on Bruker 400 MHz spectrometer using appropriate solvent and TMS as internal standard, chemical shifts are expressed in ppm

2.1. Materials and Methods

Titanium isopropoxide [Ti{OCH(CH_3)$_2$}$_4$], Ethanol (C_2H_5OH), were purchased from Sigma Aldrich Chemicals (India) Ltd. All the reagents were analytically grade and used300 spectrometer.

2.2. General procedure for the synthesis of 2,4,5-trisubstituted1H-imidazoles:

A mixture of benzil (1 mmol), substituted aryl aldehydes (1 mmol), ammonium acetate (2.5mmol), and TiO_2 (10 mmol) in ethanol (25 mL) was stirred at reflux temperature for 2-3 hr. The progress of the reaction was monitored by TLC. After completion of the reaction, the mixture was cooled to room temperature, neutralized with 10% $NaHCO_3$ solution, diluted with water and poured on crushed ice. The obtained crude solid product was filtered, dried, and recrystallized from ethanol.

2.3. Synthesis of TiO2 nanoparticles:

9 mL of TTIP (Titanium tetra isopropoxide) is added drop wise to the homogeneous mixture of 150 mL of ethanol and 3.72 mL of distilled water. This mixture is stirred for half an hour and maintained at 85°C for 4 h. This solution was kept under continuous magnetic stirring and later it was sent to centrifugation and finally dried at 60°C in a hot air oven for 30 min. Finally it was calcined at 400°C for 3 h to yield Anatase TiO2 nanoparticles.

2,4,5-triphenyl-1H-imidazole (4a):
Paleyellowsolid, Yield-86%. m.p:251-253°C. ^{1}HNMR (400MHz, $CDCl_3$,) δppm:10.541(s,1H, NH-imidazole), 8.214 (d, J = 8.8 Hz, 2H), 7.651 (d, J = 7.6 Hz, 4H), 7.45 (t, J = 8.0 Hz, 2H), 7.384 (t, J = 9.2 Hz, 6H) 7.328 (d, J = 6.8 Hz, 2H). ^{13}C NMR (100MHz, $CDCl_3$) δppm: 175.26, 135.45, 130.09, 129.68, 129.17, 128.92, 128.55, 128.08, 127.74, 127.13; Molecular weight (m/z):297.38(M+H); Formula of compound: $C_{21}H_{16}N_2$.

4-(4, 5-Diphenyl-1H-imidazol 2-yl)-phenol (4b):
Pale yellow compound; Mp: 260-262°C. ^{1}HNMR ($CDCl_3$,400MHz)δppm: 10.784 (NH, s , 1H), 9.459 (1H,s, OH), 7.874 (d, J = 8.8 Hz, 2H), 7.612-7.357 (m, 8H, Ar-H),7.241-7.025(m, 2H, Ar-H), 6.945 (d, J=8.0Hz, 2H,Ar-H); ^{13}CNMR (100 MHz, $CDCl_3$)δppm: 159.52, 147.65, 138.77, 134.60, 132.07, 130.12, 129.31, 128.89, 128.49, 128.16, 128.04, 127.84, 127.35, 127.03, 126.48, 123.41, 116.98; LCMS(m/z): 312.19(M^+): Formulae of compound $C_{21}H_{16}N_2O$,

4-(4,5-diphenyl-1H-imidazol-2-yl)-2-ethoxyphenol(4c) :
Pale yellow; Mp 258–260◦C.Yield-89%, FTIR (KBr, cm−1): 3514, 3457, 3271, 3049, 2975, 2861, 1687, and 1277;1HNMR (400 MHz, CDCl3): 11.974 (s, 1H,NH-imidazole), 9.851 (s, 1H,-OH),7.614(d,J=7.6Hz,1H),7.592–7.314(m,10H,Ar-H),7.134(d,8.0Hz,1H,Ar-H),3.917-3.784(m,2H,-CH_2-),1.125 (t,3H,-CH_3);^{13}CNMR(400MHz,$CDCl_3$):174.69, 146.72, 144.38, 136.21, 129.11, 128.83, 128.16, 127.77, 126.54, 124.09, 116.45, 113.09, 65.73, 14.27; Molecular weight (m/z): 357.76(M+H);Formula of compound: $C_{23}H_{20}N_2O_2$

2-(4-Methoxyphenyl)-4, 5-diphenylimidazole (4d)
Pale yellow compound;Mp: 232-233°C. ^{1}HNMR ($CDCl_3$, 400MHz)δppm : 10.521 (NH,s, 1H), 7.775(t, J=8.0 Hz, 2H), 7.612 (d, J = 12.4 Hz,2H),7.480(d,J=6.8Hz,Ar-H), 7.431-7.287 (m, 4H, Ar-H), 7.250-7.046(m,2H,Ar-H), 6.887 (t, J=7.6 Hz, 2H), 3.789 (s,3H,OCH_3); ^{13}C NMR (100 MHz, $CDCl_3$)δppm:160.51, 148.84, 128.83, 128.34, 127.68 127.25, 123.12, 114.88, 54..43; LCMS(m/z) :327.47(M+H); Formulae of compound : $C_{22}H_{18}N_2O$.

4,5-diphenyl-2-(3,4,5-trimethoxyphenyl)-1H-imidazole (4e):
Whitesold,Mp:231–233◦C.Yield-89%;^{1}HNMR(400MHz,$CDCl_3$):11.354(s,1H,NH-imidazole),7.482–7.287(m,10H,Ar-H),7.148(d,J=7.2Hz,2H,Ar-H),3.784(s,6H,(-OCH_3)$_2$), 3.627

(s,3H,-OCH_3), 13CNMR(400MHz,CDCl3):171.15, 155.85, 152.47, 137.07, 129.48, 128.81, 128.34, 127.55, 116.41, 112.26, 109.54,56.17,53.28; Molecular weight (m/z): 387.32(M+H) ; Formula of compound: $C_{24}H_{22}N_2O_3$.

2-(4-Chlorophenyl)-4, 5-diphenyl-1H-imidazole (4f):

Pale yellow compound;Mp:210-212°C. ^{1}HNMR ($CDCl_3$, 400MHz) δppm : 10.987 (s, 1H, NH), 8.134 (d, J= 5.2 Hz, 2H, Ar-H), 7.860-7.286 (m, 10H, Ar-H), 7.108-6.895(m,2H,Ar-H); ^{13}C NMR (100 MHz, $CDCl_3$)δppm: 147.23, 131.44, 129.46, 129.12, 128.88, 128.53, 127.18, 127.84, 127.55, 100.12; LCMS(m/z):332.66(M+H);Formulae of compound : $C_{21}H_{15}ClN_2$.

2-(3-Bromophenyl)-4, 5-diphenyl-1H-imidazole (4g):

Pale yellow compound;Mp: 218-220°C. ^{1}HNMR ($CDCl_3$, 400MHz) δppm : 11.281 (s, 1H, NH), 8.185 (t, J =10.4 Hz, 1H), 8.024 (d, J = 8.0 Hz, 1H), 7.874-7.324 (m, 12H, Ar-H); ^{13}C NMR (100 MHz, $CDCl_3$)δppm: 147.50, 138.48, 134.56, 132.67, 131.84, 131.55, 131.25, 129.75, 129.42, 128.88, 128.57, 128.24, 12805, 127.74, 127.32, 124.75, 122.36; LCMS(m/z) : 377.38 (M+2) ;Formulae of compound :$C_{21}H_{15}BrN_2$.

4-(4,5-diphenyl-1H-imidazol-2-yl)-N,N-dimethylaniline (4h):

Pale yellow compound; m.p: 235-237°C.^{1}HNMR ($CDCl_3$, 400MHz) δppm : 12.32 (s, 1H, NH), 7.91 (d, J= 9 Hz, 2H), 7.37–7.20 (m, 10H, Ar-H), 6.80 (d, J= 9 Hz, 2H), 2.98 (s, 6H $(CH_3)_2N$); ^{13}C NMR (100 MHz, $CDCl_3$)δppm: 150.7, 146.9, 129.1, 128.7, 128.6, 127.9, 127.5, 126.8, 118.8, 112.4; LCMS(m/z):340.07(M+H);Formulae of compound : $C_{23}H_{21}N_3$,

2-(2-bromo-3,4-dimethoxyphenyl)-4,5-diphenyl-1H-imidazole (4i)

Palered;Mp:248–250°C.Yield-89%;^{1}HNMR($CDCl_3$,400MHz):11.987(NH-imidazole,s,1H), 7.715–7.360(m,10H,Ar-H),7.324(d,J=8.0,1H,Ar-H),7156(d,J=8.8,1H,Ar-H),3.810(s,3H,-OCH_3),3.684(s,3H,-OCH_3),^{13}CNMR(400MHz,$CDCl_3$):170.65,151.58,148.73,136.44,132.51, 129.14, 128.88, 128.17, 127.91, 125.69, 120.24, 110.07, 58.34,55.71;Molecular weight (m/z): 436.56(M+2); Formula of compound: $C_{23}H_{19}BrN_2O_2$

2-(2-iodo-3,5-dimethoxyphenyl)-4,5-diphenyl-1H-imidazole(4j):

PaleYellow;Mp:270–272◦C.yield-88%;^{1}HNMR(400MHz,$CDCl_3$):11.520(s,1H,NH-imida zole),7.532–7.310(m,10H,Ar-H),6.947(s,1H,Ar-H),6.584(s,1H,Ar-H),3.842(s,3H,-OCH_3), 3.581(s,3H,-OCH_3), 13CNMR(400MHz,$CDCl_3$):172.47, 158.18, 155.10, 138.37, 136.54, 129.33, 128.77, 128.49, 127.68, 110.44, 107.26, 105.19,55.02,50.25;Molecular weight (m/z): 484.24(M+2) ; Formula of compound: $C_{23}H_{19}IN_2O_2$;

4-(4,5-diphenyl-1H-imidazol-2-yl) benzonitrile (4k):

Redcompound;m.p-245-247^{0}C;^{1}HNMR(CDCl3,00MHz)δppm:11.807(s,1H,NH-imidazole),8.114(2H,Ar-H,d,J=8.0Hz),7.725(2H,Ar-H,d,J=10.6Hz),7.510–7.271(m,10H,Ar-H);^{13}CNMR(400MHz, $CDCl_3$)δppm:174.65, 138.21, 136.04, 131.16,129.47, 129.02, 128.84, 128.29, 127.16,119.55,113.73;Molecularweight(m/z): 320.96(M-H);Formula of compound: $C_{22}H_{15}N_3$.

2-(3-Nitrophenyl)-4, 5-diphenyl-1H-imidazole (4l):

Pale yellow compound; Mp:268-270°C.^{1}HNMR ($CDCl_3$, 400MHz) δppm : 11.254 (s, 1H, NH), 8.387(t, J=8.4 Hz, 1H,NO_2-Ar-H), 8.247(d,J =10.2, 1H, NO_2-Ar-H), 8.123 (d, *J* =8.8 Hz, 1H, NO_2-Ar-H), 7.718(t, J= 7.2Hz, 1H, NO_2-Ar-H), 7.665-7278 (m, 10H, Ar-H); ^{13}C NMR (100 MHz, $CDCl_3$)δppm: 149.49, 144.18, 130.47, 129.56, 128.89, 128.49, 124.71, 120.99;LCMS(m/z):340.08(M-H);Formulae of compound: $C_{21}H_{15}N_3O_2$, LCMS (m/z):341.36.

2-(furan-2-yl)-4, 5-diphenyl-1H-imidazole (4m):

Pale yellow compound; Mp: 200-202°C. ^{1}HNMR (CDCl3, 400MHz) δppm: 12.83 (s, 1H NH), 7.81 (s, 1H, furan-H),7.53-7.22 (m, 10 H, Ar-H), 6.98 (s, 1H, furan-H), 6.66 (d, *J*=1.5,8 Hz 1H, furan-H); ^{13}C NMR (100 MHz, $CDCl_3$)δppm: 143.2, 128.8,112.5,107.6; LCMS(m/z) 287.51 (M+H)**;**Formulae of compound : $C_{19}H_{14}N_2O$.

2-(α-Napthyl)-4, 5-diphenylimidazole (4n):

Pale yellow compound;Mp: 215-217°C. ^{1}HNMR ($CDCl_3$,400MHz) δppm: 12.80 (s, 1H, NH), 9.19 (d, *J*=8 Hz, 1H,), 8.01 (t, *J* = 8 Hz, 2H), 7.97 (s, 1H), 7.64-7.25 (m, 13H, Ar-H); ^{13}C NMR (100 MHz, $CDCl_3$)δppm: 146, 137.6, 135.8, 135.1, 131.5, 130.8, 129.4, 129.1, 128.8, 128.7, 128.4, 128.2,

127.9, 127.7, 127.2, 127, 126.6, 125.7;HRMS: LCMS(m/z): 346.22 (M^+)Formulae of compound: $C_{25}H_{18}N_2$,

3-(4,5-diphenyl-1H-imidazol-2-yl)pyridine(4p):

Pale yellow compound; Mp: 214-216°C;^{1}HNMR ($CDCl_3$, 400MHz) δppm:12.083 (s,1H NH),8.671(1H,s,pyridine),8.257(d,J=7.6Hz,1H,pyridine),8.112(1H,pyridine,J=8.0Hz,d),7.984(t,J=12 .2 Hz,1H,Pyridine),7.614-7.348 (m, 10 H, Ar-H); ^{13}C NMR (100 MHz, $CDCl_3$)δppm:156.29,149.08,145.87,167.44,133.84,130.77,129.38,128.91,128.54,128.17,127.45,124. 87; LCMS(m/z) :298.24(M+H); Formulae of compound : $C_{20}H_{15}N_3$.

2.4. Assay of in vitro antibacterial activity:

The synthesized compounds (4a–4n) were tested for their minimum inhibitory concentrations (MIC) using the broth dilution method, which is advised by the National Committee for Clinical Laboratory (NCCL) standards [28]. Gram-positive organisms tested included Staphylococcus aureus (MTCC 1478) and Bacillus subtilis (MTCC 2184), while gram-negative bacteria included Escherichia coli (MTCC 987) and Proteus vulgaris (MTCC 1580). Test compounds were serially diluted in a 200 mL culture medium final volume, with concentrations ranging from 150 to 0.56 μg mL−1. Following that, a 50 μL microbial solution with 0.5 MacFarland densities was added to each well. Both a sterility control (negative) and a microbiological culture control were run for every test. The plates were incubated at 37 °C for 24 hours. The minimum inhibitory concentration (MIC) was defined as the lowest concentration that prevented detectable microbial growth.

3. Results and discussion:

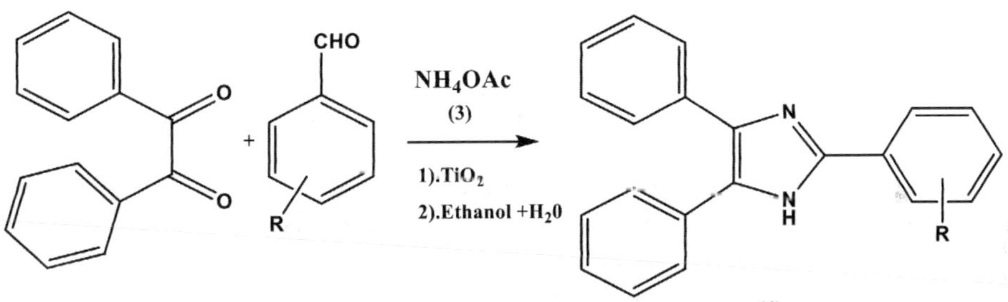

(1) (2) (4)

(4a-4p)

R= H, 4-OH, 3-OC_2H_5-4-OH, 4-OCH_3, 3,4,5-$(OCH_3)_3$, 4-Cl, 4-Br,

4-$N(CH_3)_2$, 2-Br-3,4$(OCH_3)_2$, 2-I-3,5$(OCH_3)_2$, 4-CN, 4-NO_2,

1-Napthyl, Furyl, pyradyl

(scheme-1)

Initially, the first step of our experiment involved heating benzil (1 mmol), benzaldehyde (1 mmol), and ammonium acetate (2 mmol) in the presence of supported TiO_2 nanoparticles in an ambient environment. This reaction yielded the appropriate 2, 4, 5-trisubstituted imidazole (Scheme-2).Due to its high activity, non-toxicity, ease of availability, reusability, and potent oxidizing power, and long-term stability, nano-TiO_2 has emerged as one of the most effective Nano metal oxides. In addition, titanium dioxide is widely used in many different industrial catalysis applications, such as the selective reduction of nitrite or nitrate ions in stationary sources, photocatalysis for the removal of pollutants or organic synthesis, and photovoltaic devices, sensors, and paints. The present study presents TiO_2 nanoparticles as a novel and effective low loading catalyst for the synthesis of imidazoles.

3.1. XRD analysis:

The preliminary characterization of TiO_2 NPs was confirmed with XRD peaks (Figure 1). XRD peaks for obtained adsorbent with diffraction angles at 2θ = 25.28°, 37.5°, 47.5°, 55°, 62.4°, 70°, and 76.5° respectively. Sharp peaks at diffraction angles of 25.28°, 37.5°, 47.5°, and 55° in XRD pattern confirmed highly Nano crystalline structure of Anatase a mineral form of TiO_2 NPs.

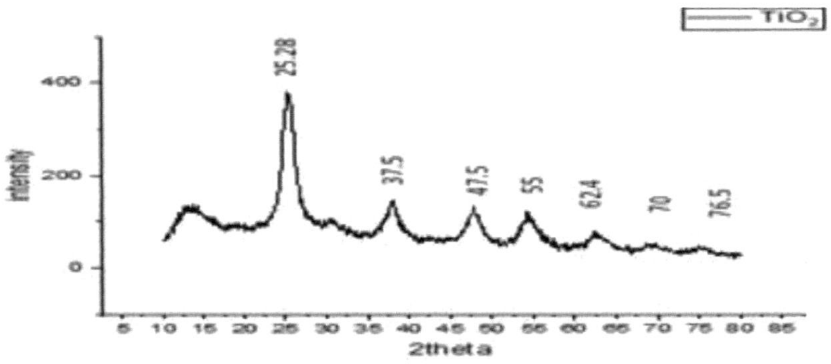

Figure 1: The XRD of TiO_2 NPs

To find the ideal conditions, the reaction was run using various solvents, temperatures, and catalyst concentrations. Table 1 provides an overview of the findings. Without the catalyst, the aforementioned reaction would not have happened at all. In order to optimize the catalyst and reaction conditions, the reaction was first conducted without a catalyst and without a solvent, but even after increasing the reaction time, the intended product was not produced. Therefore, the catalyst's use is necessary for the reaction to proceed. We choose TiO_2Nano catalyst because it is mild, affordable, nontoxic, extremely stable, and easily accessible, taking into account the significance of catalyst.

In table-1 represented that the time of the reaction and yield of the product, the various substituted aromatic aldehydes and hetero aromatic aldehydes is treated with benzil and ammonium acetate scaffold titled derivatives of 2,4,5-triphenyl imidazoles, the aromatic aldehydes bearing electron donating groups and also electron attracting groups. The derivative "4e" was obtained high yield of the product that having highly electron releasing groups. The compounds 4b,4c,4d having electron donating groups that having lower yield than the compound "5" while the derivatives 9,10 bearing EDG and EWG but also such derivatives obtained moderate to good percentage of product obtained. Whereas heteroatom bearing imidazoles such as pyridine and furan generate good yield of product.

Table-1: Synthesis of series of titled derivatives in the presence of TiO_2

Entry	R	Product	Time(min)	Yield (%)
1	H	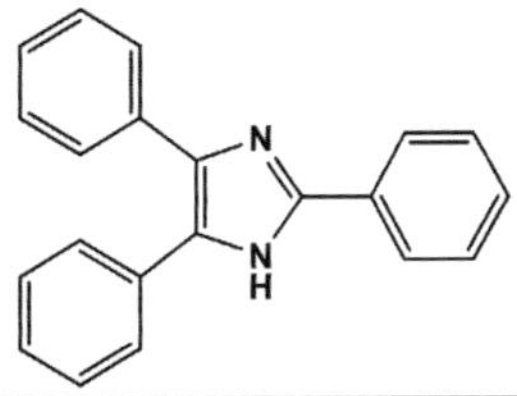	180	84
2	4-OH	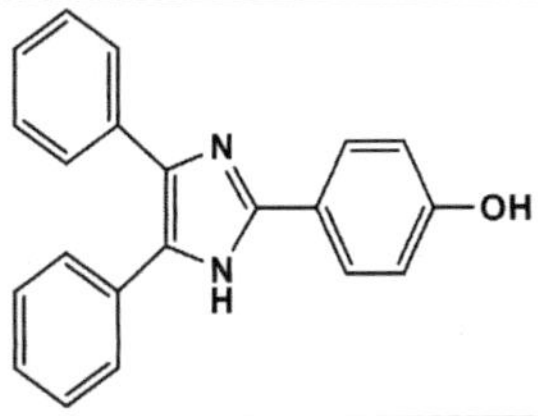	150	89
3	3-OC_2H_5-4-OH	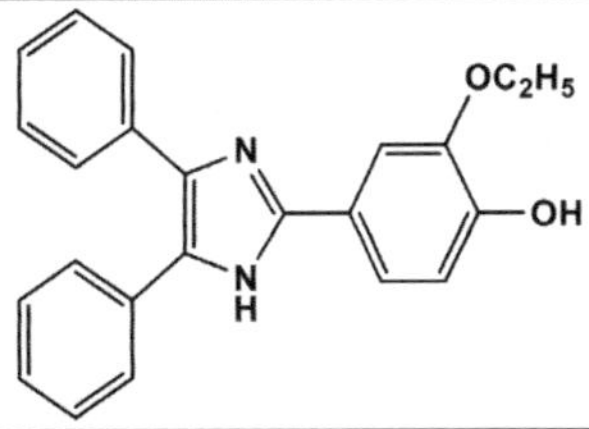	120	90
4	4-OCH_3	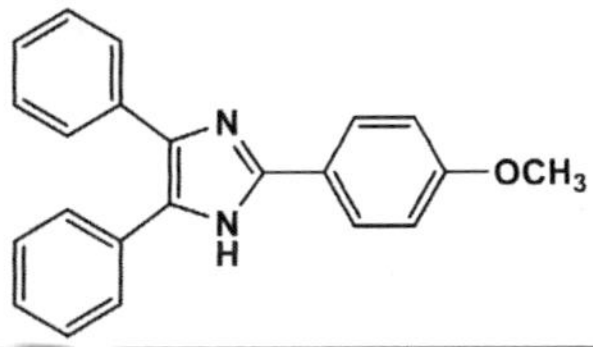	120	86
5	3,4,5(OCH_3)$_3$	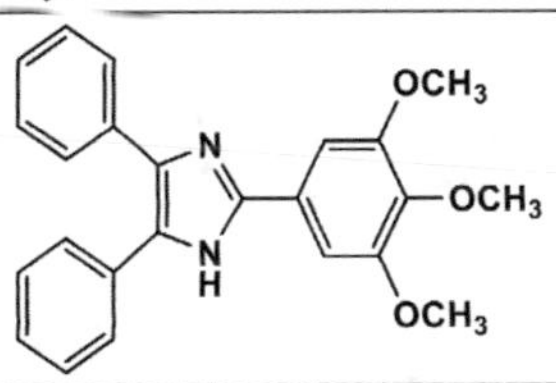	120	92
6	4-Cl	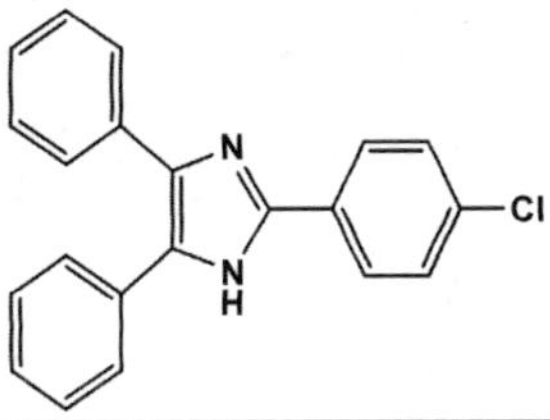	210	87
7	4-Br	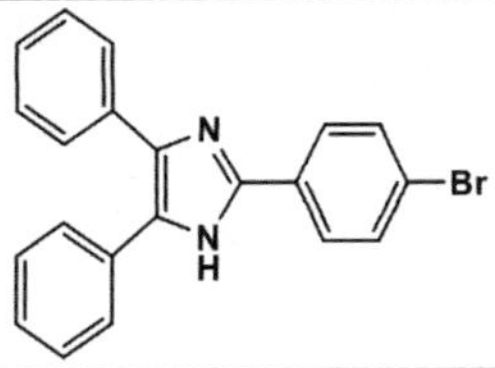	180	87

8	$N(CH_3)_2$	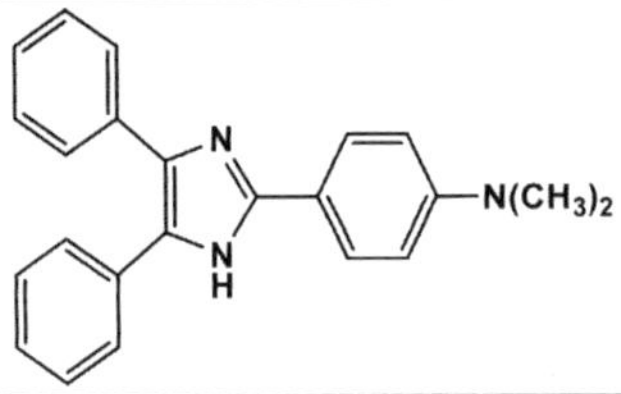	210	85
9	2-Br-3,4$(OCH_3)_3$	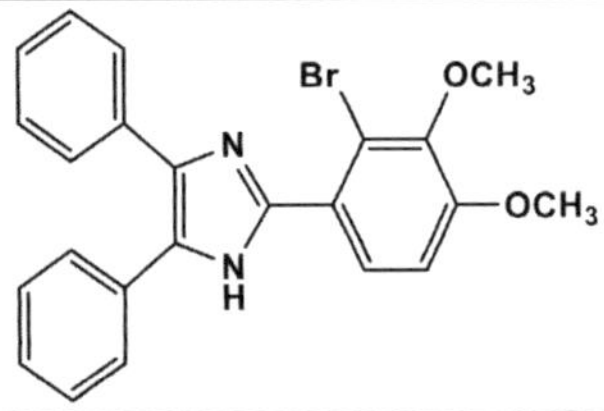	240	87
10	2-I-3,5$(OCH_3)_2$	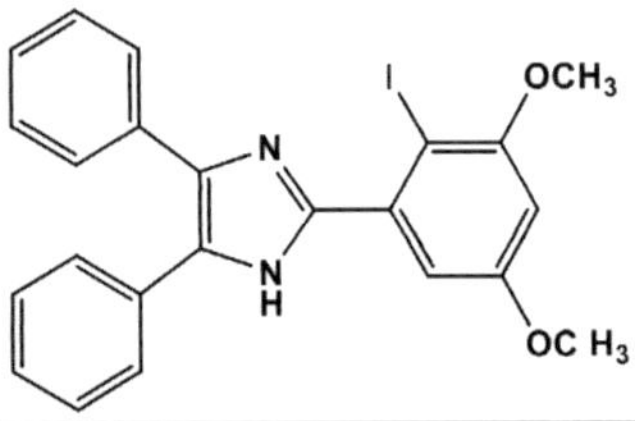	240	86
11	4-CN	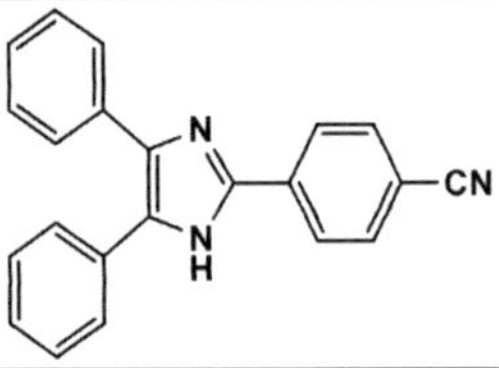	210	85
12	3-NO_2	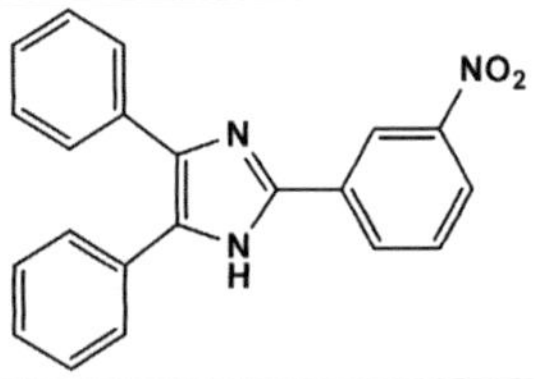	180	83
13	furyl	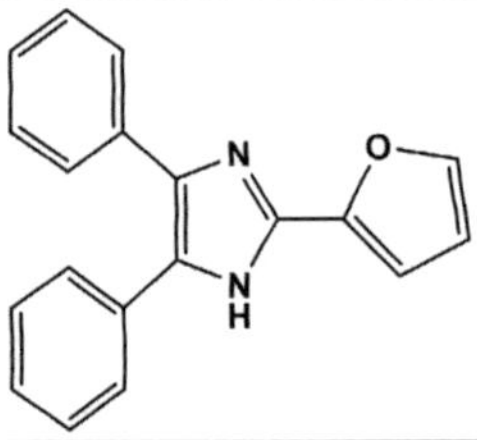	180	86
14	pyridine	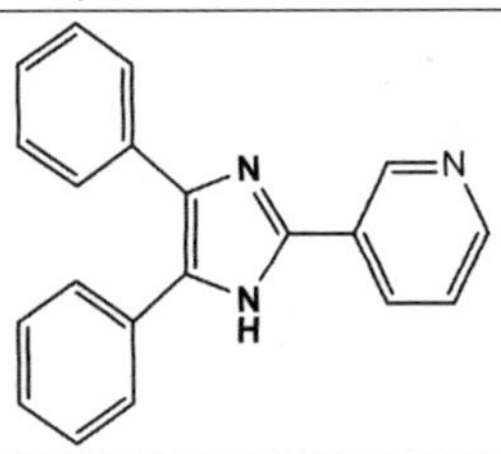	240	86

15	Napthyl	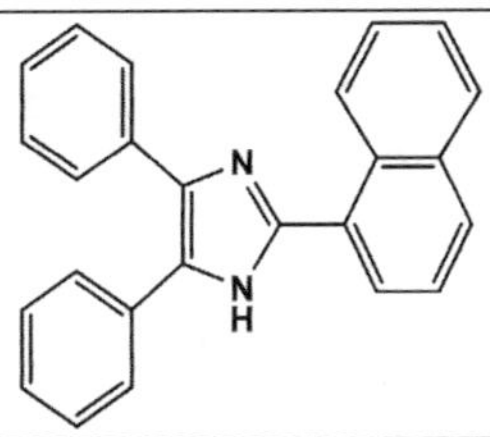	300	82

3.2. NanoTiO₂ catalyzed synthesis for derivatives of imidazole

Several parameters were evaluated in order to optimize the reaction conditions. The quantity (mmol %) of TiO_2 Nano particles had a significant impact on their catalytic efficiency. Consequently, a model reaction involving isatin, ammonium acetate, and benzaldehyde was conducted with varying concentrations ofTiO_2 NPs (Table-1). It has been observed that the catalyst has a significant impact on product yield; after 120 minutes, only a low yield was produced without the catalyst (Entry 1, Table 1). It was also discovered that as catalyst concentration increases, product yield graduallywas improved l. In 60 minutes, 60% yield could be achieved with only 5 mol% of catalyst (Entry 2, Table-1). With 15 mol% of TiO_2NPs, the highest yield of 88% was achieved (Entry -5) With 15 mol% TiO_2 NPs, the highest yield of 88% was achieved (Entry-5).

Table-2: Optimization of various catalyst of synthesis of titled derivatives (4e)

Entry	**Catalyst**	**Time(min)**	**Yield (%)**
1	$NiFeO_4$	120	48
2	Fe_2O_3	120	68
3	CaO	120	56
4	MgO	120	51
5	TiO_2	120	92
6	SnO_2	120	71
7	ZnO	120	76

At first, different several mole ratios ranges from 5 to 25% were used to screen the catalyst in ethanol as solvent and the absence of a catalyst, there was no traces amount ofproduct yield at room temperature. However, a trace amount of product was obtained by raising the reaction temperature to increases slowly70°C. As the amount of catalyst is increased, the yield of product gradually rises to 20%mole at 70°C. The reaction rate did not significantly improve even when the catalyst was increased by 25% mole. Therefore, the observation indicated that 20% mole equivalent of catalyst is sufficient to bring the reaction to a close (Table-1). Following catalyst optimization, we used different solvents, including, EtOH, MeOH, DCM, Toluene, CH_3CN, and THF, to screen for the effects of the solvents. The observation represented that the ideal reaction condition is one that is solvent. The effect of solvent observed thatthe significant impact onyield of thedesired product. Themaximum yield was obtained with polar solvent (Entry 1, 2, 3, 6 and 7) than nonpolar solvent (Entry 4 and 5) even though a highest yield was obtained without the solvent in a shorter amount of time. Entry-2, Table-2, suggests that the solvent's competitive adsorption of the solvent with the substrate molecule surface on the catalyst is responsible for the remarkable yield.

Table -3:. Effect of solvent for Nano TiO_2 catalyzed synthesis of 2, 4, 5-triaryl imidazole (4e)

Entry	**solvent**	**Time (min)**	**Yield (%)**
1	H_2O	150	Traces
2	EtOH	150	92
3	MeOH	150	58
4	CH_3CN	150	45
5	DCM	150	50
6	THF	150	60
7	Acetone	150	35

In order to determine the excellent reaction conditions, we studied the reaction of benzil with substituted aromatic aldehyde in the presence of different amounts of nano-TiO_2 at 70^0 C under different conditions. As shown in Table-3, the best result was obtained with 10 mol% of nano-TiO_2 under solvent conditions (entry 2).

Table-4: Optimization of the good reaction conditions by using different mmol of nano-TiO_2 (4e)

Entry	**Catalyst**	**Solvent**	**Time (min)**	**Yield (%)**
1	Absence of TiO2	C_2H_5OH and H_2O	Prolong	Traces
2	1mmol	C_2H_5OH and H_2O	90	10
3	2.5mmol	C_2H_5OH and H_2O	90	33
4	5.0mmol	C_2H_5OH and H_2O	90	54
5	10mmol	C_2H_5OH and H_2O	90	92
6	15mmol	C_2H_5OH and H_2O	90	92
7	20mmol	C_2H_5OH and H_2O	90	92

The structures of the new derivatives were evaluated by LCMS, IR, ^{1}HNMR and ^{13}CNMR as well as elemental analysis.IR spectra of compounds 1a-d showed CN bands at 2224 cm–1 and a stretching NH band at 3465cm^{-1}. The IR spectra of =C-H strencing of benzene nucleus all compounds showed the values 1602, 1547, 1524, 1496 cm^{-1}. The 1HNMR spectra exhibited a peak at 12.083-10.541 δppm shown the presence of N-H proton, a peak between 8.384-6.945 δppm showed aromatic proton, a peak exhibited between 3.842-3.581 δppm represents methoxy protons, a peak shown between 9.459 δppm indicates hydroxyl proton(-OH). The LCMS spectra recorded the molecular weight of halogen substituted such as 4f,4g,4i and 4Jshowed (M+2) peak. The derivatives "4a,4c,4d,4e,4h,4m and4p" exhibited (M+H) peaks, the derivatives "4k and 4l" showed at (M-H) peaks and the derivatives "4b and 4n "showed M+ peak.

3.3. In vitro antibacterial activity:

The minimum inhibitory concentrations (MIC) of synthesized compounds (4a–4p) were tested against Gram-positive and Gram negative bacteria. The antibacterial activity screening results (Table- 5) indicated that, compounds possessing 2,4,6-triphenyl imidazole ring (4b and 4p) have shown excellent activity against *E. coli and B. Subtilis* when comparing with the standard drug Streptomycin. Compound 4d against Aeruginosa 4k against S. aureus, and compound 31 against S. aureus and B. subtilis have shown promising antibacterial activity.

Table-5: The MIC of 2,4,6-triphenyl imidazole derivatives (4a–4n) on different bacterial strains.

Derivatives	***E. coli***	***S. aureus***	***B. Subtilis***	**P.aeruginosa**
4a	04	03	07	09
4b	14	16	17	19
4c	13	16	15	18
4d	15	20	22	20
4e	**21**	**21**	**22**	**23**
4f	15	18	19	20
4g	**20**	**19**	**22**	**21**
4h	19	18	21	23
4i	17	20	22	19
4j	17	15	21	23
4k	19	18	18	20
4l	12	19	20	18
4m	11	21	22	12
4n	14	16	23	10
4p	12	15	10	13
Streptomycin	25	25	27	27
DMSO	-	-	-	-

The standard "Streptomycin" as gathered in (Table-IV) was compared with the in vitro antibacterial activity of the compounds titled (4a-4p). The majority of the synthetic derivatives typically showed strong activity against all of the tested bacterial strains, as shown in Table IV. Because compound "4d, 4i, and 4n" contains halogen atoms, it demonstrated strong antibacterial action against gram-positive bacterial strains (*E. coli, P. aeruginosa*) and gram-negative bacterial strains (*B. subtilis, Staphylococcus aureus*), respectively. When tested against bacterial strains, the derivatives 4a, 4b, 4c, 4f, 4g, 4h, 4k, 4l, and 4m showed moderate to good active potential. The limited action of compounds "4e, 4j, and 4o" against bacterial stains can be attributed to their strongly electron withdrawing groups. The compounds with electron-releasing groups exhibited moderate to good activity, as demonstrated by these data, compared to those with electron-withdrawing groups. The compounds containing halogen atoms showed excellent active potential.

Conclusions:

We presented a practical and effective conventional synthesis method for the synthesis of a range of transition metal oxides. Through the use of nanotechnology, substituted aromatic aldehydes, ammonium acetate, and benzil were reacted in a three-component reaction to efficiently synthesize derivatives of 2, 4, and 5 trisubstituted-1H-imidazoles. This process has several benefits, such as a good to excellent yield of products, the use of TiO_2, a cheap, reusable, nontoxic, and readily available catalyst, and a simple experimental setup process that makes it a valuable method for the synthesis of titled derivatives. Additionally, the altered experimental protocol for the separation and purification of the unreacted dimedone was examined, as was the initial reactant from the reaction's end products. Furthermore, we have evaluated antibacterial activity successfully. In comparison to compounds screened against Gram-(-ve) organisms, the antibacterial and antifungal activity data of the acquired named compounds showed very good anti-bacterial activity against Gram-(+ve) organisms. Positive outcomes were shown by the targeted products' strong activity.

References:

1) .Mahnaz Sakhdari, Ali Amoozadeh and Eskandar Kolvari, 2021, *Heterocyclic Communications* , **27**: 71–78.

2) Rana S.Ahmed ,2019, *Journal of Global Pharma Technology*, **11**(7):437-444.
3) M. SVS Lahari, D. Sankarao, S. Shyam, Ch. Syamala and Dr. N. Krishnarao, 2022, *World Journal of Pharmaceutical Research*,**11**,5.
4) Iftikhar Ahsan and K.K. Sharma, 2015, *The Pharma Innovation Journal* **4**(3): 68-73
5) Ankit Siwach & Prabhakar Kumar Verma, 2021, *BMC Chemistry*, **15**, 12.
6) MilindBhitre,2013, *Journal of Current Pharma Research*, **3**,889-900.
7) Swati D. Burungale and M. J. Bhitre, 2013, *IJPSR*, **4**(10): 4051-4057.
8) Mohamed Reda Aouad,2015, *Acta Pharm.* **65**, 117–132.
9) M. Ganga, K.R. Sankaran, 2020, *Chemical Data Collections*, **28**, 100412.
10) Iván Córdova-Guerrero, 2020, *Appl. Sci.* **10**(8), 2889;
11) AdelA. Marzouk, AmrK.A. Bass, Montaser Sh. Ahmed, Antar A. Abdelhamid, Yaseen A.M.M. Elshaier, AsmaaM.M. Salman f, Omar. Aly, 2020, *Bioorganic*,**101**,104020.
12) Rutuja Sonawane, Chandrakant Magdum, 2015, *Asian Journal of Biomedical and Pharmaceutical Sciences*, **5**(49) 1-4.
13) B Somashekara, C, B Thippeswamy,C And G R Vijayakumar,2019, *J. Chem. Sci.* **131**:62.
14) L. Athishu Anthony, P. Nethaji, G. Sundararajan C, D. Rajaraman,2022, *Journal of Molecular Structure*, **1250**,131892.
15) Avishek Ghatak, 2021, *Journal of Molecular Structure* **1242**:130823.
16) Abbas Teimouri a,∗ , Alireza Najafi Chermahini, 2011, *Journal of Molecular Catalysis A: Chemical* **346** , 39–45.
17) Leila Zamani, Bi Bi Fatemeh Mirjalili*, Mansoor Namazian,2013, *chemija.* **24**.4. 312–319.
18) Javad Safaei Ghomi, Ali Kareem Abbas, Marzieh Shahpir, 2020, Organic Preparations and Procedures International, **36**, 2004
19) FatemehMostaghni, FereshtehTaat ,2020, *Eurasian Chem. Commun.* 427-432.
20) Ghodsi Mohammadi Ziarani, Zeina Dashtianeh , Monireh Shakiba Nahad, Alireza Badiei , 2015, *Arabian Journal of Chemistry* **8**,5,692-697.
21) Maryam Banazadeh Sara Amirnejat Shahrzad Javanshir, 2020, *Front. Chem., Sec.CatalysisandPhotocatalysis*,**8**,2020.
22) Somayeh Zolfagharinia, EskandarKolvari, NadiyaKoukabi, Maliheh M. Hosseini, 2020,*Arabian Journal of Chemistry*,**13**, 1, 227-241.
23) Mehdi Bakavolia, Hossein Eshghia, Ali Mohammadi, Hassanali Moradia, Javad Ebrahimi, 2015, *Iranian Journal of Catalysis* **5**(3), 237-243
24) Bhosale, S. V.; Kalyankar, M. B.; Nalage, S. V.; Bhosale, D. S.; Pandhare, S. L.; Kotbagi, T. V.; Umbarkar, S. B.; Dongare, M. K. 2011 *Synth.Commun.* **41**(5), 762-769
25)Bamoniri, A.; Rohani, S.,2012, *J. Nanostruct.,* **2**, 221-226.

AUTHOR INDEX